McGRAW-HILL BOOK COMPANY

New York
St. Louis
San Francisco
Düsseldorf
Johannesburg
Kuala Lumpur
London
Mexico
Montreal
New Delhi
Panama
Rio de Janeiro
Singapore
Sydney
Toronto

CLIVE L. DYM
Carnegie-Mellon University

IRVING H. SHAMES
State University of New York at Buffalo

Solid Mechanics

A VARIATIONAL APPROACH

This book was set in Times Roman.
The editor was B. J. Clark;
and the production supervisors were
Alan Chapman and Thomas J. Lo Pinto.
The printer and binder was Kingsport Press Inc.

Library of Congress Cataloging in Publication Data

Dym, Clive L
    Solid mechanics.

    (Advanced engineering series)
    Includes bibliographies.
    1. Elastic analysis (Theory of structures)
2. Calculus of variations. I. Shames, Irving
Herman, 1923–   joint author. II. Title.
TA653.D9   620.1'05   72-10186
IBSN 0-07-018556-5

SOLID MECHANICS: A VARIATIONAL APPROACH

1234567890 KPKP 79876543

CONTENTS

FOREWORD

In this text we shall employ a number of mathematical techniques and methods. We shall introduce these techniques and methods at places where it is felt maximum understanding can be achieved. The physical aspects of the concepts will be stressed where possible. And, although we shall present most of this material with the purpose of immediate use in solid and structural mechanics, we shall also "open-end" the discussions where feasible to other fields of study. Such discussions are of necessity more mathematical in nature and are generally asterisked, indicating that they can be deleted with no loss in continuity.

We shall employ Cartesian tensor concepts in parts of this text and the accompanying notation will be used where it is most meaningful. (It will accordingly not be used exclusively.) For those readers not familiar with Cartesian tensors (or for those wishing a review), we have presented a self-contained treatment of this subject in Appendix I. This treatment (it includes exercises as well) will more than suffice the needs of this text.

CLIVE L. DYM
IRVING H. SHAMES

PREFACE

This text is written for senior and first year graduate students wishing to study variational methods as applied to solid mechanics. These methods are extremely useful as means of properly formulating boundary-value problems and also as a means of finding approximate analytical solutions to these boundary-value problems.

We have endeavored to make this text self-contained. Accordingly, virtually all the solid and structural mechanics needed in the text is developed as part of the treatment. Furthermore the variational considerations have been set forth in a rather general manner so that the reader should be able to apply them in fields other than solid mechanics.

A brief description of the contents of the text will now be given. For those readers not familiar with cartesian tensor notation or for those wishing a review, we have presented in Appendix I a development of this notation plus certain ancillary mathematical considerations. In Chapter 1 we present a self-contained treatment of the theory of linear elasticity that will serve our needs in this area throughout the text. Next in Chapter 2 comes a study of the calculus of variations wherein we consider the first variation of functionals under a variety of circumstances. The delta operator is carefully formulated in this discussion. The results of the first two chapters are then brought together in Chapter 3 where the key variational principles of elasticity are undertaken. Thus we consider work and energy principles, including the Reissner principle, as well as the Castigliano theorems. Serving to illustrate these various theorems and principles, there is set forth a series of truss problems. These truss problems serve simultaneously as the beginning of our efforts in structural mechanics. In developing the aforementioned energy principles and theorems, we proceeded by presenting a functional first and then, by considering a null first variation, arrived at the desired equations. We next reverse this process by presenting certain classes of differential equations and then finding the appropriate functional. This sets the stage for examining the Ritz and the closely related Galerkin approximation methods. In later chapters we shall present other approximation techniques. In Chapter 4 we continue the study of structural mechanics by applying the principles and theorems of Chapter 3 to beams, frames and rings. These problems are characterized by the fact that they involve only one independent variable—they may thus be called one-dimensional structural problems. In Chapter 5 we consider the elastic and inelastic torsion of shafts. Use is made of earlier methods in the text but now, because there are two independent variables, new approximation techniques are presented—namely the methods of Trefftz and Kantorovich. Chapter 6 dwells on the classical theory of plates. We set forth the equations of equilibrium and the appropriate boundary conditions, via variational methods, and then we find approximate solutions to various problems via the techniques presented earlier. In Chapter 7 there is covered the free vibrations of beams and plates. With time now as a variable we first present Hamilton's Principle and then go on to formulate the equations of motion of beams and plates. The methods of Ritz and Rayleigh–Ritz are then employed for generating approximate natural frequencies of free vibration as well as mode shapes. To put these methods on firm foundation we then examine the eigenvalue–eigenfunction problem in general and this leads to the Rayleigh quotient that will be used in the study of stability in Chapter 9. Also developed is the maximum–minimum principle of the calculus of variations—thus providing a continuation of the variational calculus in Chapter 2. Up to this point only small deformation has been considered in our undertakings (the non-linear considerations thus far have been in the constitutive laws) and so in Chapter 9 we consider large deformation theory. In particular the principle of virtual work and the method of total potential energy are presented. The climax of the chapter is the presentation of the von Kármán plate theory. The closing chapter considers the elastic stability of columns and plates. Various approaches are set forth including the criterion of Trefftz and the asymptotic postbuckling theory of Koiter.

Note that we have not included finite element applications despite the importance of variational methods in this field. We have done this because the finite element approach has become so broad in its approach that a short treatment would not be worthwhile. We recommend accordingly that this text serve as a precursor to a study in depth of the finite element approach.

At the end of each chapter there is a series of problems that either call for applications of the theory in the chapter or augment the material in the chapter. Particularly long or difficult problems are starred.

We wish to thank Prof. T. A. Cruse of Carnegie-Mellon University for reading the entire manuscript and giving us a number of helpful comments. Also our thanks go to Prof. J. T. Oden of the University of Alabama for his valuable suggestions. Dr. A. Baker and Dr. A. Frankus helped out on calculations and we thank them for their valuable assistance. One of the authors (C.L.D.) wishes to pay tribute to three former teachers—Prof. J. Kempner of Brooklyn Polytechnic Institute and Profs. N. J. Hoff and J. Mayers of Stanford University—who inspired his interest in variational methods. The other author (I.H.S.) wishes to thank his colleague Prof. R. Kaul of State University of New York at Buffalo for many stimulating and useful conversations concerning several topics in this text. Finally we both wish to thank Mrs. Gail Huck for her expert typing.

CLIVE L. DYM
IRVING H. SHAMES

1

THEORY OF LINEAR ELASTICITY

1.1 INTRODUCTION

In much of this text we shall be concerned with the study of elastic bodies. Accordingly, we shall now present a brief treatment of the theory of elasticity. In developing the theory we shall set forth many concepts that are needed for understanding the variational techniques soon to be presented.

We will not attempt solutions of problems directly using the full theory of elasticity; indeed very few such analytical solutions are available. Generally we will work with special simplifications of the theory wherein a priori assumptions are made as to

(*a*) the stress field (plane stress problems for example)

(b) the strain field (plane strain problems for example)

(c) the deformation field (structural mechanics of beams, plates and shells).

The significance of these simplications and the limitations of their use can best be understood in terms of the general theory.

Specifically in this chapter we will examine the concepts of stress, strain, constitutive relations, and various forms of energy. This will permit us to present the equations of linear elasticity and to consider the question of uniqueness of solutions to these equations. The plane stress simplification will be considered in this chapter; examples of simplification (c) will be presented in subsequent chapters.

Although we shall present certain introductory notions pertaining to finite deformations in this chapter so as to view small deformation in the proper perspective, we shall defer detailed considerations of finite deformation to Chap. 8 wherein we consider geometrically nonlinear elasticity.

Part A
STRESS

1.2 FORCE DISTRIBUTIONS

In the study of continuous media, we are concerned with the manner in which forces are transmitted through a medium. At this time, we set forth two classes of forces that will concern us. The first is the so-called *body-force distribution* distinguished by the fact that it acts directly on the distribution of matter in the domain of specification. Accordingly, it is represented as a function of position and time and will be denoted as $\mathbf{B}(x,y,z,t)$ or, in index notation, as $B_i(x_1,x_2,x_3,t)$. The body force distribution is an intensity function and is generally evaluated per unit mass or per unit volume of the material acted on. (In the study of fluids, the basis of measure is usually per unit mass while in the study of solids the basis of measure is usually per unit volume.)

In discussing a continuum there may be some apparent physical boundary that encloses the domain of interest such as, for example, the outer surface of a beam. On the other hand, we may elect to specify a domain of interest and thereby generate a "mathematical" boundary. In either case, we will be concerned with the force distribution that is applied to such boundaries directly from material outside the domain of interest. We call such force distributions *surface tractions* and denote them as $\mathbf{T}(x,y,z,t)$ or $T_i(x_1,x_2,x_3,t)$. The surface traction is again an intensity, given on the basis of per unit area.

Now consider an infinitesimal area element on a boundary (see Fig. 1.1) over which we have a surface traction distribution $\mathbf{T}$ at some time t. The force df_i trans-

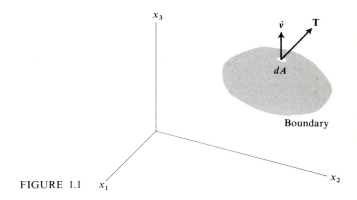

FIGURE 1.1

mitted across this area element can then be given as follows:

$$df_i = T_i \, dA$$

Note that **T** need not be normal to the area element and so this vector and the unit outward normal vector $\hat{\mathbf{v}}$ may have any orientation whatever relative to each other. We have not brought the unit normal $\hat{\mathbf{v}}$ into consideration thus far, but we will find it useful, as we proceed, to build into the notation for the surface traction a superscript referring to the direction of the area element at the point of application of the surface traction. Thus we will give the traction vector as

$$\mathbf{T}^{(v)}(x,y,z,t)$$

or as

$$T_i^{(v)}(x_1,x_2,x_3,t)$$

where $^{(v)}$ is not to be considered as a power. If the area element has the unit normal in the x_j direction, then we would express the traction vector on this element as $\mathbf{T}^{(j)}$ or as $T_i^{(j)}$. In the following section we shall show how we can use the superscript to good advantage.

1.3   STRESS

Consider now a vanishingly small rectangular parallelepiped taken at some time t from a continuum. Choose reference x_1, x_2, x_3 so as to be parallel to the edges of this rectangular parallelepiped as has been shown in Fig. 1.2. We have shown surface tractions on three rectangular boundary surfaces of the body. Note that we have

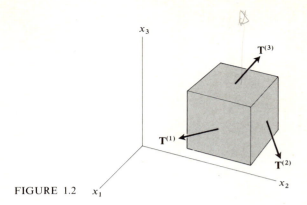

FIGURE 1.2

employed the superscript to identify the surfaces. The cartesian components of the vector $\mathbf{T}^{(1)}$ are then $T_1^{(1)}$, $T_2^{(1)}$, $T_3^{(1)}$. We shall now represent these components by employing τ in place of T and moving the superscript down to be the first subscript while deleting the enclosing parenthesis. Hence, the components of $\mathbf{T}^{(1)}$ are then given as $\tau_{11}, \tau_{12}, \tau_{13}$. In general we have

$$T_i^{(1)} \equiv \tau_{11}, \tau_{12}, \tau_{13}$$
$$T_i^{(2)} \equiv \tau_{21}, \tau_{22}, \tau_{23}$$
$$T_i^{(3)} \equiv \tau_{31}, \tau_{32}, \tau_{33}$$

In a more compact manner we have

$$T_j^{(i)} \equiv \tau_{ij}$$

where the nine quantities comprising τ_{ij} are called *stresses* and are forces per unit area wherein the first subscript gives the coordinate direction of the normal of the area element and the second subscript gives the direction of the force intensity itself. These nine force intensities are shown in Fig. 1.3 on three orthogonal faces of an infinitesimal rectangular parallelepiped with faces parallel to the coordinate planes of x_1, x_2, x_3.

Representing the set τ_{ij} as an array we have

$$\begin{vmatrix} \tau_{11} & \tau_{12} & \tau_{13} \\ \tau_{21} & \tau_{22} & \tau_{23} \\ \tau_{31} & \tau_{32} & \tau_{33} \end{vmatrix}$$

where the terms in the main diagonal are called normal stresses, since the force intensities corresponding to these stresses are normal to the surface, while the off-diagonal terms are the shear stresses.

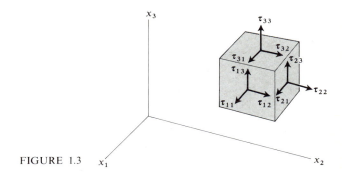

FIGURE 1.3

We shall employ the following sign convention for stresses. A normal stress directed outward from the interface is termed a tensile stress and is taken by definition as positive. A normal stress directed toward the surface is called a compressive stress (it is exactly the same as pressure on the surface) and is, by definition, negative. For shear stresses we employ the following convention:

> A shear stress is positive if (*a*) *both* the stress itself and the unit normal point in positive coordinate directions (the coordinates need not be the same), or (*b*) both point in the negative coordinate directions (again the coordinates need not be the same). A mixture of signs for coordinate directions corresponding to shear stress and the unit normal indicates a negative value for this shear stress.[1]

Note that all the stresses shown in Fig. 1.3 are positive stresses according to the above sign convention.

Knowing τ_{ij} for a set of axes, i.e., for three orthogonal interfaces at a point, we can determine a stress vector $\mathbf{T}^{(\nu)}$ for an interface at the point having *any* direction whatever. We shall now demonstrate this. Consider any point P in a continuum (Fig. 1.4(*a*)). Form as a free body a tetrahedron with P as a corner and with three orthogonal faces parallel to the reference planes as has been shown enlarged in Fig. 1.4(*b*). The legs of the tetrahedron are given as Δx_1, Δx_2, and Δx_3 as has been indicated in the diagram, and the inclined face $\overline{ABC}$ has a normal vector $\hat{\mathbf{\nu}}$. The stresses have been shown for the orthogonal faces as have the stress vector $\mathbf{T}^{(\nu)}$ and its components for the inclined face. We will denote by h the perpendicular distance from P to the inclined face (the "altitude" of the tetrahedron). The stresses and traction components given on the faces of the tetrahedron are average values over the surfaces on which they act. Also, the body force vector $(\mathbf{B})_{av}$ (not shown) represents the average intensity over the tetrahedron. Using these average values as well as an

---

[1] Note that normal stress actually follows this very same convention.

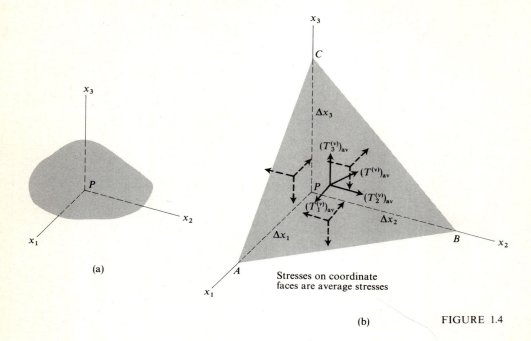

Stresses on coordinate
faces are average stresses

(b)   FIGURE 1.4

average density, ρ_{av}, we can express Newton's law for the mass center of the tetrahedron in the x_1 direction as follows

$$-(\tau_{11})_{av}\overline{PCB} - (\tau_{21})_{av}\overline{ACP} - (\tau_{31})_{av}\overline{APB} + (B_1)_{av}\rho_{av}\tfrac{1}{3}\overline{ABCh}$$
$$+ (T_1^{(v)})_{av}\overline{ABC} = \rho_{av}\tfrac{1}{3}\overline{ABC}ha_1 \tag{1.1}$$

where a_1 is the acceleration of the mass center in the x_1 direction. We can next replace
the average values by values taken at P itself, plus a small increment which goes to
zero as Δx_i goes to zero.[1] Thus we have

$$\begin{aligned}
(\tau_{11})_{av} &= (\tau_{11})_P + e_1 & (B_1)_{av} &= (B_1)_P + e_B \\
(\tau_{21})_{av} &= (\tau_{21})_P + e_2 & (T_1^{(v)})_{av} &= (T_1^{(v)})_P + e_T \\
(\tau_{31})_{av} &= (\tau_{31})_P + e_3 & \rho_{av} &= \rho_P + e_\rho
\end{aligned} \tag{1.2}$$

where e_1, e_2, e_3, e_B, e_T and e_ρ go to zero with Δx_i. Thus we have for Eq. (1.1):

$$-[(\tau_{11})_P + e_1]\overline{PCB} - [(\tau_{21})_P + e_2]\overline{ACP} - [(\tau_{31})_P + e_3]\overline{APB}$$
$$+ [(B_1)_P + e_B](\rho_P + e_\rho)(\tfrac{1}{3})(\overline{ABCh}) + [(T_1^{(v)})_P + e_T]\overline{ABC}$$
$$= \tfrac{1}{3}(\rho_P + e_\rho)(\overline{ABC})ha_1 \tag{1.3}$$

[1] We are thus tacitly assuming that the quantities in Eq. 1.1 vary in a continuous manner.

Next dividing through by area $\overline{ABC}$ and noting that $\overline{PCB}/\overline{ABC} = v_1$, $\overline{ACP}/\overline{ABC} = v_2$, etc., we have:

$$-[(\tau_{11})_P + e_1]v_1 - [(\tau_{21})_P + e_2]v_2 - [(\tau_{31})_P + e_3]v_3 +$$

$$[(B_1)_P + e_B](\rho_P + e_\rho)\frac{h}{3} + [(T_1^{(v)})_P + e_T] = (\rho_P + e_\rho)\frac{h}{3}a_1$$

Now go to the limit as $\Delta x_i \to 0$ in such a manner as to keep v_i constant. Clearly, the e's disappear and $h \to 0$. We then get in the limit:

$$-(\tau_{11})_P v_1 - (\tau_{21})_P v_2 - (\tau_{31})_P v_3 + (T_1^{(v)})_P = 0$$

Since P is any point of the domain we need no longer carry along the subscript. Rearranging terms we get:

$$T_1^{(v)} = \tau_{11}v_1 + \tau_{21}v_2 + \tau_{31}v_3 \qquad (1.4)$$

For any coordinate direction i we accordingly get:

$$T_i^{(v)} = \tau_{1i}v_1 + \tau_{2i}v_2 + \tau_{3i}v_3$$

And so using the summation convention, the stress vector $T_i^{(v)}$ can be given as

$$T_i^{(v)} = \tau_{ji}v_j$$

where you will note that v_j can be considered to give the direction cosines of the unit normal of the interface on which the traction force is desired. We will soon show that the stress terms τ_{ij} form a symmetric array and, accordingly the above equation can be put in the following form:

$$\boxed{T_i^{(v)} = \tau_{ij}v_j} \qquad (1.5)$$

Thus, knowing τ_{ij} we can get the traction vector for any interface at the point. The above is called *Cauchy's formula* and may be used to relate tractions on the boundary with stresses directly next to the boundary.

1.4 EQUATIONS OF MOTION

Consider an element of the body of mass dm at any point P. Newton's law requires that

$$d\mathbf{f} = dm\dot{\mathbf{V}}$$

where $d\mathbf{f}$ is the sum of the total traction force on the element and the total body force on the element. Integrating the above equation over some arbitrary spatial domain having a volume D and a boundary surface S, we note as a result of Newton's third law that only tractions on the bounding surface do not cancel out so that we have, using indicial notation:

$$\oiint_S T_i^{(v)} \, dA + \iiint_D B_i \, dv = \iiint_D \dot{V}_i \rho \, dv \qquad (1.6)$$

Now employ Eq. (1.5) to replace the stress vector $T_i^{(v)}$ by stresses. Thus:

$$\oiint_S \tau_{ji} v_j \, dv + \iiint_D B_i \, dv = \iiint_D \dot{V}_i \rho \, dv$$

Next employ Gauss' theorem for the first integral and collect terms under one integral. We get:

$$\iiint_D [\tau_{ji,j} + B_i - \dot{V}_i \rho] \, dv = 0$$

Since the domain D is arbitrary we conclude from above that at any point the following must hold:

$$\boxed{\tau_{ji,j} + B_i = \rho \dot{V}_i} \qquad (1.7)$$

This is the desired equation of motion.

Suppose next we consider an integral form of the moment of momentum equation derivable from Newton's law, i.e., $\mathbf{M} = \dot{\mathbf{H}}$. Thus we may say (considering Fig. 1.5):

$$\oiint_S \mathbf{r} \times \mathbf{T}^{(v)} \, dA + \iiint_D \mathbf{r} \times \mathbf{B} \, dv = \iiint_D \mathbf{r} \times \frac{d}{dt}(\mathbf{V} \, dm)$$

Considering the continuum to be composed of elements whose mass dm is constant but whose shape may be changing, we can express the integrand of the last expression as $\mathbf{r} \times \dot{\mathbf{V}} \, dm$. Thus we have for the above equation in tensor notation

$$\oiint_S \epsilon_{ijk} x_j T_k^{(v)} \, dA + \iiint_D \epsilon_{ijk} x_j B_k \, dv - \iiint_D \epsilon_{ijk} x_j \dot{V}_k \rho \, dv = 0$$

where we replace dm by $\rho \, dv$. Now replace $T_k^{(v)}$ by $\tau_{lk} v_l$ and employ Gauss' theorem. Thus we may write the above equation as:

$$\iiint_D \epsilon_{ijk}[(x_j \tau_{lk})_{,l} + x_j B_k - \rho x_j \dot{V}_k] \, dv = 0$$

Since the above is true for any domain D, we can set the integrand equal to zero. Carrying out differentiation of the first expression in the bracket and collecting terms

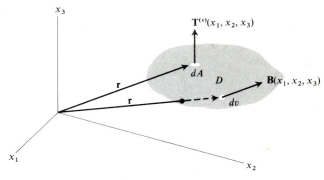

FIGURE 1.5

we then get:

$$\epsilon_{ijk}x_j[\tau_{lk,l} + B_k - \rho\dot{V}_k] + \epsilon_{ijk}x_{j,l}\tau_{lk} = 0$$

Because of Eq. (1.7) we can set the first expression equal to zero and so we get:

$$\epsilon_{ijk}x_{j,l}\tau_{lk} = 0$$

Noting that $x_{j,l} = \delta_{jl}$ we have

$$\epsilon_{ijk}\delta_{jl}\tau_{kl} = \epsilon_{ijk}\tau_{kj} = 0 \qquad (1.8)$$

It should be clear by considering the above equations for each value of the free index i, that the stresses with reversed indices with respect to each other are equal. That is:

$$\tau_{kj} = \tau_{jk} \qquad (1.9)$$

The stress components form a symmetric array.[1]

1.5 TRANSFORMATION EQUATIONS FOR STRESS

We have thus far shown how a stress vector $\mathbf{T}^{(v)}$ on an arbitrarily oriented interface at a point P in a continuum can be related to the set of nine stress components on a set of orthogonal interfaces at the point. Furthermore, we have found through consideration of Newton's law, that the stress terms form a symmetric array. We shall

---

[1] We have only considered body force distributions here. If one assumes *body-couple* distributions that may occur as a result possibly of magnetic or electric fields on certain kinds of dielectric and magnetic materials, we will have in Eq. (1.8) the additional integral $\iiint \mathbf{M}\, dv$ where $\mathbf{M}$ is the couple-moment vector per unit volume. The result is that Eq. (1.8) becomes $\epsilon_{ijk}\tau_{kj} + M_i = 0$. The stress tensor is now no longer a symmetric tensor. Such cases are beyond the scope of this text.

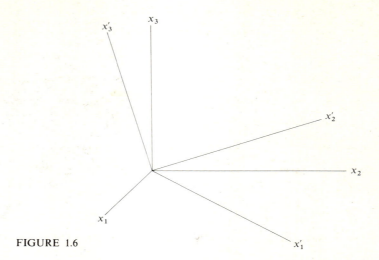

FIGURE 1.6

now show that the stress components for a set of Cartesian axes (i.e., for three ortho-gonal interfaces at a point) transform as a second-order tensor.

For this purpose consider Fig. 1.6 showing axes x_1, x_2, x_3 and x'_1, x'_2, x'_3 rotated arbitrarily relative to each other. Suppose we know the set of stresses for the un-primed reference, i.e., for a set of orthogonal interfaces having $x_1 x_2 x_3$ as edges, and wish to determine stresses for reference $x'y'z'$. Accordingly, in Fig. 1.7, we have shown a vanishingly small rectangular parallelepiped having $x'_1 x'_2 x'_3$ as edges so as to present the set of orthogonal interfaces for the desired stresses. The stress vectors for these interfaces have been shown as well as the corresponding stresses.

Suppose we wish to evaluate the stress component τ'_{31}. We can do this by first computing the stress vector on the interface corresponding to this stress (this interface is denoted as (3′)). The direction cosines v'_1, v'_2, and v'_3 for this interface can be given in our familiar notation as a_{31}, a_{32}, and a_{33} where the first subscript identifies the normal to interface (x'_3) and the second subscript identifies the proper *unprimed axis*. Thus we can say, using Eq. (1.5):

$$T_i^{(3')} = \tau_{ij} a_{3j} \qquad (1.10)$$

You must remember that the stress vector *components* for surface (3′) are given above as components along the *unprimed* axes. Hence, to get τ'_{31} we must project each one of the above components in a direction along the x'_1 direction. This is accomplished by taking an inner product using a_{1i} in the above equation.[1] Thus

$$\tau'_{31} = T_i^{(3')} a_{1i} = (\tau_{ij} a_{3j}) a_{1i} = a_{3j} a_{1i} \tau_{ij} = a_{3j} a_{1i} \tau_{ji}$$

[1] a_{1i} represents the set of direction cosines between the x'_1 axis and the x_1, x_2 and x_3 axes. It is therefore a *unit vector* in the x'_1 direction.

and they are called the *first, second,* and *third tensor invariants.* Thus

$$\tau_{11} + \tau_{22} + \tau_{33} = I_\tau$$
$$\tau_{11}\tau_{22} + \tau_{11}\tau_{33} + \tau_{22}\tau_{33} - \tau_{12}{}^2 - \tau_{23}{}^2 - \tau_{31}{}^2 = II_\tau$$
$$\tau_{11}\tau_{22}\tau_{33} - \tau_{11}\tau_{23}{}^2 - \tau_{22}\tau_{13}{}^2 - \tau_{33}\tau_{12}{}^2 + 2\tau_{12}\tau_{23}\tau_{13} = III_\tau \qquad (1.21)$$

The *first tensor invariant* is simply the sum of the terms of the left to right diagonal from τ_{11} to τ_{33}—the so-called *principal diagonal.* This sum is called the *trace* of the tensor and can be given in tensor notation as

$$I_\tau = \tau_{ii} \qquad (1.22)$$

The *second tensor invariant* is the sum of three sub-determinants formed from the matrix representation of the stress tensor. These submatrices form the minors of terms of the principal diagonal, and so II_τ is the sum of the principal minors. Thus we have:

$$II_\tau = \begin{vmatrix} \tau_{11} & \tau_{12} \\ \tau_{21} & \tau_{22} \end{vmatrix} + \begin{vmatrix} \tau_{22} & \tau_{23} \\ \tau_{32} & \tau_{33} \end{vmatrix} + \begin{vmatrix} \tau_{11} & \tau_{13} \\ \tau_{31} & \tau_{33} \end{vmatrix} \qquad (1.23)$$

Finally, the *third tensor invariant* can be seen to be simply the determinant of the tensor itself. Thus:

$$III_\tau = \begin{vmatrix} \tau_{11} & \tau_{12} & \tau_{13} \\ \tau_{21} & \tau_{22} & \tau_{23} \\ \tau_{31} & \tau_{32} & \tau_{33} \end{vmatrix} \qquad (1.24)$$

These tensor invariants to a great extent characterize a tensor, just as the invariant length, $V_i V_i$, of a vector characterizes to some extent a vector.

As a next step in this section, we shall show that the principal axes correspond to the directions for which the normal stress forms an *extremum* when compared to normal stresses in neighboring directions. Consider an arbitrary interface having a normal $\hat{\mathbf{v}}$ as shown in Fig. 1.9. The surface traction vector $\mathbf{T}^{(v)}$ is shown for the interface with its components parallel to the reference x_1, x_2, x_3. We can give the traction vector in terms of stresses associated with reference $x_1 x_2 x_3$ as follows using Cauchy's formula:

$$T_i^{(v)} = \tau_{ij} v_j \qquad (1.25)$$

To get the traction component, τ_{nn}, normal to the interface, we have

$$\tau_{nn} = T_i^{(v)} v_i = \tau_{ij} v_j v_i \qquad (1.26)$$

We now ask what direction $\hat{\mathbf{v}}$ will extremize τ_{nn}? A necessary condition for an extremum would be that the partials of the quantity $(\tau_{ij} v_i v_j)$ be zero with respect to v_1, v_2, and v_3 were these independent variables. However the variables v_1, v_2, and v_3 are not independent since there is

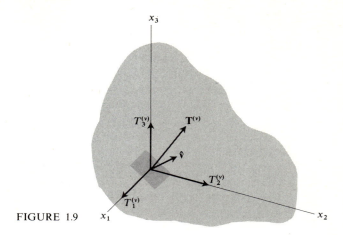

FIGURE 1.9

the constraining equation

$$v_i v_i = 1$$

that must always be satisfied. Accordingly, using λ as a Lagrange multiplier, we must extremize the function $[\tau_{ij}v_i v_j + \lambda(v_i v_i - 1)]$. Thus

$$\frac{\partial}{\partial v_s}[\tau_{ij}v_i v_j + \lambda(v_i v_i - 1)] = 0$$

We have here three equations. Noting that τ_{ij} is constant with respect to v_s:

$$\tau_{ij}v_i \frac{\partial v_j}{\partial v_s} + \tau_{ij}\frac{\partial v_i}{\partial v_s}v_j + 2\lambda v_i \frac{\partial v_i}{\partial v_s} = 0 \qquad s = 1, 2, 3$$

But

$$\frac{\partial v_j}{\partial v_s} = \delta_{js} \quad \text{and} \quad \frac{\partial v_i}{\partial v_s} = \delta_{is}$$

We get, on making these substitutions:

$$\tau_{ij}v_i \delta_{js} + \tau_{ij}\delta_{is}v_j + 2\lambda v_i \delta_{is} = 0$$

This becomes:

$$\tau_{is}v_i + \tau_{sj}v_j + 2\lambda v_i \delta_{is} = 0$$

Using the symmetry of the stress tensor in the second term and changing repeated indices j to i, we can put the above equation into the form:

$$(\tau_{is} + \lambda\delta_{is})v_i = 0 \qquad (1.27)$$

Notice that λ, the Lagrange multiplier, has the dimensions of stress and, accordingly, represents a stress quantity. Note further, that Eq. (1.27) is of identically the same form as Eq. (1.13) for determining the principal stresses σ_x. Accordingly, the three values of λ that you get on carrying

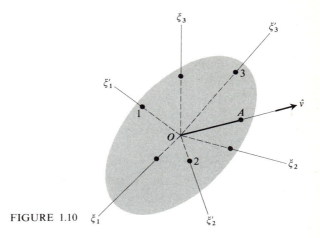

FIGURE 1.10

through the above calculations, $(\lambda_1, \lambda_2, \lambda_3)$, and their corresponding sets of direction cosines $[\overset{1}{v}_1, \overset{1}{v}_2, \overset{1}{v}_3], [\overset{2}{v}_1, \overset{2}{v}_2, \overset{2}{v}_3], [\overset{3}{v}_1, \overset{3}{v}_2, \overset{3}{v}_3]$ will be respectively identical to $\sigma_1, \sigma_2, \sigma_3$ and the corresponding sets of direction cosines. Thus the directions that emerge for extremizing the normal stress are the directions corresponding to the principal stresses.

Now consider the variations of normal stress τ_{nn} as the direction $\hat{\mathbf{v}}$ is varied at a point to cover all directions (Fig. 1.9). Using Cauchy's formula:

$$T_i^{(v)} v_i = \tau_{nn} = \tau_{ij} v_i v_j \qquad (1.28)$$

For convenience we lay off, using a second reference $\xi_1\xi_2\xi_3$, a distance $\overline{OA}$ along direction $\hat{\mathbf{v}}$ (see Fig. 1.10) such that

$$(\overline{OA})^2 = \frac{\pm d^2}{\tau_{nn}} \qquad (1.29)$$

where d is an arbitrary constant with dimensions to render $\overline{OA}$ dimensionless. The plus sign obviously must be used if τ_{nn} is positive (i.e., tensile stress) and the negative sign is used when τ_{nn} is negative (compression). We can then give the direction cosines v_i as follows:

$$v_i = \frac{\xi_i}{\overline{OA}} = \frac{\xi_i}{\sqrt{\pm d^2/\tau_{nn}}} \qquad (1.30)$$

Hence Eq. (1.28) can be given as:

$$\tau_{nn} = \tau_{ij}\left(\frac{\xi_i \xi_j}{\pm d^2/\tau_{nn}}\right) \qquad (1.31)$$

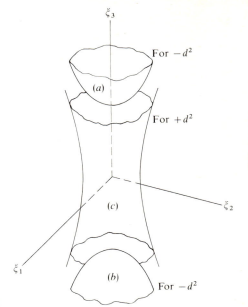

FIGURE 1.11

Cancelling τ_{nn} we get the following result on rearranging terms in the above equation

$$\xi_i \xi_j \tau_{ij} = \pm d^2 \qquad (1.32)$$

For a particular choice of sign of d^2, the above represents a real second-order surface in a particular region of $\xi_1 \xi_2 \xi_3$. We call this surface the *stress quadric*. The distance from the origin to this surface in some direction $\hat{\mathbf{v}}$ is inversely proportional to the square root of the normal stress τ_{nn} (see Eq. (1.29)) for the same direction $\hat{\mathbf{v}}$ in physical space (reference x_1, x_2, x_3). For instance with τ_{11}, τ_{22}, and τ_{33} positive, $+d^2$ in Eq. (1.32) gives the only real surface—that of an ellipsoid as has been shown in Fig. 1.10. It is apparent that for the three symmetrical semi-axes (ξ'_1, ξ'_2, ξ'_3) the distances from origin to surface are local extrema and so these directions correspond to the principal axes in accordance with the previous remarks. Since one of these distances (03) is a maximum for the ellipsoid, the corresponding principal stress must be the minimum normal stress at the point in the body and since the distance (01) is a minimum for the ellipsoid, the corresponding normal stress must be the maximum normal stress at the point in the body. The third principal stress corresponding to direction (02) must then have some intermediate value such that the sum of the principal stresses gives the proper first tensor invariant at the point in the body. Other kinds of second-order surfaces are possible if the signs of τ_{11}, τ_{22}, and τ_{33} are not all positive or all negative. Thus for the particular case where $\sigma_1 \geq \sigma_2 > 0$, $\sigma_3 < 0$ we get the surfaces shown in Fig. 1.11 where $+d^2$ is needed in Eq. (1.32) to generate surface (c) as a real surface and $-d^2$ is needed in Eq. (1.32) to generate (a) and (b) as real surfaces. The stress for any direction is found by measuring $\overline{OA}$ for that direction and employing Eq. (1.29) using the same sign with d^2 as is associated with that part of the surface intercepted by $\overline{OA}$. Thus the proper sign of τ_{nn} is then determined for use in Eq. (1.29) for that direction. The earlier conclusion that the largest (algebraically) normal stress and the smallest (algebraically) normal stress at a point are principal stresses still holds. We thus can see why principal stresses are so important in engineering work.

We will not formally use the so-called stress quadric. However, it does serve as a graphical representation of stress (or any other second-order symmetric tensor) just as an arrow is a graphical representation of a vector.

It should be clearly understood that the main ingredient in arriving at the conclusions in this section was the fact that stress transforms according to the formula

$$\tau'_{ij} = a_{ik}a_{jl}\tau_{kl}$$

and the fact that $\tau_{ij} = \tau_{ji}$. Thus, all the conclusions made concerning principal stresses, tensor-invariants, etc., apply to any *second-order symmetric tensor*. We shall make ample use of these results in studies to follow.

Part B

STRAIN

1.7 STRAIN COMPONENTS

We shall now propose means of expressing the deformation of a body. Accordingly we consider an undeformed body as shown in Fig. 1.12. If this body is given a rigid-body motion we know that each and every line segment in the body undergoes no change in length. Accordingly the change in length of line segments in the body or, more appropriately, the distance between points in the body, can serve as a measure of the deformation (change of shape and size) of the body. We therefore consider two points A and B in the body at positions x_i and $x_i + dx_i$ as shown in Fig. 1.12. The distance between these points is given by

$$(\overline{AB})^2 = (ds)^2 = dx_i\,dx_i \qquad (1.33)$$

When external forces are applied, the body will deform so that points A and B move to points A^* and B^* respectively. It is convenient now to consider that the x_i reference is labeled the ξ_i reference when considering the deformed state as has been

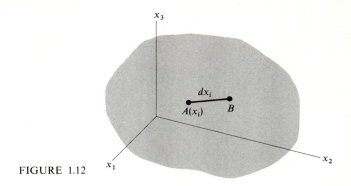

FIGURE 1.12

shown in Fig. 1.13. That is, we may consider the deformation as depicted by a mapping of each point from coordinate x_i to coordinate ξ_i. We can then say for a deformation

$$\xi_i = \xi_i(x_1, x_2, x_3) \qquad (1.34)$$

And since the mapping must be one-to-one we can expect a unique inverse to the above formulation in the form:

$$x_i = x_i(\xi_1, \xi_2, \xi_3) \qquad (1.35)$$

We can accordingly express the differentials dx_i and $d\xi_i$ as follows by making use of the above relations:

$$dx_i = \left(\frac{\partial x_i}{\partial \xi_j}\right) d\xi_j$$

$$d\xi_i = \left(\frac{\partial \xi_i}{\partial x_j}\right) dx_j \qquad (1.36)$$

This permits us to express $(ds)^2$ in Eq. (1.33) in the following way:

$$ds^2 = dx_i\, dx_i = \frac{\partial x_i}{\partial \xi_m} \frac{\partial x_i}{\partial \xi_k} d\xi_m\, d\xi_k \qquad (1.37)$$

It should then be clear that in the deformed state we have for the line segment $\overline{A^*B^*}$:

$$\overline{A^*B^*}^2 = (ds^*)^2 = d\xi_i\, d\xi_i$$

$$= \frac{\partial \xi_i}{\partial x_k} \frac{\partial \xi_i}{\partial x_l} dx_k\, dx_l \qquad (1.38)$$

Now we will examine the change in length of the segment—that is, we investigate the measure of deformation mentioned earlier. We may do this in either of two ways

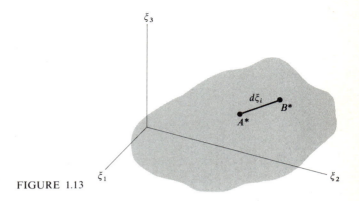

FIGURE 1.13

by using either of the above formulations (1.37) or (1.38). Thus:

$$(ds^*)^2 - (ds)^2 = \left(\frac{\partial \xi_k}{\partial x_i} \frac{\partial \xi_k}{\partial x_j} - \delta_{ij}\right) dx_i\, dx_j \qquad (1.39)$$

$$(ds^*)^2 - (ds)^2 = \left(\delta_{ij} - \frac{\partial x_k}{\partial \xi_i} \frac{\partial x_k}{\partial \xi_j}\right) d\xi_i\, d\xi_j \qquad (1.40)$$

We now rewrite the above equations as follows

$$
\begin{aligned}
(ds^*)^2 - (ds)^2 &= 2\epsilon_{ij}\, dx_i\, dx_j \quad &(a)\\
(ds^*)^2 - (ds)^2 &= 2\eta_{ij}\, d\xi_i\, d\xi_j \quad &(b)
\end{aligned}
\qquad (1.41)
$$

where we have introduced so-called strain terms:

$$
\begin{aligned}
\epsilon_{ij} &= \frac{1}{2}\left(\frac{\partial \xi_k}{\partial x_i} \frac{\partial \xi_k}{\partial x_j} - \delta_{ij}\right) \quad &(a)\\
\eta_{ij} &= \frac{1}{2}\left(\delta_{ij} - \frac{\partial x_k}{\partial \xi_i} \frac{\partial x_k}{\partial \xi_j}\right) \quad &(b)
\end{aligned}
\qquad (1.42)
$$

The set of terms ϵ_{ij} contains the implicit assumption that they are expressed as functions of the coordinates in the undeformed state—i.e., the so called *Lagrange coordinates*. The set of terms ϵ_{ij}, formulated by Green and St. Venant is called Green's strain tensor. The second set, η_{ij}, is formulated as a function of the coordinate for the deformed state—the so-called *Eulerian coordinates*. This form was introduced by Cauchy for infinitesimal strain (soon to be discussed) and by Almansi and Hamel for finite strains. It is often called the Almansi measure of strain.

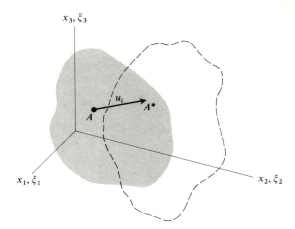

FIGURE 1.14

We now introduce the *displacement field* u_i defined such that

$$u_i = \xi_i - x_i \qquad (1.43)$$

Thus u_i gives the displacement of each point in the body from the initial undeformed configuration to the deformed configuration as has been shown in Fig. 1.14. We may express u_i as a function of the Lagrangian coordinates x_i, in which case, it expresses the displacement from the position x_i in the undeformed state to the deformed position ξ_i. On the other hand u_i can equally well be expressed in terms of ξ_i, the Eulerian coordinates, in which case it expresses the displacement that must have taken place to get to the position ξ_i from some undeformed configuration. The following relations can then be written from Eq. (1.43).

$$\frac{\partial x_i}{\partial \xi_j} = \delta_{ij} - \frac{\partial u_i}{\partial \xi_j} \qquad (a)$$

$$\frac{\partial \xi_i}{\partial x_j} = \frac{\partial u_i}{\partial x_j} + \delta_{ij} \qquad (b)$$

$$(1.44)$$

Substituting these results into Eq. (1.42) we obtain the following:

$$\epsilon_{ij} = \frac{1}{2}\left(\frac{\partial u_i}{\partial x_j} + \frac{\partial u_j}{\partial x_i} + \frac{\partial u_k}{\partial x_i}\frac{\partial u_k}{\partial x_j}\right) \qquad (a)$$

$$\eta_{ij} = \frac{1}{2}\left(\frac{\partial u_i}{\partial \xi_j} + \frac{\partial u_j}{\partial \xi_i} - \frac{\partial u_k}{\partial \xi_i}\frac{\partial u_k}{\partial \xi_j}\right) \qquad (b)$$

$$(1.45)$$

The Green strains ϵ_{ij} are thus referred to the initial undeformed geometry and indicate what must occur during a given deformation; the other strain terms η_{ij} are referred to the deformed or instantaneous geometry of the body and indicate what must have occurred to reach this geometry from an earlier undeformed state.

Up to this point in the discussion we have made no restriction on the magnitude of deformations for which Eq. (1.45) is valid. We now restrict ourselves to what is commonly called *infinitesimal strain* wherein the derivatives of the displacement components are small compared to unity. Thus:

$$\frac{\partial u_i}{\partial x_j} \ll 1 \qquad \frac{\partial u_i}{\partial \xi_j} \ll 1$$

With this in mind consider the following operator acting on an arbitrary function $J(x_i)$:

$$\frac{\partial J}{\partial \xi_i} = \frac{\partial J}{\partial x_j}\left(\frac{\partial x_j}{\partial \xi_i}\right) = \frac{\partial J}{\partial x_j}\left[\frac{\partial}{\partial \xi_i}(\xi_j - u_j)\right]$$

$$= \left[\delta_{ij} - \frac{\partial u_j}{\partial \xi_i}\right]\frac{\partial J}{\partial x_j}$$

For infinitesimal strain we may drop the term $\partial u_j/\partial \xi_i$ above to reach the result:[1]

$$\frac{\partial}{\partial \xi_i} = \frac{\partial}{\partial x_i}$$

and we conclude that we need no longer distinguish between Eulerian and Lagrangian coordinates in expressing strain. A further simplification occurs when we note that the product of the derivatives of displacement components in Eq. (1.45) can now be considered negligible compared to the linear terms. We then get the following formulation for strain[2]:

$$\epsilon_{ij} = \eta_{ij} = \frac{1}{2}\left(\frac{\partial u_i}{\partial x_j} + \frac{\partial u_j}{\partial x_i}\right) = \frac{1}{2}(u_{i,j} + u_{j,i}) \qquad (1.46)$$

---

[1] This assumes that $\partial J/\partial x_i$, $i = 1,2,3$ are all of the same order of magnitude. Thus consider

$$\frac{\partial J}{\partial \xi_1} = \left(1 - \frac{\partial u_1}{\partial x_1}\right)\frac{\partial J}{\partial x_1} - \frac{\partial u_2}{\partial x_1}\frac{\partial J}{\partial x_2} - \frac{\partial u_3}{\partial x_1}\frac{\partial J}{\partial x_3}$$

It is clear we can neglect $\partial u_1/\partial x_1$ compared to unity. To neglect the last two terms means that

$$\frac{\partial u_2}{\partial x_1}\frac{\partial J}{\partial x_2} \quad \text{and} \quad \frac{\partial u_3}{\partial x_1}\frac{\partial J}{\partial x_3}$$

must be small compared to $\partial J/\partial x_1$. For this to be true the $\partial J/\partial x_i$ must be of the same order of magnitude.

[2] You will recall that in developing the equations of equilibrium we employed only a single reference, and the tacit assumption taken was that the geometry employed for the equation was the *deformed* geometry. For infinitesimal deformation we may use the undeformed geometry rather than the deformed geometry for expressing equations of equilibrium. We shall discuss this question and other related questions in Chap. 8

Accordingly we shall now use only ϵ_{ij} for infinitesimal strain. In unabridged notation we have:

$$\epsilon_{xx} = \frac{\partial u_x}{\partial x} \qquad \epsilon_{xy} = \frac{1}{2}\left(\frac{\partial u_x}{\partial y} + \frac{\partial u_y}{\partial x}\right) = \frac{1}{2}\gamma_{xy}$$

$$\epsilon_{yy} = \frac{\partial u_y}{\partial y} \qquad \epsilon_{yz} = \frac{1}{2}\left(\frac{\partial u_y}{\partial z} + \frac{\partial u_z}{\partial y}\right) = \frac{1}{2}\gamma_{yz}$$

$$\epsilon_{zz} = \frac{\partial u_z}{\partial z} \qquad \epsilon_{xz} = \frac{1}{2}\left(\frac{\partial u_x}{\partial z} + \frac{\partial u_z}{\partial x}\right) = \frac{1}{2}\gamma_{xz} \qquad (1.47)$$

where the γ_{ij} are called the engineering shear strains.

1.8 PHYSICAL INTERPRETATIONS OF STRAIN TERMS

In the previous section we introduced the strain terms ϵ_{ij} by considering changes in length between points separated by infinitesimal distances ("adjacent points"). Clearly ϵ_{ij} also affords us directly some measure of the deformation of each vanishingly small element of the body and thus gives us a means of describing the deformation of the body as a whole.

We shall examine this local deformation in this section and we will find it helpful to employ for this purpose a small rectangular parallelepiped at point P in the body as shown in Fig. 1.15. Notice we have placed a Cartesian reference at P. Let us imagine next that the body has some deformation and let us focus on line $\overline{PQ} = \Delta y$. As shown in Fig. 1.16 point P moves to P^* and point Q moves to Q^* as a result of the deformation. The projection of $\overline{P^*Q^*}$ in the y direction, which we denote as $(\overline{P^*Q^*})_y$, is computed in terms of the original length Δy and in terms of the displacement of point P and Q in the following way:

$$(\overline{P^*Q^*})_y = \Delta y + (u_y)_Q - (u_y)_P$$

Now express $(u_y)_Q$ as a Taylor series in terms of $(u_y)_P$ in the above equation. We then get

$$(\overline{P^*Q^*})_y = \Delta y + \left[(u_y)_P + \left(\frac{\partial u_y}{\partial y}\right)_P \Delta y + \cdots\right] - (u_y)_P$$

$$= \Delta y + \left(\frac{\partial u_y}{\partial y}\right)_P \Delta y + \cdots \qquad (1.48)$$

The net y component of elongation of the segment $\overline{PQ}$ can then be given as follows:

$$(\overline{P^*Q^*})_y - \Delta y = \left(\frac{\partial u_y}{\partial y}\right)_P \Delta y + \cdots$$

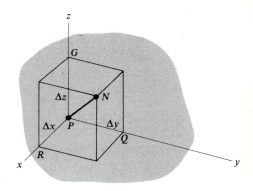

FIGURE 1.15

Dividing through by Δy and taking the limit of each term as $\Delta y \to 0$ we get

$$\frac{\overline{(P^*Q^*)}_y - \Delta_y}{\Delta y} = \frac{\partial u_y}{\partial y} = \epsilon_{yy}$$

where with the coalescence of points P and Q we may now drop the subscript P. We can conclude that the normal strain ϵ_{yy} at a point is the change in length in the y direction per unit original length of vanishingly small line segment originally in the y direction. For small deformations we can, in the above formulation, take $\overline{(P^*Q^*)}$ $= \overline{(P^*Q^*)}_y$. This permits us to say that the strain ϵ_{yy} is simply *the elongation per unit original length of a vanishingly small line segment originally in the y direction.* We can make corresponding interpretations for ϵ_{xx} and ϵ_{zz} or, given a direction p, for ϵ_{pp}.

Now let us consider respectively line segments $\overline{PR}$ of length Δx and $\overline{PQ}$ of length Δy in Fig. 1.17. In the deformed state P, Q, and R move to P^*, Q^*, and R^* respectively as we have shown in Fig. 1.17. We shall be interested in the projection of $\overline{P^*R^*}$ and $\overline{P^*Q^*}$ onto the xy plane—i.e., onto the plane of these line segments in the

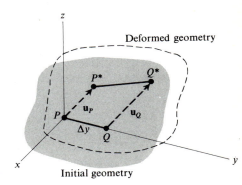

FIGURE 1.16

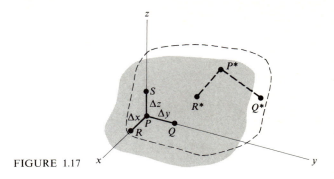

FIGURE 1.17

undeformed state. We have shown this projection in Fig. 1.18. Note that θ is the angle between the projection of $\overline{P^*Q^*}$ and the y axis while β is the angle between the projection of $\overline{P^*R^*}$ and the x axis. The displacement component of point P in the x direction has been shown simply as $(u_x)_P$ and the displacement component of point Q in the x direction has been given with the aid of a Taylor series expansion in the form

$$\left[(u_x)_P + \left(\frac{\partial u_x}{\partial y} \right)_P \Delta y + \cdots \right]$$

Finally the component of the projection of $\overline{P^*Q^*}$ taken in the y direction has been given as $\Delta y + \delta_y^2$ where δ_y^2 is a second-order increment for a small deformation. Now we can give $\tan \theta$ as follows:

$$\tan \theta = \frac{(\partial u_x/\partial y)_P \Delta y + \cdots}{\Delta y + \delta_y^2}$$

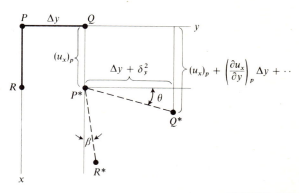

FIGURE 1.18

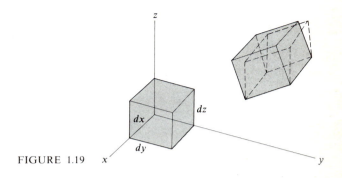

FIGURE 1.19

Take the limit as $\Delta y \rightarrow 0$. Higher-order terms in the numerator vanish. Also we delete the second-order increment $\delta_y{}^2$. We then have

$$\tan\theta = \theta = \left(\frac{\partial u_x}{\partial y}\right)$$

at any point P. Similarly we have for β:

$$\beta = \left(\frac{\partial u_y}{\partial x}\right)$$

The sum $(\theta + \beta)$ then can be directly related to the shear strain as follows:

$$(\theta + \beta) = \frac{\partial u_x}{\partial y} + \frac{\partial u_y}{\partial x} = 2\epsilon_{xy} = \gamma_{xy}$$

The sum of the angles $(\theta + \beta)$ and hence the engineering shear γ_{xy}, is the decrease in right angle of a pair of infinitesimal line segments originally in the x and y directions at P when we project the aforementioned deformed pair onto the xy plane in the undeformed geometry. Because of the infinitesimal deformation restriction, however, the change in right angle of the line segments themselves can be taken as equal to that of the aforementioned projections of these line segments. In general then, γ_{ij} *gives the change in right angle of vanishingly small line segments originally in the i and j directions at a point.*

Now consider the effects of strain on an infinitesimal rectangular parallelepipted in the undeformed geometry. With zero shear strain, the sides must remain orthogonal on deformation. However, the position and orientation of the element may change as may the length of sides and volume. This has been shown using full lines in Fig. 1.19. The shear strain can now be applied (the effects of normal strain and shear strain superpose in an uncoupled manner in accordance with Eq. (1.47). The result is that the orientation of the sides may lose their mutual perpendicularity so that we have parallelograms instead of rectangles for the sides (see dashed lines in Fig. 1.19). In short we can say that *the size of the rectangular parallelepiped is changed by normal strain while the basic shape is changed by the shear strain.* In fact you are asked to show (in the exercises) that for infinitesimal deformation the volume change per unit volume is simply ϵ_{ii}.

1.9   THE ROTATION TENSOR

In the previous sections we considered stretching of a line element to generate the strain tensor ϵ_{ij}, and then used the deformation of a vanishingly small rectangular parallelepiped to give physical interpretations to the components of the strain tensor.

We shall now go through a similar discussion so as to introduce the rotation tensor. This time, rather than considering just the stretch of a vanishingly small line element we now consider the *complete* mutual relative motion of the endpoints of the line element (thus we include rotation as well as stretching of the element). For this purpose consider line element $\overline{PN}$ in Fig. 1.15. The relative movement of the end points can be given using the displacement field as follows

$$\mathbf{u}_N - \mathbf{u}_P = \left[\mathbf{u}_P + \left(\frac{\partial \mathbf{u}}{\partial x_j} \right)_P \Delta x_j + \cdots \right] - \mathbf{u}_P \qquad (1.49)$$

where we have expanded $\mathbf{u}_N$ as a Taylor series about point P. In the limit as $\Delta x_i \to 0$ we have from the above equation:

$$d\mathbf{u} = \frac{\partial \mathbf{u}}{\partial x_j} dx_j$$

In index notation we get:

$$du_i = \frac{\partial u_i}{\partial x_j} dx_j \qquad (1.50)$$

Thus the relative movement du_i between the two adjacent points dx_i apart is determined by the tensor $u_{i,j}$. But this tensor can be written as follows:

$$u_{i,j} = \tfrac{1}{2}(u_{i,j} + u_{j,i}) + \tfrac{1}{2}(u_{i,j} - u_{j,i}) \qquad (1.51)$$

The first expression on the right side of the equation is the strain tensor ϵ_{ij}. The second expression is denoted as ω_{ij} and is called the *rotation tensor* for reasons soon to be

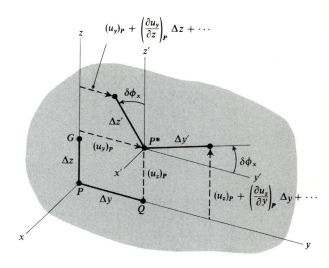

FIGURE 1.20

made clear. Note that the rotation tensor ω_{ij} is a *skew-symmetric* tensor. We can then express the above equation in the following manner

$$u_{i,j} = \epsilon_{ij} + \omega_{ij} \qquad (1.52)$$

where

$$\omega_{ij} = \frac{1}{2}(u_{i,j} - u_{j,i}) \qquad (1.53)$$

Thus the general relative movement between adjacent points is a result of the strain tensor plus a second effect which we will now interpret by examining again an infinitesimal parallelepiped having $\overline{PN}$ as a diagonal, as has been shown in Fig. 1.15. For now we assume that there is no deformation of the body—only rigid-body motion. Then line segments $\overline{PQ}$ and $\overline{PG}$ of the rectangular parallelepiped each undergo the same rotation $\delta\phi_x$ about the x axis. We have shown these lines (Fig. 1.20) after rigid body motion, projected onto plane $y'z'$ which is parallel to plane yz. We can then give $\sin \delta\phi_x$ as follows

$$\sin[\delta\phi_x] = \frac{[(u_z)_P + (\partial u_z/\partial y)_P \, \Delta y + \cdots] - (u_z)_P}{\Delta y'}$$

where $\Delta y'$ is the projection of $\overline{P^*Q^*}$ onto plane $y'z'$. Now take the limit as $\Delta y \to 0$. Noting for small rotations that we can replace the sine of the angle by the angle itself and that $\Delta y'$ may be replaced by Δy, we get for the above equation in the limit

$$\delta\phi_x = \frac{\partial u_z}{\partial y}$$

at any point P. Similarly considering line element PG we get for $\sin \delta\phi_x$,

$$\sin \delta\phi_x = \frac{(u_y)_P - [(u_y)_P + (\partial u_y/\partial z)_P \, \Delta z + \cdots]}{\Delta z'}$$

In the limit considering small deformations, we get:

$$\delta\phi_x = -\frac{\partial u_y}{\partial z}$$

Thus we may express $\delta\phi_x$ in the following manner

$$\delta\phi_x = \frac{1}{2}\left(\frac{\partial u_z}{\partial y} - \frac{\partial u_y}{\partial z}\right) \qquad (1.54)$$

For the other two components of rotation we have, on permuting subscripts,

$$\delta\phi_y = \frac{1}{2}\left(\frac{\partial u_x}{\partial z} - \frac{\partial u_z}{\partial x}\right)$$

$$\delta\phi_z = \frac{1}{2}\left(\frac{\partial u_y}{\partial x} - \frac{\partial u_x}{\partial y}\right) \qquad (1.55)$$

The expressions on the right sides of Eqs. (1.54) and (1.55) are clearly the off-diagonal terms of the rotation tensor ω_{ij}. Thus we can say

$$\delta\phi_x = \omega_{32} = -\omega_{23}$$

$$\delta\phi_y = \omega_{13} = -\omega_{31}$$

$$\delta\phi_z = \omega_{21} = -\omega_{12} \qquad (1.56)$$

We see that for a rigid-body movement of the element the non-zero components of the rotation tensor give the infinitesimal rotation components of the element. What does ω_{ij} represent when the rectangular parallelepiped is undergoing a movement including deformation of the element and not just rigid-body rotation? For this case each line segment in the rectangular volume has its own angle of rotation and we have shown in Appendix II that ω_{ij} for such a situation gives the *average rotation* components of all the line segments in the body. We shall term the components of ω_{ij}, however, the rigid-body rotation components.

In summary, by considering the general relative motion between adjacent points and by then going to the infinitesimal rectangular parallelepiped we have by the first step introduced ω_{ij} and by the second step given a physical interpretation for ω_{ij}. Now going back to the adjacent point approach we would expect that the strain tensor should bear direct relation to the force intensities (i.e., the stress tensor) at a point. (This is to be expected from your studies in atomic physics where inter-atomic forces were considered as a function of the separation of atoms.) Indeed we know from experiment that it is the ϵ_{ij} portion of Eq. (1.52) that is related to the stress τ_{ij} at a point. We shall discuss constitutive laws shortly.

1.10 TRANSFORMATION EQUATIONS FOR STRAIN

Since $u_{i,j}$ must be a second-order tensor (see the conclusions of Section I.7 of Appendix I on taking the partial derivative of a tensor), clearly ϵ_{ij} must be a second-order tensor. However, we shall show this formally here since this is a simple in-structive procedure. We shall first consider the normal strain at a point P in the direction $\hat{\alpha}$ (see Fig. 1.21). For this purpose a line segment $\varDelta \alpha$ has been shown con-necting points P and Q in the undeformed geometry. The positions of P and Q in the deformed geometry have been shown as P^* and Q^* respectively. We express the displacement component of point Q in the α direction in terms of the displacement component of point P in the α direction as follows:

$$(u_\alpha)_Q = (u_\alpha)_P + \left(\frac{\partial u_\alpha}{\partial x_i}\right)_P \varDelta x_i + \frac{1}{2!}\left(\frac{\partial^2 u_\alpha}{\partial x_i\,\partial x_j}\right)_P \varDelta x_i\, \varDelta x_j + \cdots$$

Therefore

$$\frac{(u_\alpha)_Q - (u_\alpha)_P}{\varDelta \alpha} = \left(\frac{\partial u_\alpha}{\partial x_i}\right)_P\left(\frac{\varDelta x_i}{\varDelta \alpha}\right) + \frac{1}{2!}\left(\frac{\partial^2 u_\alpha}{\partial x_i\,\partial x_j}\right)_P\frac{\varDelta x_i}{\varDelta \alpha}\varDelta x_j + \cdots$$

Take the limit as $\varDelta x_i$ and $\varDelta \alpha$ go to zero. The left side clearly is the normal strain $\epsilon_{\alpha\alpha}$. On the right side only the first term remains, so that we have

$$\epsilon_{\alpha\alpha} = \left(\frac{\partial u_\alpha}{\partial x_i}\right)\left(\frac{dx_i}{d\alpha}\right) = \left(\frac{\partial u_\alpha}{\partial x_i}\right)a_{\alpha i} \qquad (1.57)$$

where $a_{\alpha i}$ is the direction cosine between the α axis and the x_i axis. Now express u_α in terms of u_i as follows by projecting components u_i in the direction α:

$$u_\alpha = a_{\alpha j} u_j$$

Substituting into Eq. (1.57) we get

$$\epsilon_{\alpha\alpha} = u_{j,i}a_{\alpha i}a_{\alpha j} = [\tfrac{1}{2}(u_{j,i} + u_{i,j})]a_{\alpha i}a_{\alpha j}$$

$$= \epsilon_{ij}a_{\alpha i}a_{\alpha j} = a_{\alpha i}a_{\alpha j}\epsilon_{ij}$$

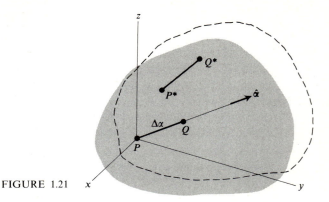

FIGURE 1.21

where we have interchanged dummy indices i and j in expanding $u_{j,i}$ above. We can imagine that the α direction coincides with the axes of a primed reference x_i'. Accordingly we may then say:

$$\epsilon_{\alpha\alpha}' = a_{\alpha i}a_{\alpha j}\epsilon_{ij} \qquad (1.58)$$

Now consider two segments $\overline{PQ}$ and $\overline{PR}$ at right angles to each other in the directions $\hat{n}$ and $\hat{s}$ in the undeformed geometry as has been shown in Fig. 1.22. The points P^*, R^*, and Q^* correspond to the deformed geometry. Considering n and s to be coordinate axes, the shear strain ϵ_{ns} for these axes is given as:

$$\epsilon_{ns} = \frac{1}{2}\left(\frac{\partial u_n}{\partial s} + \frac{\partial u_s}{\partial n}\right)$$

Expressing u_n and u_s as follows using vector projections,

$$u_n = a_{nj}u_j$$
$$u_s = a_{sj}u_j \qquad (1.59)$$

we can write ϵ_{ns} in the following way:

$$\epsilon_{ns} = \frac{1}{2}\left(\frac{\partial u_j}{\partial s}a_{nj} + \frac{\partial u_j}{\partial n}a_{sj}\right) \qquad (1.60)$$

Next, using the chain rule we note that:

$$\frac{\partial u_j}{\partial s} = \left(\frac{\partial u_j}{\partial x_k}\right)\left(\frac{\partial x_k}{\partial s}\right) = \frac{\partial u_j}{\partial x_k}a_{sk}$$

$$\frac{\partial u_j}{\partial n} = \left(\frac{\partial u_j}{\partial x_k}\right)\left(\frac{\partial x_k}{\partial n}\right) = \frac{\partial u_j}{\partial x_k}a_{nk} \qquad (1.61)$$

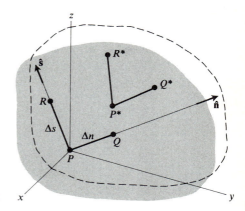

FIGURE 1.22

Substituting Eq. (1.61) into Eq. (1.60) we get:

$$\epsilon_{ns} = \frac{1}{2}\left(\frac{\partial u_j}{\partial x_k} a_{sk} a_{nj} + \frac{\partial u_j}{\partial x_k} a_{nk} a_{sj}\right)$$

In the second expression on the right side of the equation interchange the dummy symbols j and k. We thus get:

$$\epsilon_{ns} = \frac{1}{2}\left(\frac{\partial u_j}{\partial x_k} + \frac{\partial u_k}{\partial x_j}\right) a_{sk} a_{nj} = a_{nj} a_{sk} \epsilon_{jk}$$

We may think of n and s as a pair of primed axes and so we have:

$$\epsilon'_{ns} = a_{nj} a_{sk} \epsilon_{jk} \qquad (1.62)$$

And if $n = s = \alpha$ in the above formulation, we get back to Eq. (1.58). Thus we have reached the transformation equation of strain at a point from an unprimed reference to a primed rotated reference and the transformation equation is that of a second-order tensor.

The various properties set forth for the stress tensor stemming from the transformation equations then apply to the strain tensor.

1.11 COMPATIBILITY EQUATIONS

Let us consider further the strain-displacement relations

$$\epsilon_{ij} = \tfrac{1}{2}(u_{i,j} + u_{j,i}) \qquad (1.63)$$

If the displacement field is given, we can readily compute the strain-tensor field by substituting u_i into the above equations. The inverse problem of finding the displacement field from a strain field is not so simple. Here the displacement field, composed of *three* functions u_i, must be determined by integration of *six* partial differential equations given by Eq. (1.63). In order to insure *single-valued*, *continuous* solutions u_i, we must impose certain restrictions on ϵ_{ij}. That is, we cannot set forth any strain field ϵ_{ij} and expect it automatically to be associated with a single-valued continuous displacement field. But actual deformations must have single-valued displacement fields. Furthermore the deformations of interest to us will be those having continuous displacement fields. Hence the restrictions we will reach in rendering u_i single-valued and continuous apply to our formulations. The resulting equations are called the *compatibility equations*.

We shall first set forth necessary conditions on ϵ_{ij} for single-valuedness and continuity of u_i. The use of Eq. (1.63) to give ϵ_{ij} requires that u_i have the aforestated properties. Accordingly we shall insure these properties for u_i by working directly with these equations. We now form the following derivatives from these equations.

$$\epsilon_{ij,kl} = \tfrac{1}{2}(u_{i,jkl} + u_{j,ikl})$$

$$\epsilon_{kl,ij} = \tfrac{1}{2}(u_{k,lij} + u_{l,kij})$$

$$\epsilon_{lj,ki} = \tfrac{1}{2}(u_{j,lik} + u_{l,jik})$$

$$\epsilon_{ki,lj} = \tfrac{1}{2}(u_{k,ijl} + u_{i,kjl})$$

By adding the first two equations and then subtracting the last two equations we may eliminate the u_i components and thus arrive at a set of relations involving only strains. That is:

$$\boxed{\epsilon_{ij,kl} + \epsilon_{kl,ij} - \epsilon_{lj,ki} - \epsilon_{ki,lj} = 0} \qquad (1.64)$$

These form a set of eighty-one equations known as the *compatibility equations* that a strain field must satisfy if it is to be related to u_i via Eq. (1.63) which in turn means that u_i is single-valued and continuous. These equations are thus *necessary requirements*.

We shall now consider the *sufficiency* of these equations for generating a single-valued, continuous displacement field for the case of simply-connected regions.[1]

Let $P^0(x_1, x_2, x_3)$ be a point at which displacement $u_i{}^0$ and rotation terms $\omega_{ij}{}^0$ are known. Then the displacement of *any other point* P^* in the domain is representable by a line integral along a continuous curve C from P^0 to P^* in the following way:

$$u_i{}^* = u_i{}^0 + \int_{P^0}^{P^*} du_i = u_i{}^0 + \int_{P^0}^{P^*} u_{i,j}\, dx_j$$

---

[1] A simply-connected region is one for which each and every curve can be shrunk to a point without cutting a boundary.

We now employ Eq. (1.52) in the last expression to give us:

$$u_i^* = u_i^0 + \int_{P^0}^{P^*} \epsilon_{ij}\, dx_j + \int_{P^0}^{P^*} \omega_{ij}\, dx_j \qquad (1.65)$$

Since x_j^* is a given fixed point, as far as integration is concerned, we can employ $d(x_j - x_j^*)$ in the last line integral. Integrating this expression by parts we then may say:

$$\int_{P^0}^{P^*} \omega_{ij}\, d(x_j - x_j^*) = \omega_{ij}(x_j - x_j^*)\Big|_{P^0}^{P^*} - \int_{P^0}^{P^*} (x_j - x_j^*)\, d\omega_{ij}$$

$$= -\omega_{ij}^0(x_j^0 - x_j^*) - \int_{P^0}^{P^*} (x_j - x_j^*)\omega_{ij,k}\, dx_k \qquad (1.66)$$

Now the expression $\omega_{ij,k}$ can be written as follows

$$\omega_{ij,k} = \tfrac{1}{2}(u_{i,jk} - u_{j,ik}) = \tfrac{1}{2}(u_{i,jk} + u_{k,ij}) - \tfrac{1}{2}(u_{j,ik} + u_{k,ij})$$

where we have added and subtracted $\tfrac{1}{2}u_{k,ij}$. From this we conclude:

$$\omega_{ij,k} = \tfrac{1}{2}(u_{i,kj} + u_{k,ij}) - \tfrac{1}{2}(u_{j,ki} + u_{k,ji})$$

$$= (\epsilon_{ik,j} - \epsilon_{jk,i}) \qquad (1.67)$$

Substituting from Eqs. (1.67) and (1.66) into Eq. (1.65) we get

$$u_i^* = u_i^0 - \omega_{ij}^0(x_j^0 - x_j^*) + \int_{P^0}^{P^*} U_{ik}\, dx_k \qquad (a)$$

where

$$U_{ik} = \epsilon_{ik} - (x_j - x_j^*)(\epsilon_{ik,j} - \epsilon_{jk,i}) \qquad (b) \quad (1.68)$$

If u_i^* is to be single-valued and continuous, the integral of Eq. (1.68(a)) must be independent of the path. This in turn means that the integrand $U_{ik}\, dx_k$ must be an exact differential. In a simply-connected domain the necessary and sufficient condition for ($U_{ik}\, dx_k$) to be an exact differential is that:

$$U_{ik,l} - U_{il,k} = 0 \qquad (1.69)$$

Now using Eq. (1.68(b)) in the above equation and noting that $x_{j,l} = \delta_{jl}$, we find that:

$$\varepsilon_{ik,l} - \delta_{jl}(\epsilon_{ik,j} - \epsilon_{jk,i}) - (x_j - x_j^*)(\epsilon_{ik,jl} - \epsilon_{jk,il}) - \epsilon_{il,k}$$
$$+ \delta_{jk}(\epsilon_{il,j} - \epsilon_{jl,i}) + (x_j - x_j^*)(\epsilon_{il,jk} - \epsilon_{jl,ik}) = 0$$

Rearranging the equation and simplifying we get:

$$\epsilon_{ik,l} - \epsilon_{il,k} - (\epsilon_{ik,l} - \epsilon_{lk,i}) + (\epsilon_{il,k} - \epsilon_{kl,i})$$
$$+ (x_j - x_j^*)(\epsilon_{il,jk} + \epsilon_{jk,il} - \epsilon_{ik,jl} - \epsilon_{jl,ik}) = 0$$

The first six terms cancel each other, while the expression multiplied by $(x_j - x_j^*)$, set equal to zero, gives us the compatibility equation presented earlier. We have thus proved the sufficiency requirements for the compatibility equations.

Actually only six of the eighty-one equations of compatibility are independent; the rest are either identities or repetitions due to the symmetry of ϵ_{ij}. The six independent equations of compatibility are given as follows in unabridged notation:

$$\frac{\partial^2 \epsilon_{xx}}{\partial y^2} + \frac{\partial^2 \epsilon_{yy}}{\partial x^2} = \frac{\partial^2 \gamma_{xy}}{\partial x \, \partial y} \qquad (a)$$

$$\frac{\partial^2 \epsilon_{yy}}{\partial z^2} + \frac{\partial^2 \epsilon_{zz}}{\partial y^2} = \frac{\partial^2 \gamma_{yz}}{\partial y \, \partial z} \qquad (b)$$

$$\frac{\partial^2 \epsilon_{zz}}{\partial x^2} + \frac{\partial^2 \epsilon_{xx}}{\partial z^2} = \frac{\partial^2 \gamma_{zx}}{\partial z \, \partial x} \qquad (c)$$

$$2\frac{\partial^2 \epsilon_{xx}}{\partial y \, \partial z} = \frac{\partial}{\partial x}\left(-\frac{\partial \gamma_{yz}}{\partial x} + \frac{\partial \gamma_{xz}}{\partial y} + \frac{\partial \gamma_{xy}}{\partial z} \right) \quad (d)$$

$$2\frac{\partial^2 \epsilon_{yy}}{\partial z \, \partial x} = \frac{\partial}{\partial y}\left(-\frac{\partial \gamma_{zx}}{\partial y} + \frac{\partial \gamma_{yx}}{\partial z} + \frac{\partial \gamma_{yz}}{\partial x} \right) \quad (e)$$

$$2\frac{\partial^2 \epsilon_{zz}}{\partial x \, \partial y} = \frac{\partial}{\partial z}\left(-\frac{\partial \gamma_{xy}}{\partial z} + \frac{\partial \gamma_{zy}}{\partial x} + \frac{\partial \gamma_{zx}}{\partial y} \right) \quad (f) \quad (1.70)$$

Part C
GENERAL CONSIDERATIONS

1.12 ENERGY CONSIDERATIONS

We have described the stress tensor arising from equilibrium considerations and the strain tensor arising from kinematical considerations. These tensors are related to each other, as noted earlier, by laws that are called *constitutive laws*. In general such relations include as other variables temperature and time and, in addition, often require knowledge of the history of the deformations leading to the instantaneous condition of interest in order to properly relate stress and strain. In this text we shall consider that the constitutive laws relate stress and strain directly and uniquely. That is:

$$\tau_{ij} = \tau_{ij}(\epsilon_{11}, \epsilon_{12}, \ldots, \epsilon_{33}) \qquad (1.71)$$

We will discuss specific constitutive laws later in the text. Now it will be of interest to us to consider such constitutive laws that render the following integral

$$U_0 = \int_0^{\epsilon_{ij}} \tau_{ij} \, d\epsilon_{ij} \qquad (1.72)$$

a point function of the upper limit ϵ_{ij}. For U_0 to be a point function the integral must be independent of the path and this in turn means that $\tau_{ij} \, d\epsilon_{ij}$ must be a perfect

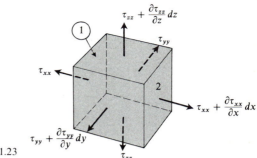

FIGURE 1.23

differential. When such is the case we can say:

$$dU_0 = \tau_{ij}\,d\epsilon_{ij} \quad (a)$$

therefore

$$\frac{\partial U_0}{\partial \epsilon_{ij}} = \tau_{ij} \quad (b) \quad (1.73)$$

The function U_0 under such circumstances is called the *strain energy density function*. We now pose two queries—what physical attributes can be ascribed to this function and when do they exist? As to the first query consider an infinitesimal rectangular parallelepiped under the action of normal stresses as shown in Fig. 1.23. The displacements on faces (1) and (2) in the x direction are given as u_x and $u_x + (\partial u_x/\partial x)\,dx$ respectively so that the increment of mechanical work done by the stresses on the element during deformation is

$$-\tau_{xx}\,du_x\,dy\,dz + \left(\tau_{xx} + \frac{\partial \tau_{xx}}{\partial x}\,dx\right)d\left(u_x + \frac{\partial u_x}{\partial x}\,dx\right)dy\,dz$$

$$+ B_x\,dx\,dy\,dz\,d\left(u_x + \kappa\frac{\partial u_x}{\partial x}\,dx\right)$$

where κ is some factor between 0 and 1. Collecting terms and deleting higher-order expressions we then get for the above expressions:

$$\left\{\tau_{xx}\,d\left(\frac{\partial u_x}{\partial x}\right) + \left(\frac{\partial \tau_{xx}}{\partial x} + B_x\right)du_x\right\}dx\,dy\,dz$$

We may employ equilibrium considerations to delete the second expression in the bracket leaving us with the following expression for the increment of mechanical

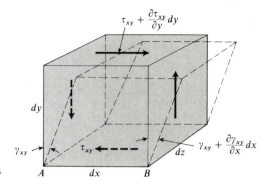

FIGURE 1.24

work:

$$\tau_{xx} \, d\left(\frac{\partial u_x}{\partial x}\right) dx \, dy \, dz = \tau_{xx} \, d\epsilon_{xx} \, dv$$

By considering normal stress and strain in the y and z directions we may form similar expressions for the element. Thus, for normal stresses on an element, the increment of mechanical work is then

$$(\tau_{xx} \, d\epsilon_{xx} + \tau_{yy} \, d\epsilon_{yy} + \tau_{zz} \, d\epsilon_{zz}) \, dv$$

We shall denote w as the mechanical work per unit volume. From the above expression we see that for normal stresses only we have:

$$dw = \tau_{xx} \, d\epsilon_{xx} + \tau_{yy} \, d\epsilon_{yy} + \tau_{zz} \, d\epsilon_{zz}$$

Next consider a case of pure shear strain such as is shown in Fig. 1.24. The mechanical increment of work can be given here as follows

$$\left[\left(\tau_{xy} + \frac{\partial \tau_{xy}}{\partial y} dy\right) dz \, dx\right]\left\{d\left[\gamma_{xy} + \beta\left(\frac{\partial \gamma_{xy}}{\partial x}\right) dx\right] dy\right\}$$

$$+ B_x \, dx \, dy \, dz \, d\left(\gamma_{xy} + \eta \frac{\partial \gamma_{xy}}{\partial x} dx\right)(\kappa \, dy)$$

where β, η, and κ are factors between 0 and 1. Now carrying out arithmetical operations and dropping higher-order terms we get for the above expression:

$$\tau_{xy} \, d\gamma_{xy} \, dx \, dy \, dz = 2\tau_{xy} \, d\epsilon_{xy} \, dv$$

Thus for pure shear stresses on all faces we get the following result for the increment of mechanical work:

$$2(\tau_{xy}\, d\epsilon_{xy} + \tau_{xz}\, d\epsilon_{xz} + \tau_{yz}\, d\epsilon_{yz})\, dv$$

Accordingly the mechanical work increment per unit volume at a point for a general state of stress is then given as:

$$dw = \tau_{ij}\, d\epsilon_{ij} \qquad (1.74)$$

It should be clear that the above result is valid only for infinitesimal deformations. And now integrating from 0 to some strain level ϵ_{ij} we get

$$w = \int_0^{\epsilon_{ij}} \tau_{ij}\, d\epsilon_{ij} = U_0$$

in accordance with Eq. (1.72). Thus we can conclude that U_0, the strain energy density, is the mechanical work performed *on* an element per unit volume at a point during a deformation. The total strain energy U then becomes

$$U = \iiint_V U_0\, dv = \iiint_V \left[\int_0^{\epsilon_{ij}} \tau_{ij}\, d\epsilon_{ij} \right] dv \qquad (a)$$

where

$$U_0 = \int_0^{\epsilon_{ij}} \tau_{ij}\, d\epsilon_{ij} \qquad (b) \quad (1.75)$$

As to the second query (when does U_0 exist) we may proceed using two different arguments. First we may rely on a strictly mathematical approach whereby it is known from the calculus that for $\tau_{ij}\, d\epsilon_{ij}$ to be a perfect differential (and thus to permit existence for U_0) it is necessary and sufficient that:

$$\frac{\partial \tau_{ij}}{\partial \epsilon_{kl}} = \frac{\partial \tau_{kl}}{\partial \epsilon_{ij}} \qquad (1.76)$$

On the other hand, we may resort to arguments from *thermodynamics*. The *first law of thermodynamics* states that

$$du = d'q + d'w_k$$

where u is the internal energy per unit volume (specific internal energy), q is the heat transfer per unit volume, and w_k is the work done on the system per unit volume by the surroundings. The primes associated with the differentials of q and w_k are to indicate that these quantities are not *perfect differentials*—they represent simply vanishingly small increments of these quantities. In other words, u is a *point function* depending on the state of the system whereas q and w_k are *path functions* depending on the process.

We can say for a solid that the first law using Eq. (1.74) may be given as follows:

$$du = d'q + \tau_{ij}\, d\epsilon_{ij} \qquad (1.77)$$

The second law of thermodynamics indicates that no process is possible in an isolated system which would decrease a property called the *entropy* of the system. The specific entropy s in a reversible process is related to q and the absolute temperature T according to the following equation:

$$d'q = T\, ds \qquad (1.78)$$

Combining Eqs. (1.77) and (1.78) to eliminate $d'q$ we get a form of the *combined first and second law* given as:

$$du = T\, ds + \tau_{ij}\, d\epsilon_{ij} \qquad (1.79)$$

Let us now examine a reversible *adiabatic process* (no heat transfer). For such a process there can be no change in the entropy and the above equation becomes:

$$du = \tau_{ij}\, d\epsilon_{ij}$$

Since u is a point function we can conclude for this process that $\tau_{ij}\, d\epsilon_{ij}$ is a perfect differential. Indeed, noting Eq. (1.73(a)) we can conclude that the strain energy density is simply the specific internal energy for such processes (isentropic processes).

Let us now consider another thermodynamic function used in thermodynamics called the *Helmholtz function*, F, defined as a specific value in the following way:

$$F = u - Ts$$

Since u, T, and s are point functions, clearly F must be a point function. Now consider a reversible *isothermal process*. The above equation in differential form becomes for such a process:

$$dF = du - T\, ds \qquad (1.80)$$

Now replacing du in Eq. (1.79) using Eq. (1.80) we arrive at the result:

$$dF = \tau_{ij}\, d\epsilon_{ij} \qquad (1.81)$$

Once again $\tau_{ij}\, d\epsilon_{ij}$ is a perfect differential. Indeed for such processes we conclude from Eq. (1.73(a)) that the strain energy density function equals the Helmholtz function.[1]

We thus see that the energy density function U_0 exists for certain reversible processes. Because of the reversibility requirement we see that the existence of U_0

---

[1] From our discussion thus far, we can say that the Helmholtz function represents the energy that can be converted to mechanical work in a reversible isothermal process. This is often called *free energy*. This is one reason for its importance in thermodynamics.

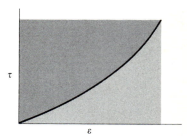

$$\tau$$

FIGURE 1.25 ε

calls for *elastic behavior* of the bodies in that the body must return identically to its original condition when the loads are released. In most situations in elasticity, furthermore, the process is somewhere between isothermal and adiabatic. Since in solid mechanics, unlike fluid mechanics, the difference between the adiabatic and isothermal processes is usually not great, the strain energy function U_0 is assumed to exist for most processes involving elastic behavior.

Furthermore it can be shown from thermodynamics considerations that the strain energy density functions are *positive-definite* functions[1] of the strains for small deformation. We will ask you to demonstrate this in Problem 1.18 for the case of linear elastic materials.

Let us next consider a stress–strain diagram for a tensile test in a case where the energy density function exists (Fig. 1.25). For τ to be a single-valued function of ϵ the curve τ vs ϵ must be monotonically increasing as has been shown in the diagram. Furthermore the value of U_0 is simply the area under the curve for a given strain.

Now just as the curve generates a single-valued function U_0 as the area between the curve and the ϵ axis, so does it generate a single-valued function, giving the area between the curve and the τ axis (shown darkened in the diagram). This function is called the *complementary energy density function* and is denoted as U_0^*. For the one-dimensional case we have:

$$dU_0^* = \epsilon \, d\tau$$

therefore

$$U_0^* = \int_0^\tau \epsilon \, d\tau$$

For a general state of stress at a point we may generalize the above result as follows:

$$dU_0^* = \epsilon_{ij} \, d\tau_{ij} \quad (a)$$

[1] A positive definite function has the property of never being less than zero for the range of the variables involved and is zero only when the variables are zero. This means $U_0 = 0$ at a point when the strains are zero at the point.

therefore

$$U_0^* = \int_0^{\tau_{ij}} \epsilon_{ij}\, d\tau_{ij} \qquad (b) \quad (1.82)$$

It is apparent from Eq. (1.82) (a) that:

$$\frac{\partial U_0^*}{\partial \tau_{ij}} = \epsilon_{ij} \qquad (1.83)$$

For linear elastic behavior (to be discussed in the next section) the one-dimensional stress–strain curve is that of a straight line and the strain energy density function (the area between the curve and the ϵ axis) equals the complementary strain energy density function (the area between the curve and the τ axis). It is similarly true, for a general state of stress, that for linear-elastic behavior:

$$U_0 = U_0^* \qquad (1.84)$$

We shall find that both functions U_0 and U_0^* will be of considerable use in this text.

1.13   HOOKE'S LAW

In the previous section we pointed out and used the fact that the stress tensor and the strain tensor are related. These relations depend on the nature of the material and are called constitutive laws. We shall be concerned in most of this text with *linear elastic behavior* wherein each stress component is linearly related, in the general case, to all the strains by equations of the form

$$\tau_{ij} = C_{ijkl}\epsilon_{kl} \qquad (1.85)$$

where the C's are at most functions only of position. The above law is called the *generalized Hooke's Law*. Because ϵ_{ij} and τ_{ij} are second-order tensor fields, one may employ a quotient rule of Appendix I to show that C_{ijkl} is a fourth-order tensor field. Since τ_{ij} is symmetric it should be clear that C_{ijkl} must also be symmetric in i, j. That is:

$$C_{ijkl} = C_{jikl} \qquad (1.86)$$

And since ϵ_{kl} is symmetric we can always express the components C_{ijkl} in a form symmetric in kl without violating Eq. (1.85). That is, we will stipulate that:

$$C_{ijkl} = C_{ijlk} \qquad (1.87)$$

We next assume the existence of a strain energy function U_0 as discussed in the previous section. Now in order to satisfy Eq. (1.73) and Hooke's law simultaneously this function must take the following form:

$$U_0 = \tfrac{1}{2} C_{ijkl} \epsilon_{ij} \epsilon_{kl} \qquad (1.88)$$

where the tensor C_{ijkl} must have the following symmetry property[1]

$$C_{ijkl} = C_{klij} \qquad (1.89)$$

Thus starting with eighty-one terms for C_{ijkl} we may show, using the three afore-mentioned symmetry relations for C_{ijkl}, that only twenty-one of these terms are independent. We will assume now that the material is homogeneous[2] so we may consider C_{ijkl} to be a set of *constants* for a given reference.

We next consider the property of *isotropy* for the mechanical behavior of the body. This property requires that the mechanical properties of a material at a point are not dependent on direction. Thus a stress such as τ_{xx} must be related to all the strains ϵ_{ij} for reference xyz exactly as the stress $\tau_{x'x'}$ is related to all the strains ϵ'_{ij} for a reference $x'y'z'$ rotated relative to xyz. Accordingly C_{ijkl} must have the *same components* for all references. A tensor such as C_{ijkl} whose components are invariant with respect to a rotation of axes is called an *isotropic* tensor. For an isotropic second-order tensor:

$$A_{ij} = A'_{ij} \qquad (1.90)$$

Transforming the right side of the above equation according to the rules of second-order tensors we have:

$$A_{ij} = a_{il} a_{jk} A_{lk} \qquad (1.91)$$

You may demonstrate (Problem 1.15) that the following set satisfies the above requirements and is the most general second-order isotropic tensor

$$A_{ij} = \alpha \delta_{ij} \qquad (1.92)$$

where α is a scalar constant. One can show that the above is the only such second-order tensor. As for a fourth-order isotropic tensor we require that:

$$D_{mnop} = D'_{mnop} \qquad (1.93)$$

---

[1] To understand why this is so, consider the computation of τ_{12} from U_0. That is:

$$\frac{\partial U_0}{\partial \epsilon_{12}} = \frac{\partial}{\partial \epsilon_{12}} [\tfrac{1}{2} C_{ijkl} \epsilon_{ij} \epsilon_{kl}] = \tfrac{1}{2} C_{12kl} \epsilon_{kl} + \tfrac{1}{2} C_{ij12} \epsilon_{ij} = C_{12kl} \epsilon_{kl} = \tau_{12}$$

Thus by using Eq. (1.88) for U_0 in conjunction with Eq. (1.89) we are able to compute stresses from U_0 properly and still have the basic relation for Hooke's law intact.

[2] A homogeneous material has the same composition throughout.

Using the fourth-order transformation formula on the right side of the above equation we get:

$$D_{mnop} = a_{mi}a_{nj}a_{ok}a_{pl}D_{ijkl}$$

You may demonstrate (Problem 1.16) that the following set is a fourth-order isotropic tensor

$$D_{ijkl} = \lambda\,\delta_{ij}\,\delta_{kl} + \beta\,\delta_{ik}\,\delta_{jl} + \gamma\,\delta_{il}\,\delta_{jk} \qquad (1.94)$$

where λ, β, and γ are scalar constants. It is convenient to express the constants β and γ as follows

$$\beta = G + B$$
$$\gamma = G - B$$

where G and B are constants determined by the above equations. Then we can give Eq. (1.94) as follows

$$
\begin{aligned}
D_{ijkl} &= \lambda\,\delta_{ij}\,\delta_{kl} + (G + B)(\delta_{ik}\,\delta_{jl}) + (G - B)(\delta_{il}\,\delta_{jk}) \\
&= \lambda\,\delta_{ij}\,\delta_{kl} + G(\delta_{ik}\,\delta_{jl} + \delta_{il}\,\delta_{jk}) \\
&\qquad\qquad + B(\delta_{ik}\,\delta_{jl} - \delta_{il}\,\delta_{jk}) \qquad (1.95)
\end{aligned}
$$

Clearly the tensor associated with G is symmetric in the indices i, j while the tensor associated with B is skew-symmetric in these indices. For the tensor D_{ijkl} to be used for isotropic elastic behavior it should be clear that the constant B must be zero (see Eq. (1.86)). Setting $B = 0$, we can consider the above tensor as the most general isotropic tensor that may be used for Hooke's law. Thus we can say:

$$\tau_{ij} = [\lambda\,\delta_{ij}\,\delta_{kl} + G(\delta_{ik}\,\delta_{jl} + \delta_{il}\,\delta_{jk})]\epsilon_{kl}$$

Carrying out the operations with the Kronecker delta terms we get:

$$\tau_{ij} = \lambda\,\delta_{ij}\epsilon_{ll} + 2G\epsilon_{ij} \qquad (1.96)$$

This is the general form of Hooke's law giving stress components in terms of strain components for isotropic materials. The constants λ and G are the so-called *Lamé constants*. We see that as a result of isotropy the number of independent elastic moduli have been reduced from twenty-one to two. The inverse form of Hooke's law yielding strain components in terms of stress components may be given as follows:

$$\epsilon_{ij} = \frac{1}{2G}\tau_{ij} - \left[\frac{\lambda + G}{2G(3\lambda + G)}\tau_{kk}\right]\delta_{ij} \qquad (1.97)$$

In terms of the commonly used constants E and ν, respectively, the Young's modulus and the Poisson ratio stemming from one-dimensional test data, you may readily show (see Problem 1.21) that:[1]

$$G = \frac{E}{2(1 + v)} \qquad (a)$$

$$\lambda = \frac{Ev}{(1 + v)(1 - 2v)} \qquad (b) \quad (1.98)$$

The inverse forms of the above results are given as:

$$E = \frac{G(3\lambda + 2G)}{\lambda + G}$$

$$v = \frac{\lambda}{2(\lambda + G)} \qquad (1.99)$$

Using Eq. (1.98) we express Eq. (1.97) as follows:

$$\epsilon_{ij} = \frac{1 + v}{E} \tau_{ij} - \frac{v}{E} \tau_{kk} \delta_{ij} \qquad (1.100)$$

In unabridged notation we then have the following familiar relations:

$$\epsilon_{xx} = \frac{1}{E}[\tau_{xx} - v(\tau_{yy} + \tau_{zz})]$$

$$\epsilon_{yy} = \frac{1}{E}[\tau_{yy} - v(\tau_{xx} + \tau_{zz})]$$

$$\epsilon_{zz} = \frac{1}{E}[\tau_{zz} - v(\tau_{xx} + \tau_{yy})]$$

$$\epsilon_{xy} = \frac{1 + v}{E}\tau_{xy} = \frac{1}{2G}\tau_{xy}$$

$$\epsilon_{yz} = \frac{1 + v}{E}\tau_{yz} = \frac{1}{2G}\tau_{yz}$$

$$\epsilon_{xz} = \frac{1 + v}{E}\tau_{xz} = \frac{1}{2G}\tau_{xz} \qquad (1.101)$$

In closing this section it is well to remember that a material may be both isotropic and inhomogeneous or, conversely, anisotropic and homogeneous. These two characteristics are independent.

---

[1] G is the shear modulus and is often represented by the letter μ in the literature.

1.14 BOUNDARY-VALUE PROBLEMS FOR LINEAR ELASTICITY

The complete system of equations of linear elasticity for homogeneous, isotropic solids includes the equilibrium equations:

$$\tau_{ij,j} + B_i = 0 \qquad (1.102)$$

the stress-strain law:

$$\tau_{ij} = \lambda \epsilon_{ll}\, \delta_{ij} + 2G\epsilon_{ij} \qquad (1.103)$$

and the strain-displacement relations:

$$\epsilon_{ij} = \tfrac{1}{2}(u_{i,j} + u_{j,i}) \qquad (1.104)$$

We have here a complete system of fifteen equations for fifteen unknowns. When explicit use of the displacement field is not made we must be sure that the compatibility equations are satisfied. It must also be understood that B_i and $T_i^{(v)}$ on the boundary have resultants that satisfy equilibrium relations for the body as dictated by rigid-body mechanics. We shall say in this regard that B_i and $T_i^{(v)}$ must be *statically compatible*.

We may pose three classes of boundary value problems.

1 Determine the distribution of stresses and displacements in the interior of the body under a given body force distribution and a *given surface traction over the boundary*. This is called a boundary-value problem of the *first kind*.

2 Determine the distribution of stresses and displacements in the interior of the body under the action of a given body-force distribution and *a prescribed displacement distribution over the entire boundary*. This is called a boundary value problem of the *second kind*.

3 Determine the distribution of stresses and displacements in the interior of a body under the action of a given body force distribution *with a given traction distribution over part of the boundary denoted as S_1 and a prescribed displacement distribution over the remaining part of the boundary S_2*. This is called a *mixed* boundary value problem.

It should be noted that on the surfaces where the $T_i^{(v)}$ are prescribed, Cauchy's formula $T_i^{(v)} = \tau_{ij}\nu_j$ must apply.

For boundary-value problems of the *first kind* we find it convenient to express the basic equations in terms of stresses. To do this we substitute for ϵ_{ij} using Eq. (1.100) in the compatibility equation (1.64). Using the equilibrium equation we can then arrive (see Problem 1.24) at the following system of equations.

$$\nabla^2 \tau_{ij} + \frac{1}{1+v}\kappa_{,ij} - \frac{v}{1+v}\delta_{ij}\nabla^2\kappa = -(B_{i,j} + B_{j,i}) \qquad (1.105)$$

where $\kappa = \tau_{kk}$. These are the *Beltrami-Michell* equations. The solution of these equations, subject to the satisfaction of Cauchy's formula on the boundary for simply connected domains, will lead to a set of stress components that both satisfy the equilibrium equations and are derivable from a single-valued continuous displacement field.

As for boundary-value problems of the *second kind*, we employ Eq. (1.103) and (1.104) in the equilibrium equation (1.102) to yield differential equations with the displacement field as the dependent variable. By straightforward substitutions we then get the well-known *Navier* equations of elasticity.

$$G\nabla^2 u_i + (\lambda + G)u_{j,ji} + B_i = 0 \qquad (1.106)$$

For dynamic conditions we need only employ Eq. (1.7) in place of the equilibrium equations. The result is the addition of the term $\rho\ddot{u}_i$ on the right side of the above equation. If the above equation can be solved in conjunction with the prescribed displacements on the surface and if the resulting solution is single-valued and continuous the problem may be considered solved.

Solutions for *mixed boundary-value* problems will be investigated throughout the text with emphasis on techniques stemming from the variational approach.

1.15 ST. VENANT'S PRINCIPLE

Sometimes it is advantageous to simplify the specification of surface tractions in the boundary-value problems mentioned in the previous section. Thus, as in rigid body mechanics where point forces are used to replace a force distribution, we have the analogous situation in elasticity where, as a result of St. Venant's principle, a surface traction distribution over a comparatively small part of a boundary may be replaced by a statically equivalent[1] system without altering the stress distribution at points sufficiently far away from the surface traction. Thus we can say that the effects of surface tractions over a part of the boundary that are felt far into the interior of an elastic solid are dependent only on the rigid-body resultant of the applied tractions over this part of the boundary. By this principle we can (Fig. 1.26) replace the complex supporting force distribution exerted by the wall on the cantilever beam by a single force and couple as shown in the diagram, for the purpose of simplifying the computation of stress and strain in the domain to the right of the support.

Although we shall not present them here, it is to be pointed out that mathematical justifications have been advanced for St. Venant's principle.[2]

---

[1] Having the same rigid-body resultant.

[2] Goodier, J. N., "A General Proof of St. Venant's Principle," *Philosophy Magazine*, 7, No. 23, 637 (1937). Hoff, N. J., "The Applicability of Saint Venant's Principle to Airplane Structures," *J. of Aero. Sciences*, **12**, 455 (1945). Fung, Y. C., "Foundation of Solid Mechanics," Prentice-Hall Inc., 1965, p. 300.

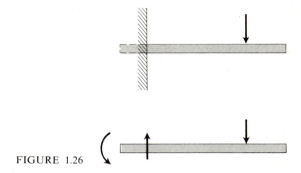

FIGURE 1.26

1.16 UNIQUENESS

We shall present a proof, due to the German mathematician Franz Neumann, that the solution to the mixed boundary-value problem of classical elasticity is unique—that is, all the stresses and strains can be found without ambiguity. We shall consider the linearized *dynamic* case where the boundary conditions are expressed as functions of time as follows

$$\text{On } S_1: \quad T_i^{(v)}(x,y,z,t) = f_i(x,y,z,t) \qquad \text{for} \quad t \geq 0$$
$$\text{On } S_2: \quad u_i(x,y,z,t) = g_i(x,y,z,t) \qquad \text{for} \quad t \geq 0$$

where f_i and g_i are known functions. Furthermore we must also express *initial conditions* here of the form:

$$u_i(x,y,z,0) = h_i(x,y,z) \qquad \text{in} \quad V$$
$$\dot{u}_i(x,y,z,0) = k_i(x,y,z) \qquad \text{in} \quad V$$

Let us assume that there are two solutions u_i' and u_i'' for the displacement field of a given mixed boundary-value problem. The corresponding stress and strain fields are given respectively as τ_{ij}', ϵ_{ij}' and τ_{ij}'', ϵ_{ij}''. Now form a new displacement field given as:

$$u_i = u_i' - u_i'' \tag{1.107}$$

Because of the *linearity* of the basic equations this new function must be a solution to these equations for the situation where we have zero body forces, zero surface tractions on S_1 and zero displacements u_i on S_2. Furthermore the initial values of u_i and $\dot{u}_i$ are zero. The equation of motion for such a case is then:

$$\tau_{ij,j} - \rho \frac{\partial^2 u_i}{\partial t^2} = 0$$

Using the strain energy function U_0 this becomes:

$$\frac{\partial}{\partial x_j}\left(\frac{\partial U_0}{\partial \epsilon_{ij}}\right) - \rho \frac{\partial^2 u_i}{\partial t^2} = 0$$

Now multiply by $\partial u_i/\partial t$ and integrate first over the entire volume and then with respect to time from t_0 to t, where for t_0 we take $U_0 = \dot{u}_i = 0$.

$$\int_{t_0}^{t} dt \iiint_V \left[\frac{\partial}{\partial x_j}\left(\frac{\partial U_0}{\partial \epsilon_{ij}}\right) - \rho \frac{\partial^2 u_i}{\partial t^2}\right] \frac{\partial u_i}{\partial t} \, dv = 0 \tag{1.108}$$

Consider the second expression in the integrand. We may express it as follows:

$$\int_{t_0}^{t} dt \iiint_V \rho \frac{\partial^2 u_i}{\partial t^2} \frac{\partial u_i}{\partial t} \, dv = \iiint_V \left[\int_{t_0}^{t} \left(\rho \frac{\partial^2 u_i}{\partial t^2} \frac{\partial u_i}{\partial t} \right) dt \right] dv$$

$$= \iiint_V \left[\int_{t_0}^{t} \frac{1}{2} \rho \frac{\partial}{\partial t} \left(\frac{\partial u_i}{\partial t} \right)^2 dt \right] dv$$

$$= \frac{1}{2} \iiint_V \rho \left(\frac{\partial u_i}{\partial t} \right)^2 \bigg|_{t_0}^{t} dv = \frac{1}{2} \iiint_V \rho \left(\frac{\partial u_i}{\partial t} \right)^2 dv \qquad (1.109)$$

where we have noted that $\dot{u}_i(t_0) = 0$. Now consider the first expression in the integrand of Eq. (1.108). We may express this as follows

$$\int_{t_0}^{t} dt \iiint_V \left[\frac{\partial}{\partial x_j} \left(\frac{\partial U_0}{\partial \epsilon_{ij}} \right) \right] \frac{\partial u_i}{\partial t} \, dv = \int_{t_0}^{t} dt \iiint_V \frac{\partial}{\partial x_j} \left(\frac{\partial U_0}{\partial \epsilon_{ij}} \frac{\partial u_i}{\partial t} \right) dv$$

$$- \int_{t_0}^{t} dt \iiint_V \frac{\partial U_0}{\partial \epsilon_{ij}} \frac{\partial^2 u_i}{\partial x_j \partial t} \, dv \qquad (1.110)$$

Examine next the expression

$$\left(\frac{\partial U_0}{\partial \epsilon_{ij}} \frac{\partial^2 u_i}{\partial x_j \partial t} \right)$$

in the second integral on the right side of the above equation. We may express it as follows

$$\frac{\partial U_0}{\partial \epsilon_{ij}} \frac{\partial^2 u_i}{\partial x_j \partial t} = \frac{\partial U_0}{\partial \epsilon_{ij}} \frac{\partial}{\partial t} (u_{i,j}) = \frac{1}{2} \frac{\partial U_0}{\partial \epsilon_{ij}} \frac{\partial}{\partial t} u_{i,j} + \frac{1}{2} \frac{\partial U_0}{\partial \epsilon_{ji}} \frac{\partial}{\partial t} (u_{j,i})$$

wherein we have simply decomposed the expression into two equal parts while interchanging the dummy indices. Now making use of the fact that ϵ_{ij} is symmetric and employing Eq. (1.46) we then get:

$$\frac{\partial U_0}{\partial \epsilon_{ij}} \frac{\partial^2 u_i}{\partial x_j \partial t} = \frac{\partial U_0}{\partial \epsilon_{ij}} \frac{\partial \epsilon_{ij}}{\partial t} = \frac{\partial U_0}{\partial t} \qquad (1.111)$$

Employing the above result and, in addition, using Gauss' theorem for the first integral on the right side, Eq. (1.110) becomes:

$$\int_{t_0}^{t} dt \iiint_V \left[\frac{\partial}{\partial x_j} \left(\frac{\partial U_0}{\partial \epsilon_{ij}} \right) \right] \frac{\partial u_i}{\partial t} \, dv = \int_{t_0}^{t} dt \oiint_S \frac{\partial U_0}{\partial \epsilon_{ij}} \frac{\partial u_i}{\partial t} v_j \, dA - \int_{t_0}^{t} dt \iiint_V \frac{\partial U_0}{\partial t} \, dv \qquad (1.112)$$

Noting that $\partial U_0 / \partial \epsilon_{ij} = \tau_{ij}$ is zero on S_1 and that $\partial u_i / \partial t$ is zero on S_2 we conclude that the surface integral vanishes. After interchanging the order of integration in the last integral we get for that expression on noting that we have taken $U_0(t_0) = 0$:

$$- \iiint_V U_0 \, dv.$$

Now substitute this result and that of Eq. (1.109) into Eq. (1.108). We get:

$$\iiint_V \left[U_0 + \frac{1}{2} \rho \left(\frac{\partial u_i}{\partial t} \right)^2 \right] dv = 0$$

Since both expressions in the integrand are positive definite we must conclude each must vanish everywhere in the domain. This means that $U_0 = 0$ and hence the strain field is zero throughout the body. Accordingly, except possibly for rigid-body motions which we may ignore, the displacement fields u' and u'' are identical, proving uniqueness of the solution.

It is well to note that this proof hinged on the *linearity* of the governing equations. No such uniqueness theorem is possible in nonlinear elasticity. Indeed we shall see in our discussion of stability later that a multiplicity of solutions is possible when nonlinearities are involved.

Part D
PLANE STRESS

1.17   INTRODUCTION

The rigorous solutions of three-dimensional problems of elasticity are few. Consequently there is a need to simplify problems so that a mathematical solution is feasible wherein the solution is reasonably close to representing the actual physical problem. As an illustration of such an approach we now present briefly the plane stress problem. Other ways of simplifying the problem using variational approaches will be presented later and the present undertaking will provide a basis for comparison.

1.18   EQUATIONS FOR PLANE STRESS

We define plane stress in the z plane as a stress distribution where:

$$\tau_{xz} = \tau_{yz} = \tau_{zz} = 0 \qquad (1.113)$$

Thin plates acted on by loads lying in the plane of symmetry of a plate (Fig. 1.27(a)) can often be considered to be in a state of plane stress with the z direction taken normal to the plate. Clearly, with no loads normal to lateral surfaces of the plate, τ_{zz} must be zero there, and since the plate is thin we consider τ_{zz} to be zero inside. If the loads are distributed so as to have a constant intensity over the thickness of the plate then the shearing stresses τ_{zx} and τ_{zy} in addition to being zero on the faces of the plate will be zero throughout the thickness. If however, the peripheral load is symmetrically distributed over the thickness (see Fig. 1.27(b)) then the shear stresses τ_{zx} and τ_{zy} may have considerable magnitudes at points along the thickness of the plate such as is shown along a–a in the diagram. This is possible in spite of the fact that these shear stresses must be zero at the lateral faces of the plate. However, the shear stresses must be antisymmetric over the thickness with a net area of zero for the curve τ vs z. Thus these shear stresses have an average value of zero over the thickness of the plate. This motivates the theory of *generalized plane stress* where the stress tensor is replaced by the average values of stress over the plate thickness. We shall consider for simplicity that the surface loads are applied so as to have uniform intensity over the plate thickness and so we can assume the shear stress τ_{xz} and τ_{yz} are zero everywhere. The

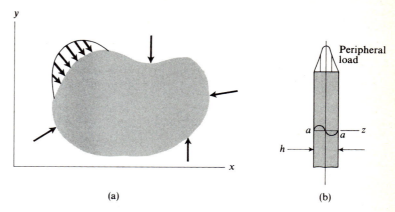

(a)   (b)

FIGURE 1.27

equations of equilibrium then become

$$\frac{\partial \tau_{xx}}{\partial x} + \frac{\partial \tau_{xy}}{\partial y} = 0$$

$$\frac{\partial \tau_{yx}}{\partial x} + \frac{\partial \tau_{yy}}{\partial y} = 0 \qquad (1.114)$$

We may permanently satisfy these equations by expressing the stresses in terms of a function Φ, called the *Airy stress function*, as follows:

$$\tau_{xx} = \frac{\partial^2 \Phi}{\partial y^2} \qquad \tau_{yy} = \frac{\partial^2 \Phi}{\partial x^2} \qquad \tau_{xy} = -\frac{\partial^2 \Phi}{\partial x\,\partial y} \qquad (1.115)$$

A substitution for stresses in Eq. (1.114) using the Airy stress function renders these equations to be trivial identities. Next, Hooke's law for strain gives us:[1]

$$\epsilon_{xx} = \frac{1}{E}(\tau_{xx} - v\tau_{yy}) \qquad\qquad \gamma_{xy} = \frac{1}{G}\tau_{xy}$$

$$\epsilon_{yy} = \frac{1}{E}(\tau_{yy} - v\tau_{xx}) \qquad\qquad \gamma_{xz} = 0$$

$$\epsilon_{zz} = -\frac{v}{E}(\tau_{xx} + \tau_{yy}) \qquad\qquad \gamma_{yz} = 0 \qquad (1.116)$$

[1] Hooke's law for plane stress giving stress in terms of strain will be used later in the text and can be found from Eq. (1.116) by straightforward algebraic steps to be:

$$\tau_{xx} = \frac{E}{1 - v^2}(\epsilon_{xx} + v\epsilon_{yy})$$

$$\tau_{yy} = \frac{E}{1 - v^2}(\epsilon_{yy} + v\epsilon_{xx})$$

$$\tau_{xy} = 2G\epsilon_{xy} \qquad (1.118)$$

Now replace the stresses using Airy's function in the above equations. We get:

$$\epsilon_{xx} = \frac{1}{E}\left[\frac{\partial^2 \Phi}{\partial y^2} - v\frac{\partial^2 \Phi}{\partial x^2}\right] \qquad \gamma_{xy} = -\frac{1}{G}\frac{\partial^2 \Phi}{\partial x\,\partial y}$$

$$\epsilon_{yy} = \frac{1}{E}\left[\frac{\partial^2 \Phi}{\partial x^2} - v\frac{\partial^2 \Phi}{\partial y^2}\right] \qquad \gamma_{yz} = \gamma_{xz} = 0$$

$$\epsilon_{zz} = -\frac{v}{E}\left[\frac{\partial^2 \Phi}{\partial x^2} + \frac{\partial^2 \Phi}{\partial y^2}\right] \tag{1.118}$$

Because we will be working with stress quantities (and not displacements) we shall have to satisfy the compatibility equations. Accordingly we now turn to the compatibility equations (1.70). Examine the first of these. Substituting from the preceding equations we get:

$$\frac{1}{E}\left[\frac{\partial^4 \Phi}{\partial y^4} - v\frac{\partial^4 \Phi}{\partial y^2\,\partial x^2} + \frac{\partial^4 \Phi}{\partial x^4} - v\frac{\partial^4 \Phi}{\partial x^2\,\partial y^2}\right] = -\frac{1}{G}\left(\frac{\partial^4 \Phi}{\partial y^2\,\partial x^2}\right)$$

Multiply through by E and replace E/G by $2(1 + v)$ in accordance with Eq. (1.98(a)). We then get:

$$\frac{\partial^4 \Phi}{\partial x^4} + 2\frac{\partial^4 \Phi}{\partial y^2\,\partial x^2} + \frac{\partial^4 \Phi}{\partial y^4} = \nabla^4 \Phi = 0 \tag{1.119}$$

We see that Φ must satisfy the so-called *biharmonic* equation and is thus a biharmonic function as a result of compatibility. As for the other compatibility equations, we see on inspection that Eqs. (1.70(d)) and (1.70(e)) are satisfied identically. The remaining equations are not satisfied but involve only the strain ϵ_{zz} which is of no interest to us here. We shall accordingly disregard these latter equations.[1] Equation (1.119) incorporates equations of equilibrium, Hooke's law, and satisfies those aspects of compatibility that are of concern to us. It thus forms the key equation for plane stress problems. The function Φ desired must be a biharmonic function that satisfies, within the limits of St. Venant's principle, the given surface tractions of the problem. We now consider the boundary conditions that we must impose on Φ. We confine ourselves here to the case where surface tractions are prescribed. Applying Cauchy's formula to plane stress (Eq. (1.5)) we get:

$$T_x^{(v)} = \tau_{xx}\cos(v,x) + \tau_{xy}\cos(v,y)$$

$$T_y^{(v)} = \tau_{yx}\cos(v,x) + \tau_{yy}\cos(v,y) \tag{1.120}$$

---

[1] One can show that the error incurred using Eq. (1.119) to determine Φ leads to results that are very close to the correct result for thin plates. (Timoshenko and Goodier, "Theory of Elasticity," McGraw-Hill Book Co.)

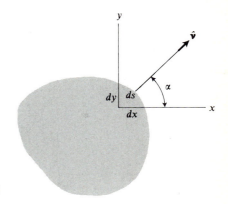

FIGURE 1.28

Now employing the Airy stress function we have:

$$T_x^{(v)} = \frac{\partial^2 \Phi}{\partial y^2} \cos (v,x) - \frac{\partial^2 \Phi}{\partial x \, \partial y} \cos (v,y)$$

$$T_y^{(v)} = -\frac{\partial^2 \Phi}{\partial x \, \partial y} \cos (v,x) + \frac{\partial^2 \Phi}{\partial x^2} \cos (v,y) \qquad (1.121)$$

Observe next Fig. 1.28 showing an arbitrary boundary wherein element ds of the boundary is depicted with coordinate elements dx and dy. We can say, on inspecting this diagram, that:

$$\cos (v,x) = \cos \alpha = \frac{dy}{ds}$$

$$\cos (v,y) = \sin \alpha = -\frac{dx}{ds} \qquad (1.122)$$

Substituting the above results into Eqs. (1.121) we then get:

$$T_x^{(v)} = \frac{\partial^2 \Phi}{\partial y^2} \frac{dy}{ds} + \frac{\partial^2 \Phi}{\partial x \, \partial y} \frac{dx}{ds} = \frac{d}{ds}\left(\frac{\partial \Phi}{\partial y}\right)$$

$$T_y^{(v)} = -\frac{\partial^2 \Phi}{\partial x \, \partial y} \frac{dy}{ds} - \frac{\partial^2 \Phi}{\partial x^2} \frac{dx}{ds} = -\frac{d}{ds}\left(\frac{\partial \Phi}{\partial x}\right) \qquad (1.123)$$

where we use the chain rule of differentiation to reach the furthermost expression on the right sides of the equations. The above equations are the usual form used for the boundary conditions on Φ. We have thus posed the boundary value problem for plane stress.

We now illustrate the entire procedure in solving a plane stress problem.

1.19 PROBLEM OF THE CANTILEVER BEAM

Consider now a cantilever beam (Fig. 1.29) loaded at the tip by a force P. The cross section of the beam is that of a rectangle with height h and thickness t. The length of the beam is L. For t small compared to both h and L, we may use the plane stress analysis for determining deflections in the midplane of the beam.

We shall postulate the form of the stress function Φ for this problem by examining the results of elementary strength of materials for the beam. You will recall that

$$\tau_{xx} = -\frac{My}{I_{zz}} = -\frac{P(L-x)}{I_{zz}}y \qquad (a)$$

$$\tau_{yy} = 0 \qquad (b)$$

$$\tau_{xy} = \frac{VQ}{I_{zz}t} = \frac{P(h^2/4 - y^2)}{2I_{zz}} \qquad (c) \quad (1.124)$$

where V is the shear force and Q is the first moment about the z axis of the section area above the elevation y. Thus we can see that the stresses are of the form:

$$\tau_{xx} = C_1 y + C_2 xy$$

$$\tau_{yy} = 0$$

$$\tau_{xy} = C_3 + C_4 y^2 \qquad (1.125)$$

You may demonstrate directly that the following function Φ yields a stress distribution of the form given above:

$$\Phi = Ay^3 + Bxy^3 + Dxy \qquad (1.126)$$

It is also readily demonstrated that the above function is a biharmonic function and so this represents some plane-stress problem. Accordingly we must investigate to see whether we can satisfy the boundary conditions of the problem at hand by adjusting the constants A, B, and D. Thus we require:

$$\text{For } y = \pm\frac{h}{2}: \quad (1) \quad \tau_{yy} = 0; \quad (2) \quad \tau_{xy} = 0$$

$$\text{For } x = L: \quad (3) \quad \tau_{xx} = 0$$

$$\text{Everywhere:} \quad (4) \quad \int_{-h/2}^{h/2} \tau_{xy} t \, dy = P \qquad (1.127)$$

Applying these conditions to Φ we have:

(1) $\tau_{yy} = \dfrac{\partial^2 \Phi}{\partial x^2} = 0$ (everywhere)

(2) $(\tau_{xy})_{y = \pm h/2} = -\left(\dfrac{\partial^2 \Phi}{\partial x \partial y}\right)_{y = \pm h/2} = (3By^2 + D)_{y = \pm h/2} = 0$

therefore $D = -\frac{3}{4}Bh^2$

(a)

(3) $(\tau_{xx})_{x = L} = \left(\dfrac{\partial^2 \Phi}{\partial y^2}\right)_{x = L} = (6Ay + 6Bxy)_{x = L} = 0$

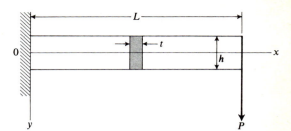

FIGURE 1.29

therefore $\quad A = -LB$ $\qquad\qquad\qquad\qquad\qquad\qquad\qquad\qquad$ (b)

(4) $\quad\displaystyle\int_{-h/2}^{h/2} \tau_{xy} t \, dy = \int_{-h/2}^{h/2} -\left(\frac{\partial^2 \Phi}{\partial x \, \partial y}\right) t \, dy = \int_{-h/2}^{h/2} -(3By^2 + D)_{x=L} t \, dy = P$

therefore $\quad -Bt\dfrac{h^3}{4} - Dth = P$ $\qquad\qquad\qquad\qquad\qquad\qquad$ (c)

Solving Eqs. (a), (b), and (c) simultaneously we get for the constants A, B, and D the values:

$$B = 2\frac{P}{th^3}$$

$$D = -\tfrac{3}{2}\frac{P}{th}$$

$$A = -2\frac{PL}{th^3}$$

Thus the following form of Φ satisfies the stipulated boundary conditions:

$$\Phi = -2\frac{PL}{th^3}y^3 + \frac{2P}{th^3}xy^3 - \tfrac{3}{2}\frac{P}{th}xy \qquad (1.128)$$

If we now compute the stresses using the above function we find results identical to those given by Eq. (1.124). Thus the stress distributions from strength of materials is identical to that from the theory of elasticity. Indeed if the load P has the parabolic distribution stipulated in Eq. (1.124 (c)) for τ_{xy} then we have an "exact" solution of the problem for regions to the right of the base support.[1] If this is not the case then the solution is valid away from the support and away from the right end.

Let us next consider the displacement field for the problem. We have, using Hooke's law and the notation $u_x = u$ and $u_y = v$:

$$\epsilon_{xx} = \frac{\partial u}{\partial x} = \frac{1}{E}\tau_{xx} = -\frac{P(L-x)}{EI_{zz}}y \qquad (a)$$

$$\epsilon_{yy} = \frac{\partial v}{\partial y} = -\frac{1}{E}(v\tau_{xx}) = \frac{Pv}{EI_{zz}}(L-x)y \qquad (b)$$

$$\gamma_{xy} = \left(\frac{\partial u}{\partial y} + \frac{\partial v}{\partial x}\right) = \frac{P}{2GI_{zz}}\left(\frac{h^2}{4} - y^2\right) \qquad (c) \;\; (1.129)$$

---

[1] One can show that when the bending moment varies linearly with x the flexure formula is exact. See: Borg, S. R., "Matrix-Tensor Methods in Continuum Mechanics," Section 5.5. D. Van Nostrand Co., Inc., 1962.

We integrate Eqs. (a) and (b) as follows

$$u = -\frac{P}{EI_{zz}}\left(Lyx - \frac{x^2}{2}y\right) + [g_1(y)]\frac{P}{EI_{zz}}$$

$$v = \frac{Pv}{EI_{zz}}\left(\frac{Ly^2}{2} - \frac{xy^2}{2}\right) + [g_2(x)]\frac{P}{EI_{zz}} \qquad (1.130)$$

where g_1 and g_2 are arbitrary with P/EI_{zz} attached for later convenience. Now substitute the above results into Eq. (1.129(c)).

$$-\frac{P}{EI_{zz}}\left(Lx - \frac{x^2}{2}\right) + g_1'\frac{P}{EI_{zz}} + \frac{Pv}{EI_{zz}}\left(-\frac{y^2}{2}\right) + g_2'\frac{P}{EI_{zz}} = \frac{P}{2GI_{zz}}\left(\frac{h^2}{4} - y^2\right)$$

Replacing G by $E/\{2(1 + v)\}$ (see Eq. (1.98(a)) we may cancel out the term P/EI_{zz}. We then have on rearranging terms:

$$-\left(Lx - \frac{x^2}{2}\right) + g_2'(x) = \frac{vy^2}{2} + (1 + v)\left(\frac{h^2}{4} - y^2\right) - g_1'(y)$$

Note that each side is a function of a different variable and hence each side must equal a constant which we denote as K. Thus:

$$-\left(Lx - \frac{x^2}{2}\right) + g_2' = K$$

$$\frac{vy^2}{2} + (1 + v)\left(\frac{h^2}{4} - y^2\right) - g_1' = K$$

We may integrate the above equations now to get g_1 and g_2 as follows:

$$g_2 = Kx + \frac{Lx^2}{2} - \frac{x^3}{6} + C_1$$

$$g_1 = -Ky + \frac{vy^3}{6} + (1 + v)\left(\frac{h^2y}{4} - \frac{y^3}{3}\right) + C_2$$

We then have for u and v:

$$u = -\frac{P}{EI_{zz}}\left\{Lxy - \frac{x^2y}{2} - Ky + \frac{vy^3}{6} + (1 + v)\left(\frac{h^2y}{4} - \frac{y^3}{3}\right) + C_2\right\} \qquad (a)$$

$$v = \frac{P}{EI_{zz}}\left\{v\left(\frac{Ly^2}{2} - \frac{xy^2}{2}\right) + Kx + \frac{Lx^2}{2} - \frac{x^3}{6} + C_1\right\} \qquad (b) \quad (1.131)$$

We have three arbitrary constants C_1, C_2, and K that must now be determined. We shall find these constants by fixing the support in some way. Let us say that point 0 (see Fig. 1.29) is stationary. Then $u = v = 0$ when $x = y = 0$. It is clear that the constants C_1 and C_2 must then be zero. Next let us say that the centerline of the beam at 0 remains horizontal. That is, $\partial v/\partial x = 0$ when $x = y = 0$. We see on inspection of Eq. (1.131(b)) that K is also zero. The displacement field for this case is then:

$$u = -\frac{P}{EI_{zz}}\left\{Lxy - \frac{x^2y}{2} + \frac{vy^3}{6} + (1 + v)\left(\frac{h^2y}{4} - \frac{y^3}{3}\right)\right\} \qquad (a)$$

$$v = \frac{P}{EI_{zz}}\left\{v\left(\frac{Ly^2}{2} - \frac{xy^2}{2}\right) + \frac{Lx^2}{2} - \frac{x^3}{6}\right\} \qquad (b) \quad (1.132)$$

In the one-dimensional study of beams to be soon undertaken we center attention on the vertical deflection of the centerline of the beam. We may get this result from above by setting $y = 0$ in the formulation of v. Thus:

$$(v)_{y=0} = \frac{P}{2EI_{zz}}\left[Lx^2 - \frac{x^3}{3}\right] \qquad (1.133)$$

This coincides with the result from strength of materials.[1]

1.20   CLOSURE

In this chapter we have presented a self-contained treatment of classical linear elasticity that will serve our needs in this area. This does not mean that the text is restricted to infinitesimal linear-elastic behavior. In Chap. 8, in anticipation of the study of elastic stability, we will consider finite deformations for Hookean materials: we will thus introduce geometric nonlinearities. And in various problems, such as the torsion problem of Chap. 5, we consider a nonlinear constitutive law.

In the next chapter we continue to lay the foundation for this text by considering certain salient features of the calculus of variations.

READING

BORESI, A. P.: "Theory of Elasticity," Prentice-Hall Inc., 1965.

FILONENKO AND BORODITCH: "Theory of Elasticity," Peace Publishers, Moscow.

FUNG, Y. C.: "Foundations of Solid Mechanics," Prentice-Hall Inc., N.J., 1965.

HODGE, P. G.: "Continuum Mechanics," McGraw-Hill Book Co., N.Y., 1970.

SHAMES, I. H.: "Mechanics of Deformable Solids," Prentice-Hall Inc., N.J., 1964.

SOKOLNIKOFF, I. S.: "Mathematical Theory of Elasticity," McGraw-Hill Book Co., N.Y., 1956.

TIMOSHENKO, S., AND GOODIER, J. N.: "Theory of Elasticity," McGraw-Hill Book Co., N.Y., 1951.

PROBLEMS

1.1   Label the stresses shown on the infinitesimal rectangular parallelepiped in Fig. 1.30 and give their correct signs.

---

[1] It is to be pointed out that if we fix the beam in another manner at the end (for example we can reasonably assume that $\partial u/\partial y = 0$ at 0 which is a case you will be asked to investigate) there results a different deflection curve for the centerline. Thus we see that, whereas St. Venant's principle assumes that the *stresses will not be affected* away from the support for such changes, *this does not hold for the deflection curve.*

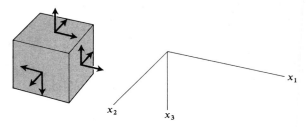

FIGURE 1.30

1.2 The stress components at orthogonal interfaces parallel to $x_1 x_2 x_3$ at a point are known:

$$\tau_{11} = \quad 1000 \text{ psi} \qquad \tau_{12} = \tau_{21} = \quad 200 \text{ psi}$$

$$\tau_{22} = -6000 \text{ psi} \qquad \tau_{13} = \tau_{31} = \quad 0$$

$$\tau_{33} = \quad 0 \qquad \tau_{23} = \tau_{32} = -400 \text{ psi}$$

(a) Find the components of the surface traction vector for an interface whose normal vector is:

$$\hat{v} = 0.11\hat{\imath} + 0.35\hat{\jmath} + 0.93\hat{k}$$

(b) What is the component of this force intensity in the $\hat{\epsilon}$ direction?

$$\hat{\epsilon} = 0.33\hat{\imath} + 0.90\hat{\jmath} + 0.284\hat{k}$$

1.3 Suppose you had a body-couple distribution given as $\mathbf{M}$ in-lb/in^3. What is the relation between the stresses and $\mathbf{M}$ at a point? In particular if

$$M_1 = 200 \text{ lb-in/in}^3; \qquad \tau_{11} = 100 \text{ psi}; \qquad \tau_{22} = -200 \text{ psi}$$

$$\tau_{33} = 0 \text{ psi}; \qquad \tau_{13} = 500 \text{ psi}; \qquad \tau_{31} = \quad 100 \text{ psi}$$

$$\tau_{23} = 100 \text{ psi}; \qquad \tau_{21} = 50 \text{ psi}; \qquad \tau_{12} = \quad 80 \text{ psi.}$$

what is the stress τ_{32}? What are the body couple components M_2 and M_3?

1.4 The state of stress at a point in a given reference $(x_1 x_2 x_3)$ is given by

$$\begin{pmatrix} 200 & 100 & 0 \\ 100 & 0 & 0 \\ 0 & 0 & 500 \end{pmatrix}$$

What is the array of stress terms for a new set of axes $(x_1' x_2' x_3')$ formed by rotating $(x_1 x_2 x_3)$ 60° about the x_3 axis?

1.5 Given the following stress tensor:

$$\tau_{ij} = \begin{pmatrix} -1000 & 1000 & 0 \\ 1000 & 1000 & 0 \\ 0 & 0 & 0 \end{pmatrix}$$

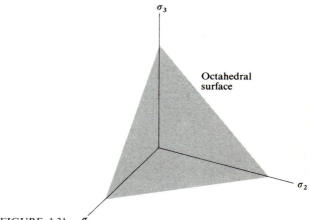

FIGURE 1.31

Determine:

(a) The principal stresses $\sigma_1, \sigma_2, \sigma_3$.

(b) The three stress tensor invariants. Show that they are indeed invariant by calculating them from the given τ_{ij} and from the array of principal values.

(c) The principal directions $\overset{1}{v}_j, \overset{2}{v}_j, \overset{3}{v}_j$ and check them for orthogonality.

1.6 Show that the stress quadric becomes:

(a) For *uniaxial tension* a pair of plane surfaces.

(b) For *plane stress* (i.e., $\tau_{zz} = \tau_{yz} = \tau_{xz} = 0$) a cylinder.

(c) For *simple shear* a set of four surfaces forming as traces rectangular hyperbolae on a plane normal to the surfaces.

1.7 The *octahedral shear stress* is the maximum shear stress on a plane equally inclined toward the principal axes at a point (Fig. 1.31). Using the transformation of stress equations show that the normal stress on the octahedral plane is related to the principal stresses σ_i as follows:

$$(\tau_{nn})_{\text{oct}} = \tfrac{1}{3}(\sigma_1 + \sigma_2 + \sigma_3)$$

Now using Fig. 1.31 show from Newton's law that:

$$\tau_{\text{oct}}{}^2 = \tfrac{1}{9}[(\sigma_1 - \sigma_2)^2 + (\sigma_2 - \sigma_3)^2 + (\sigma_1 - \sigma_3)^2]$$

The octahedral shear stress is often considered the stress to observe to predict *yielding* in a general state of stress at a point. If the yield stress from a one-dimensional test is Y, what is then the octahedral stress for yielding under general conditions according to the preceding test? This test for yielding is the so-called *Mises–Hencky* criterion.

1.8 Show that

$$9\tau_{\text{oct}}{}^2 = 2(\text{I}_\tau)^2 - 6(\text{II}_\tau)$$

Thus the tensor invariants are shown to be simply related to a physically meaningful stress measure.

1.9 Given the descriptions below, of various types of deformation, determine the various Green tensor representations (Eq. 1.42(a)):
   (a) Simple dilatation: $\xi_1 = \lambda x_1, \xi_2 = x_2, \xi_3 = x_3$
   (b) Pure deformation: $\xi_1 = \lambda_1 x_1, \xi_2 = \lambda_2 x_2, \xi_3 = \lambda_3 x_3$
   (c) Cubical dilatation: $\lambda_1 = \lambda_2 = \lambda_3 = \lambda$ in (b)
   (d) Simple shear: $\xi_1 = x_1 + \Gamma x_2, \quad \xi_2 = x_2, \quad \xi_3 = x_3$.

1.10 Determine the displacements ($u_i = \xi_i - x_i$) corresponding to the deformations of Problem 1.9, and from them compute the infinitesimal strain components. Can any physical meanings be deduced for the parameters λ and Γ involved?

1.11 For the case of pure deformation (see Problem 1.9) show that the volume V of an element $dx_1\, dx_2\, dx_3$ in the deformed state is represented as $V^* = \lambda_1 \lambda_2 \lambda_3\, dx_1\, dx_2\, dx_3$. Further, show that if the strains are infinitesimal, that $V^*/V = 1 + \epsilon_{ii}$.
   Hint: Consider dS^{*2}/dS^2 for an element such as dx_1. Use Eq. (1.41(a)) and the fact that $(dS^{*2}/dx_1{}^2) = d\xi_1{}^2/dx_1{}^2$.

1.12 Working in rectangular coordinates, show that integration of the equations $\epsilon_{ij} = 0$ (in terms of displacements) leads to rigid body rotations. Consider infinitesimal strain.

1.13 The strain tensor ϵ_{ij} can be written as follows:

$$\epsilon_{ij} = \bar{\epsilon}_{ij} + \frac{\epsilon_{kk}}{3}\delta_{ij} = \bar{\epsilon}_{ij} + \frac{e}{3}\delta_{ij}$$

where

$$\bar{\epsilon}_{ij} = \begin{pmatrix} \epsilon_{11} - e/3 & \epsilon_{12} & \epsilon_{13} \\ \epsilon_{21} & \epsilon_{22} - e/3 & \epsilon_{23} \\ \epsilon_{31} & \epsilon_{32} & \epsilon_{33} - e/3 \end{pmatrix}$$

The tensor $\bar{\epsilon}_{ij}$ is called the strain *deviator* tensor. Explain why $\bar{\epsilon}_{ij}$ characterizes the distortion of an element with no change in volume. (This is the shearing distortion.) Show that the first, second, and third-tensor invariants for the strain deviator are given in terms of principal strains as follows:

$$\mathrm{I}_\epsilon = 0$$
$$\mathrm{II}_\epsilon = -\tfrac{1}{6}[(\epsilon_1 - \epsilon_2)^2 + (\epsilon_2 - \epsilon_3)^2 + (\epsilon_1 - \epsilon_3)^2]$$
$$\mathrm{III}_\epsilon = (\epsilon_1 - e/3)(\epsilon_2 - e/3)(\epsilon_3 - e/3)$$

The strain deviator invariants (as well as those for stress) play an important role in the theory of plasticity since for the plastic state we have primarily shear distortion deformation.

1.14 Show that $\lambda\delta_{ij}$ satisfies the equation for a second-order isotropic tensor.

1.15 Show that $\lambda\delta_{ij}$ is the most general isotropic, second-order tensor by starting with a second order tensor A_{ij} and considering the following rotation of axes while employing the

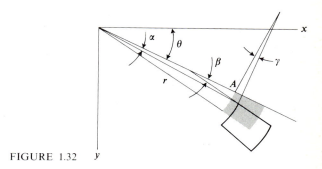

FIGURE 1.32 y

isotropic condition (Eq. 1.91). Thus with a rotation of 180° about the z axis show that:

$$A_{23} = A_{32} = 0$$

$$A_{13} = A_{31} = 0$$

For a rotation about the x axis of 180° show that $A_{12} = A_{21} = 0$. Thus the non-diagonal terms must be zero. Next, by considering separately, rotation of 90° about two axes, show that the diagonal terms are equal. But this means that $A_{ij} = A \, \delta_{ij}$ and so the proposed tensor is the most general isotropic second-order tensor.

1.16 Show that D_{ijkl} as given by Eq. (1.94) is an isotropic fourth-order tensor.

1.17 Consider u_r to be the displacement component in the radial direction and u_θ to be the displacement component in the transverse (θ) direction. Show that:

$$\epsilon_{\theta\theta} = \frac{1}{r} \frac{\partial u_\theta}{\partial \theta} + \frac{u_r}{r}$$

As for $\gamma_{r\theta}$ observe that the shaded element (Fig. 1.32) is the undeformed geometry while the unshaded element is in the deformed geometry. What is the change in right angles at corner A in terms of angles γ, β, and α? Show that:

$$\gamma_{r\theta} = \frac{\partial u_\theta}{\partial r} - \frac{u_\theta}{r} + \frac{1}{r} \frac{\partial u_r}{\partial \theta}$$

What is ϵ_{rr}?

1.18 We want to find the conditions on the elastic constants to make the strain energy density U_0 *positive-definite*. Thus integrate Eq. (1.75), using the stress-strain law (1.96), to obtain

$$U_0 = \frac{\lambda}{2} e^2 + G \epsilon_{ij} \epsilon_{ij}, \qquad e = \epsilon_{kk}$$

Then introducing the strain deviator (Problem 1.13), one can write

$$U_0 = \tfrac{1}{2} K e^2 + G \bar{\epsilon}_{ij} \bar{\epsilon}_{ij}, \qquad K = \lambda + \tfrac{2}{3} G$$

From this result, deduce appropriate conditions on E, for $U_0 > 0$. Also notice that $e \equiv I_\epsilon$, and $\bar{\epsilon}_{ij}\bar{\epsilon}_{ij} = 2II_\epsilon$ (Problem 1.13 extended). Hence U_0 is an invariant!

1.19 The stress-strain law for an isotropic elastic body undergoing a temperature change ΔT is:

$$\tau_{ij} = \lambda \epsilon_{kk} \delta_{ij} + 2G\epsilon_{ij} - (3\lambda + 2G)\Delta T \, \delta_{ij}$$

where α is the *thermal expansion coefficient*. Show for *plane stress* ($\tau_{yz} = \tau_{xz} = \tau_{zz} = 0$) that:

$$\tau_{xx} = \frac{E}{1-v^2}(\epsilon_{xx} + v\epsilon_{yy}) - \frac{E}{1+v}\alpha \Delta T$$

$$\tau_{yy} = \frac{E}{1-v^2}(v\epsilon_{xx} + \epsilon_{yy}) - \frac{E}{1+v}\alpha \Delta T$$

$$\tau_{xy} = 2G\epsilon_{xy}$$

Construct next the strain energy density function for this case (see Eq. (1.72)). Is it equal to the simple expression $\frac{1}{2}\tau_{ij}\epsilon_{ij}$? Explain why you should expect your conclusion when you consider the basic forms of constitutive law for the isothermal and nonisothermal cases.

1.20 For a two-dimensional (plane-stress) *orthotropic* continuum, the constitutive law is usually given as:

$$\tau_{xx} = C_{11}\epsilon_{xx} + C_{12}\epsilon_{yy}$$

$$\tau_{yy} = C_{21}\epsilon_{xx} + C_{22}\epsilon_{yy}$$

$$\tau_{xy} = G_{12}\epsilon_{xy}$$

Construct an appropriate strain energy density. Also, verify that with no loss in generality we can make $C_{12} = C_{21}$. Check your constitutive law by computing stresses τ_{xx}, τ_{yy}, and τ_{xy}.

1.21 Using the stress-strain law (1.96), consider a one-dimensional (uniaxial) tensile test in the z direction (i.e., $\tau_{zz} \neq 0$, $\tau_{xx} = \tau_{yy} = 0$). Relate ϵ_{xx} and ϵ_{yy} to ϵ_{zz}, and τ_{zz} to ϵ_{zz}. Next, compare these "analytical" results with the following "experimental data",

$$\epsilon_{xx} = \epsilon_{yy} = -v\epsilon_{zz}, \qquad \tau_{zz} = E\epsilon_{zz}$$

and hence derive Eqs. (1.98).

1.22 Using only the linearity of the constitutive (stress-strain) law, derive the *Betti reciprocal theorem*:

$$\iiint_V \tau_{ij}^{(1)}\epsilon_{ij}^{(2)} \, dv = \iiint_V \tau_{ij}^{(2)}\epsilon_{ij}^{(1)} \, dv$$

The superscripts represent different states of loading on the same body with the same kinematic constraints. Hint: make use of symmetry condition (Eq. 1.89).

1.23 Consider the expression

$$\iiint_V \tau_{ij}^{(1)}\epsilon_{ij}^{(2)} \, dv,$$

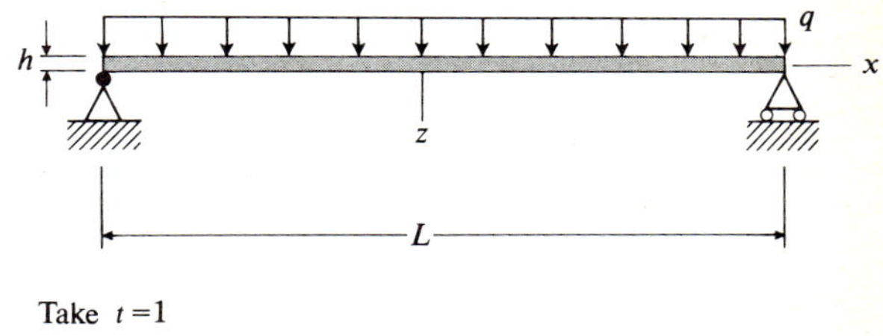

Take $t = 1$

FIGURE 1.33

where the superscripts refer to different states. Show that it can be written as follows:

$$\iiint_V [(\tau_{ij}{}^{(1)}u_i{}^{(2)})_{,j} - \tau_{ij,j}{}^{(1)}u_i{}^{(2)}]\,dv$$

Now using the divergence theorem, the equations of equilibrium, Cauchy's formula as well as the Betti reciprocal theorem (Problem 1.22) show that:

$$\iint_S T_i{}^{(v)(1)}u_i{}^{(2)}\,dS + \iiint_V B_i{}^{(1)}u_i{}^{(2)}\,dv = \iint_S T_i{}^{(v)(2)}u_i{}^{(1)}\,dS + \iiint_V B_i{}^{(2)}u_i{}^{(1)}\,dv$$

1.24 To develop the *Beltrami–Michell Equation* substitute ϵ_{ij} in Eq. (1.64) using Eq. (1.100), with τ_{kk} replaced by κ. Perform a contraction over indices k and l. Use Eq. (1.102) to replace terms of the form $\tau_{ij,j}$ by body forces $-B_i$. Show that the result is that given by Eq. (1.105).

1.25 (*a*) In the example of the cantilever beam show that plane sections do not in fact remain plane but distort slightly as a third-order curve.

(*b*) Find the deflection curve for the case where we fix the end so that $\partial u/\partial y = 0$. For a steel $(E = 30 \times 10^6 \text{ psi})$ beam of length $L = 10\,\text{ft}$, $h = (1/2)\,\text{ft}$, $t = 1''$ and $F = 10,000\,\text{lb}$, what is the difference between the deflection at the end for this case and for the case where $\partial v/\partial x = 0$ at the base?

1.26 Show that the following function is the proper *Airy function* for the simply-supported rectangular beam problem carrying a uniform load as shown in Fig. 1.33.

$$\Phi = \frac{Az^3}{6} + \frac{Bx^2z^3}{6} - \frac{C}{2}x^2z - \frac{qx^2}{4} - \frac{B}{30}z^5$$

where

$$B = -\frac{6q}{h^3}; \qquad C = -\frac{3}{2}\frac{q}{h}; \qquad A = \frac{12}{h^3}\left(\frac{qL^2}{8} - \frac{qh^2}{20}\right)$$

Check the boundary conditions. Show that:

$$\tau_{xx} = -\frac{q}{2I}\left[x^2 - \left(\frac{L}{2}\right)^2\right]z + \frac{q}{2I}\left(\frac{2}{3}z^3 - \frac{h^2}{10}z\right)$$

$$\tau_{zz} = -\frac{q}{2I}\left(\frac{1}{3}z^3 - \frac{h^2}{4}z + \frac{h^3}{12}\right)$$

$$\tau_{xz} = -\frac{q}{2I}\left(\frac{h^2}{4} - z^2\right)x$$

1.27 Show for the beam of the previous problem that the deflection of the centerline is given as:

$$w(x, 0) = \frac{qL^4}{24EI}\left[\frac{5}{16} - \frac{3}{2}\left(\frac{x}{L}\right)^2 + \left(\frac{x}{L}\right)^4\right] + \frac{qL^4}{24EI}\left(\frac{h}{L}\right)^2\left[\left(\frac{12}{5} + \frac{3v}{2}\right)\left(\frac{1}{4} - \left(\frac{x}{L}\right)^2\right)\right]$$

INTRODUCTION TO THE CALCULUS OF VARIATIONS

2.1 INTRODUCTION

In dealing with a function of a single variable, $y = f(x)$, in the ordinary calculus, we often find it of use to determine the values of x for which the function y is a *local maximum* or a *local minimum*. By a local maximum at position x_1, we mean that f at position x *in the neighborhood* of x_1 is less than $f(x_1)$ (see Fig. 2.1). Similarly for a local minimum of f to exist at position x_2 (see Fig. 2.1) we require that $f(x)$ be larger than $f(x_2)$ for all values of x in the *neighborhood of* x_2. The values of x in *the neighborhood of* x_1 *or* x_2 may be called the *admissible values* of x relative to which x_1 or x_2 is a maximum or minimum position.

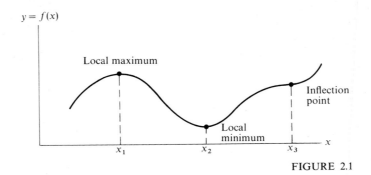

FIGURE 2.1

To establish the condition for a local extremum (maximum or minimum), let us expand the function f as a Taylor series about a position $x = a$. Thus assuming that $f(x)$ has continuous derivatives at position $x = a$ we have:

$$f(x) = f(a) + \left(\frac{df}{dx}\right)_{x=a}(x-a) + \frac{1}{2!}\left(\frac{d^2f}{dx^2}\right)_{x=a}(x-a)^2$$

$$+ \frac{1}{3!}\left(\frac{d^3f}{dx^3}\right)_{x=a}(x-a)^3 + \cdots$$

We next rearrange the series and rewrite it in the following more compact form:

$$f(x) - f(a) = [f'(a)](x-a) + \frac{1}{2!}[f''(a)](x-a)^2$$

$$+ \frac{1}{3!}[f'''(a)](x-a)^3 + \cdots \tag{2.1}$$

For $f(a)$ to be a minimum it is necessary that $[f(x) - f(a)]$ be a positive number for all values of x in the neighborhood of "a". Since $(x - a)$ can be positive or negative for the admissible values of x, then clearly the term $f'(a)$ must then be zero to prevent the dominant term in the series from yielding positive and negative values for the admissible values of x. That is, a necessary condition for a local minimum at "a" is that $f'(a) = 0$. By similar reasoning we can conclude that the same condition prevails for a local maximum at "a". Considering the next term in the series, we see that there will be a constancy in sign for admissible values of x and so the sign of $f''(a)$ will determine whether we have a local minimum or a local maximum at position "a". Thus with $f'(a) = 0$, the sign of $f''(a)$ (assuming $f''(a) \neq 0$) supplies the information for establishing a local minimum or a local maximum at position "a".

Suppose next that both $f'(a)$ and $f''(a)$ are zero but that $f'''(a)$ does not equal zero. Then the third term of the series of Eq. (2.1) becomes the dominant term, and for admissible values of x there must be a change in sign of $[f(x) - f(a)]$ as we move across point "*a*". Such a point is called an *inflection point* and is shown in Fig. 2.1 at position x_3.

Thus we see that point "*a*", for which $f'(a) = 0$, may correspond to a local minimum point, to a local maximum point, or to an inflection point. Such points as a group are often of much physical interest[1] and they are called *extremal* positions of the function.

We have presented a view of elements of the theory of local extrema in order to set the stage for the introduction of the calculus of variations[2] which will be of considerable use in the ensuing studies of elastic structures. In place of the function of the preceding discussion we shall be concerned now with *functionals* which are, plainly speaking, functions of functions. Specifically, a functional is an expression that takes on a particular value which is dependent on the function used in the functional. A form of functional that is employed in many areas of applied mathematics is the integral of $F(x,y,y')$ between two points (x_1, y_1) and (x_2, y_2) in two-dimensional space. Denoting this functional as I we have:

$$I = \int_{x_1}^{x_2} F(x,y,y') \, dx \qquad (2.2)$$

Clearly, the value of I for a given set of end points x_1 and x_2 will depend on the function $y(x)$. Thus, just as $f(x)$ depends on the value of x, so does the value of I depend on the form of the function $y(x)$. And, just as we were able to set up necessary conditions for a local extreme of f at some point "*a*" by considering *admissible values* of x (i.e., x in the neighborhood of "*a*") so can we find necessary conditions for extremizing I with respect to an *admissible set of functions* $y(x)$. Such a procedure, forming one of the cornerstones of the calculus of variations, is considerably more complicated than the corresponding development in the calculus of functions and we shall undertake this in a separate section.

In this text we shall only consider necessary conditions for establishing an extreme. We usually know a priori whether this extreme is a maximum or a minimum by physical arguments. Accordingly the complex arguments[3] needed in the calculus of variations for giving sufficiency conditions for maximum or minimum states of I will be omitted.

We may generalize the functional I in the following ways:

(*a*) The functional may have many independent variables other than just x;
(*b*) The functional may have many functions (dependent variables) of these independent variables other than just $y(x)$;

---

[1] See Courant: "Differential and Integral Calculus," Interscience Press.
[2] For a rigorous study of this subject refer to "Calculus of Variations" by Gelfand and Fomin, Prentice-Hall Inc., or to "An Introduction to the Calculus of Variations," by Fox, Oxford University Press.
[3] See the references cited earlier.

(*c*) The functional may have higher-order derivatives other than just first-order.

We shall examine such generalizations in subsequent sections.

In the next section we set forth some very simple functionals.

2.2 EXAMPLES OF SIMPLE FUNCTIONALS

Historically the calculus of variations became an independent discipline of mathematics at the beginning of the eighteenth century. Much of the formulation of this mathematics was developed by the Swiss mathematician Leonhard Euler (1707–83). It is instructive here to consider three of the classic problems that led to the growth of the calculus of variations.

(a)   The Brachistochrone

In 1696 Johann Bernoulli posed the following problem. Suppose one were to design a frictionless chute between two points (1) and (2) in a vertical plane such that a body sliding under the action of its own weight goes from (1) to (2) in the *shortest* interval of time. The time for the descent from (1) to (2) we denote as I and it is given as follows

$$I = \int_{(1)}^{(2)} \frac{ds}{V} = \int_{(1)}^{(2)} \frac{\sqrt{dx^2 + dy^2}}{V} = \int_{x_1}^{x_2} \frac{\sqrt{1 + (y')^2}}{V}\, dx$$

where V is the speed of the body and s is the distance along the chute. Now employ the conservation of energy for the body. If V_1 is the initial speed of the body we have at any position y:

$$\frac{mV_1^{\,2}}{2} + mgy_1 = \frac{mV^2}{2} + mgy$$

therefore

$$V = [V_1^{\,2} - 2g(y - y_1)]^{1/2}$$

We can then give I as follows:

$$I = \int_1^2 \frac{\sqrt{1 + (y')^2}}{[V_1^{\,2} - 2g(y - y_1)]^{1/2}}\, dx \qquad (2.3)$$

We shall later show that the chute (i.e., $y(x)$) should take the shape of a *cycloid*.[1]

---

[1] This problem has been solved by both Johann and Jacob Bernoulli, Sir Isaac Newton, and the French mathematician L'Hôpital.

(b)   Geodesic Problem

The problem here is to determine the curve on a given surface $g(x,y,z) = 0$ having the *shortest length* between two points (1) and (2) on this surface. Such curves are called *geodesics*. (For a spherical surface the geodesics are segments of the so-called great circles.) The solution to this problem lies in determining the extreme values of the integral.

$$I = \int_{(1)}^{(2)} ds = \int_{x_1}^{x_2} \sqrt{1 + (y')^2 + (z')^2}\, dx \qquad (2.4)$$

We have here an example where we have two functions y and z in the functional, although only one independent variable, i.e., x. However, y and z are not independent of each other but must have values satisfying the equation:

$$g(x,y,z) = 0 \qquad (2.5)$$

The extremizing process here is analogous to the *constrained* maxima or minima problems of the calculus of functions and indeed Eq. (2.5) is called a constraining equation in connection with the extremization problem.

If we are able to solve for z in terms of x and y, or for y in terms of z and x in Eq. (2.5), we can reduce Eq. (2.4) so as to have only one function of x rather than two. The extremization process with the one function of x is then no longer constrained.

(c)   Isoperimetric Problem

The original isoperimetric problem is given as follows: of all the closed non-intersecting plane curves having a given fixed length L, which curve encloses the *greatest* area A?

The area A is given from the calculus by the following line integral:

$$A = \frac{1}{2} \oint (x\, dy - y\, dx)$$

Now suppose that we can express the variables x and y parametrically in terms of τ. Then we can give the above integral as follows

$$A = I = \frac{1}{2} \int_{\tau_1}^{\tau_2} \left(x \frac{dy}{d\tau} - y \frac{dx}{d\tau} \right) d\tau \qquad (2.6)$$

where τ_1 and τ_2 correspond to beginning and end of the closed loop. The constraint on x and y is now given as follows:

$$L = \oint ds = \oint (dx^2 + dy^2)^{1/2}$$

Introducing the parametric representation we have:

$$L = \int_{\tau_1}^{\tau_2} \left[\left(\frac{dx}{d\tau} \right)^2 + \left(\frac{dy}{d\tau} \right)^2 \right]^{1/2} d\tau \qquad (2.7)$$

We have thus one independent variable, τ, and two functions of τ, x and y, constrained this time by the *integral relationship* (2.7).

In the historical problems set forth here we have shown how functionals of the form given by Eq. (2.2) may enter directly into problems of interest. Actually the extremization of such functionals or their more generalized forms is equivalent to solving certain corresponding differential equations, and so this approach may give alternate viewpoints of various areas of mathematical physics. Thus in optics the extremization of the time required for a beam of light to go from one point to another in a vacuum relative to an admissible family of light paths is equivalent to satisfying Maxwell's equations for the radiation paths of light. This is the famous Fermat principle. In the study of particle mechanics the extremization of the difference between the kinetic energy and the potential energy (i.e., the Lagrangian) integrated between two points over an admissible family of paths yields the correct path as determined by Newton's law. This is Hamilton's principle.[1] In the theory of elasticity, which will be of primary concern to us in this text, we shall amongst other things extremize the so-called total potential energy of a body with respect to an admissible family of displacement fields to satisfy the equations of equilibrium for the body. We can note similar dualities in other areas of mathematical physics and engineering science, notably electromagnetic theory and thermodynamics. Thus we conclude that the extremization of functionals of the form of Eq. (2.2) or their generalizations affords us a different view of many fields of study. We shall have ample opportunity in this text to see this new viewpoint as it pertains to solid mechanics. One important benefit derived by recasting the approach as a result of variational considerations is that some very powerful *approximate procedures* will be made available to us for the solution of problems of engineering interest. Such considerations will form a significant part of this text.

It is to be further noted that the series of problems presented required respectively: a *fastest* time of descent, a *shortest* distance between two points on a surface, and a *greatest* area to be enclosed by a given length. These problems are examples of what are called *optimization* problems.[2]

---

[1] We shall consider Hamilton's principle in detail in Chap. 7.

[2] In seeking an *optimal solution* in a problem we strive to attain, subject to certain given constraints, that solution, amongst other possible solutions, that satisfies or comes closest to satisfying a certain criterion or certain criteria. Such a solution is then said to be optimal relative to this criterion or criteria, and the process of arriving at this solution is called optimization.

We now examine the extremization process for functionals of the type described in this section.

2.3 THE FIRST VARIATION

Consider a functional of the form

$$I = \int_{x_1}^{x_2} F(x,y,y')\, dx \qquad (2.8)$$

where F is a known function, twice differentiable for the variables x, y, and y'. As discussed earlier, the value of I between points (x_1, y_1) and (x_2, y_2) will depend on the path chosen between these points, i.e., it will depend on the function $y(x)$ used. We shall assume the existence of a path, which we shall henceforth denote as $y(x)$, having the property of extremizing I with respect to other neighboring paths which we now denote collectively as $\tilde{y}(x)$.[1] We assume further that $y(x)$ is twice differentiable. We shall for simplicity refer henceforth to $y(x)$ as the *extremizing path* or the *extremizing function* and to $\tilde{y}(x)$ as the *varied paths*.

We will now introduce a single-parameter family of varied paths as follows

$$\tilde{y}(x) = y(x) + \epsilon\eta(x) \qquad (2.9)$$

where ϵ is a small parameter and where $\eta(x)$ is a differentiable function having the requirement that:

$$\eta(x_1) = \eta(x_2) = 0$$

We see that an infinity of varied paths can be generated for a given function $\eta(x)$ by adjusting the parameter ϵ. All these paths pass through points (x_1, y_1) and (x_2, y_2). Furthermore for any $\eta(x)$ the varied path becomes coincident with the extremizing path when we set $\epsilon = 0$.

With the agreement to denote $y(x)$ as the extremizing function, then I in Eq. (2.8) becomes the extreme value of the integral

$$\int_{x_1}^{x_2} F(x,\tilde{y},\tilde{y}')\, dx.$$

We can then say:

$$\tilde{I} = \int_{x_1}^{x_2} F(x,\tilde{y},\tilde{y}')\, dx = \int_{x_1}^{x_2} F(x, y + \epsilon\eta, y' + \epsilon\eta')\, dx \qquad (2.10)$$

---

[1] Thus $y(x)$ will correspond to "a" of the early extremization discussion of $f(x)$ while $\tilde{y}(x)$ corresponds to the values of x in the neighborhood of "a" of that discussion.

By having employed $y + \epsilon\eta$ as the admissible functions we are able to use the extremization criteria of simple function theory as presented earlier since $\tilde{I}$ is now, for the desired extremal $y(x)$, a function of the parameter ϵ and thus it can be expanded as a power series in terms of this parameter. Thus

$$\tilde{I} = (\tilde{I})_{\epsilon=0} + \left(\frac{d\tilde{I}}{d\epsilon}\right)_{\epsilon=0}\epsilon + \left(\frac{d^2\tilde{I}}{d\epsilon^2}\right)_{\epsilon=0}\frac{\epsilon^2}{2!} + \cdots$$

Hence:

$$\tilde{I} - I = \left(\frac{d\tilde{I}}{d\epsilon}\right)_{\epsilon=0}\epsilon + \left(\frac{d^2\tilde{I}}{d\epsilon^2}\right)_{\epsilon=0}^{'}\frac{\epsilon^2}{2!} + \cdots$$

For $\tilde{I}$ to be extreme when $\epsilon = 0$ we know from our earlier discussion that

$$\left(\frac{d\tilde{I}}{d\epsilon}\right)_{\epsilon=0} = 0$$

is a *necessary condition*. This, in turn, means that

$$\left\{\int_{x_1}^{x_2}\left(\frac{\partial F}{\partial \tilde{y}}\frac{d\tilde{y}}{d\epsilon} + \frac{\partial F}{\partial \tilde{y}'}\frac{d\tilde{y}'}{d\epsilon}\right)dx\right\}_{\epsilon=0} = 0$$

Noting that $d\tilde{y}/d\epsilon = \eta$ and that $d\tilde{y}'/d\epsilon = \eta'$, and realizing that deleting the tilde for $\tilde{y}$ and $\tilde{y}'$ in the derivatives of F is the same as setting $\epsilon = 0$ as required above, we may rewrite the above equation as follows:

$$\int_{x_1}^{x_2}\left(\frac{\partial F}{\partial y}\eta + \frac{\partial F}{\partial y'}\eta'\right)dx = 0 \qquad (2.11)$$

We now integrate the second term by parts as follows:

$$\int_{x_1}^{x_2}\frac{\partial F}{\partial y'}\eta'\,dx = \frac{\partial F}{\partial y'}\eta\bigg|_{x_1}^{x_2} - \int_{x_1}^{x_2}\left[\frac{d}{dx}\left(\frac{\partial F}{\partial y'}\right)\right]\eta\,dx$$

Noting that $\eta = 0$ at the end points, we see that the first expression on the right side of the above equation vanishes. We then get on substituting the above result into Eq. (2.11):

$$\int_{x_1}^{x_2}\left[\frac{\partial F}{\partial y} - \frac{d}{dx}\left(\frac{\partial F}{\partial y'}\right)\right]\eta\,dx = 0 \qquad (2.12)$$

With $\eta(x)$ arbitrary between end points, a basic lemma of the calculus of variations[1]

---

[1] For a particular function $\phi(x)$ continuous in the interval (x_1,x_2), if $\int_{x_1}^{x_2}\phi(x)\eta(x)\,dx = 0$ for every continuously differentiable function $\eta(x)$ for which $\eta(x_1) = \eta(x_2) = 0$, then $\phi \equiv 0$ for $x_1 \leq x \leq x_2$.

indicates that the bracketed expression in the above integrand is zero. Thus:

$$\frac{d}{dx}\frac{\partial F}{\partial y'} - \frac{\partial F}{\partial y} = 0 \qquad (2.13)$$

This the famous *Euler–Lagrange equation.* It is the condition required for $y(x)$ in the role we have assigned it of being the extremizing function. Substitution of $F(x,y,y')$ will result in a second-order ordinary differential equation for the unknown function $y(x)$. In short, the variational procedure has resulted in an ordinary differential equation for getting the function $y(x)$ which we have "tagged" and handled as the extremizing function.

We shall now illustrate the use of the Euler–Lagrange equation by considering the brachistochrone problem presented earlier.

✓

EXAMPLE 2.1 BRACHISTOCHRONE PROBLEM   We have from the earlier study of the *brachistochrone* problem of Sect. 2.2 the requirement to extremize:

$$I = \int_1^2 \frac{\sqrt{1 + (y')^2}}{[V_1{}^2 - 2g(y - y_1)]^{1/2}} \, dx \qquad (a)$$

If we take the special case where the body is released from rest at the origin the above functional becomes:

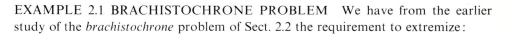

$$I = \frac{1}{\sqrt{2g}} \int_1^2 \sqrt{\frac{1 + (y')^2}{y}} \, dx \qquad (b)$$

The function F can be identified as $\{[1 + (y')^2]/y\}^{1/2}$. We go directly to the Euler–Lagrange equation to substitute for F. After some algebraic manipulation we obtain:

$$y'' = -\frac{1 + (y')^2}{2y}$$

Now make the substitution $u = y'$. Then we can say

$$u\frac{du}{dy} = -\frac{1 + u^2}{2y}$$

Separating variables and integrating we have:

$$y(1 + u^2) = C_1$$

therefore

$$y[1 + (y')^2] = C_1$$

We may arrange for another separation of variables and perform another quadrature as follows:

$$x = \int \frac{\sqrt{y}}{\sqrt{C_1 - y}} \, dy + x_0$$

Next make the substitution

$$y = C_1\left[\sin^2 \frac{t}{2}\right] \qquad (c)$$

We then have:

$$x = C_1 \int \sin^2 \frac{t}{2} \, dt + x_0 = C_1\left[\frac{t - \sin t}{2}\right] + x_0$$

Since at time $t = 0$, we have $x = y = 0$ then $x_0 = 0$ in the above equation. We then have as results

$$x = \frac{C_1}{2}(t - \sin t) \qquad (d)$$

$$y = \frac{C_1}{2}(1 - \cos t) \qquad (e)$$

wherein we have used the double-angle formula to arrive at Eq. (e) from Eq. (c). These equations represent a cycloid which is a curve generated by the motion of a point fixed to the circumference of a rolling wheel. The radius of the wheel here is $C_1/2$.   ////

2.4 THE DELTA OPERATOR

We now introduce an operator δ, termed the *delta operator*, in order to give a certain formalism to the procedure of obtaining the first variation. We define $\delta[y(x)]$ as follows:

$$\delta[y(x)] = \tilde{y}(x) - y(x) \qquad (2.14)$$

Notice that the delta operator represents a small arbitrary change in the dependent variable y for a *fixed* value of the independent variable x. Thus in Fig. 2.2 we have shown extremizing path $y(x)$ and some varied paths $\tilde{y}(x)$. At the indicated position x any of the increments $a-b$, $a-c$ or $a-d$ may be considered as δy—i.e., as a variation of y. Most important, note that we *do not associate a δx with each δy*. This is in contrast to the differentiation process wherein a dy is associated with a given dx. We can thus say that δy is simply the vertical distance between points on different curves at the

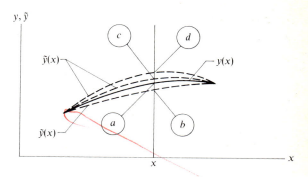

FIGURE 2.2

same value of x whereas dy is the vertical distance between points on the *same* curve at positions dx apart. This has been illustrated in Fig. 2.3.

We may generalize the delta operator to represent a small (usually infinitesimal) change of a function wherein the independent variable is kept fixed. Thus we may take the variation of the function dy/dx. We shall agree here, however, to use as varied function the derivatives $d\tilde{y}/dx$ where the $\tilde{y}$ are varied paths for y. We can then say:

$$\delta\left[\frac{dy}{dx}\right] = \left(\frac{d\tilde{y}}{dx}\right) - \left(\frac{dy}{dx}\right) = \frac{d}{dx}(\tilde{y} - y) = \frac{d(\delta y)}{dx} \qquad (2.15)$$

As a consequence of this arrangement we conclude that the δ operator is *commutative* with the differential operator. In a similar way, by agreeing that $\int \tilde{y}(x)\, dx$ is a varied function for $\int y(x)\, dx$, we can conclude that the variation operator is commutative with the integral operator.

Henceforth we shall make ample use of the δ operator and its associated notation. It will encourage the development of mechanical skill in carrying out the varia-

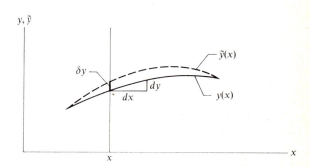

FIGURE 2.3

tional process and will aid in developing "physical" feel in the handling of problems.

It will be well to go back now to Eq. (2.10) and re-examine the first variation using the delta-operator notation. First note that for the one-parameter family of varied functions, $y + \epsilon\eta$, it is clear that

$$\delta y = \tilde{y} - y = \epsilon\eta \quad (a)$$

and that: (2.16)

$$\delta y' = \tilde{y}' - y' = \epsilon\eta' \quad (b)$$

Accordingly we can give F along a varied path as follows using the δ notation:

$$F(x, y + \delta y, y' + \delta y')$$

Now at any position x we can expand F as a Taylor series about y and y' in the following manner

$$F(x, y + \delta y, y' + \delta y') = F(x,y,y') + \left[\frac{\partial F}{\partial y}\delta y + \frac{\partial F}{\partial y'}\delta y'\right] + O(\delta^2)$$

therefore

$$F(x, y + \delta y, y' + \delta y') - F(x,y,y') = \left[\frac{\partial F}{\partial y}\delta y + \frac{\partial F}{\partial y'}\delta y'\right] + O(\delta^2) \quad (2.17)$$

where $O(\delta^2)$ refers to terms containing $(\delta y)^2$, $(\delta y')^2$, $(\delta y)^3$, etc., which are of negligibly higher order. We shall denote the left side of the equation as the *total variation* of F and denote it as $\delta^{(T)}F$. The bracketed expression on the right side of the equation with δ's to the *first power* we call the *first variation*, $\delta^{(1)}F$. Thus:

$$\delta^{(T)}F = F(x, y + \delta y, y' + \delta y') - F(x,y,y') \quad (a)$$

(2.18)

$$\delta^{(1)}F = \left[\frac{\partial F}{\partial y}\delta y + \frac{\partial F}{\partial y'}\delta y'\right] \quad (b)$$

On integrating Eq. (2.17) from x_1 to x_2 we get:

$$\int_{x_1}^{x_2} F(x, y + \delta y, y' + \delta y')\,dx - \int_{x_1}^{x_2} F(x,y,y')\,dx$$

$$= \int_{x_1}^{x_2} \left[\frac{\partial F}{\partial y}\delta y + \frac{\partial F}{\partial y'}\delta y'\right]dx + O(\delta^2)$$

This may be written as:

$$\tilde{I} - I = \int_{x_1}^{x_2} \left[\frac{\partial F}{\partial y}\delta y + \frac{\partial F}{\partial y'}\delta y'\right]dx + O(\delta^2)$$

We shall call $\tilde{I} - I$ the total variation of I and denote it as $\delta^{(T)}I$, while as expected, the first expression on the right side of the equation is the first variation of I, $\delta^{(1)}I$. Hence:

$$\delta^{(T)}I = \tilde{I} - I$$

$$\delta^{(1)}I = \int_{x_1}^{x_2} \left(\frac{\partial F}{\partial y} \, \delta y + \frac{\partial F}{\partial y'} \, \delta y' \right) dx$$

Integrating the second expression in the above integral by parts we get for $\delta^{(1)}I$:

$$\delta^{(1)}I = \int_{x_1}^{x_2} \left[\frac{\partial F}{\partial y} - \frac{d}{dx} \frac{\partial F}{\partial y'} \right] \delta y \, dx + \left[\frac{\partial F}{\partial y'} \, \delta y \right] \Bigg|_{x_1}^{x_2}$$

Since all $\tilde{y}(x)$ must take the specific values corresponding to those of $y(x)$ at x_1 and x_2, clearly the variation in $y(x)$ must be zero at these points—i.e., $\delta y = 0$ at x_1 and x_2. Thus we have for $\delta^{(1)}I$:

$$\delta^{(1)}I = \int_{x_1}^{x_2} \left[\frac{\partial F}{\partial y} - \frac{d}{dx} \frac{\partial F}{\partial y'} \right] \delta y \, dx \qquad (2.19)$$

We can then say for $\delta^{(T)}I$:

$$\delta^{(T)}I = \delta^{(1)}I + O(\delta^2) + \cdots$$

$$= \int_{x_1}^{x_2} \left[\frac{\partial F}{\partial y} - \frac{d}{dx} \frac{\partial F}{\partial y'} \right] \delta y \, dx + O(\delta^2)$$

In order for I to be a maximum or a minimum it must retain the same sign for all possible variations δy over the interval. Thus δy at any position x could be $\pm K$ where K is a small number. For this to be possible the bracketed expression in the integrand on the right side of the equation has to be zero which in turn leads to the familiar Euler–Lagrange equations. We may thus conclude that

$$\delta^{(1)}I = 0 \qquad (2.20)$$

Accordingly, the requirement for extremization of I is that its *first variation be zero*. Suppose now that $F = F(\epsilon_{ij})$. Then for independent variables x, y, z we have for the first variation of I, on extrapolating from Eq. (2.18(b))

$$\delta^{(1)}I = \iiint_V \delta^{(1)}F \, dx \, dy \, dz$$

$$= \iiint_V \frac{\partial F}{\partial \epsilon_{ij}} \delta \epsilon_{ij} \, dx \, dy \, dz$$

where we observe the summation convention of the repeated indices. If F is a function of ϵ as a result of using a one-parameter family approach then one might suppose that:

$$\delta^{(1)}I = \int_{x_1}^{x_2} \frac{\partial F}{\partial \epsilon} \delta\epsilon \, dx$$

But the dependent variable represents the extremal function in the development, and for the one-parameter development $\epsilon = 0$ corresponds to this condition. Hence we must compute $\partial F/\partial \epsilon$ at $\epsilon = 0$ in the above formulation and $\delta\epsilon$ can be taken as ϵ itself. Hence we get:

$$\delta^{(1)}I = \int_{x_1}^{x_2} \left(\frac{\partial \tilde{F}}{\partial \epsilon}\right)_{\epsilon=0} \epsilon \, dx \qquad (2.21)$$

If we use a two-parameter family (as we soon shall) then it should be clear for ϵ_1 and ϵ_2 as parameters that:

$$\delta^{(1)}I = \int_{x_1}^{x_2} \left[\left(\frac{\partial \tilde{F}}{\partial \epsilon_1}\right)_{\substack{\epsilon_1=0 \\ \epsilon_2=0}} \epsilon_1 + \left(\frac{\partial \tilde{F}}{\partial \epsilon_2}\right)_{\substack{\epsilon_1=0 \\ \epsilon_2=0}} \epsilon_2 \right] dx \qquad (2.22)$$

Now going back to Eq. (2.12) and its development we can say:

$$\left(\frac{d\tilde{I}}{d\epsilon}\right)_{\epsilon=0} = \int_{x_1}^{x_2} \left[\frac{\partial F}{\partial y} - \frac{d}{dx}\left(\frac{\partial F}{\partial y'}\right) \right] \eta \, dx \qquad (2.23)$$

Next rewriting Eq. (2.19) we have:

$$\delta^{(1)}I = \int_{x_1}^{x_2} \left[\frac{\partial F}{\partial y} - \frac{d}{dx}\left(\frac{\partial F}{\partial y'}\right) \right] \delta y \, dx$$

Noting that $\delta y = \epsilon\eta$ for a single-parameter family approach we see by comparing the right sides of the above equations that:

$$\delta^{(1)}I \equiv \left(\frac{d\tilde{I}}{d\epsilon}\right)_{\epsilon=0} \epsilon \qquad (2.24)$$

Accordingly setting $(d\tilde{I}/d\epsilon)_{\epsilon=0} = 0$, as we have done for a single-parameter family approach, is tantamount to setting the first variation of I equal to zero.[1] We shall use both approaches in this text for finding the extremal functions.

As a next step, we examine certain simple cases of the Euler–Lagrange equation to ascertain first integrals.

---

[1] Note that for a two parameter family we have from Eq. (2.22) the result:

$$\delta^{(1)}I = \left(\frac{\partial \tilde{I}}{\partial \epsilon_1}\right)_{\substack{\epsilon_1=0 \\ \epsilon_2=0}} \epsilon_1 + \left(\frac{\partial \tilde{I}}{\partial \epsilon_2}\right)_{\substack{\epsilon_1=0 \\ \epsilon_2=0}} \epsilon_2 \qquad (2.25)$$

2.5 FIRST INTEGRALS OF THE EULER–LAGRANGE EQUATION

We now present four cases where we can make immediate statements concerning first integrals of the Euler–Lagrange equation as presented thus far.

Case (a). *F* is not a function of *y*—i.e., *F* = *F*(*x*, *y*′)

In this case the Euler–Lagrange equation degenerates to the form:

$$\frac{d}{dx}\left(\frac{\partial F}{\partial y'}\right) = 0$$

Accordingly, we can say for a first integral that:

$$\frac{\partial F}{\partial y'} = \text{Const.} \qquad (2.26)$$

Case (b). *F* is only a function of *y*′—i.e., *F* = *F*(*y*′)

Equation 2.26 still applies because of the lack of presence of the variable *y*. However, now we know that the left side must be a function only of *y*′. Since this function must equal a constant at all times we conclude that *y*′ = const. is a possible solution. This means that for this case an extremal path is simply that of a straight line.

Case (c). *F* is independent of the independent variable *x*—i.e., *F* = *F*(*y*, *y*′)

For this case we begin by presenting an identity which you are urged to verify. Thus noting that *d*/*dx* may be expressed here as

$$\left(\frac{\partial}{\partial x} + y'\frac{\partial}{\partial y} + y''\frac{\partial}{\partial y'}\right)$$

we can say:

$$\frac{d}{dx}\left(y'\frac{\partial F}{\partial y'} - F\right) = -y'\left[\frac{\partial F}{\partial y} - \frac{d}{dx}\left(\frac{\partial F}{\partial y'}\right)\right] - \frac{\partial F}{\partial x}$$

If *F* is not explicitly a function of *x*, we can drop the last term. A satisfaction of the Euler–Lagrange equation now means that the right side of the equation is zero so that we may conclude that

$$y'\frac{\partial F}{\partial y'} - F = C_1 \qquad (2.27)$$

for the extremal function. We thus have next to solve a first-order differential equation to determine the extremal function.

Case (d). F is the total derivative of some function $g(x, y)$—i.e., $F = dg/dx$

It is easy to show that when $F = dg/dx$, it must satisfy identically the Euler–Lagrange equation. We first note that:

$$F = \frac{\partial g}{\partial x} + \frac{\partial g}{\partial y} y' \qquad (2.28)$$

Now substitute the above result into the Euler–Lagrange equation. We get:

$$\frac{\partial}{\partial y}\left(\frac{\partial g}{\partial x} + \frac{\partial g}{\partial y} y'\right) - \frac{d}{dx}\left(\frac{\partial g}{\partial y}\right) = 0$$

Carry out the various differentiation processes:

$$\frac{\partial^2 g}{\partial x\, \partial y} + \frac{\partial^2 g}{\partial y^2} y' - \frac{\partial^2 g}{\partial x\, \partial y} - \frac{\partial^2 g}{\partial y^2} y' = 0$$

The left side is clearly identically zero and so we have shown that a *sufficient* condition for F to satisfy the Euler–Lagrange equation identically is that $F = dg(x,y)/dx$. It can also be shown that this is a *necessary* condition for the identical satisfaction of the Euler–Lagrange equation. It is then obvious that we can always add a term of the form dg/dx to the function F in I without changing the Euler–Lagrange equations for the functional I. That is, for any Euler–Lagrange equation there are an *infinity of functionals differing from each other in these integrals by terms of the form dg/dx.*

We shall have ample occasion to use these simple solutions. Now we consider the geodesic problem, presented earlier, for the case of the sphere.

EXAMPLE 2.2   Consider a sphere of radius R having its center at the origin of reference xyz. We wish to determine the shortest path between two points on this sphere. Using spherical coordinates R, θ, ϕ (see Fig. 2.4) a point P on the sphere has the following coordinates:

$$x = R \sin\theta \cos\phi$$
$$y = R \sin\theta \sin\phi$$
$$z = R \cos\theta \qquad (a)$$

The increment of distance ds on the sphere may be given as follows:

$$ds^2 = dx^2 + dy^2 + dz^2 = R^2[d\theta^2 + \sin^2\theta\, d\phi^2]$$

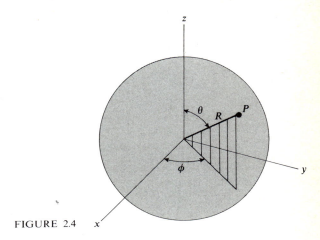

FIGURE 2.4

Hence the distance between point P and Q can be given as follows:

$$I = \int_P^Q ds = \int_P^Q R[d\theta^2 + \sin^2\theta \, d\phi^2]^{1/2} \qquad (b)$$

With ϕ and θ as independent variables, the transformation equations (a) depict a sphere. If ϕ is related to θ, i.e., it is a function of θ, then clearly x, y, and z in Eq. (a) are functions of a single parameter θ and this must represent some curve on the sphere. Accordingly, since we are seeking a curve on the sphere we shall assume that ϕ is a function of θ in the above equation so that we can say:

$$I = R \int_P^Q \left[1 + \sin^2\theta \left(\frac{d\phi}{d\theta}\right)^2 \right]^{1/2} d\theta \qquad (c)$$

We have here a functional with θ as the independent variable and ϕ as the function, with the derivative of ϕ appearing explicitly. With ϕ not appearing explicitly in the integrand, we recognize this to be case (a) discussed earlier. We may then say, using Eq. (2.26):

$$\frac{\partial}{\partial \phi'}[1 + (\sin^2\theta)(\phi')^2]^{1/2} = C_1 \qquad (d)$$

This becomes:

$$\frac{(\sin^2\theta)(\phi')}{[1 + (\sin^2\theta)(\phi')^2]^{1/2}} = C_1$$

Solving for ϕ' we get:

$$\phi' = \frac{C_1}{\sin\theta[\sin^2\theta - C_1{}^2]^{1/2}}$$

Integrating we have:

$$\phi = C_1 \int \frac{d\theta}{\sin\theta[\sin^2\theta - C_1{}^2]^{1/2}} + C_2 \qquad (e)$$

We make next the following substitution for θ:

$$\theta = \tan^{-1}\frac{1}{\eta} \qquad (f)$$

This gives us:

$$\phi = C_1 \int \frac{\dfrac{1}{1+\eta^2}\,d\eta}{\left(\dfrac{1}{1+\eta^2}\right)^{1/2}\left[\dfrac{1}{1+\eta^2} - C_1{}^2\right]^{1/2}} + C_2$$

$$= \int \frac{d\eta}{\left[\left(\dfrac{1}{C_1{}^2} - 1\right) - \eta^2\right]^{1/2}} + C_2$$

Denoting $(1/C_1{}^2 - 1)^{1/2}$ as $1/C_3$ we get on integrating:

$$\phi = \sin^{-1}(C_3\eta) + C_2 \qquad (g)$$

Replacing η from Eq. (f) we get:

$$\phi = \sin^{-1}\left[C_3\frac{1}{\tan\theta}\right] + C_2$$

Hence we can say:

$$\sin(\phi - C_2) = \frac{C_3}{\tan\theta}$$

This equation may next be written as follows:

$$\sin\phi\cos C_2 - \cos\phi\sin C_2 = C_3\frac{\cos\theta}{\sin\theta}$$

Hence:

$$\sin\phi\sin\theta\cos C_2 - \sin\theta\cos\phi\sin C_2 = \cos\theta$$

Observing the transformation equations (a) we may express the above equation in terms of Cartesian coordinates in the following manner:

$$y \cos C_2 - x \sin C_2 = zC_3 \qquad (h)$$

wherein we have cancelled the term R. This is the equation of a plane surface going through the origin. The intersection of this plane surface and the sphere then gives the proper locus of points on the sphere that forms the desired extremal path. Clearly this curve is the expected great circle. ////

2.6 FIRST VARIATION WITH SEVERAL DEPENDENT VARIABLES

We now consider the case where we may have any number of functions with still a single *independent* variable. We shall denote the functions as $q_1, q_2, \ldots, q_n$ and the independent variable we shall denote as t. (This is a notation that is often used in particle mechanics.) The functional for this case then is given as follows:

$$I = \int_{t_1}^{t_2} F(q_1, q_2, \ldots, q_n, \dot{q}_1, \dot{q}_2, \ldots, \dot{q}_n, t) \, dt \qquad (2.29)$$

where $\dot{q}_1 = dq_1/dt$, etc. We wish to determine a set of functions $q_1(t), q_2(t) \ldots q_n(t)$ which are twice differentiable and which extremize the functional I with respect to a broad class of admissible functions. We shall denote the varied functions as $\tilde{q}_i(t)$ and, as before, we shall henceforth consider the notation $q_i(t)$ to identify the extremizing functions we are seeking.

We shall use the following single-parameter family of admissible functions

$$\tilde{q}_1(t) = q_1(t) + \epsilon \eta_1(t)$$
$$\tilde{q}_2(t) = q_2(t) + \epsilon \eta_2(t)$$
$$\vdots$$
$$\tilde{q}_n(t) = q_n(t) + \epsilon \eta_n(t) \qquad (2.30)$$

where $\eta_1(t), \eta_2(t), \ldots, \eta_n(t)$ are arbitrary functions having proper continuity and differentiability properties for the ensuing steps. Also, these functions are equal to zero at the end points t_1 and t_2. Finally, when we set $\epsilon = 0$ we get back to the designated extremizing functions $q_1(t), \ldots, q_n(t)$.

We now form $\tilde{I}(\epsilon)$ as follows:

$$\tilde{I}(\epsilon) = \int_{t_1}^{t_2} F(\tilde{q}_1, \tilde{q}_2 \cdots \tilde{q}_n, \dot{\tilde{q}}_1, \dot{\tilde{q}}_2 \cdots \dot{\tilde{q}}_n, t) \, dt$$

In order for the $q_i(t)$ to be the extremal paths we now require that:

$$\left[\frac{d\tilde{I}(\epsilon)}{d\epsilon}\right]_{\epsilon=0} = 0$$

Hence:

$$\left\{\int_{t_1}^{t_2}\left(\frac{\partial F}{\partial \tilde{q}_1}\eta_1 + \cdots \frac{\partial F}{\partial \tilde{q}_n}\eta_n + \frac{\partial F}{\partial \dot{\tilde{q}}_1}\dot{\eta}_1 + \cdots \frac{\partial F}{\partial \dot{\tilde{q}}_n}\dot{\eta}_n\right)dt\right\}_{\epsilon=0} = 0$$

Now setting $\epsilon = 0$ in the above expression is the same as removing the tildes from the q's and $\dot{q}$'s. Thus we have:

$$\int_{t_1}^{t_2}\left(\frac{\partial F}{\partial q_1}\eta_1 + \cdots \frac{\partial F}{\partial q_n}\eta_n + \frac{\partial F}{\partial \dot{q}_1}\dot{\eta}_1 + \cdots \frac{\partial F}{\partial \dot{q}_n}\dot{\eta}_n\right)dt = 0$$

Now the functions $\eta_i(t)$ are arbitrary and so we can take all $\eta_i(t)$ except $\eta_1(t)$ equal to zero. We then have on integrating $(\partial F/\partial \dot{q}_1)\dot{\eta}_1$ by parts:

$$\int_{t_1}^{t_2}\left[\frac{\partial F}{\partial q_1} - \frac{d}{dt}\frac{\partial F}{\partial \dot{q}_1}\right]\eta_1\, dt = 0$$

Using the fundamental lemma of the calculus of variations, we conclude that:

$$\frac{\partial F}{\partial q_1} - \frac{d}{dt}\left(\frac{\partial F}{\partial \dot{q}_1}\right) = 0$$

We may similarly assume that η_2 is the only non-zero function and so forth to lead us to the conclusion that

$$\boxed{\begin{array}{l}\dfrac{\partial F}{\partial q_i} - \dfrac{d}{dt}\left(\dfrac{\partial F}{\partial \dot{q}_i}\right) = 0 \\[2ex] \qquad\qquad i = 1, 2, \cdots, n\end{array}} \qquad (2.31)$$

are necessary conditions for establishing the $q_i(t)$ as the extremal functions. These are again the Euler–Lagrange equations which lead us on substitution for F to a system of second-order ordinary differential equations for establishing the $q_i(t)$. These equations may be coupled (i.e., simultaneous equations) or uncoupled, depending on the variables q_i chosen to be used in forming the functional I in Eq. (2.29).

We now illustrate the use of the Euler–Lagrange equations for a problem in particle mechanics.

EXAMPLE 2.3   We will present for use here the very important Hamilton principle in order to illustrate the multi-function problem of this section. Later we shall take the time to consider this principle in detail.

For a system of particles acted on by conservative forces, Hamilton's principle states that the proper paths taken from a configuration at time t_1 to a configuration at time t_2 are those that extremize the following functional

$$I = \int_{t_1}^{t_2} (T - V)\, dt \qquad (a)$$

where T is the kinetic energy of the system and therefore a function of velocity $\dot{x}_i$ of each particle while V is the potential energy of the system and therefore a function of the coordinates x_i of the particles. We thus have here a functional of many dependent variables x_i with the presence of a single independent variable, t.

Consider two identical masses connected by three springs as has been shown in Fig. 2.5. The masses can only move along a straight line as a result of frictionless constraints. Two independent coordinates are needed to locate the system; they are shown as x_1 and x_2. The springs are unstretched when $x_1 = x_2 = 0$.

From elementary mechanics we can say for the system:

$$T = \tfrac{1}{2}M\dot{x}_1^2 + \tfrac{1}{2}M\dot{x}_2^2$$
$$V = \tfrac{1}{2}K_1 x_1^2 + \tfrac{1}{2}K_2(x_2 - x_1)^2 + \tfrac{1}{2}K_1 x_2^2 \qquad (b)$$

Hence we have for I:

$$I = \int_{t_1}^{t_2} \left\{ \tfrac{1}{2}M(\dot{x}_1^2 + \dot{x}_2^2) - \tfrac{1}{2}[K_1 x_1^2 + K_2(x_2 - x_1)^2 + K_1 x_2^2] \right\} dt \qquad (c)$$

To extremize I we employ Eq. (2.31) as follows:

$$\frac{\partial F}{\partial x_1} - \frac{d}{dt}\left(\frac{\partial F}{\partial \dot{x}_1}\right) = 0$$

$$\frac{\partial F}{\partial x_2} - \frac{d}{dt}\left(\frac{\partial F}{\partial \dot{x}_2}\right) = 0 \qquad (d)$$

Substituting we get:

$$-K_1 x_1 + K_2(x_2 - x_1) - \frac{d}{dt}(M\dot{x}_1) = 0$$

$$-K_2 x_2 - K_2(x_2 - x_1) - \frac{d}{dt}(M\dot{x}_2) = 0 \qquad (e)$$

Rearranging we then have:

$$M\ddot{x}_1 + K_1 x_1 - K_2(x_2 - x_1) = 0$$
$$M\ddot{x}_2 + K_2 x_2 + K_2(x_2 - x_1) = 0 \qquad (f)$$

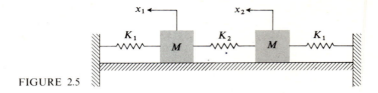

FIGURE 2.5

These are recognized immediately to be equations obtainable directly from Newton's laws. Thus the Euler–Lagrange equations lead to the basic equations of motion for this case. We may integrate these equations and, using initial conditions of $\dot{x}_1(0)$, $\dot{x}_2(0)$, $x_1(0)$ and $x_2(0)$, we may then fully establish the subsequent motion of the system.

In this problem we could have more easily employed Newton's law directly. There are many problems, however, where it is easier to proceed by the variational approach to arrive at the equations of motion. Also, as will be seen in the text, other advantages accrue to the use of the variational method.   ////

2.7   THE ISOPERIMETRIC PROBLEM

We now investigate the isoperimetric problem in its simplest form whereby we wish to extremize the functional

$$I = \int_{x_1}^{x_2} F(x,y,y')\,dx \qquad (2.32)$$

subject to the restriction that $y(x)$ have a form such that:

$$J = \int_{x_1}^{x_2} G(x,y,y')\,dx = \text{Const.} \qquad (2.33)$$

where G is a given function.

We proceed essentially as in Sec. 2.3. We take $y(x)$ from here on to represent the extremizing function for the functional of Eq. (2.32). We next introduce a system of varied paths $\tilde{y}(x)$ for computation of $\tilde{I}$. Our task is then to find conditions required to be imposed on $y(x)$ so as to extremize $\tilde{I}$ with respect to the admissible varied paths $\tilde{y}(x)$ while satisfying the isoperimetric constraint of Eq. (2.33). To facilitate the computations we shall require that the admissible varied paths also satisfy Eq. (2.33). Thus we have:

$$\tilde{I} = \int_{x_1}^{x_2} F(x,\tilde{y},\tilde{y}')\,dx \qquad (a)$$

$$\tilde{J} = \int_{x_1}^{x_2} G(x,\tilde{y},\tilde{y}')\,dx = \text{Const.} \qquad (b) \quad (2.34)$$

Because of this last condition on $y(x)$ we shall no longer use the familiar single-parameter family of admissible functions, since varying ϵ alone for a single parameter family of functions may mean that the constraint for the corresponding paths $y(x)$ is not satisfied. To allow for enough flexibility to carry out extremization while maintaining *intact* the constraining equations, we employ a two-parameter family of admissible functions of the form,

$$\tilde{y}(x) = y(x) + \epsilon_1 \eta_1(x) + \epsilon_2 \eta_2(x) \qquad (2.35)$$

where $\eta_1(x)$ and $\eta_2(x)$ are arbitrary functions which vanish at the end points x_1, x_2, and where ϵ_1 and ϵ_2 are two small parameters. Using this system of admissible functions it is clear that $\tilde{I}$ and $\tilde{J}$ are functions of ϵ_1 and ϵ_2. Thus:

$$\tilde{I}(\epsilon_1, \epsilon_2) = \int_{x_1}^{x_2} F(x, \tilde{y}, \tilde{y}') \, dx \qquad (a)$$

$$\tilde{J}(\epsilon_1, \epsilon_2) = \int_{x_1}^{x_2} G(x, \tilde{y}, \tilde{y}') \, dx \qquad (b) \quad (2.36)$$

To extremize $\tilde{I}$ when $\tilde{y} \to y$ we require (see Eq. 2.25) that:

$$\delta^{(1)}\tilde{I} = \left(\frac{\partial \tilde{I}}{\partial \epsilon_1}\right)_{\substack{\epsilon_1 = 0 \\ \epsilon_2 = 0}} \epsilon_1 + \left(\frac{\partial \tilde{I}}{\partial \epsilon_2}\right)_{\substack{\epsilon_1 = 0 \\ \epsilon_2 = 0}} \epsilon_2 = 0 \qquad (2.37)$$

If ϵ_1 and ϵ_2 were independent of each other, we could then set each of the partial derivatives in the above equation equal to zero separately to satisfy the above equation. However, ϵ_1 and ϵ_2 are related by the requirement that $\tilde{J}(\epsilon_1, \epsilon_2) = \text{const.}$ The first variation of $\tilde{J}$ must be zero because of the constancy of its value and we have the equation:

$$\left(\frac{\partial \tilde{J}}{\partial \epsilon_1}\right)_{\substack{\epsilon_1 = 0 \\ \epsilon_2 = 0}} \epsilon_1 + \left(\frac{\partial \tilde{J}}{\partial \epsilon_2}\right)_{\substack{\epsilon_1 = 0 \\ \epsilon_2 = 0}} \epsilon_2 = 0 \qquad (2.38)$$

At this time we make use of the method of the *Lagrange multiplier*. That is, we multiply Eq. (2.38) by an undetermined constant λ (the Lagrange multiplier) and add Eq. (2.37) and (2.38) to get:

$$\left[\left(\frac{\partial \tilde{I}}{\partial \epsilon_1}\right)_{\substack{\epsilon_1 = 0 \\ \epsilon_2 = 0}} + \lambda \left(\frac{\partial \tilde{J}}{\partial \epsilon_1}\right)_{\substack{\epsilon_1 = 0 \\ \epsilon_2 = 0}}\right] \epsilon_1 + \left[\left(\frac{\partial \tilde{I}}{\partial \epsilon_2}\right)_{\substack{\epsilon_1 = 0 \\ \epsilon_2 = 0}} + \lambda \left(\frac{\partial \tilde{J}}{\partial \epsilon_2}\right)_{\substack{\epsilon_1 = 0 \\ \epsilon_2 = 0}}\right] \epsilon_2 = 0 \qquad (2.39)$$

We now choose λ so that one of the two bracketed quantities is zero. We choose here the second bracketed quantity so that:

$$\left(\frac{\partial \tilde{I}}{\partial \epsilon_2}\right)_{\substack{\epsilon_1 = 0 \\ \epsilon_2 = 0}} + \lambda \left(\frac{\partial \tilde{J}}{\partial \epsilon_2}\right)_{\substack{\epsilon_1 = 0 \\ \epsilon_2 = 0}} = 0 \qquad (2.40)$$

Now we may consider that ϵ_2 is the dependent variable and that ϵ_1 is the independent variable. We must then conclude from Eq. (2.39) that the coefficient of ϵ_1, is zero. Thus:

$$\left(\frac{\partial \tilde{I}}{\partial \epsilon_1}\right)_{\substack{\epsilon_1 = 0 \\ \epsilon_2 = 0}} + \lambda \left(\frac{\partial \tilde{J}}{\partial \epsilon_1}\right)_{\substack{\epsilon_1 = 0 \\ \epsilon_2 = 0}} = 0 \qquad (2.41)$$

Thus, Eq. (2.40) and (2.41) with the use of the multiplier λ give us the necessary conditions for an extreme of $\tilde{I}$ while maintaining the constraint condition intact.

We can now shorten the notation by introducing $\tilde{I}*$ as follows:

$$\tilde{I}* = \tilde{I} + \lambda \tilde{J} \qquad (2.42)$$

so that Eqs. (2.40) and (2.41) become

$$\left(\frac{\partial \tilde{I}*}{\partial \epsilon_1}\right)_{\substack{\epsilon_1 = 0 \\ \epsilon_2 = 0}} = \left(\frac{\partial \tilde{I}*}{\partial \epsilon_2}\right)_{\substack{\epsilon_1 = 0 \\ \epsilon_2 = 0}} = 0 \qquad (2.43)$$

In integral form we have for $\tilde{I}*$

$$\tilde{I}* = \int_{x_1}^{x_2} F(x,\tilde{y},\tilde{y}')\, dx + \lambda \int_{x_1}^{x_2} G(x,\tilde{y},\tilde{y}')\, dx$$

$$= \int_{x_1}^{x_2} [F(x,\tilde{y},\tilde{y}') + \lambda G(x,\tilde{y},\tilde{y}')]\, dx = \int_{x_1}^{x_2} F*(x,\tilde{y},\tilde{y}')\, dx \qquad (2.44)$$

where:

$$F* = F + \lambda G \qquad (2.45)$$

We now apply the conditions given by Eq. (2.43) using Eq. (2.44) to replace $\tilde{I}*$.

$$\left(\frac{\partial \tilde{I}*}{\partial \epsilon_i}\right)_{\substack{\epsilon_1 = 0 \\ \epsilon_2 = 0}} = \left\{ \int_{x_1}^{x_2} \left(\frac{\partial F*}{\partial \tilde{y}}\eta_i + \frac{\partial F*}{\partial \tilde{y}'}\eta_i'\right) dx \right\}_{\substack{\epsilon_1 = 0 \\ \epsilon_2 = 0}} = 0 \qquad i = 1, 2 \qquad (2.46)$$

We thus get a statement of the extremization process without the appearance of the constraint and this now becomes the starting point of the extremization process. Removing the tildes from $\tilde{y}$ and $\tilde{y}'$ is equivalent to setting $\epsilon_i = 0$ and so we may say:

$$\left(\frac{\partial \tilde{I}*}{\partial \epsilon_i}\right)_{\substack{\epsilon_1 = 0 \\ \epsilon_2 = 0}} = \int_{x_1}^{x_2} \left(\frac{\partial F*}{\partial y}\eta_i + \frac{\partial F*}{\partial y'}\eta_i'\right) dx = 0 \qquad i = 1, 2 \qquad (2.47)$$

Integrating the second expression in the integrand by parts and noting that $\eta_i(x_1) = \eta_i(x_2) = 0$ we get:

$$\int_{x_1}^{x_2} \left[\frac{\partial F*}{\partial y} - \frac{d}{dx}\frac{\partial F*}{\partial y'}\right]\eta_i\, dx = 0 \qquad i = 1, 2$$

Now using the fundamental lemma of the calculus of variations we conclude that:

$$\boxed{\frac{\partial F^*}{\partial y} - \frac{d}{dx}\left(\frac{\partial F^*}{\partial y'}\right) = 0} \qquad (2.48)$$

Thus the Euler–Lagrange equation is again a necessary condition for the desired extremum, this time applied to F^* and thus including the Lagrange multiplier. This leads to a second-order differential equation for $y(x)$, the extremizing function. Integrating this equation then leaves us two constants of integration plus the Lagrange multiplier. These are determined from the specified values of y at the end points plus the constraint condition given by Eq. (2.33).

We have thus far considered only a single constraining integral. If we have "n" such integrals

$$\int_{x_1}^{x_2} G_k(x,y,y')\,dx = C_k \qquad k = 1, 2, \cdots, n$$

then by using an $n + 1$ parameter family of varied paths $\tilde{y} = y + \epsilon_1\eta_1 + \epsilon_2\eta_2 + \cdots$ $\epsilon_{n+1}\eta_{n+1}$ we can arrive as before at the following requirement for extremization:

$$\frac{\partial F^*}{\partial y} - \frac{d}{dt}\left(\frac{\partial F^*}{\partial y'}\right) = 0$$

where:

$$F^* = F + \sum_{k=1}^{n} \lambda_k G_k$$

The λ's are again the Lagrange multipliers. Finally, for p dependent variables, i.e.,

$$I = \int_{t_1}^{t_2} F(q_1, q_2, \ldots, q_p, \dot{q}_1, \ldots, \dot{q}_p, t)\,dt$$

with n constraints

$$\int_{t_1}^{t_2} G_k(q_1, \ldots, q_p, \dot{q}_1, \ldots, \dot{q}_p, t)\,dt = C_k \qquad k = 1, 2, \ldots, n$$

the extremizing process yields

$$\frac{\partial F^*}{\partial q_i} - \frac{d}{dt}\left(\frac{\partial F^*}{\partial q_i}\right) = 0 \qquad i = 1, \ldots, p \qquad (2.49)$$

where

$$F^* = F + \sum_{1}^{n} \lambda_k G_k \qquad (2.50)$$

We now illustrate the use of these equations by considering in detail the isoperimetric problem presented earlier.

EXAMPLE 2.4   Recall from Sec. 2.2 that the isoperimetric problem asks us to find the particular curve $y(x)$ which for a given length L encloses the largest area A. Expressed parametrically we have a functional with two functions y and x; the independent variable is τ. Thus:

$$I = A = \frac{1}{2} \int_{\tau_1}^{\tau_2} \left(x\frac{dy}{d\tau} - y\frac{dx}{d\tau} \right) d\tau \qquad (a)$$

The constraint relation is:

$$L = \int_{\tau_1}^{\tau_2} \sqrt{\left(\frac{dx}{d\tau}\right)^2 + \left(\frac{dy}{d\tau}\right)^2}\, d\tau \qquad (b)$$

We now form F^* for this case. Thus:

$$F^* = \tfrac{1}{2}(x\dot{y} - y\dot{x}) + \lambda(\dot{x}^2 + \dot{y}^2)^{1/2}$$

where we use the dot superscript to represent $(d\ /d\tau)$. We now set forth the Euler–Lagrange equations:

$$\frac{\dot{y}}{2} - \frac{d}{d\tau}\left[-\frac{y}{2} + \frac{\lambda\dot{x}}{(\dot{x}^2 + \dot{y}^2)^{1/2}} \right] = 0 \qquad (c)$$

$$-\frac{\dot{x}}{2} - \frac{d}{d\tau}\left[\frac{x}{2} + \frac{\lambda\dot{y}}{(\dot{x}^2 + \dot{y}^2)^{1/2}} \right] = 0 \qquad (d)$$

We next integrate Eqs. (c) and (d) with respect to τ to get:

$$y - \frac{\lambda\dot{x}}{(\dot{x}^2 + \dot{y}^2)^{1/2}} = C_1 \qquad (e)$$

$$x + \frac{\lambda\dot{y}}{(\dot{x}^2 + \dot{y}^2)^{1/2}} = C_2 \qquad (f)$$

After eliminating λ between the equations we will reach the following result:

$$(x - C_2)\, dx + (y - C_1)\, dy = 0 \qquad (g)$$

The last equation is easily integrated to yield:

$$\frac{(x - C_2)^2}{2} + \frac{(y - C_1)^2}{2} = C_3{}^2 \qquad (h)$$

where C_3 is a constant of integration. Thus we get as the required curve a circle— a result which should surprise no one. The radius of the circle is $\sqrt{2}C_3$ which you may readily show (by eliminating C_1 and C_2 from Eq. (h) using Eqs. (e) and (f) and solving for λ) is the value of the Lagrange multiplier[1] λ. The constants C_1 and C_2 merely position the circle.   ////

2.8   FUNCTIONAL CONSTRAINTS

We now consider the functional

$$I = \int_{t_1}^{t_2} F(q_1, q_2, \ldots, q_n, \dot{q}_1, \ldots, \dot{q}_n, t)\, dt \qquad (2.51)$$

with the following m constraints on the n functions q_i:[2]

$$G_1(q_1, \ldots, q_n, \dot{q}_1, \ldots, \dot{q}_n, t) = 0$$
$$\vdots$$
$$G_m(q_1, \ldots, q_n, \dot{q}_1, \ldots, \dot{q}_n, t) = 0 \qquad (2.52)$$

We assume $m < n$. To extremize the functional I we may proceed by employing n one-parameter families of varied functions of the form

$$\tilde{q}_i(t) = q_i(t) + \epsilon\eta_i(t) \qquad i = 1, \ldots, n \qquad (2.53)$$

where $\eta_i(t_1) = \eta_i(t_2) = 0$. Furthermore we assume that the varied functions $\tilde{q}_i$ satisfy the constraining equation (2.52). That is,

$$G_j(\tilde{q}_1, \ldots, \tilde{q}_n, \dot{\tilde{q}}_1, \ldots, \dot{\tilde{q}}_n, t) = 0 \qquad j = 1, \ldots, m \qquad (2.54)$$

We now set the first variation of I equal to zero.

$$\delta^{(1)}I = 0 = \left(\frac{\partial \tilde{I}}{\partial \epsilon}\right)_{\epsilon=0} \epsilon$$

Setting $(\partial \tilde{I}/\partial \epsilon)_{\epsilon=0}$ equal to zero we have:

$$\left[\int_{t_1}^{t_2}\left[\frac{\partial F}{\partial \tilde{q}_1}\eta_1 + \cdots + \frac{\partial F}{\partial \tilde{q}_n}\eta_n + \frac{\partial F}{\partial \dot{\tilde{q}}_1}\dot{\eta}_1 + \cdots + \frac{\partial F}{\partial \dot{\tilde{q}}_n}\dot{\eta}_n\right] dt\right]_{\epsilon=0} = 0$$

Dropping the tildes and subscript $\epsilon = 0$ we have:

$$\int_{t_1}^{t_2}\left(\frac{\partial F}{\partial q_1}\eta_1 + \cdots + \frac{\partial F}{\partial q_n}\eta_n + \frac{\partial F}{\partial \dot{q}_1}\dot{\eta}_1 + \cdots + \frac{\partial F}{\partial \dot{q}_n}\dot{\eta}_n\right) dt = 0 \qquad (2.55)$$

---

[1] The Lagrange multiplier is usually of physical significance.
[2] In dynamics of particles, if the constraining equations do not have derivatives the constraints are called *holomonic*.

Since the varied q's satisfy the constraining equations, by assumption, we can conclude since $G_i = 0$ that:

$$\delta^{(1)}(G_i) = 0 = \left(\frac{\partial G}{\partial \epsilon}\right)_{\epsilon=0} \epsilon$$

Setting $(\partial G_i/\partial \epsilon)_{\epsilon=0} = 0$ here we have:

$$\left\{\frac{\partial G_i}{\partial \tilde{q}_1}\eta_1 + \cdots + \frac{\partial G_i}{\partial \tilde{q}_1}\eta_n + \frac{\partial G_i}{\partial \dot{\tilde{q}}_1}\dot{\eta}_1 + \cdots + \frac{\partial G_i}{\partial \dot{\tilde{q}}_n}\dot{\eta}_n\right\}_{\epsilon=0} = 0$$

$$i = 1, 2, \ldots, m$$

Dropping the tildes and $\epsilon = 0$ we then have:

$$\frac{\partial G_i}{\partial q_1}\eta_1 + \cdots + \frac{\partial G_i}{\partial q_n}\eta_n + \frac{\partial G_i}{\partial \dot{q}_1}\dot{\eta}_1 + \cdots + \frac{\partial G_i}{\partial \dot{q}_n}\dot{\eta}_n = 0 \qquad i = 1, \ldots, m$$

We now multiply each of the above m equations by an *arbitrary time function*, $\lambda_i(t)$, which we may call a *Lagrange multiplier function*. Adding the resulting equations we have:

$$\sum_{i=1}^{m} \lambda_i(t)\left[\frac{\partial G_i}{\partial q_1}\eta_1 + \cdots + \frac{\partial G_i}{\partial q_n}\eta_n + \frac{\partial G_i}{\partial \dot{q}_1}\dot{\eta}_1 + \cdots + \frac{\partial G_i}{\partial \dot{q}_n}\dot{\eta}\right] = 0$$

Now integrate the above sum from t_1 to t_2 and then add the results to Eq. (2.55). We then have:

$$\int_{t_1}^{t_2} \left\{\frac{\partial F}{\partial q_1}\eta_1 + \cdots + \frac{\partial F}{\partial q_n}\eta_n + \frac{\partial F}{\partial \dot{q}_1}\dot{\eta}_1 + \cdots + \frac{\partial F}{\partial \dot{q}_n}\dot{\eta}_n \right.$$
$$\left. + \sum_{i=1}^{m} \lambda_i\left(\frac{\partial G_i}{\partial q_1}\eta_1 + \cdots + \frac{\partial G_i}{\partial q_n}\eta_n + \frac{\partial G_i}{\partial \dot{q}_1}\dot{\eta}_1 + \cdots + \frac{\partial G_i}{\partial \dot{q}_n}\dot{\eta}_n\right)\right\} dt = 0$$

Integrating by parts the terms with $\dot{\eta}_i$ and regrouping the results we then have:

$$\int_{t_1}^{t_2} \left[\left\{\left(\frac{\partial F}{\partial q_1} - \frac{d}{dt}\left(\frac{\partial F}{\partial \dot{q}_1}\right)\right) + \sum_{i=1}^{m}\left[\lambda_i\frac{\partial G_i}{\partial q_1} - \frac{d}{dt}\left(\lambda_i\frac{\partial G_i}{\partial \dot{q}_1}\right)\right]\right\}\eta_1\right.$$
$$+ \cdots\cdots\cdots\cdots\cdots\cdots\cdots\cdots\cdots\cdots\cdots\cdots$$
$$+ \cdots\cdots\cdots\cdots\cdots\cdots\cdots\cdots\cdots\cdots\cdots\cdots$$
$$\left. + \left\{\left(\frac{\partial F}{\partial q_n} - \frac{d}{dt}\left(\frac{\partial F}{\partial \dot{q}_n}\right)\right) + \sum_{i=1}^{m}\left[\lambda_i\frac{\partial G_i}{\partial q_n} - \frac{d}{dt}\left(\lambda_i\frac{\partial G_i}{\partial \dot{q}_n}\right)\right]\right\}\eta_n\right] dt = 0 \qquad (2.56)$$

Now introduce F^* defined as:

$$F^* = F + \sum_{i=1}^{m} \lambda_i(t)G_i \qquad (2.57)$$

We can then rewrite Eq. (2.56) as follows:

$$\int_{t_1}^{t_2} \left\{ \left(\frac{\partial F^*}{\partial q_1} - \frac{d}{dt} \frac{\partial F^*}{\partial \dot{q}_1} \right) \eta_1 + \cdots + \left(\frac{\partial F^*}{\partial q_n} - \frac{d}{dt} \frac{\partial F^*}{\partial \dot{q}_n} \right) \eta_n \right\} dt = 0$$

Now the η's are not independent (they are related through m equations (2.54)) and so we cannot set the coefficients of the bracketed expressions equal to zero separately. However, we can say that $(n - m)$ of the η's are independent. For the remaining m η's we now assume that the time functions $\lambda_i(t)$ are so chosen that the coefficients of the η's are zero. Then, since we are left with only independent η's, we can take the remaining coefficients equal to zero. In this way we conclude that:

$$\frac{\partial F^*}{\partial q_i} - \frac{d}{dt} \frac{\partial F^*}{\partial \dot{q}_i} = 0 \qquad i = 1, 2, \ldots, n$$

We thus arrive at the Euler–Lagrange equations once again. However, we have now m unknown time functions λ_i to be determined with the aid of the original m constraining equations.

We shall have ample opportunity of using the formulations of this section in the following chapter when we consider the Reissner functional.

2.9   A NOTE ON BOUNDARY CONDITIONS

In previous efforts at extremizing I we specified the end points (x_1, y_1) and (x_2, y_2) through which the extremizing function had to proceed. Thus, in Fig. 2.2 we asked for the function $y(x)$ going through (x_1, y_1) and (x_2, y_2) to make it extremize I relative to neighboring varied paths *also* going through the aforestated end points. The boundary conditions specifying y at x_1 and at x_2 are called the *kinematic* or *rigid boundary* conditions of the problem.

We now pose a different query. Suppose only x_1 and x_2 are given as has been shown in Fig. 2.6 and we ask what is the function $y(x)$ that extremizes the functional $\int_{x_1}^{x_2} F(x, y, y') \, dx$ between these limits. Thus, we *do not specify* y at x_1 and x_2 for the extremizing function.

As before, we denote the extremizing function in the discussion as $y(x)$ and we have shown it so labeled at some position in Fig. 2.6. A system of nearby admissible "paths" $\tilde{y}(x)$ has also been shown. Some of these paths go through the endpoints of the extremizing path while others do not. We shall extremize I with respect to such a family of admissible functions to obtain certain necessary requirements for $y(x)$ to maintain the role of the extremizing function. Thus with the δ operator we arrive at the following necessary condition using the same steps of earlier discussions

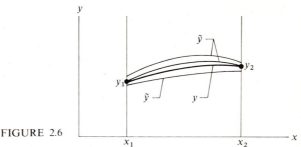

FIGURE 2.6

and noting that δy need no longer always be zero at the end points:

$$\delta^{(1)}I = 0 = \int_{x_1}^{x_2} \left[\frac{\partial F}{\partial y} - \frac{d}{dx}\left(\frac{\partial F}{\partial y'}\right) \right] \delta y \, dx + \frac{\partial F}{\partial y'} \delta y \bigg|_{x=x_1} - \frac{\partial F}{\partial y'} \delta y \bigg|_{x=x_2} \qquad (2.58)$$

There are admissible functions that go through the end points (x_1, y_1) and (x_2, y_2) of the designated extremizing function $y(x)$. For such admissible functions $\delta y_1 = \delta y_2 = 0$ and we conclude that a necessary condition for an extremum is:

$$\int_{x_1}^{x_2} \left[\frac{\partial F}{\partial y} - \frac{d}{dx}\left(\frac{\partial F}{\partial y'}\right) \right] \delta y \, dx = 0$$

Following the familiar procedures of previous computations we readily then arrive at the Euler–Lagrange equations as a necessary condition for $y(x)$ to be an extreme:

$$\frac{\partial F}{\partial y} - \frac{d}{dx}\left(\frac{\partial F}{\partial y'}\right) = 0$$

There are now, however, additional necessary requirements for extremizing I if δy is not zero at the end points of the extremizing function. Accordingly from Eq. (2.58) we conclude that for such circumstances we require:

$$\frac{\partial F}{\partial y'} \bigg|_{x=x_1} = 0 \qquad (a)$$

$$\frac{\partial F}{\partial y'} \bigg|_{x=x_2} = 0 \qquad (b) \quad (2.59)$$

The conditions (2.59) are termed the *natural boundary conditions*. They are the boundary conditions that must be prescribed if both the values $y(x_1)$ and $y(x_2)$ are not specified.[1] However, it is also possible to assign one natural and one kinematic

---

[1] In problems of solid mechanics dealing with the total potential energy we will see that the kinematic boundary conditions involve displacement conditions of the boundary while natural boundary conditions involve force conditions at the boundary.

boundary condition to satisfy the requirements for $y(x)$ to fulfill its assigned role as extremizing function. In more general functionals I set forth earlier and those to be considered in following sections, we find natural boundary conditions in much the same way as set forth in this section. In essence we proceed with the variation process *without* requiring the η functions (or by the same token the variations δy) to be zero at the boundaries. Rather we set equal to zero all expressions established on the boundary by the integration by parts procedures. The resulting conditions are the natural boundary conditions.

Note that for all various possible boundary conditions, we work in any particular case with the same differential equation. However, the extremal function will eventually depend on the particular permissible combination of boundary conditions we choose to employ. Usually certain kinematic boundary conditions are imposed by the constraints present in a particular problem. We must use these boundary conditions or else our extremal functions will not correspond to the problem at hand. The remaining boundary conditions then are natural ones that satisfy the requirements for the extremizing process. We shall illustrate these remarks in Example 2.5 after a discussion of higher-order derivatives.

2.10 FUNCTIONALS INVOLVING HIGHER-ORDER DERIVATIVES

We have thus far considered only first-order derivatives in the functionals. At this time we extend the work by finding extremal functions $y(x)$ for functionals having higher-order derivatives. Accordingly we shall consider the following functional:

$$I = \int_{x_1}^{x_2} F(x,y,y',y'',y''')\,dx \qquad (2.60)$$

The cases for lower or higher-order derivatives other than y''' are easily attainable from the procedure that we shall follow. Using the familiar one-parameter family of admissible functions for extremizing I we can then say:

$$\left(\frac{d\tilde{I}}{d\epsilon}\right)_{\epsilon=0} = \left[\frac{d}{d\epsilon}\int_{x_1}^{x_2} F(x,\tilde{y},\tilde{y}',\tilde{y}'',\tilde{y}''')\,dx\right]_{\epsilon=0} = 0$$

This becomes:

$$\left\{\int_{x_1}^{x_2}\left[\frac{\partial F}{\partial \tilde{y}}\eta + \frac{\partial F}{\partial \tilde{y}'}\eta' + \frac{\partial F}{\partial \tilde{y}''}\eta'' + \frac{\partial F}{\partial \tilde{y}'''}\eta'''\right]dx\right\}_{\epsilon=0} = 0$$

We may drop the subscript $\epsilon = 0$ along with the tildes as follows:

$$\int_{x_1}^{x_2} \left[\frac{\partial F}{\partial y} \eta + \frac{\partial F}{\partial y'} \eta' + \frac{\partial F}{\partial y''} \eta'' + \frac{\partial F}{\partial y'''} \eta''' \right] dx = 0$$

We now carry out the following series of integration by parts:

$$\int_{x_1}^{x_2} \frac{\partial F}{\partial y'} \eta' \, dx = \left. \frac{\partial F}{\partial y'} \eta \right|_{x_1}^{x_2} - \int_{x_1}^{x_2} \frac{d}{dx}\left[\frac{\partial F}{\partial y'} \right] \eta \, dx$$

$$\int_{x_1}^{x_2} \frac{\partial F}{\partial y''} \eta'' \, dx = \left. \frac{\partial F}{\partial y''} \eta' \right|_{x_1}^{x_2} - \int_{x_1}^{x_2} \frac{d}{dx}\left[\frac{\partial F}{\partial y''} \right] \eta' \, dx$$

$$= \left. \frac{\partial F}{\partial y''} \eta' \right|_{x_1}^{x_2} - \left. \frac{d}{dx}\left(\frac{\partial F}{\partial y''} \right) \eta \right|_{x_1}^{x_2} + \int_{x_1}^{x_2} \frac{d^2}{dx^2}\left(\frac{\partial F}{\partial y''} \right) \eta \, dx$$

$$\int_{x_1}^{x_2} \frac{\partial F}{\partial y'''} \eta''' \, dx = \left. \frac{\partial F}{\partial y'''} \eta'' \right|_{x_1}^{x_2} - \int_{x_1}^{x_2} \frac{d}{dx}\left(\frac{\partial F}{\partial y'''} \right) \eta'' \, dx$$

$$= \left. \frac{\partial F}{\partial y'''} \eta'' \right|_{x_1}^{x_2} - \left. \frac{d}{dx}\left(\frac{\partial F}{\partial y'''} \right) \eta' \right|_{x_1}^{x_2} + \int_{x_1}^{x_2} \frac{d^2}{dx^2}\left(\frac{\partial F}{\partial y'''} \right) \eta' \, dx$$

$$= \left. \frac{\partial F}{\partial y'''} \eta'' \right|_{x_1}^{x_2} - \left. \frac{d}{dx}\left(\frac{\partial F}{\partial y'''} \right) \eta' \right|_{x_1}^{x_2} + \left. \frac{d^2}{dx^2}\left(\frac{\partial F}{\partial y'''} \right) \eta \right|_{x_1}^{x_2}$$

$$- \int \frac{d^3}{dx^3}\left(\frac{\partial F}{\partial y'''} \right) \eta \, dx$$

Now combining the results we get:

$$- \int_{x_1}^{x_2} \left[\frac{d^3}{dx^3}\left(\frac{\partial F}{\partial y'''} \right) - \frac{d^2}{dx^2}\left(\frac{\partial F}{\partial y''} \right) + \frac{d}{dx}\left(\frac{\partial F}{\partial y'} \right) - \frac{\partial F}{\partial y} \right] \eta \, dx$$

$$+ \left. \frac{\partial F}{\partial y'''} \eta'' \right|_{x_1}^{x_2} - \left. \left\{ \frac{d}{dx} \frac{\partial F}{\partial y'''} - \frac{\partial F}{\partial y''} \right\} \eta' \right|_{x_1}^{x_2}$$

$$+ \left. \left\{ \frac{d^2}{dx^2}\left(\frac{\partial F}{\partial y'''} \right) - \frac{d}{dx}\left(\frac{\partial F}{\partial y''} \right) + \frac{\partial F}{\partial y'} \right\} \eta \right|_{x_1}^{x_2} = 0 \qquad (2.61)$$

Since the function η having the properties such that $\eta(x_1) = \eta'(x_1) = \eta''(x_1) = \eta(x_2) = \eta'(x_2) = \eta''(x_2) = 0$ are admissible no matter what the boundary conditions may be, it is clear that a necessary condition for an extremum is:

$$\int_{x_1}^{x_2} \left[\frac{d^3}{dx^3}\left(\frac{\partial F}{\partial y'''} \right) - \frac{d^2}{dx^2}\left(\frac{\partial F}{\partial y''} \right) + \frac{d}{dx}\left(\frac{\partial F}{\partial y'} \right) - \frac{\partial F}{\partial y} \right] \eta \, dx = 0$$

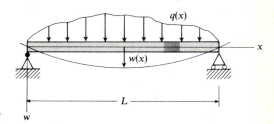

FIGURE 2.7 w

And we can then arrive at the following Euler–Lagrange equation:

$$\frac{d^3}{dx^3}\left(\frac{\partial F}{\partial y'''}\right) - \frac{d^2}{dx^2}\left(\frac{\partial F}{\partial y''}\right) + \frac{d}{dx}\left(\frac{\partial F}{\partial y'}\right) - \frac{\partial F}{\partial y} = 0 \qquad (2.62)$$

If, specifically, the conditions $y(x_1)$, $y(x_2)$, $y'(x_1)$, $y'(x_2)$, $y''(x_1)$ and $y''(x_2)$ are *specified* (these are the kinematic boundary conditions of this problem), the admissible functions must have these prescribed values and these prescribed derivatives at the end points. Then clearly we have for this case:

$$\eta(x_1) = \eta(x_2) = \eta'(x_1) = \eta'(x_2) = \eta''(x_1) = \eta''(x_2) = 0$$

Thus we see that giving the kinematic end conditions and satisfying the Euler–Lagrange equations permits the resulting $y(x)$ to be an extremal. If the kinematic conditions are not specified then we may satisfy the *natural boundary conditions* for this problem. Summarizing, we may use either of the following overall sets of requirements (see Eq. (2.61)) at the end points:

Kinematic	*Natural*	
y'' specified or	$\dfrac{\partial F}{\partial y'''} = 0$	(a)
y' specified or	$\dfrac{d}{dx}\left(\dfrac{\partial F}{\partial y'''}\right) - \dfrac{\partial F}{\partial y''} = 0$	(b)
y specified or	$\dfrac{d^2}{dx^2}\left(\dfrac{\partial F}{\partial y'''}\right) - \dfrac{d}{dx}\left(\dfrac{\partial F}{\partial y''}\right) + \dfrac{\partial F}{\partial y'} = 0$	(c) (2.63)

Clearly combinations of kinematic and natural boundary conditions are also permissible.

It should be apparent that by disregarding the terms involving y''' we may use the results for a functional whose highest-order derivative is second-order. And, by disregarding both terms with y''' and with y'', we get back to the familiar expressions used earlier where y' was the highest-order term in the functional. Furthermore the pattern of growth from first order derivatives on up is clearly established by the results so that one can readily extrapolate the formulations for orders higher than three. Finally we may directly extend the results of this problem to include function I with more than one function y. We simply get equations of the form (2.62) for each function. Thus for t as the independent variable and $q_1, q_2, \ldots, q_n$ as the functions in I we have for derivatives up to order three for all functions:

$$\frac{d^3}{dt^3}\left(\frac{\partial F}{\partial q_1''}\right) - \frac{d^2}{dt^2}\left(\frac{\partial F}{\partial q_1''}\right) + \frac{d}{dt}\left(\frac{\partial F}{\partial q_1'}\right) - \frac{\partial F}{\partial q_1} = 0$$

$$\vdots$$

$$\frac{d^3}{dt^3}\left(\frac{\partial F}{\partial q_n''}\right) - \frac{d^2}{dt^2}\left(\frac{\partial F}{\partial q_n''}\right) + \frac{d}{dt}\left(\frac{\partial F}{\partial q_n'}\right) - \frac{\partial F}{\partial q_n} = 0 \qquad (2.64)$$

Similarly for each q we have a set of conditions corresponding to Eq. (2.63).

EXAMPLE 2.5   We now consider the deflection w of the centerline of a simply-supported beam having a rectangular cross section and loaded in the plane of symmetry of the cross section (see Fig. 2.7). In Chap. 4 we will show through the principal of minimum total potential energy that the following functional is to be extremized in order to insure equilibrium:

$$I = \int_0^L \left[\frac{EI}{2}\left(\frac{d^2w}{dx^2}\right)^2 - qw\right] dx \qquad (a)$$

where I is the second moment of the cross section about the horizontal centroidal axis, and q is the transverse loading function.

The Euler–Lagrange equation for this case may be formed from Eq. (2.62) as follows:

$$-\frac{d^2}{dx^2}\left(\frac{\partial F}{\partial w''}\right) + \frac{d}{dx}\left(\frac{\partial F}{\partial w'}\right) - \frac{\partial F}{\partial w} = 0 \qquad (b)$$

where

$$F = \frac{EI}{2}(w'')^2 - qw \qquad (c)$$

We then get for the Euler–Lagrange equation:

$$-\frac{d^2}{dx^2}(EIw'') + \frac{d}{dx}(0) + q = 0$$

therefore

$$EIw^{IV} = q \qquad (d)$$

The possible boundary conditions needed for extremization of functional (*a*) can be readily deduced from Eq. (2.63). Thus at the end points:

$$w' \text{ specified} \quad \text{or} \quad \frac{\partial F}{\partial w''} = 0 \qquad (e)$$

$$w \text{ specified} \quad \text{or} \quad -\frac{d}{dx}\left(\frac{\partial F}{\partial w''}\right) + \frac{\partial F}{\partial w'} = 0 \qquad (f)$$

It is clear from Fig. 2.7 that we *must* employ the kinematic boundary condition $w = 0$ at the ends. In order then to carry out the extremization process properly we must require additionally that:

$$\frac{\partial F}{\partial w''} = 0 \qquad \text{at ends} \qquad (g)$$

Substituting for F we get:

$$w''(0) = w''(L) = 0 \qquad (h)$$

You may recall that this indicates that the bending moments are zero at the ends—a condition needed for the frictionless pins there. We thus see here that the natural boundary conditions have to do with forces in structural problems.   ////

In the following section we consider the case where we have higher-order derivatives involving more than one independent variable.

2.11   A FURTHER EXTENSION

As we shall demonstrate later, the total potential energy in a plate is generally expressible as a double integral of the displacement function w and partial derivatives of w in the following form:

$$I = \int\int_S F\left(x,y,w,\frac{\partial w}{\partial x},\frac{\partial w}{\partial y},\frac{\partial^2 w}{\partial x^2},\frac{\partial^2 w}{\partial x\,\partial y},\frac{\partial^2 w}{\partial y\,\partial x},\frac{\partial^2 w}{\partial y^2}\right) dx\,dy \qquad (2.65)$$

where S is the area over which the integration is carried out. Using the notation $\partial w/\partial x = w_x$, $\partial^2 w/\partial x^2 = w_{xx}$, etc., the above functional may be given as follows:

$$I = \int\int_S F(x,y,w,w_x,w_y,w_{xy},w_{yx},w_{yy},w_{xx})\,dx\,dy \qquad (2.66)$$

We have here a functional with *more than one independent variable*. The procedure for finding the extremizing function, which we consider now to be represented as

$w(x,y)$ is very similar to what we have done in the past. We use as admissible functions, $\tilde{w}(x,y)$, a one parameter family of functions defined as follows:

$$\tilde{w}(x,y) = w(x,y) + \epsilon\eta(x,y) \qquad (2.67)$$

Accordingly we have for the extremization process:

$$\left(\frac{d\tilde{I}}{d\epsilon}\right)_{\epsilon=0} = 0 = \left\{\int\int_S \left\{\frac{\partial F}{\partial \tilde{w}}\eta + \frac{\partial F}{\partial \tilde{w}_x}\eta_x + \frac{\partial F}{\partial \tilde{w}_y}\eta_y + \frac{\partial F}{\partial \tilde{w}_{xx}}\eta_{xx} + \frac{\partial F}{\partial \tilde{w}_{xy}}\eta_{xy}\right.\right.$$

$$\left.\left. + \frac{\partial F}{\partial \tilde{w}_{yx}}\eta_{yx} + \frac{\partial F}{\partial \tilde{w}_{yy}}\eta_{yy}\right\} dx\,dy\right\}_{\epsilon=0}$$

We may remove the tildes and the subscript notation $\epsilon = 0$ simultaneously to get the following statement:

$$\int\int_S \left[\frac{\partial F}{\partial w}\eta + \frac{\partial F}{\partial w_x}\eta_x + \frac{\partial F}{\partial w_y}\eta_y + \frac{\partial F}{\partial w_{xx}}\eta_{xx}\right.$$

$$\left. + \frac{\partial F}{\partial w_{xy}}\eta_{xy} + \frac{\partial F}{\partial w_{yx}}\eta_{yx} + \frac{\partial F}{\partial w_{yy}}\eta_{yy}\right] dx\,dy = 0 \qquad (2.68)$$

The familiar integration by parts process can be carried out here by employing Green's theorem (see Appendix I) in two dimensions which is given as follows:

$$\int\int_S G\frac{\partial \eta}{\partial x} dx\,dy = -\int\int_S \frac{\partial G}{\partial x}\eta\,dx\,dy + \oint \eta G \cos{(n,x)}\,dl \qquad (2.69)$$

Recall that n is the normal to the bounding curve C of the domain S. We apply the above theorem once to the second and third terms in the integrand of Eq. (2.68) and twice to the last four terms to get:

$$\int\int_S \left[\frac{\partial F}{\partial w} - \frac{\partial}{\partial x}\left(\frac{\partial F}{\partial w_x}\right) - \frac{\partial}{\partial y}\left(\frac{\partial F}{\partial w_y}\right) + \frac{\partial^2}{\partial x^2}\left(\frac{\partial F}{\partial w_{xx}}\right)\right.$$

$$\left. + \frac{\partial^2}{\partial x\,\partial y}\left(\frac{\partial F}{\partial w_{xy}}\right) + \frac{\partial^2}{\partial y\,\partial x}\left(\frac{\partial F}{\partial w_{yx}}\right) + \frac{\partial^2}{\partial y^2}\left(\frac{\partial F}{\partial w_{yy}}\right)\right]\eta\,dx\,dy$$

$$+ \text{System of Line Integrals} = 0$$

We can conclude, as in earlier cases, that a necessary requirement is that:

$$\frac{\partial F}{\partial w} - \frac{\partial}{\partial x}\left(\frac{\partial F}{\partial w_x}\right) - \frac{\partial}{\partial y}\left(\frac{\partial F}{\partial w_y}\right) + \frac{\partial^2}{\partial x^2}\left(\frac{\partial F}{\partial w_{xx}}\right)$$

$$+ \frac{\partial^2}{\partial y\,\partial x}\left(\frac{\partial F}{\partial w_{yx}}\right) + \frac{\partial^2}{\partial x\,\partial y}\left(\frac{\partial F}{\partial w_{xy}}\right) + \frac{\partial^2}{\partial y^2}\left(\frac{\partial F}{\partial w_{yy}}\right) = 0 \qquad (2.70)$$

This is the Euler–Lagrange equation for this case. We may deduce the natural boundary conditions from the line integrals. However, we shall exemplify this step later when we study the particular problem of classical plate theory, since the results require considerable discussion.

If we have several functions in F, we find that additional equations of the form given by Eq. (2.70) then constitute additional Euler–Lagrange equations for the problem.

2.12 CLOSURE

We have set forth a brief introduction in this chapter into elements of the calculus of variations. In short, we have considered certain classes of functionals with the view toward establishing necessary conditions for finding functions that extremize the functionals. The results were ordinary or partial differential equations for the extremizing functions (the Euler–Lagrange equations) as well as the establishment of the dualities of kinematic (or rigid) and natural boundary conditions. We will see later that the natural boundary conditions are not easily established without the use of the variational approach. And since these conditions are often important for properly posing particular boundary value problems, we can then conclude that the natural boundary conditions are valuable products of the variational approach.

We also note that by the nature of the assumption leading to Eq. (2.9) we have been concerned with varied paths, $\tilde{y}(x)$, which not only are close to the extremal $y(x)$ but which have derivatives $\tilde{y}'(x)$ close to $y'(x)$. Such variations are called *weak variations*. There are, however, variations which do not require the closeness of the derivatives of the varied paths to that of the extremal function (see Fig. 2.8 showing such a varied path). When this is the case we say that we have *strong variations*. A more complex theory is then needed beyond the level of this text. The reader is referred to more advanced books such as those given in the footnote on page 67.

In the remainder of the text we shall employ the formulations presented in this chapter for Euler–Lagrange equations and boundary conditions when there is only a single independent variable present. The equations, you should recall, are then ordinary differential equations and the boundary conditions are prescriptions at end points. For more than one independent variable, we shall work out the Euler–Lagrange equations (now partial differential equations) as well as the boundary conditions (now line or surface integrals) from first principles using the formulation $\delta^{(1)}(I) = 0$ or $(\partial \tilde{I}/\partial \epsilon)_{\epsilon=0} = 0$ as the basis of evaluations.

It is to be pointed out that this chapter does not terminate the development of the calculus of variations. In the later chapters we shall investigate the process of going from a boundary value problem to a functional for which the differential

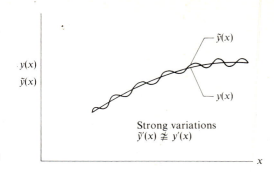

$y(x)$
$\tilde{y}(x)$

$\tilde{y}(x)$

$y(x)$

Strong variations
$\tilde{y}'(x) \not\approx y'(x)$

FIGURE 2.8

equation corresponds to the Euler–Lagrange equations. This is inverse to the process set forth in this chapter and leads to the useful quadratic functional. In Chap. 7 we shall set forth two basic theorems, namely the maximum theorem and the "mini-max" theorem which form the basis for key approximation procedures used for estimating eigenvalues needed in the study of vibrations and stability of elastic bodies. Additionally, in Chap. 9 we shall examine the second variation. Clearly there will be a continued development of the calculus of variations as we proceed.

Our immediate task in Chap. 3 is to set forth functionals whose Euler–Lagrange equation and boundary conditions form important boundary value problems in solid mechanics. We shall illustrate the use of such functionals for problems of simple trusses; we thus launch the study of structural mechanics as a by-product. Additionally we shall introduce certain valuable approximation techniques associated with the variational process that form the cornerstone of much such work throughout the text.

It may be apparent from these remarks that Chap. 3 is one of the key chapters in this text.

READING

COURANT, R., AND HILBERT, D.: "Methods of Mathematical Physics," Vol. I, Interscience, 1953.

ELSGOLC, L. E.: "Calculus of Variations," Addison-Wesley Book Co., 1961.

FOX, C.: "An Introduction to the Calculus of Variations," Oxford University Press, 1954.

GELFAND, I. M., and FOMIN, S. V.: "Calculus of Variations," Prentice-Hall, Inc., 1963.

SCHECHTER, R. S.: "The Variational Method in Engineering," McGraw-Hill Book Co., 1967.

SMIRNOV, V. I.: "Integral Equations and Partial Differential Equations," Vol. IV Chapter II, Pergamon Press, 1964.

WEINSTOCK, R.: "Calculus of Variations," McGraw-Hill Book Co., 1952.

PROBLEMS

2.1   Given the functional:

$$I = \int_{x_1}^{x_2} (3x^2 + 2x(y')^2 + 10xy)\, dx$$

What is the Euler–Lagrange equation?

2.2   What is the first variation of F in the preceding problem? What is the first variation of I in this problem?

2.3   What is the Euler–Lagrange equation for the functional

$$I = \int_{x_1}^{x_2} (3x^2 + 2xy' + 10)\, dx$$

2.4   Justify the identity given at the outset of Case (c) in Sec. 2.5.

2.5   Consider the functional:

$$I = \int_{x_1}^{x_2} [3(y')^2 + 4x]\, dx$$

What is the extremal function? Take $y = 0$ at the end points.

2.6   Fermat's principle states that the path of light is such as to minimize the time passage from one point to another. What is the proper path of light in a plane medium from point (1) to point (2) (Fig. 2.9) wherein the speed of light is given as: (a) $v = Cy$ (b) $v = C/y$?

2.7   Demonstrate that the shortest distance between two points in a plane is a straight line.

2.8   Consider a body of revolution moving with speed V through rarefied gas outside the earth's atmosphere (Fig. 2.10). We wish to design the shape of the body, i.e., get $y(x)$, so as to minimize the drag. Assume that there is no friction on the body at contact with the molecules of the gas. Then one can readily show that the pressure on the body is given as:

$$p = 2\rho V^2 \sin^2 \theta$$

(This means that the molecules are reflected *specularly*.) Show that the drag F_D for a length L of the body is given as:

$$F_D = \int_0^L 4\pi\rho V^2 (\sin^3 \theta) y \left[1 + \left(\frac{dy}{dx} \right)^2 \right]^{1/2} dx$$

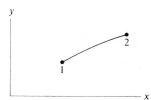

FIGURE 2.9

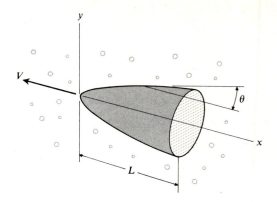

FIGURE 2.10

Assume now that dy/dx is small so that

$$\sin \theta = \frac{dy/dx}{[1 + (dy/dx)^2]^{1/2}} \cong \left(\frac{dy}{dx}\right)$$

Show that

$$F_D = 4\pi\rho V^2 \int_0^L \left(\frac{dy}{dx}\right)^3 y \, dx$$

What is the Euler–Lagrange equation for $y(x)$ in order to minimize the drag?

2.9   Consider the problem wherein a curve is desired between points (x_1, y_1) and (x_2, y_2) (see Fig. 2.11) which upon rotation about the x axis produces a surface of revolution having a minimum area. Show that the solution is of the form:

$$y = C_1 \cosh t, \qquad t = \frac{x + C_2}{C_1}$$

which is the parametric representation of a *catenary*.

2.10   Use the δ operator on the functional given in the text to find the Euler–Lagrange equation for the brachistochrone problem.

2.11   Using Hamilton's principle find the equations of motion for the system shown in Fig. 2.12 for small oscillations.

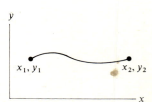

FIGURE 2.11

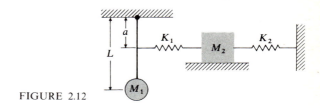

FIGURE 2.12

2.12 Using Hamilton's principle show that

$$(mR^2\dot{\phi}\sin^2\theta) = \text{Const.}$$

for the *spherical pendulum* shown in Fig. 2.13. Gravity acts in the $-z$ direction. Neglect friction and the weight of the connecting link R. Use spherical coordinates θ and ϕ as shown in the diagram and take the lowest position of m as the datum for potential energy.

2.13 Use the δ operator on the functional given in the text to find the Euler–Lagrange equations for the spring-mass system in Example 2.3.

2.14 Extend the results of Problem 2.7 to any two points in three-dimensional space.

2.15 Consider the case of a rope hanging between two points (x_1, y_1) and (x_2, y_2) as shown in Fig. 2.14. Find the Euler–Lagrange equation for the rope by minimizing the potential energy of the rope *for a given length L*. The rope is perfectly flexible. Show that

$$y = \frac{-\lambda}{gw} + \frac{C_1}{wg}\cosh\frac{(x + C_2)}{C_1}$$

How are C_1, C_2, and λ obtained?

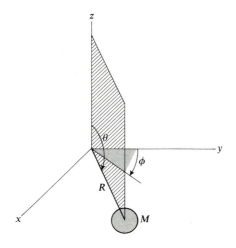

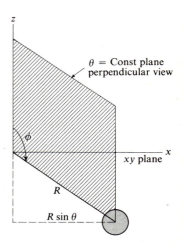

FIGURE 2.13

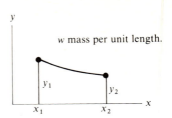

FIGURE 2.14

2.16 Show that the curve (Fig. 2.15) from position (x_1, y_1) to (x_2, y_2) which has *a given length L* and which maximizes the first moment of inertia of the cross-hatched area about the x axis has a differential equation given as:

$$\frac{y^2}{2} + \frac{\lambda}{\sqrt{1 + (y')^2}} = C_1$$

where λ is a Lagrange multiplier. Solve for $y(x)$ in the form of a quadrature but do not carry out the integration.

2.17 Consider two points A and B and a straight line joining these points. Of all the curves of length L connecting these points what is the curve that maximizes the area between the curve and the line AB? What is the physical interpretation of the Lagrange multiplier λ?

2.18 Consider a uniform rod *fixed* at points (x_0, y_0) and (x_1, y_1) as has been shown in Fig. 2.16. The distance s measured along the rod will be considered a coordinate and the length $\int_A^B ds$ of the rod is L. Let $\theta(s)$ be the angle between the tangent to the curve and the x axis. Clearly $\theta(0) = \theta(L) = 0$ due to the constraints. We will later show that the strain energy is proportional to the integral.

$$\int_0^L \left(\frac{d\theta}{ds}\right)^2 ds$$

Express two isoperimetric constraints that link the length L and the positions (x_0, y_0) and (x_1, y_1). Extremize the above functional using the aforementioned constraints and show for a first integral of Euler's equation that we get:

$$(\theta')^2 = C + \lambda_1 \cos \theta + \lambda_2 \sin \theta$$

We may thus determine the centerline curve of the rod constrained as shown in Fig. 2.16.

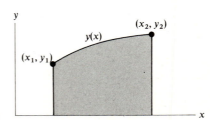

FIGURE 2.15

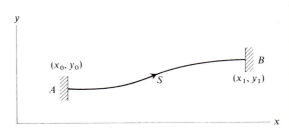

FIGURE 2.16

2.19 Consider the problem of extremizing the following functional

$$I = \int_{t_1}^{t_2} F(y,\dot{y},\ddot{y},t)\, dt \qquad (a)$$

where y and $\dot{y}$ are specified at the end points. Reformulate the functional as one having only *first-order* derivatives and a *constraining equation*. Using Lagrange multiplier functions express the Euler–Lagrange equations. Eliminate the Lagrange multiplier function and write the resulting Euler–Lagrange equation.

2.20 In the preceding problem, take

$$F = [y + 2(\ddot{y})^2 - ty]$$

What is the extremal function if $y = \dot{y} = 0$ at $t = 0$ and $y = \dot{y} = 1$ at 1? What is the Lagrange multiplier function?

2.21 Consider the brachistochrone problem in a *resistive medium* where the drag force is expressed as a given function of speed, $R(V)$, per unit mass of particle. Note there are now two functions y and V to be considered. What is the constraining equation between these functions? Show that using $\lambda(x)$, a Lagrange multiplier function, that one of the Euler–Lagrange equations is

$$\frac{V\lambda'(x)}{\sqrt{1 + (y')^2}} = \frac{dH}{dV}$$

and an integral of the other is

$$\frac{Hy'}{\sqrt{1 + (y')^2}} = C + \lambda(x)g$$

where C is a constant of integration, g is the acceleration of gravity, and H is given as:

$$H = \frac{1}{V} + \lambda(x)R(V)$$

2.22 Derive the Euler–Lagrange equation and the natural boundary conditions for the following functional:

$$I = \int_{x_1}^{x_2} F(x, d^4y/dx^4)\, dx$$

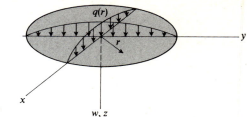

FIGURE 2.17

2.23   Do Problem 2.19 using the given functional F (i.e., with second-order derivative).

2.24   Derive the Euler–Lagrange equation and natural and geometric boundary conditions of beam theory using the δ operator on the functional given in the text (Example 2.5).

2.25   In our study of plates we will see (Chap. 6) that to find the equation of equilibrium for a circular plate (see Fig. 2.17) loaded symmetrically by a force per unit area $q(r)$, and fixed at the edges, we must extremize the following functional:

$$I = D\pi \int_0^a \left[r\left(\frac{d^2w}{dr^2}\right)^2 + \frac{1}{r}\left(\frac{dw}{dr}\right)^2 + 2v\frac{dw}{dr}\frac{d^2w}{dr^2} - \frac{2q}{D}rw \right] dr$$

where D and v are elastic constants, and w is the deflection (in the z direction) of the center-plane of the plate. Show that the following is the proper governing differential equation for w:

$$r\frac{d^4w}{dr^4} + 2\frac{d^3w}{dr^3} - \frac{1}{r}\frac{d^2w}{dr^2} + \frac{1}{r^2}\frac{dw}{dr} = \frac{qr}{D}$$

VARIATIONAL PRINCIPLES OF ELASTICITY

3.1 INTRODUCTION

This chapter is subdivided into four parts. In part A we shall set forth certain key principles which are related to or directly involve variational approaches. Specifically we set forth the principles of virtual work and complementary virtual work, and from these respectively derive the principles of total potential energy and total complementary energy. Reissner's principle is also derived in this part of the chapter. Serving to illustrate certain of the aforementioned principles, we have employed a number of simple truss problems. These problems also serve the purpose of the beginnings of our work in this text in structural mechanics. In Part B of the chapter we concentrate on formulations which are derivable from the principles of part A and which are particularly useful for the study of structural mechanics. These are the well-known Castigliano theorems. We continue the discussion of trusses begun in Part A, thereby presenting in this chapter a reasonably complete discussion of simple trusses. In the following chapter we extend the structural considerations begun here to beams, frames, and rings.

Whereas in Part A of the chapter we set forth certain functionals and show that the corresponding Euler–Lagrange equations are of critical importance in solid mechanics, we take the opposite approach in Part C of the chapter. Here, under certain prescribed conditions, we start with a differential equation and establish the functional for which the equation is the Euler–Lagrange equation. This reverse approach is much more difficult. All the results of this discussion apply to linear elasticity and, importantly, have relevance to other fields of study. They set the stage for the very useful approximation techniques that we consider in Part D of this chapter, namely the Ritz method and the Galerkin method.

Part A
KEY VARIATIONAL PRINCIPLES

3.2 VIRTUAL WORK

You will recall from particle mechanics that *virtual work* is defined as the work done on a particle by all the forces acting on the particle as this particle is given a small hypothetical displacement, a *virtual displacement*, which is consistent with the constraints present. The applied forces are kept constant during the virtual displacement. The concept of virtual work will now be extended to the case of a deformable body by specifying a continuous displacement field which falls within the category of infinitesimal deformation and which does not violate any of the constraints of the problem under consideration. The *applied forces* are again *kept constant* during such a displacement. It should be clear that we can conveniently denote a virtual displacement by employing the variational operator δ. Thus, δu_i may represent a virtual displacement field from a given configuration u_i. The constraints present are taken into account by imposing proper conditions on the variation. For example, consider the case of a beam clamped at both ends as shown in Fig. 3.1. The constraints for the beam are zero displacement and zero slope at the ends, $x = 0$ and $x = L$. We can give the virtual displacement for points along the neutral axis of the beam (this we will see suffices to characterize important aspects of the deformation of the beam under usual circumstances) as follows:

$$\delta w = A\left(1 - \cos\frac{2\pi x}{L}\right)$$

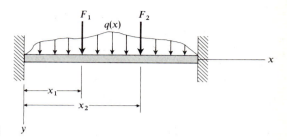

FIGURE 3.1

Notice that both δw and $(d/dx)(\delta w)$ are zero at the ends. The virtual work δW_{virt} for the indicated applied loads then becomes:

$$\delta W_{\text{virt}} = \int_0^L q(x)\left[A\left(1 - \cos\frac{2\pi x}{L}\right)\right] dx$$
$$+ F_1 A\left(1 - \cos\frac{2\pi x_1}{L}\right) + F_2 A\left(1 - \cos\frac{2\pi x_2}{L}\right)$$

In a more general situation we would have as load possibilities a body-force distribution B_i as well as surface tractions $T_i^{(v)}$ over part of the boundary, S_1, of the body. Over the remaining point of the boundary, S_2, we have prescribed the displacement field u_i—in which case to avoid violating the constraints we must be sure that $\delta u_i = 0$ on S_2. It should then be clear that the virtual work for such a general situation would be given as follows:

$$\delta W_{\text{virt}} = \iiint_V B_i\,\delta u_i\,dv + \iint_S T_i^{(v)}\,\delta u_i\,dA \qquad (3.1)$$

We note again that B_i and $T_i^{(v)}$ must not depend on δu_i in the computation of δW_{virt}. Also since $\delta u_i = 0$ on S_2, we have extended the surface integral to cover the entire surface $S = S_1 + S_2$.

We now develop the principle of virtual work for a deformable body. Using successively Cauchy's formula (Eq. (1.5)) and the divergence theorem, we rewrite the right side of the above equation as follows:

$$\iiint_V B_i\,\delta u_i\,dv + \iint_S T_i^{(v)}\,\delta u_i\,dA = \iiint_V B_i\,\delta u_i\,dv + \iint_S \tau_{ij}v_j\,\delta u_i\,dA$$

$$= \iiint_V B_i\,\delta u_i\,dv + \iiint_V (\tau_{ij}\,\delta u_i)_{,j}\,dv$$

$$= \iiint_V (B_i + \tau_{ij,j})\,\delta u_i\,dv$$

$$+ \iiint_V \tau_{ij}\,(\delta u_i)_{,j}\,dv \qquad (3.2)$$

We now introduce a *kinematically compatible strain field variation*,[1] $\delta\epsilon_{ij}$, in the last integral as follows:

$$(\delta u_i)_{,j} = \delta(u_{i,j}) = \delta(\epsilon_{ij} + \omega_{ij}) = \delta\epsilon_{ij} + \delta\omega_{ij}$$

Because of the skew-symmetry of the rotation tensor ω_{ij} and the symmetry of the stress tensor, $\tau_{ij}\delta\omega_{ij} = 0$, and we can conclude that:

$$\tau_{ij}(\delta u_i)_{,j} = \tau_{ij}\,\delta\epsilon_{ij}$$

Accordingly Eq. (3.2) becomes:

$$\iiint_V B_i\,\delta u_i\,dv + \iint_S T_i^{(v)}\,\delta u_i\,dA = \iiint_V (\tau_{ij,j} + B_i)\,dv + \iiint_V \tau_{ij}\,\delta\epsilon_{ij}\,dv$$

We now impose the condition that we have static equilibrium. This means in the above equation that

(a) The external loads B_i and $T_i^{(v)}$ are such that there is overall equilibrium for the body from the viewpoint of rigid-body mechanics. We say that B_i and $T_i^{(v)}$ are *statically compatible*. And, consequently,

(b) at any point in the body $\tau_{ij,j} + B_i = 0$ so that the first term on the right side of the above equation vanishes.

The resulting equation is given as follows:

$$\boxed{\iiint_V B_i\,\delta u_i\,dv + \iint_S T_i^{(v)}\,\delta u_i\,dS = \iiint_V \tau_{ij}\,\delta\epsilon_{ij}\,dv} \qquad (3.3)$$

This is the principle of virtual work for a deformable body. Another way of viewing this equation is to consider the left side as *external virtual work* and the right side as *internal virtual work*. We can then say that a necessary condition for equilibrium is that for any kinematically compatible deformation field (δu_i, $\delta\epsilon_{ij}$), the external virtual work, with statically compatible body forces and surface tractions, must equal the internal virtual work. (We will soon show that this is sufficient for equilibrium.) As for using the principle of virtual work directly we can view it as mathematically *relating any kinematically compatible deformation* (as embodied in δu_i and $\delta\epsilon_{ij}$) *with a stress field* τ_{ij} *which, for a given statically compatible force system* B_i *and* $T_i^{(v)}$, *must satisfy equilibrium requirements everywhere in the body.* This relation is in the form of an *integral equation* for the unknowns τ_{ij}, assuming $T_i^{(v)}$ and B_i are known and δu_i

---

[1] The strain field variation is kinematically compatible because it is formed directly from the displacement field variation.

and $\delta_{\epsilon_{ij}}$ are chosen. Most significant is the fact that this mathematical relation between a deformation field and a stress field is *independent of any constitutive* law and applies to all materials within the limitations of small deformation.[1]

We may use the principle of virtual work directly in structural problems when we can completely specify a virtual displacement field by giving the variation of a finite number of parameters. Thus for trusses we need only specify the virtual displacements of the joints; the virtual strain in the members is then readily determined by geometrical considerations. Considering a plane truss, suppose we institute a virtual displacement on the qth pin in the x direction δu_q, but keep all other pins stationary. The method of virtual work then stipulates that:

$$(P_q)_x \, \delta u_q = \int \int \int_V \tau_{ij} \, \delta \epsilon_{ij} \, dv \qquad (3.4)$$

where P_q is the external force at pin q. The virtual strains are computed from δu_q so as to give a kinematically compatible deformation field. We shall now give stresses τ_{ij} in the members through the appropriate constitutive law of the problem in terms of displacement components $\bar{u}$, $\bar{v}$ of *all* the pins of the truss that are not fixed by external constraints. Because we are using *actual* external loads, the stresses τ_{ij} and hence the displacement components $\bar{u}$, $\bar{v}$ of the movable pins must then represent respectively the *actual stresses* in the members and the *actual displacement* components of the pins. This is so because Eq. (3.3) requires that for a kinematically compatible deformation field the internally developed stresses and external loads must be statically compatible—i.e., they must satisfy the requirements of equilibrium. And since we are using actual external loads, then the stresses, etc., must be the actual equilibrium values. Now by choosing different virtual displacements we can form a sufficient number of equations to solve the displacements for $\bar{u}$, $\bar{v}$ of the movable pins of the truss (the virtual displacement terms cancel out) and thus we can determine the deformation of the truss and the forces in its members.

It is important to note here that we may consider by this method materials having any constitutive law and may also consider statically indeterminate trusses.[2] In succeeding discussions on trusses we will at times impose restriction as to linearity of the constitutive law and statical determinacy of the system.

We now illustrate the approach for a relatively simple problem.

---

[1] We shall extend the principle of virtual work to large deformations in Chap. 8.

[2] You will recall from earlier studies that a *necessary* condition for statical determinacy of a *plane truss* is that:

$$m = 2j - 3 \qquad \text{where} \qquad \begin{matrix} m = \text{number of members} \\ j = \text{number of pins} \end{matrix}$$

And for a *space truss* we must have

$$m = 3j - 6$$

These conditions are not, however, sufficient.

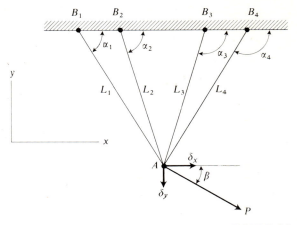

FIGURE 3.2

EXAMPLE 3.1   Consider a system of n pin-connected rods supporting a force P as has been shown in Fig. 3.2. Each rod q is inclined at an angle α_q with the horizontal, and load P is inclined at an angle β with the horizontal. We shall neglect body forces in this problem and determine the forces in the members for linearly elastic behavior.

Notice that the movement of joint A characterizes, through the use of trigonometry, the entire deformation of the system. Thus we now impose a virtual displacement of joint A in the x direction and we denote this virtual displacement as δ_x. The rods then change length and rotate about the fixed ends B_q. (The constraints of the problem are then in no way violated.) Each rod q has a virtual change in length given as:

$$\delta L_q = \delta_x \cos \alpha_q \qquad (a)$$

and so the virtual strain in each rod is simply given as:

$$\delta \epsilon_q = \frac{\delta_x \cos \alpha_q}{L_q} \qquad (b)$$

The only pin that is not constrained is pin A and so we assume it has displacement components $\bar{u}$ and $\bar{v}$. Upon satisfaction of the virtual work principle these will be the *actual* displacement components of pin A resulting from load P. The strain $\bar{\epsilon}_q$ in the qth rod from these displacement components is then:

$$\bar{\epsilon}_q = \frac{\bar{u} \cos \alpha_q}{L_q} + \frac{\bar{v} \sin \alpha_q}{L_q} \qquad (c)$$

Hence the stress $\bar{\tau}_q$ from $\bar{u}$ and $\bar{v}$ in the qth rod, with elastic modulus E_q, is

$$\bar{\tau}_q = E_q \frac{\bar{u} \cos \alpha_q}{L_q} + E_q \frac{\bar{v} \sin \alpha_q}{L_q} \qquad (d)$$

We now employ the principle of virtual work using the actual external load and stresses $\bar{\tau}_q$ along with the aforestated virtual displacement and deformation field. The only surface traction force that does virtual work is P. Thus we have for Eq. (3.3) applied to this problem:

$$P \cos \beta \delta_x = \sum_{q=1}^{4} \iiint_V [\bar{\tau}_q][\delta\epsilon_q] \, dv$$

$$= \sum_{q=1}^{4} [\bar{\tau}_q][\delta\epsilon_q] L_q A_q$$

Substituting from Eqs. (b) and (d) we get:

$$P \cos \beta \, \delta_x = \sum_{q=1}^{4} \left[E_q \frac{\bar{u} \cos \alpha_q}{L_q} \left(\delta_x \frac{\cos \alpha_q}{L_q} \right) + E_q \frac{\bar{v} \sin \alpha_q}{L_q} \left(\delta_x \frac{\cos \alpha_q}{L_q} \right) \right] L_q A_q$$

Canceling δ_x we get:

$$P \cos \beta = \sum_{q=1}^{4} [\bar{u} \cos^2 \alpha_q + \bar{v} \cos \alpha_q \sin \alpha_q] \frac{A_q}{L_q} E_q \qquad (e)$$

We have here a single algebraic equation with two unknowns, namely, $\bar{u}$ and $\bar{v}$, which we know from the previous discussion are the actual displacement components of joint A due to the load P. By giving joint A second virtual displacement δ_y we generate in a similar manner the following equation:

$$P \sin \beta = \sum_{q=1}^{4} [\bar{u} \sin \alpha_q \cos \alpha_q + \bar{v} \sin^2 \alpha_q] \frac{A_q}{L_q} E_q \qquad (f)$$

We may solve Eqs. (e) and (f) simultaneously for the unknowns $\bar{u}$ and $\bar{v}$. The actual stresses $\bar{\tau}_q$ and actual forces in the rods are then readily computed.   ////

The method presented, stemming from the principle of virtual work, is called the *dummy displacement* method. It is one of several related approaches called, in general, *displacement* methods, wherein deformation parameters become the key factors in the approach. We will examine other displacement approaches in this chapter. There are also other approaches which are analogous to the *displacement methods*, namely the *force methods*, wherein force parameters rather than deformation parameters become the key factors. We shall also examine the force methods in this chapter and the analogy alluded to above will then be quite clear.

We have shown that the satisfaction of the principle of virtual work is a *necessary* relation between the external loads and the stresses in a body which is in equilibrium. We can also show that the satisfaction of the principle of virtual work is *sufficient* to satisfy the equilibrium requirements of a body. We start by assuming that Eq. (3.3) is valid for a body. Rewrite the last integral of this equation in the following manner:

$$\iiint_V \tau_{ij}\,\delta\epsilon_{ij}\,dv = \iiint_V \tau_{ij}\,\delta\left(\frac{u_{i,j}+u_{j,i}}{2}\right)dv$$

$$= \iiint_V \tau_{ij}\frac{(\delta u_i)_{,j}}{2}\,dv + \iiint_V \tau_{ij}\frac{(\delta u_j)_{,i}}{2}\,dv$$

$$= \iiint_V \tau_{ij}(\delta u_i)_{,j}\,dv$$

wherein we have made use of the symmetry of τ_{ij} in the last equation. We can rewrite this last expression as follows:

$$\iiint_V \tau_{ij}(\delta u_i)_{,j}\,dv = \iiint_V (\tau_{ij}\,\delta u_i)_{,j}\,dv - \iiint_V \tau_{ij,j}\,\delta u_i\,dv$$

Now using the divergence theorem we get:

$$\iiint_V \tau_{ij}(\delta u_i)_{,j}\,dv = \iint_S \tau_{ij}\,\delta u_i v_j\,ds - \iiint_V \tau_{ij,j}\,\delta u_i\,dv$$

$$= \iint_{S_1} \tau_{ij}\,\delta u_i v_j\,ds - \iiint_V \tau_{ij,j}\,\delta u_i\,dv$$

where we have made use of the fact that $\delta u_i = 0$ on S_2. Now substitute these results for the last integral in Eq. (3.3). We get, on rearranging and consolidating the terms:

$$\iiint_V (\tau_{ij,j}+B_i)\delta u_i\,dv + \iint_{S_1} (T_i^{(v)}-\tau_{ij}v_j)\delta u_i\,ds = 0$$

Since δu_i is *arbitrary*, we can assume for now that $\delta u_i = 0$ on S_1 but is not zero inside the body. We must then conclude that:

$$\tau_{ij,j}+B_i = 0 \quad \text{in } V \qquad (a)$$

By the same reasoning we can also conclude that:

$$T_i^{(v)}-\tau_{ij}v_j = 0 \quad \text{on } S_1 \qquad (b)$$

therefore

$$T_i^{(v)} = \tau_{ij}v_j \quad \text{on } S_1 \qquad (c) \quad (3.5)$$

We have thus generated Newton's law for equilibrium at any point inside the boundary and Cauchy's formula which insures equilibrium at the boundary. We accordingly conclude that satisfaction of the principle of virtual work is both *necessary* and *sufficient* for equilibrium.

3.3   THE METHOD OF TOTAL POTENTIAL ENERGY

The method of virtual work is very valuable since it is valid for any constitutive law. We shall now develop from it the very useful concept of *total potential energy* which applies only to elastic bodies. Accordingly, we shall now restrict the discussion to elastic continua (not necessarily linear elastic continua). As discussed in Chap. 2, for such cases there exists a specific strain energy (i.e., strain energy per unit volume) function U_0 such that:

$$\tau_{ij} = \frac{\partial U_0}{\partial \epsilon_{ij}} \qquad (3.6)$$

where U_0 is a function of the strains at a point and is assumed to be a positive-definite function.[1] Whereas in the previous section the virtual displacement field δu_i was a priori not related to the stress field τ_{ij} when applying the principle of virtual work, we now *link these fields* through a constitutive law which is in accord with Eq. (3.6). Thus replacing τ_{ij} in Eq. (3.3) via Eq. (3.6) we get

$$\iiint_V B_i \, \delta u_i \, dv + \iint_{S_1} T_i^{(v)} \, \delta u_i \, ds = \iiint_V \frac{\partial U_0}{\partial \epsilon_{ij}} \delta \epsilon_{ij} \, dv$$

$$= \iiint_V \delta^{(1)} U_0 \, dv$$

$$= \delta^{(1)} \left[\iiint_V U_0 \, dv \right] = \delta^{(1)} U \qquad (3.7)$$

where we have interchanged the variation and the integration operations. Equation (3.7) equates the virtual work on the body with the variation of the strain energy of a body. We now *define* the potential energy V of the applied loads as a function of displacement fields u_i and the applied loads as follows:

$$V = -\iiint_V B_i u_i \, dv - \iint_{S_1} T_i^{(v)} u_i \, ds \qquad (3.8)$$

The first variation of V is given as follows for constant loads:

$$\delta^{(1)} V = -\iiint_V B_i \, \delta u_i \, dv - \iint_{S_1} T_i^{(v)} \, \delta u_i \, ds \qquad (3.9)$$

where δu_i are virtual displacements. Combining Eqs. (3.7) and (3.9) we then get:

$$\boxed{\delta^{(1)}(U + V) = 0} \qquad (3.10)$$

The quantity $(U + V)$, which we shall denote as π, is called the *total potential energy of the body* and is given as:

$$\pi = U - \iiint_V B_i u_i \, dv - \iint_{S_1} T_i^{(v)} u_i \, ds \qquad (3.11)$$

---

[1] Recall that U_0 represents the specific internal energy for an isentropic process and the specific free energy for an isothermal process. It can be shown through thermodynamic considerations to be a positive-definite function.

The statement (3.10) is known as the *Principle of Stationary Potential Energy.* An interpretation of this statement may be given as follows. *A kinematically admissible displacement field, being related through some constitutive law to a stress field satisfying equilibrium requirements in a body acted on by statically compatible external loads, must extremize the total potential energy with respect to all other kinematically admissible displacement fields.* Now if we employ the usual variational processes for extremizing the total potential energy functional we can arrive at the equations of equilibrium and Cauchy's formula (as you are asked to verify in the exercises).

We have shown that the extremization of the total potential energy with respect to admissible displacement fields is necessary for equilibrium to exist between the forces and the stresses in a body. Just as in the method of virtual work, it can readily be shown to be a sufficient condition for equilibrium.

We shall now show that the total potential energy (π) is actually a local *minimum* for the equilibrium configuration under loads B_i and $T_i^{(v)}$ compared with the total potential energy corresponding to neighboring admissible configurations with the same B_i and $T_i^{(v)}$. We accordingly examine now the difference between the total potential energies of the equilibrium state and an admissible neighboring state having displacement field ($u_i + \delta u_i$) and a corresponding strain field ($\epsilon_{ij} + \delta\epsilon_{ij}$). Thus:

$$(\pi)_{\epsilon_{ij}+\delta\epsilon_{ij}} - (\pi)_{\epsilon_{ij}} = \iiint_V [U_0(\epsilon_{ij} + \delta\epsilon_{ij}) - U_0(\epsilon_{ij})]\, dv$$

$$- \iiint_V B_i\, \delta u_i\, dv - \iint_{S_1} T_i^{(v)}\, \delta u_i\, ds \qquad (3.12)$$

Now expand $U_0(\epsilon_{ij} + \delta\epsilon_{ij})$ as a Taylor series:

$$U_0(\epsilon_{ij} + \delta\epsilon_{ij}) = U_0(\epsilon_{ij}) + \left(\frac{\partial U_0}{\partial \epsilon_{ij}}\right)\delta\epsilon_{ij} + \frac{1}{2!}\left(\frac{\partial^2 U_0}{\partial \epsilon_{ij}\partial \epsilon_{kl}}\right)\delta\epsilon_{ij}\,\delta\epsilon_{kl} + \cdots \qquad (3.13)$$

Substituting into Eq. (3.12) and rearranging terms we get:

$$(\pi)_{\epsilon_{ij}+\delta\epsilon_{ij}} - (\pi)_{\epsilon_{ij}} = \iiint_V \left(\frac{\partial U_0}{\partial \epsilon_{ij}}\right)\delta\epsilon_{ij}\, dv - \iiint_V B_i\, \delta u_i\, dv$$

$$- \iint_{S_1} T_i^{(v)}\, \delta u_i\, ds + \iiint_V \frac{1}{2!}\left(\frac{\partial^2 U_0}{\partial \epsilon_{ij}\partial \epsilon_{kl}}\right)\delta\epsilon_{ij}\,\delta\epsilon_{kl}\, dv + \cdots$$

The first three terms on the right side clearly give a net value of zero for all admissible configurations as a direct result of the principle of total potential energy. Now considering only the expression in the series having products of terms $\delta\epsilon_{ij}$ and calling this the second variation of π, denoted at $\delta^{(2)}\pi$, we get:[1]

$$\delta^{(2)}\pi = \frac{1}{2}\iiint_V \frac{\partial^2 U_0}{\partial \epsilon_{ij}\partial \epsilon_{kl}}\delta\epsilon_{ij}\,\delta\epsilon_{kl}\, dv \qquad (3.14)$$

We will demonstrate that the integrand of the expression on the right side of the above equation is $U_0(\delta\epsilon_{ij})$ for the case where $\epsilon_{ij} = 0$. Consider for this purpose that $\epsilon_{ij} = 0$ in Eq. (3.13). Then

---

[1] We shall consider the second variation in Chap. 9 where it will have more immediate use.

the first term of the series is a constant throughout the body and is taken to be zero, so as to have the strain energy vanish in the unstrained state. By definition, $\partial U_0/\partial \epsilon_{ij}$ in the next expression of the series is τ_{ij}. Since τ_{ij} corresponds here to the unstrained state, we may take it to be zero everywhere in the body. We see then that up to second-order terms:

$$U_0(\delta \epsilon_{ij}) = \frac{1}{2!} \left(\frac{\partial^2 U_0}{\partial \epsilon_{ij} \partial \epsilon_{kl}} \right)_{\epsilon_{ij} = 0} \delta \epsilon_{ij} \, \delta \epsilon_{kl}$$

We may thus express Eq. (3.14) as follows for the case $\epsilon_{ij} = 0$:

$$\delta^{(2)}(\pi) = \iiint_V U_0(\delta \epsilon_{ij}) \, dv$$

But we have indicated earlier that U_0 is a positive-definite function of strain and so the second variation of the total potential energy is positive. We thus may conclude that the total potential energy is a minimum for the equilibrium state $\epsilon_{ij} = 0$ when compared to all other neighboring admissible deformation fields. We now extend this conclusion by stating, without proof,[1] that the total potential energy is minimum for *any* equilibrium state ϵ_{ij} when compared to neighboring admissible deformation fields.

In the previous section we employed the dummy displacement method for a simple truss problem stemming from the principle of virtual work. We now examine the same problem to illustrate the use of a second displacement method, namely one stemming from the total potential energy principle.

EXAMPLE 3.2   Find the deflection of pin A in Fig. 3.2 from the force P assuming the same condition as given in Example 3.1.

We may set forth kinematically compatible deformation fields in the problem by giving pin A virtual displacements δ_x and δ_y. The potential energy of the external forces for such displacements then becomes:

$$V = - (P \cos \beta) \, \delta_x - (P \sin \beta) \, \delta_y \qquad (a)$$

The strain energy U is a function of δ_x and δ_y and may be given as follows:

$$U = \iiint_V \left(\int \tau_{ij} \, d\epsilon_{ij} \right) dv$$

$$= \sum_{p=1}^{4} \left[\int (E_p \epsilon_p \, d\epsilon_p) \right] A_p L_p = \sum_{p=1}^{4} E_p \frac{\epsilon_p^2}{2} A_p L_p \qquad (b)$$

From simple trigonometric relations we may express ϵ_p for small deformation as follows:

$$\epsilon_p = \frac{\delta_x \cos \alpha_p}{L_p} + \frac{\delta_y \sin \alpha_p}{L_p} \qquad (c)$$

[1] In Problem 3.7 you will be asked to show the proof of this minimum for linear elastic materials without assuming $\epsilon_{ij} = 0$.

Hence

$$U = \sum_{p=1}^{4} \frac{E_p A_p}{2 L_p} (\delta_x \cos \alpha_p + \delta_y \sin \alpha_p)^2 \qquad (d)$$

We may give π then as follows:

$$\pi = \sum_{p=1}^{4} \frac{E_p A_p}{2 L_p} (\delta_x \cos \alpha_p + \delta_y \sin \alpha_p)^2 - P(\cos \beta \, \delta_x + \sin \beta \, \delta_y) \qquad (e)$$

The principle of total potential energy then requires that π be an extremum with respect to the kinetically admissible deformation fields characterized by δ_x and δ_y. Hence:

$$\delta^{(1)} \pi = \frac{\partial \pi}{\partial \delta_x} \delta_x + \frac{\partial \pi}{\partial \delta_y} \delta_y = 0$$

Since δ_x and δ_y are independent of each other we conclude that:

$$\frac{\partial \pi}{\partial \delta_x} = 0 \qquad \frac{\partial \pi}{\partial \delta_y} = 0 \qquad (f)$$

Substitute Eq. (e) for π in the above equations (f). We then get two simultaneous equations for δ_x and δ_y. According to the principle of total potential energy these must then be the *actual* deflections of the pin A and hence must correspond to $\bar{u}$ and $\bar{v}$ of the dummy displacement method. With this in mind you may readily show that the aforementioned equations for δ_x and δ_y are identical to Eq. (e) and (f) of Example 3.1.

The procedure presented here can readily be extended to apply to more complex trusses. We simply set forth kinematically compatible deformations of the truss by imagining that every movable pin (i.e., pins not constrained) has deflection components in the x and y directions like δ_x and δ_y for pin A of this problem. We shall not illustrate such a case at this time. In part B of the chapter we shall consider a third displacement method involving Castigliano's first theorem and at this time we shall consider such a truss. The formulations are very similar, as you will then see, to the procedure we would follow here.   ////

We will now interrupt our series of examples of truss problems to consider the problem of the equilibrium configuration of a flexible string. This example will illustrate an important use of the total potential energy functional that will pervade much of the text. Later in Sec. 3.4 when we resume our discussion of trusses, we shall begin a discussion of force methods for trusses.

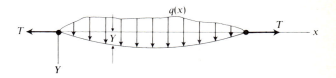

FIGURE 3.3

EXAMPLE 3.3   We now demonstrate the use of the method of total potential energy to derive the equation of equilibrium for a perfectly flexible string having a high value of initial tension T and loaded transversely by a loading $q(x)$ small enough so as not to cause large deflections (see Fig. 3.3).   Under these conditions we can assume that there is a uniform tension T all along the string.   The string is of length L and horizontal before the load $q(x)$ is applied (i.e., we neglect gravity).   It is simply supported at the ends.

We first compute the strain energy for the string.   For convenience we shall consider that U is composed of two parts; a strain U_1 due to the original tension in the string before application of loading $q(x)$ and also a strain U_2 due to the stretching of the string as a result of the transverse loading.   The term U_2 can be easily computed by multiplying the tensile force in the string, which we assume is constant, by the total elongation of the string as a result of the transverse loading.   Thus using Y to denote the deflection of the string:

$$U = T \int_0^L (ds - dx) = T \int_0^L (\sqrt{dx^2 + dY^2} - dx)$$

$$= T \int_0^L \left[\sqrt{1 + \left(\frac{dY}{dx}\right)^2} - 1 \right] dx$$

Since $(dY/dx)^2$ is less than unity here we can expand the root as a power series in the last expression.   Because dY/dx is small we need only retain the first two terms for the accuracy desired.   We then have:

$$U_2 = T \int_0^L \left\{ \left(1 + \frac{1}{2}\left(\frac{dY}{dx}\right)^2 + \cdots \right) - 1 \right\} dx \approx \frac{T}{2} \int_0^L \left(\frac{dY}{dx}\right)^2 dx$$

The strain energy for the string can then be given as follows:

$$U = U_1 + \frac{T}{2} \int_0^L \left(\frac{dY}{dx}\right)^2 dx$$

Now we can employ the principle of total potential energy as follows:

$$\delta^{(1)}\left[U_1 + \frac{T}{2}\int_0^L \left(\frac{dY}{dx}\right)^2 dx - \int_0^L qY\,dx \right] = 0$$

Since $\delta^{(1)}U_1 = 0$ we get:

$$\delta^{(1)}\left\{ \int_0^L \left[\frac{T}{2}\left(\frac{dY}{dx}\right)^2 - qY \right] dx \right\} = 0$$

We thus arrive at the classic variational problem discussed at length in the previous chapter. Employing the Euler–Lagrange equation $\partial F/\partial Y - (d/dx)(\partial F/\partial Y') = 0$ we get:

$$\frac{d^2 Y}{dx^2} = -\frac{q(x)}{T}$$

Since Y, being single-valued and continuous (as required by the operations we have carried out on it), represents a kinematically compatible deformation, it is clear in accordance with the principle of total potential energy that the above equation represents a requirement for equilibrium for the simple constitutive law we have used. Indeed it is the well-known equation of equilibrium for the loaded string.   ////

The preceding problem is a simple illustration of a use of the method of total potential energy—simple primarily because a tacitly assumed vanishingly small cross-section of the string rendered the specification of Y as sufficient for fully specifying the deformed state of the string. Then using in effect a constitutive law that maintains the tension in the string independent of the elongation, for reasons set forth in the example, we were able to arrive at a functional in terms of Y valid for small deformation of the string. The extremization of this functional then gave the differential equation of equilibrium for small deformation of the simply supported string.[1]

In the ensuing chapters on beams and plates we shall make use of the method of total potential energy in a manner paralleling this simple example. We shall propose by physical arguments a simplified mode of deformation that permits the

---

[1] Actually the variation process in this case also yields another set of permissible boundary conditions for the energies used in arriving at the total potential energy, namely the natural boundary conditions. In this case, it is thus also permitted to have:

$$\frac{dY}{dx} = 0 \qquad \text{at ends}$$

This case actually corresponds to a string connected at the ends by frictionless rings to vertical supports (see Fig. 3.4). Now because no external vertical force can be considered at the ends (we would have had to otherwise include a virtual work contribution at the ends) it is necessary that $q(x)$ have a zero resultant in order to have a statically compatible external load system required by the method of total potential energy. Physically it is obvious that this must be so to maintain equilibrium.

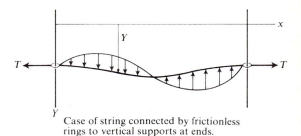

FIGURE 3.4

Case of string connected by frictionless
rings to vertical supports at ends.

expression of the three displacement field components u_1, u_2, and u_3 in terms of a single function. For beams this function will be the deflection $w(x)$ of the neutral axis of the beam (as per the discussion in Sec. 3.2); for plates it will be the deflection $w(x,y)$ of the midplane of the plate. This corresponds to using $Y(x)$ for the string. Then using a constitutive law, which may be some approximation of the actual one as in the case of the string, we may express the total potential energy functional in terms of w. Extremization of the total potential energy then yields a differential equation for w insuring equilibrium, plus the kinematic and natural boundary conditions for the problem. We thus will use the method of total potential energy to generate a boundary value problem for the particular structural body and loads in terms of a single function w.

But, after the insertions of physically inspired arguments to get u_i in terms of w (assuming for the moment we have used the correct constitutive law), what is the meaning and significance of w satisfying the resulting Euler–Lagrange equations plus boundary conditions? To best answer this, note that in expressing u_i in terms of w in the total potential energy we are *restricting* the admissible displacements u_i for the problem to a *subclass* of all those possible and therefore the extremization is carried out with respect to a smaller class of functions than if the arguments leading to w were not employed. This in turn means we are tacitly including in the body certain constraints that, because they do not appear in the calculations, do no work during deformation of the body. The resulting solutions w of the equations of equilibrium and boundary conditions from the variational process then represent equilibrium states for the body with the aforementioned constraints present. Thus by going over from u_i to w to facilitate handling of the problem we are effectively considering a *different* system—actually a *stiffer* (more highly constrained) system. Now if the simplification of the deformation process leading to w is wisely made then many computations for the stiffer system are very close under certain conditions to the actual system.

As for the use of an incorrect constitutive law in reaching the total potential energy functional, it simply means that the solution for the stiffer system corresponds to a state of equilibrium for that constitutive law. One must then demonstrate that some or all of the results of the generated boundary-value problems are reasonably close to the actual case.

With these remarks, the reader has a preview of much that will be presented in later chapters. We will show later in this chapter that the total potential energy not only permits means of arriving at equations of equilibrium and of establishing proper boundary conditions but also affords means of developing approximate solutions.

3.4 COMPLEMENTARY VIRTUAL WORK

In the previous sections we focused our attention on varying the displacement field while keeping the external forces fixed to arrive at useful variational principles. Now we investigate the process of varying in some way the stress field and external forces while holding displacement fields fixed. Specifically we allow here as admissible variations of stress and external loads only those which satisfy the equations of equilibrium inside the body and on the boundary. That is, we require:

$$(\delta\tau_{ij})_{,j} + \delta B_i = 0 \qquad \text{in } V \qquad (a)$$

$$(\delta\tau_{ij})v_j = \delta T_i^{(v)} \qquad \text{on } S_2 \qquad (b) \quad (3.15)$$

Also we assume that since $T_i^{(v)}$ is specified on S_1 of the boundary then $\delta T_i^{(v)} = 0$ there.

We now define the *complementary virtual work* δW^* as follows:

$$\delta W^* = \int\!\!\int\!\!\int_V u_i\,\delta B_i\,dv + \int\!\!\int_{S_2} u_i\,\delta T_i^{(v)}\,ds \qquad (3.16)$$

where u_i is any displacement field and δB_i and $\delta T_i^{(v)}$ satisfy Eqs. (3.15). Note because $\delta T_i^{(v)} = 0$ on S_1, we can extend the surface integrals above to cover the entire boundary S. The principle of complementary virtual work can be developed from the above equations in a manner analogous to that taken for the principle of virtual work. Thus:

$$
\begin{aligned}
\int\!\!\int\!\!\int_V u_i\,\delta B_i\,dv + \int\!\!\int_S u_i\,\delta T_i^{(v)}\,ds &= \int\!\!\int\!\!\int_V u_i\,\delta B_i\,dv + \int\!\!\int_S u_i(\delta\tau_{ij})v_j\,ds \\
&= \int\!\!\int\!\!\int_V u_i\,\delta B_i\,dv + \int\!\!\int\!\!\int_V (u_i\,\delta\tau_{ij})_{,j}\,dv \\
&= \int\!\!\int\!\!\int_V u_i[\delta B_i + (\delta\tau_{ij})_{,j}]\,dv \\
&\quad + \int\!\!\int\!\!\int_V u_{i,j}\,\delta\tau_{ij}\,dv
\end{aligned}
$$

Now employing Eq. (3.15(a)) we may drop the first integral on the right side of the equation. Next if we replace $u_{i,j}$ by $(\epsilon_{ij} + \omega_{ij})$ we *insure that the strain field ϵ_{ij} is a kinematically compatible one.* Noting the symmetry and skew-symmetry properties of τ_{ij} and ω_{ij} respectively we then arrive at the following desired statement which is the principle of complementary work:

$$\iiint_V u_i\, \delta B_i\, dv + \iint_S u_i\, \delta T_i^{(v)}\, ds = \iiint_V \epsilon_{ij}\, \delta\tau_{ij}\, dv \qquad (3.17)$$

Note as in the principle of virtual work the above relation in no way involves a constitutive law of any kind. We have here *a relation in the form of an integral equation between a kinematically compatible deformation field*[1] *and any statically compatible stress and force field* $(\delta\tau_{ij}, \delta T_i^{(v)}$ and $\delta B_i)$. This principle thus forms a necessary condition for a kinematically compatible deformation. (We will soon show that it is also sufficient for simply-connected domains.)

As in the case of the principle of virtual work we can use the principle of complementary virtual work for solving certain structural problems directly. In this case we can very easily find deflections of any particular joint in any particular direction for statically determinate structures with no restrictions as to the constitutive law. The following example illustrates the straightforward procedure that one follows for such problems and is the first of the *force methods* that we will examine.

EXAMPLE 3.4   Consider a simple pin-connected structure as shown in Fig. 3.5. We wish to determine the deflection of joint (a) in the direction $\hat{s}$ as a result of the loads P_1 and P_2. The materials in the structure follow linear elastic constitutive laws.

In approaching this problem keep in mind that δB_i, $\delta T_i^{(v)}$ and $\delta\tau_{ij}$ represent *any* system of loads and stresses that satisfy Eqs. 3.15. Accordingly we shall choose for external loading a unit force on joint (a) in the direction of $\hat{s}$ with supporting forces A_x, A_y, and B_y computed from equilibrium considerations of rigid-body mechanics for this unit load (see Fig. 3.6). The left side of Eq. (3.17) then becomes simply u_α since $\delta T_i^{(v)}$ is just the unit force at joint "a" (we have not included body forces in our loading).

Now going to the right side of the equation, we need $\delta\tau_{ij}$ for each member. Since we have a statically determinate truss here this is readily established by first

---

[1] Note that for the principle of virtual work we relate a stress field and an external load distribution that are statically compatible with *any* kinematically compatible displacement field, while for the principle of complementary virtual work we relate a kinematically compatible deformation field with *any* statically compatible stress field and external force distribution.

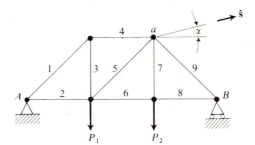

FIGURE 3.5

computing the virtual forces $(F_V)_i$ in each member from the unit force at "a" using equilibrium considerations of rigid-body mechanics. We have then for each member:

$$\delta\tau_q = \frac{(F_V)_q}{A_q}$$

As for ϵ_{ij} we shall here employ *actual* strains computed from the *actual* loads P_1 and P_2 and the appropriate reactions. Since the members are tension and compression members each with a uniform tensile or compressive strain along the axis of the member, we can say from elementary strength of materials for such strains ϵ:

$$\epsilon_q = \frac{F_q}{A_q E_q}$$

where F_q is the actual force in the qth member computed by rigid-body mechanics.

We may employ then Eq. (3.17) to compute u_α as follows:

$$u_\alpha = \sum_{q=1}^{8} \int_0^{L_q} \frac{F_q}{A_q E_q}\left(\frac{(F_V)_q}{A_q}\right)(dx\, A_q) = \sum_{q=1}^{8} \frac{F_q(F_V)_q}{A_q E_q} L_q$$

Since we have used *actual strains* on the right side of the equation, we conclude that u_α must be the *actual displacement* associated with these strains as a result of the principle of complementary virtual work. Thus by using a convenient varied force system designed to "isolate" the desired displacement, and by using easily computed actual strains within the structure, we are able to compute directly the desired actual displacement. Understandably this method is called the *unit (dummy) load method*. ////

The dummy load method is analogous to the dummy displacement method in this respect. In the former we imposed certain convenient loadings on the structure and used the method of complementary virtual work, while in the latter, you will recall, we imposed certain deflections on the structure and used the method of virtual work.

We could have done this problem without the use of energy methods by first computing the elongation or compression of each member and then by employing

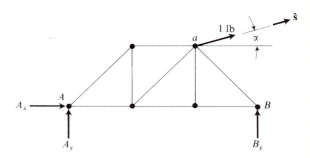

FIGURE 3.6

geometrical and trigonometric considerations, proceeding to ascertain the resulting deflection of the joint. An effort to do this yourself will show most effectively the vast superiority of the energy approach.

We have shown that the principle of complementary virtual work is a *necessary* condition for a kinematically compatible deformation. That is, a kinematically compatible deformation field (u_i, ϵ_{ij}) must satisfy Eq. (3.17) for any admissible load and stress field. We can also show that the principle of complementary energy is *sufficient* for kinematic compatibility. We shall demonstrate this for the special case of plane stress where you will recall from Chap. 1 the use of an Airy stress function Φ defined as:

$$\tau_{xx} = \frac{\partial^2 \Phi}{\partial y^2} \qquad \tau_{yy} = \frac{\partial^2 \Phi}{\partial x^2} \qquad \tau_{xy} = -\frac{\partial^2 \Phi}{\partial x\, \partial y} \qquad (3.18)$$

was sufficient for satisfying the equations of equilibrium. We shall assume that the principle of complementary virtual work applies and we shall for simplicity use as admissible force and stress distributions those for which B_i and $T_i^{(v)}$ are zero and where only the internal stresses are varied by varying the stress function Φ. That is, we shall use a subset of the admissible load and stress functions. Then expressing Eq. (3.17) for this case we get:

$$0 = \iiint_V \epsilon_{ij}\, \delta\tau_{ij}\, dv = \iiint_V \left[\epsilon_{xx} \frac{\partial^2\, \delta\Phi}{\partial y^2} - 2\epsilon_{xy} \frac{\partial^2\, \delta\Phi}{\partial x\, \partial y} + \epsilon_{yy} \frac{\partial^2\, \delta\Phi}{\partial x^2} \right] dv$$

We now employ Green's theorem (see Appendix I) to accomplish integrations by parts to reach the following result:

$$\iiint_V \left[\frac{\partial^2 \epsilon_{xx}}{\partial y^2} - 2\frac{\partial^2 \epsilon_{xy}}{\partial x\, \partial y} + \frac{\partial^2 \epsilon_{yy}}{\partial x^2} \right] \delta\Phi\, dv + \text{surface integrals} = 0$$

Since $\delta\Phi$ is arbitrary we can conclude from the above that:

$$\frac{\partial^2 \epsilon_{xx}}{\partial y^2} - 2\frac{\partial^2 \epsilon_{xy}}{\partial x\, \partial y} + \frac{\partial^2 \epsilon_{yy}}{\partial x^2} = 0 \qquad (3.19)$$

But this equation is recognized as the kinematic compatibility equation for plane stress. We thus conclude that Eq. (3.17) is sufficient for the satisfaction of kinematic compatibility at least for simply connected domains, and may be used in place of the familiar compatibility equations.

Just as we were able to formulate the principle of virtual work as a variational principle involving the total potential energy so will we next formulate the principle of complementary virtual work as a variational principle in terms of the so-called total complementary energy.

3.5   PRINCIPLE OF TOTAL COMPLEMENTARY ENERGY

You will recall from Chap. 1 the concept of the complementary energy density function U_0^* defined for elastic bodies as a function of stress such that

$$\frac{\partial U_0^*}{\partial \tau_{ij}} = \epsilon_{ij} \qquad (3.20)$$

Up to this time we have not related a priori the admissible stress and force fields with the deformation field in the principle of complementary energy. We shall now *link these fields* in Eq. (3.17) via a constitutive law embodied in Eq. (3.20) and thus limit the forthcoming results to elastic (not necessarily linearly elastic) bodies. Thus we express Eq. (3.17) as follows:

$$\iiint_V u_i \, \delta B_i \, dv + \iint_S u_i \, \delta T_i^{(v)} \, dS = \iiint_V \left(\frac{\partial U_0^*}{\partial \tau_{ij}} \right) \delta \tau_{ij} \, dv \qquad (3.21)$$

The right side of the preceding equation is simply $\delta^{(1)} U^*$, the first variation of the complementary energy for the body. Next we introduce a potential function V^* defined such that:

$$V^* = -\iiint_V u_i B_i \, dv - \iint_S u_i T_i^{(v)} \, dS \qquad (3.22)$$

for which the first variation $\delta^{(1)} V^*$ is given as follows:

$$\delta^{(1)} V^* = -\iiint_V u_i \, \delta B_i \, dv - \iint_S u_i \, \delta T_i^{(v)} \, dS \qquad (3.23)$$

Note that the variation of the traction force on S_1, where the traction is prescribed, is taken as zero. Also, in order to use $\delta^{(1)} V^*$ in Eq. (3.17) the varied body forces and surface tractions must satisfy overall requirements of equilibrium for the body. We can then give Eq. (3.17) in the following form:

$$\boxed{\delta^{(1)}(U^* + V^*) = 0} \qquad (3.24)$$

We shall call $(U^* + V^*)$ the *total complementary energy* and we shall denote it as π^*. Thus we can say:

$$\delta^{(1)} \pi^* = 0 \qquad (3.25)$$

where

$$\pi^* = U^* - \iiint_V B_i u_i \, dv - \iint_{S_2} T_i^{(v)} u_i \, dA \qquad (3.26)$$

The principle of total complementary energy has the following meaning. *A statically admissible stress field, being related through some constitutive law to a kinematically compatible strain field, must extremize the total complementary energy with respect to all other statically compatible stress fields.*

It may furthermore be shown, that the total complementary energy is a minimum for the proper stress field.

We now illustrate the use of this principle for finding the forces in a statically indeterminate structure. It is yet another force method.

EXAMPLE 3.5   Shown in Fig. 3.7 is a simple *statically indeterminate* plane truss. We wish to determine the forces in the members assuming linear elastic behavior.

To solve this problem we will cut member 5 and replace the cut by forces f_5 as has been shown in Fig. 3.8. Furthermore we shall delete all external loads. What is left then is a statically determinate truss with equal and opposite forces f_5 applied as has been shown in the diagram. We can readily solve for the forces in each of the uncut members (note the reaction forces are zero) and we shall denote the forces as f_i. Clearly (for a linear system) each such force will be directly proportional to f_5 such that:

$$f_i = C_i f_5 \qquad i = 1, 2, \ldots, 10 \qquad (a)$$

where $C_5 = 1$. Now delete the member 5 thus rendering the truss statically determinate and solve for the forces in the remaining nine members stemming from the applied loads P_1 and P_2. We shall call these forces F_i. We take $F_5 = 0$. Now add to these forces those given by Eq. (*a*) thus including the contribution from the omitted member 5 (Fig. 3.8). The total force in the *p*th member can then be given as follows:

$$[F_p + C_p f_5]$$

The stress in the *p*th member then is

$$\tau_p = \frac{F_p + C_p f_5}{A_p} \qquad (b)$$

The corresponding strain in the *p*th member is:

$$\epsilon_p = \frac{F_p + C_p f_5}{A_p E_p} \qquad (c)$$

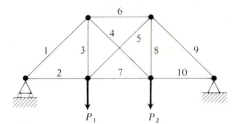

FIGURE 3.7

We may thus think of the truss as having one member cut (see Fig. 3.9) so that in addition to traction forces P_1 and P_2 and reactions, there are also two traction forces f_5 at the cut surfaces. Now *by varying f_5 we have available a family of admissible stress fields as given by Eq. (b).* The complementary strain energy function for such a family of admissible stresses then is

$$U^* = \iiint_V \int \epsilon_{ij}\, d\tau_{ij}\, dv = \frac{1}{2}\sum_{p=1}^{10} \frac{(F_p + C_p f_5)^2}{A_p E_p} L_p \qquad (d)$$

The total complementary energy principle can then be given as follows, for no body forces:

$$\delta^{(1)}U^* - \iint_{S_2} u_i\, \delta T_i^{(v)}\, dS = 0 \qquad (e)$$

The only traction force that is varied for the family of admissible stresses we have decided to use is f_5 at the two surfaces exposed by the cut (Fig. 3.8). However, since these surfaces undergo the *same* displacement and since the force f_5 on one surface is *opposite* in direction to that of the other surface, clearly the surface integral in the above equation is zero. We can accordingly say for the principle of total complementary energy:

$$\delta^{(1)}\left[\frac{1}{2}\sum_{p=1}^{10} \frac{(F_p + C_p f_5)^2}{A_p E_p} L_p\right] = 0$$

therefore

$$\frac{d}{df_5}\left[\frac{1}{2}\sum_{p=1}^{10} \frac{(F_p + C_p f_5)^2}{A_p E_p} L_p\right] = 0 \qquad (f)$$

Any value of f_5 would have led to a statically compatible stress field. Now having rendered the first variation of π^* equal to zero the resulting f_5 must be tied to a kinematically compatible deformation field. Accordingly the computed value of f_5 from Eq. (f) must then be the *actual* force in the member 5 of the truss; the actual forces in the other members can then be determined from Eq. (b).   ////

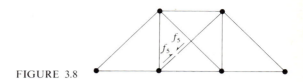

FIGURE 3.8

Although we carried out the discussion in terms of a special simple truss without involving ourselves with numbers, the reader should have no trouble using the approach for other configurations. If there are two or more redundant members, these are cut in the manner that we cut member 5 here. We may then reach a total complementary potential energy π^* having two or more force parameters. Extremizing π^* with respect to these parameters permits us to form enough equations to solve for these force parameters and thus solve for the actual forces in the truss.

Note that this method is analogous to the one presented in Example 3.2 illustrating the use of total potential energy principle. That is, in the present undertaking we extremized the total complementary energy with respect to statically compatible stress fields (found by adjusting f_5 in cut members, etc.) while in the earlier undertaking we extremized the total potential energy with respect to kinematically compatible deformations found by giving displacements to movable pins.

3.6   STATIONARY PRINCIPLES; REISSNER'S PRINCIPLE

We have presented two minimum principles, namely the principles of minimum total potential energy and the principle of minimum total complementary energy. We shall now present a principle which is set forth only as extrema—the exact nature of the extremum is not known. This and other such principles are becoming increasingly more important in structural mechanics and are called *stationary principles*.

We consider a functional given by Reissner in 1955, the extremization of which yields the equations of equilibrium, Cauchy's formula, a constitutive relation, and the appropriate boundary condition—provided only (*a*) that we employ the classical strain-displacement relations of elasticity (this assures kinematic compatibility) in the variational process and (*b*) that we take $\tau_{ij} = \tau_{ji}$ (symmetry). Thus we consider the functional:

$$I_R = \int\!\!\int\!\!\int_V [\tau_{ij}\epsilon_{ij} - U_0^*(\tau_{ij})]\, dV - \int\!\!\int\!\!\int_V [\bar{B}_i u_i]\, dV - \int\!\!\int_{S_1} \bar{T}_i^{(v)} u_i\, dS \qquad (3.27)$$

The quantities $\bar{B}_i$ and $\bar{T}_i^{(v)}$ are prescribed, the former in V, the latter on S_1. Now the stresses, strains and displacements will be varied with strain variations related

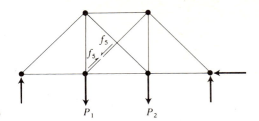

FIGURE 3.9

to the displacement variations by the equation:

$$\delta\epsilon_{ij} = \tfrac{1}{2}(\delta u_{i,j} + \delta u_{j,i}) \qquad (3.28)$$

Carrying out the variations we get

$$\delta^{(1)}I_R = \iiint_V \left[\delta\tau_{ij}\epsilon_{ij} + \tau_{ij}\,\delta\epsilon_{ij} - \frac{\partial U_0^*}{\partial\tau_{ij}}\,\delta\tau_{ij} \right]dV$$

$$- \iiint_V \bar{B}_i\,\delta u_i\,dV - \iint_{S_1} \bar{T}_i^{(v)}\,\delta u_i\,dS \qquad (3.29)$$

Note that in view of Eq. (3.28), the second term on the right of the above equation can be recast as follows (making use of the symmetry of τ_{ij}):

$$\iiint_V \tau_{ij}\,\delta\epsilon_{ij}\,dV = \iiint_V \tau_{ij}\,\delta u_{i,j}\,dV$$

$$= \iiint_V (\tau_{ij}\,\delta u_i)_{,j}\,dV - \iiint_V \tau_{ij,j}\,\delta u_i\,dV$$

If the divergence theorem is applied appropriately:

$$\iiint_V \tau_{ij}\,\delta\epsilon_{ij}\,dV = \iint_{S=S_1+S_2} \tau_{ij}v_j\,\delta u_i\,dS - \iiint_V \tau_{ij,j}\,\delta u_i\,dV$$

Thus the first variation of the Reissner functional can be written as

$$\delta^{(1)}I_R = \iiint_V \left[\left(\epsilon_{ij} - \frac{\partial U_0^*}{\partial\tau_{ij}}\right)\delta\tau_{ij} - (\tau_{ij,j} + \bar{B}_i)\,\delta u_i \right]dV$$

$$- \iint_{S_1} [\tau_{ij}v_j - \bar{T}_i^{(v)}]\,\delta u_i\,dS - \iint_{S_2} \tau_{ij}v_j\,\delta u_i\,dS. \qquad (3.30)$$

If we require $\delta^{(1)}I_R$ to vanish for independent variations $\delta\tau_{ij}$ and δu_i both in V and on $S = S_1 + S_2$, we obtain the following Euler–Lagrange equations and boundary conditions:

$$\epsilon_{ij} = \frac{\partial U_0^*}{\partial \tau_{ij}} \quad \text{in } V \qquad\qquad \tau_{ij,j} + \bar{B}_i = 0 \qquad \text{in } V \qquad (3.31)$$

$$\overline{T}_i^{(v)} = \tau_{ij} v_j \quad \text{on } S_1 \qquad\qquad \delta u_i = 0 \rightarrow u_i = \bar{u}_i \quad \text{on } S_2 \qquad (3.32)$$

Thus both the stress–strain law and the equations of equilibrium are derivable from the Reissner functional, as are stress and displacement boundary conditions. Recall that the strain-displacement relations are built into the variational process via Eq. (3.28).

Although the nature of the extremum is not known for the Reissner principle we wish to point out a distinct advantage for the Reissner principle. You will recall in the potential energy principle we vary displacements in order to arrive at an equilibrium configuration. In approximation techniques (as pointed out in Sec. 3.3) we shall employ approximate displacement fields to obtain approximate equilibrium configurations. As will be shown this can lead to results very closely resembling real displacement fields. However, the corresponding stress field may be considerably in error. This occurs because of the rapid deterioration of accuracy of an approximate solution, such as the displacement field, when differentiations are required to get other results such as the stress field. Furthermore, approximate techniques based on the complementary energy principles lead to the converse problem; while we may get good approximations as to stress fields, we often obtain poor results for displacement fields owing to the fact that we must solve further differential equations in the process. By contrast, the Reissner principle allows for the arbitrary variation of *both* stress and displacements. We can for this reason handle both quantities independently in approximation procedures to achieve good results simultaneously in both categories.[1] We shall illustrate the use of the Reissner principle in the following chapter.

Part B
THE CASTIGLIANO THEOREMS AND STRUCTURAL MECHANICS

3.7   PRELIMINARY REMARKS

In Part A we presented a number of key principles that will form the basis of study for much of this text. And to illustrate these principles in a most simple way we considered certain problems concerning both statically determinate and indeter-

---

[1] Nevertheless we note the apparent preference for the potential and complementary energy principles in the technical literature. This preference is probably due to the greater physical appeal of dealing with a *minimum* of an *energy* quantity. There is, however, a growing tendency to use the Reissner principle in finite element theory of structural mechanics.

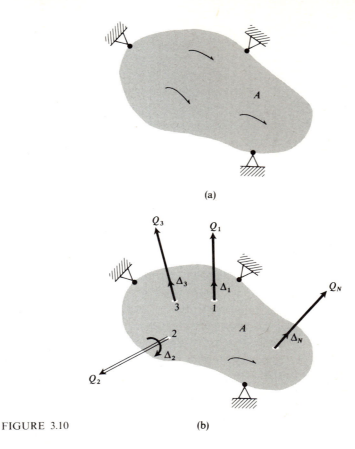

(a)

FIGURE 3.10   (b)

minate simple trusses. We have thus already launched our effort in structural mechanics.

In Part B we continue in this effort in a more formal way by deriving the Castigliano theorems directly from the principles of the previous sections and applying these theorems to various kinds of truss problems. In Chap. 4, we will find that these theorems are useful for the study of beams, frames, and rings.

3.8   THE FIRST CASTIGLIANO THEOREM

We may now employ the principle of total potential energy to formulate the first Castigliano theorem. Let us now consider an elastic body maintained at all times in a state of equilibrium by a system of rigid supports (see Fig. 3.10) against external loads. Consider a system of N external independent point-force vectors and point-couple vectors. We shall denote both force and couple vectors as Q_i. At each point force, Q_i we identify Δ_i to be the displacement component in the direction of Q_i of the movement of this point resulting from deformation of the body. And at each point of application of couple Q_i we identify Δ_i to be the rotation component of this point in the direction of Q_i resulting from deformation of the body. We shall call Δ_i the *generalized displacements*. We thus have N generalized displacements, Δ_i, and we assume next that we can give the strain energy of the body, U, in terms of these quantities. The total potential energy can then be given as follows:

$$\pi = U(\Delta_i) - \sum_{k=1}^{N} Q_k \Delta_k \qquad (3.33)$$

We see that the total potential energy under the aforestated circumstances depends on the parameters Δ_i, which characterize the deformation as far as π is concerned, and the associated external loads and couples. Accordingly for equilibrium we require that the first variation of π, found by varying Δ_i, be zero. Thus:

$$\delta^{(1)}\pi = \sum_{i=1}^{N} \left(\frac{\partial \pi}{\partial \Delta_i} \delta \Delta_i \right) = 0$$

therefore

$$\sum_{i=1}^{N} \left[\frac{\partial U}{\partial \Delta_i} \delta \Delta_i - Q_i \, \delta \Delta_i \right] = 0 \qquad (3.34)$$

We may express the above equation next as follows:

$$\sum_{i=1}^{N} \left[\frac{\partial U}{\partial \Delta_i} - Q_i \right] \delta \Delta_i = 0 \qquad (3.35)$$

We may consider the $\delta \Delta_i$ to be independent of each other and so we conclude that:

$$\boxed{Q_i = \frac{\partial U}{\partial \Delta_i}} \qquad i = 1, \ldots, N \qquad (3.36)$$

This is the well-known *first Castigliano theorem*. By computing U as a function of Δ_n from elasticity considerations we can then readily determine the required force or torque Q_n for a particular generalized displacement Δ_n. The following example illustrates how we may use this theorem.

EXAMPLE 3.6   We consider now a very simple problem shown in Fig. 3.11. We wish to determine what force P is needed to cause joint A to descend a given distance δ. We assume the members are composed of linear elastic material and are identical in every way.

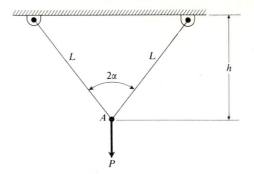

FIGURE 3.11

We could do this problem by computing the necessary elongation in each bar to achieve the deflection δ and then, using Hooke's law, we could determine the force in each member. Finally considering the pin A as a free body we could then evaluate P. We shall instead illustrate the use of the Castigliano first theorem here to accomplish the same task. Clearly δ is the sole generalized coordinate present and so we must determine U as a function of δ. This is readily done once we see that the new length of the members is given (for reasonably small *changes* in the angle α) as (see Fig. 3.12):

$$(h + \delta)/\cos \alpha \qquad (a)$$

The strain in each bar is then

$$\epsilon = \frac{[(h + \delta)/\cos \alpha - h/\cos \alpha]}{h/\cos \alpha} = \frac{\delta}{h} \qquad (b)$$

The strain energy of the system U can now easily be determined as

$$U = 2\left\{\iiint_V \int_\epsilon \tau \, d\epsilon \, dv\right\} = 2\left(\int_\epsilon E\epsilon \, d\epsilon\right) AL = E\epsilon^2 AL$$

$$= E\left(\frac{\delta}{h}\right)^2 A\left(\frac{h}{\cos \alpha}\right) = \frac{EA \, \delta^2}{h \cos \alpha} \qquad (c)$$

where A is the cross-sectional area of the members. Hence:

$$P = \frac{\partial U}{\partial \delta} = \frac{2EA \, \delta}{h \cos \alpha} \qquad (d)////$$

We will now use the Castigliano theorem to present the *stiffness method*[1] of solving *linearly elastic*, statically indeterminate trusses. Recall from Sec. 3.2 that

---

[1] The method to be developed corresponds to the dummy displacement method employed in conjunction with the principle of virtual work.

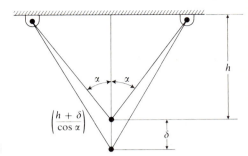

FIGURE 3.12

we could readily establish kinematically compatible deformation fields by giving those pins that are not fully constrained (we shall call them movable pins) virtual displacement components in the x and y directions. Suppose we consider n such displacement components, denoting them as $\Delta_1, \Delta_2, \ldots, \Delta_n$. It will be understood that these deflections take the truss from an undeformed geometry to a kinematically compatible deformed geometry. The strain for any member, say the pth member, can be found by superposing strains developed from the separate displacements of the movable pins.[1] We thus may say:

$$\epsilon_p = \sum_{s=1}^{n} \kappa_{ps} \Delta_s \qquad (3.37)$$

where the κ_{ps} are constants, the first subscript of which refers to the member on which the strain is being considered and the second subscript refers to the virtual displacement. (Thus κ_{12} is the strain in member 1 per unit deflection corresponding to Δ_2.) A strain energy expression can then be given in terms of the displacements Δ_p as follows for a truss with M members:

$$U = \iiint_V \int_\epsilon \tau \, d\epsilon \, dv = \sum_{p=1}^{M} \int_\epsilon (E_p \epsilon_p) \, d\epsilon_p (A_p L_p)$$

$$= \sum_{p=1}^{M} \frac{E_p \epsilon_p^{\,2}}{2} A_p L_p = \sum_{p=1}^{M} \frac{E_p}{2} \left(\sum_{s=1}^{n} \kappa_{ps} \Delta_s \right)^2 A_p L_p \qquad (3.38)$$

Now employ Castigliano's first theorem. We can then say:

$$\frac{\partial U}{\partial \Delta_r} = P_r = \sum_{p=1}^{M} E_p \left(\sum_{s=1}^{n} \kappa_{ps} \Delta_s \right) \kappa_{pr} A_p L_p$$

$$= \sum_{p=1}^{M} \sum_{s=1}^{n} E_p A_p L_p \kappa_{ps} \kappa_{pr} \Delta_s \qquad r = 1, 2, \ldots, n$$

---

[1] It is because we are using here the *superposition principle* that we are restricted to *linear elastic behavior*.

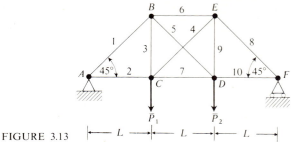

FIGURE 3.13

We have thus a set of n simultaneous equations involving the forces P_r, in the direction of the displacements Δ_r, and these displacements. These equations may be written as follows:

$$P_r = \sum_{s=1}^{n} k_{rs} \Delta_s \quad (a)$$

where

$$k_{rs} = \left(\sum_{p=1}^{M} E_p A_p L_p \kappa_{pr} \kappa_{ps} \right) \quad (b) \quad (3.39)$$

The constants k_{rs} are called *stiffness constants*. We may interpret k_{rq} as the force in the direction of P_r needed per unit deflection Δ_q. Now the forces P_r in Eq. (3.39(a)) are those needed to maintain the deflections Δ_s. If we know k_{rs} and if we insert the *known values* of the external loads we arrive at a system of simultaneous equations for directly determining the Δ_s, which clearly now must represent the *actual displacements* of the movable pins. The forces in the truss members may then be readily computed and the truss has been solved completely.

We illustrate the above procedures in the following example.

EXAMPLE 3.7 Shown in Fig. 3.13 is a statically indeterminate truss loaded by two forces $\bar{P}_1$ and $\bar{P}_2$. We wish to determine the deflection of the pins of the truss and the forces in the members. The members have the same cross-sectional area A, the same modulus of elasticity E, and are numbered for identification.

The procedure will be to reach the key equation (3.39(a)) from which, with actual loads P_r, we can solve for actual displacement components Δ_s. We could first find constants κ_{ps}, in this regard, and then get the stiffness constants k_{rs}. Rather than following this approach, we shall instead find k_{rs} by deriving Eq. (3.39(a)) directly for the problem at hand. For this purpose we assume force components P_i at all pins where there is a possible mobility present (see Fig. 3.14). Thus at joint F we have

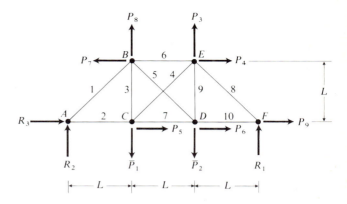

FIGURE 3.14

included a force P_9 while at pin B we have forces P_7 and P_8. The supporting forces have been denoted using the letter R. The *actual* applied loads are identified by the overbar ($\bar{P}_1$ and $\bar{P}_2$). With each of the 9 forces $\bar{P}_1, \bar{P}_2, \ldots, P_9$ we have associated respective deflection components $\Delta_1, \Delta_2, \ldots, \Delta_9$; these are the deflection components Δ_s of Eq. (3.39(*a*)). At the appropriate time we shall set all the P's equal to zero except for the applied loads $\bar{P}_1$ and $\bar{P}_2$ so as to become the appropriate set P_r for Eq. (3.39(*a*)).

The procedure we now follow is to find first the total force F_i in *each member* resulting from (and therefore in terms of) the deflection components $\Delta_1, \ldots, \Delta_9$. Then considering each joint as a *free body*, relate the force components F_i (and thus the deflection components Δ_s) to the force components P_r at each joint. Now assigning the proper values to the P's we arrive at Eq. (3.39(*a*)) which permits us to solve for the deflection components. We can then readily find the proper values of the forces in the members.

Specifically, in finding F_i in terms of Δ_i we may proceed by stipulating one displacement component at a time, keeping all others equal to zero, and computing the forces developed in the members from that single displacement component. Superposing such forces for all displacement components then gives us F_i in terms of the entire system of Δ's. We will do one such displacement calculation in detail and we will then list the results for the others. Thus consider Δ_7 (see Fig. 3.15). Note that members 1, 3, 5, and 6 are affected by this displacement. Examine member 1. For a small displacement Δ_7, there is a change in length in member 1 of $-\Delta_7 \cos 45°$ and so the strain is:

$$-\frac{\Delta_7 \cos 45°}{L/\cos 45°} = -\frac{\Delta_7}{L} \cos^2 45°$$

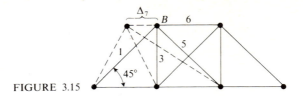

FIGURE 3.15

Using the notation whereby the force in member i due to deflection $\varDelta_j$ is $F_{i,j}$ we have for the present case:

$$F_{1,7} = -\left(\frac{\varDelta_7}{L}\cos^2 45°\right)EA = -\frac{1}{2}\frac{\varDelta_7}{L}EA \qquad (a)$$

Similarly we have for $F_{5,7}$:

$$F_{5,7} = \frac{1}{2}\frac{\varDelta_7}{L}EA \qquad (b)$$

As for $F_{3,7}$ for small deflection $\varDelta_7$ we have only a second-order contribution which we neglect. Thus:

$$F_{3,7} = 0 \qquad (c)$$

Finally it is clear that for $F_{6,7}$ we have:

$$F_{6,7} = \frac{\varDelta_7}{L}EA \qquad (d)$$

Now by symmetry we can give the results for $\varDelta_4$ using the previous results:

$$F_{8,4} = -\frac{1}{2}\frac{\varDelta_4}{L}EA \qquad F_{9,4} = 0$$

$$F_{4,4} = \frac{1}{2}\frac{\varDelta_4}{L}EA \qquad F_{6,4} = \frac{\varDelta_4}{L}EA$$

Next consider $\varDelta_8$ as has been shown in Fig. 3.16. Only members 1, 3, 5, and 6 are affected. We get:

$$F_{3,8} = \frac{\varDelta_8}{L}EA \qquad F_{5,8} = \frac{\varDelta_8}{2L}EA$$

$$F_{1,8} = \frac{\varDelta_8}{2L}EA \qquad F_{6,8} = 0$$

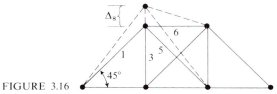

FIGURE 3.16

Again by symmetry we see for Δ_3 that:

$$F_{8,3} = \frac{\Delta_3}{2L} EA \qquad F_{9,3} = \frac{\Delta_3}{L} EA$$

$$F_{4,3} = \frac{\Delta_3}{2L} EA \qquad F_{6,3} = 0$$

We leave it for you now to show by inspection that for Δ_5 and Δ_6 we get:

$$F_{2,5} = \frac{\Delta_5}{L} AE \qquad F_{10,6} = -\frac{\Delta_6}{L} AE$$

$$F_{7,5} = -\frac{\Delta_5}{L} AE \qquad F_{7,6} = \frac{\Delta_6}{L} AE$$

$$F_{3,5} = 0 \qquad F_{9,6} = 0$$

$$F_{4,5} = -\frac{1}{2} \frac{\Delta_5}{L} AE \qquad F_{5,6} = \frac{1}{2} \frac{\Delta_6}{L} AE$$

And for Δ_1, Δ_2, and Δ_9 we get:

$$F_{2,1} = F_{7,1} = 0 \qquad F_{7,2} = F_{10,2} = 0 \qquad F_{10,9} = \frac{\Delta_9}{L} AE$$

$$F_{3,1} = \frac{\Delta_1}{L} AE \qquad F_{9,2} = \frac{\Delta_2}{L} AE \qquad F_{8,9} = \frac{1}{2} \frac{\Delta_9}{L} AE$$

$$F_{4,1} = \frac{1}{2} \frac{\Delta_1}{L} AE \qquad F_{5,2} = \frac{1}{2} \frac{\Delta_2}{L} AE$$

We now consider joint B as a free body and sum forces in the horizontal direction. We have:

$$P_7 = \underbrace{-(F_{1,7} + F_{1,8}) \cos 45°}_{\substack{\text{Forces in} \\ \text{member 1} \\ \text{from } \Delta\text{'s}}} + \underbrace{(F_{5,2} + F_{5,6} + F_{5,7} + F_{5,8}) \cos 45°}_{\substack{\text{Forces in member 5} \\ \text{from } \Delta\text{'s}}} +$$

$$(F_{6,3} + F_{6,4} + F_{6,7} + F_{6,8}) \tag{e}$$

Forces in member 6 from
Δ's

Next we get P_7 as a function of the Δ's by substituting for the forces using previous results starting with Eq. (a). Thus

$$P_7 = [-(-\tfrac{1}{2}\Delta_7 + \tfrac{1}{2}\Delta_8)(0.707) + (\tfrac{1}{2}\Delta_2 + \tfrac{1}{2}\Delta_6 + \tfrac{1}{2}\Delta_7 + \tfrac{1}{2}\Delta_8)(0.707)$$

$$+ (0 + \Delta_4 + \Delta_7 + 0)]\frac{EA}{L}$$

therefore

$$\frac{P_7 L}{EA} = 0.353\Delta_2 + \Delta_4 + 0.353\Delta_6 + 1.707\Delta_7$$

In a similar fashion, summing forces at joint B in the vertical direction, we get

$$\frac{P_8 L}{EA} = \Delta_1 + 0.353\Delta_2 + 0.353\Delta_6 + 1.707\Delta_8$$

Proceeding through the truss using the method of joints we thus arrive at a system of algebraic equations, relating the forces P_i to the displacements Δ_i, which correspond to Eq. (3.39(a)). Using matrix notation at this time we have:

$$\frac{L}{EA}\{P\} = [K]\{\Delta\}$$

where:

$\{P\}$ = column vector of loads with elements $\bar{P}_1, \bar{P}_2, \ldots, P_9$

$\{\Delta\}$ = column vector of displacements with elements $\Delta_1, \Delta_2, \ldots, \Delta_9$

$[K]$ = square matrix of stiffness constants with elements
$k_{ij}, i = 1, \ldots, 9, j = 1, \ldots, 9$

Specifically we have for $[K]$:

$$[K] = \begin{bmatrix}
1.353 & 0 & 0.353 & 0.353 & -0.353 & 0 & 0 & 1 & 0 \\
0 & 1.353 & 1 & 0 & 0 & 0.353 & 0.353 & 0.353 & 0 \\
0.353 & 1 & 1.707 & 0 & -0.353 & 0 & 0 & 0 & 0.353 \\
0.353 & 0 & 0 & 1.707 & -0.353 & 0 & 1 & 0 & -0.353 \\
-0.353 & 0 & -0.353 & -0.353 & 2.353 & -1 & 0 & 0 & 0 \\
0 & 0.353 & 0 & 0 & -1 & 2.353 & 0.353 & 0.353 & -1 \\
0 & 0.353 & 0 & 1 & 0 & 0.353 & 1.707 & 0 & 0 \\
1 & 0.353 & 0 & 0 & 0 & 0.353 & 0 & 1.707 & 0 \\
0 & 0 & 0.353 & -0.353 & 0 & -1 & 0 & 0 & -1
\end{bmatrix}$$

We may express the Δ's explicitly in terms of the P's by employing the inverse of matrix $[K]$ as follows:

$$\{\Delta\} = \frac{L}{EA}[K]^{-1}\{P\}$$

The inverse of $[K]$ is best found using a digital computer. We get for the problem at hand:

$[K]^{-1} =$

$$
\begin{bmatrix}
3.403 & 2.219 & -2.155 & 0.184 & 0.667 & 1.063 & -0.787 & -2.673 & 1.396 \\
2.219 & 3.403 & -2.673 & 0.609 & 0.333 & 0.730 & -1.213 & -2.155 & 1.396 \\
-2.135 & -2.673 & 2.840 & -0.566 & -0.333 & -0.833 & 1.046 & 1.989 & -1.50 \\
0.184 & 0.609 & -0.546 & 1.793 & 0.667 & 1.063 & -1.397 & -0.454 & 1.396 \\
0.667 & 0.333 & -0.333 & 0.667 & 1.0 & 1.0 & -0.667 & -0.667 & 1.0 \\
1.063 & 0.730 & -0.833 & 1.063 & 1.0 & 1.886 & -1.167 & -1.167 & 1.896 \\
-0.787 & -1.213 & 1.046 & -1.397 & -0.667 & -1.167 & 1.897 & 0.954 & -1.5 \\
-2.673 & -2.155 & 1.989 & -0.454 & -0.667 & -1.167 & 0.954 & 2.840 & -1.5 \\
1.396 & 1.396 & -1.5 & 1.396 & 1.0 & 1.896 & -1.5 & -1.5 & 2.896
\end{bmatrix}
$$

Assigning the P's their appropriate values (all P_i are zero except $\bar{P}_1$ and $\bar{P}_2$) we then have for the proper deflection components:

$$\Delta_1 = \frac{L}{EA}[3.403\bar{P}_1 + 2.219\bar{P}_2] \qquad \Delta_6 = \frac{L}{EA}[1.063\bar{P}_1 + 0.730\bar{P}_2]$$

$$\Delta_2 = \frac{L}{EA}[2.219\bar{P}_1 + 3.403\bar{P}_2] \qquad \Delta_7 = \frac{L}{EA}[-0.787\bar{P}_1 - 1.213\bar{P}_2]$$

$$\Delta_3 = \frac{L}{EA}[-2.155\bar{P}_1 - 2.673\bar{P}_2] \qquad \Delta_8 = \frac{L}{EA}[-2.673\bar{P}_1 - 2.155\bar{P}_2]$$

$$\Delta_4 = \frac{L}{EA}[0.184\bar{P}_1 + 0.609\bar{P}_2] \qquad \Delta_8 = \frac{L}{EA}[1.396\bar{P}_1 + 1.396\bar{P}_2]$$

$$\Delta_5 = \frac{L}{EA}[0.667\bar{P}_1 + 0.333\bar{P}_2]$$

The forces in the members are now readily computed. Thus for the ith member the total force F_i is given as:

$$F_i = \sum_{j=1}^{n} F_{i,j}\Delta_j$$

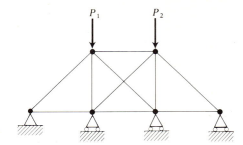

FIGURE 3.17

where $F_{i,j}$ are found from the system of formulas, starting with Eq. (*a*), in which the computed values of the Δ's are substituted.

(Suppose the truss is supported in a statically indeterminate manner such as has been shown in Fig. 3.17. How could you utilize the approach set forth here to solve for the deformation of such a system?)   ////

In order to cast Eq. (3.36) in another useful form we now introduce the so-called *generalized coordinates*, q_i, as any set of variables which *uniquely establish the configuration of a body or system of bodies in space*. These variables need not have the units of length or rotation. Suppose as a result of assumptions as to deformation that there are N variables that serve as generalized coordinates for a particular body loaded by forces $P_1, P_2, \ldots$. Then we can give the differentials of the potential energy of these forces, for use in the total potential energy functional, in the following manner:

$$dV = - \sum_{i=1}^{N} f_i(P_1, P_2, \ldots) \, dq_i$$

where f_i are functions of the external loads. We now define the *generalized force* $(Q_i)_g$ as that function f_i that accompanies dq_i in the above formulation. Clearly the generalized force may have units other than that of ordinary forces or torques. Expressing U as a function of the generalized coordinates, the total potential energy principle then becomes:

$$\delta^{(1)}\pi = \sum_{i=1}^{N} \frac{\partial U}{\partial q_i} \delta q_i - \sum_{i=1}^{N} (Q_i)_g \, \delta q_i = 0$$

As before we reach the result:

$$\frac{\partial U}{\partial q_i} = (Q_i)_g \qquad (3.40)$$

now in terms of generalized forces and generalized coordinates. We shall have occasion to employ the above result in Chap. 4 on frames.

3.9   THE SECOND CASTIGLIANO THEOREM

We may derive the second Castigliano from the principle of total complementary energy. This theorem will accordingly afford us another force method for solving structural problems.

Consider again an elastic body acted on by a system of N point forces and point couples which we represent as Q_i. This body is maintained in equilibrium by a system of rigid supports (Fig. 3.10). We suppose that we can vary Q_i *independently* to form δQ_i. We again employ the generalized displacement components Δ_i, described in the previous section, stemming from the actual deformation of the body from Q_i. The complementary potential of Q_i may then be given as:

$$V^* = -\sum_{k=1}^{N} \Delta_k Q_k \qquad (a)$$

therefore

$$\delta^{(1)}V^* = -\sum_{k=1}^{N} \Delta_k \,\delta Q_k \qquad (b)$$

Now the total complementary energy π^* becomes

$$\pi^* = U^*(Q_i) + V^* \qquad (3.41)$$

where, you will note, we have expressed the complementary energy as a function of Q_i. The principle of total complementary energy may now be used as follows:

$$\delta^{(1)}\pi^* = 0$$

therefore

$$\sum_{i=1}^{N} \left[\frac{\partial U^*}{\partial Q_i} \delta Q_i - \Delta_i \,\delta Q_i \right] = 0 \qquad (3.42)$$

Collecting terms we get:

$$\sum_{i=1}^{N} \left[\left(\frac{\partial U^*}{\partial Q_i} - \Delta_i \right) \delta Q_i \right] = 0 \qquad (3.43)$$

Since we have assumed that the δQ_i are independent of each other we can conclude that:

$$\boxed{\Delta_i = \frac{\partial U^*}{\partial Q_i}} \qquad (3.44)$$

This is the *second Castigliano theorem*. Thus if we can express U^* in terms of Q_i we can compute the values of generalized displacements. For trusses this is sufficient for determining the deformation of the entire truss.

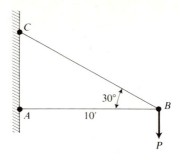

FIGURE 3.18

If the body is linearly elastic you will recall from Chap. 1 that

$$U^* = U$$

Accordingly the second Castigliano theorem for such cases may be given as follows:

$$\varDelta_i = \frac{\partial U}{\partial Q_i} \qquad (3.45)$$

We shall now illustrate the use of the second Castigliano theorem for truss problems. These are further examples of force methods.

EXAMPLE 3.8   Consider a simple structural system composed of two bars (see Fig. 3.18) and supporting a load P equal to 10,000 lb. The bars each have a cross-sectional area of 1 in² and are made of a material having a nonlinear elastic stress-strain behavior (see Fig. 3.19) that may be approximately represented by the following relation:

$$\tau = 6 \times 10^5 \epsilon^{1/2} \qquad (a)$$

We are to determine the vertical deflection $\varDelta$ of joint B. Note that $\varDelta$ is a generalized displacement for this problem.

The system is statically determinate; we can compute the forces in each member by considering joint B as a free body. Thus we get:

$$P_{(BC)} = 2P \text{ lb} \qquad \text{therefore } \tau_{BC} = 2P \text{ psi}$$
$$P_{(AB)} = 1.732P \text{ lb} \qquad \text{therefore } \tau_{AB} = 1.732P \text{ psi} \qquad (b)$$

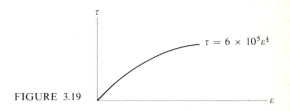

FIGURE 3.19

The complementary strain energy can now be determined as a function of P as follows:

$$U^* = \iiint_V \int_{\tau_{ij}} \epsilon_{ij}\, d\tau_{ij}\, dv = \iiint_V \left[\int_0^{2P} (\tau_{BC}/6 \times 10^5)^2\, d\tau_{BC} \right.$$
$$\left. + \int_0^{1.732P} (\tau_{AB}/6 \times 10^5)^2\, d\tau_{AB} \right] dv$$

$$= \left(\frac{1}{6 \times 10^5} \right)^2 \left\{ \int_0^{11.55} \frac{\tau_{BC}^3}{3} \bigg|_0^{2P}\, dL \right.$$
$$\left. + \int_0^{10} \frac{\tau_{AB}^3}{3} \bigg|_0^{1.732P}\, dL \right\} = 1.605 \times 10^{-9} P^3$$

Now employing the Castigliano theorem we get:

$$\Delta = \frac{\partial U^*}{\partial P} = 4.82 \times 10^{-9} P^2$$

For a load of 10,000 lb we have for Δ:

$$\Delta = 0.482 \text{ in.} \qquad ////$$

EXAMPLE 3.9   Shown in Fig. 3.20 is a statically determinate simple truss, loaded by concentrated loads at pins D and B. What is the total deflection of pin C as a result of these loads? The members are made of linear elastic material.

We shall assume a horizontal force component P and a vertical force component Q at pin C so as to permit us to use the second Castigliano theorem for the horizontal and vertical displacement components there. At the appropriate time we shall set these quantities equal to zero.

Our first step is to determine the strain energy of the system from the 1 kip load at D, the 2 kip load at B, and the loads P and Q at C. By method of joints, we

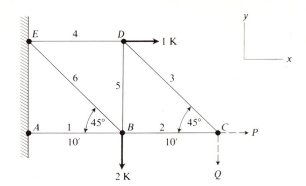

FIGURE 3.20

have for the forces in the members of the truss

$$
\begin{aligned}
AB &= P - 2(1 + Q) & \text{tension} \\
BC &= (P - Q) & \text{tension} \\
CD &= Q/0.707 & \text{tension} \\
DE &= (1 + Q) & \text{tension} \\
DB &= -Q & \text{compression} \\
EB &= \frac{2 + Q}{0.707} & \text{tension}
\end{aligned}
$$

(a)

We now determine U in the following way:

$$
U = \left[\iiint_V \int_\epsilon \tau \, d\epsilon \, dv \right] = \sum_p \int \tau_p \left(\frac{d\tau_p}{E_p} \right) A_p L_p
$$

$$
= \sum_p \frac{\tau_p{}^2}{2E_p} A_p L_p = \sum_p \frac{F_p{}^2 L_p}{2A_p E_p}
$$

(b)

Taking A_p and E_p as having the same value for each member, we get:

$$
U = \frac{1}{2AE} \left\{ [P - 2(1 + Q)]^2 (10) + (P - Q)^2 (10) + \left(\frac{Q}{0.707} \right)^2 \left(\frac{10}{0.707} \right) \right.
$$

$$
\left. + (1 + Q)^2 (10) + Q^2 (10) + \left(\frac{2 + Q}{0.707} \right)^2 \left(\frac{10}{0.707} \right) \right\}
$$

(c)

We may now compute the horizontal and vertical components of pin C by first taking partial derivatives of U with respect to P and with respect to Q respectively, and then letting P and Q equal zero. Thus for the horizontal component Δ_H, we have:

$$\Delta_H = \left(\frac{\partial U}{\partial P}\right)_{P=Q=0} = \frac{1}{2AE}[2(P-2-2Q)(10) + 2(P-Q)(10)]_{P=Q=0}$$

$$= -\frac{20}{AE} \tag{d}$$

The minus sign above indicates that the horizontal deflection component of the pin is opposite in sense to the direction of the force P shown in Fig. 3.20. Now we get Δ_V. Thus

$$\Delta_V = \left(\frac{\partial U}{\partial Q}\right)_{P=Q=0} = \frac{1}{2AE}\left[2(P-2-2Q)(10)(-2) + 2(P-Q)(10)(-1)\right.$$

$$+ \frac{2Q}{(0.707)^2}\left(\frac{10}{0.707}\right) + 2(1+Q)(10) + 20Q$$

$$\left. + \frac{2(2+Q)}{(0.707)^2}\frac{10}{0.707}\right]_{P=Q=0}$$

therefore

$$\Delta_V = \frac{106.5}{AE} \tag{e}$$

The deflection at pin C can then be given as

$$\boldsymbol{\Delta}_C = \frac{1}{AE}[-20\,\hat{\mathbf{i}} + 106.5\hat{\mathbf{j}}] \qquad (f)////$$

We can determine, by the method illustrated in the preceding example, the deformation of an entire truss. The method for this broader undertaking would then be analogous to that taken in Example 3.6 illustrating the use of the first Castigliano theorem. That is, we apply forces to the pins here in computing U whereas in the previous case we assigned generalized displacements Δ_s. Because we have worked this problem in detail in conjunction with the first Castigliano theorem, we will leave it as an exercise to develop the analogous formulations corresponding to Eqs. (3.37) and (3.39), and to apply these formulations as we did in Example 3.7. Such procedures are often called *flexibility methods*.

3.10   SUMMARY CONTENTS FOR PARTS A AND B

We have thus far set forth two analogous sets of principles and theorems, namely those generating displacement methods and those generating force methods. We now list certain aspects of these results side by side to illustrate the results in Table 3.1.

We shall not proceed further in this text with considerations of trusses. It should be obvious from Example 3.7 that matrix methods can be effectively used in truss problems. Indeed any further work in this area would most certainly center around matrix methods.[1]

In Part A we defined certain functionals and then went on to show that the Euler–Lagrange equations were important equations in solid mechanics. Thus the total potential energy functional generated the equations of equilibrium while the total complementary energy functional generated the compatibility equations. In Part C of this chapter we shall consider the reverse process in that we shall start with a boundary value problem and attempt to generate a functional whose Euler–Lagrange equations and boundary conditions comprise the given boundary value problem. This material will form additional background for later undertakings in the text. It will also help us to understand the approximation techniques that we shall introduce in Part D of the chapter.

Table 3.1

Displacement Methods	Force Methods	Comments
Principle of Virtual Work	*Principle of Complementary Virtual Work*	No restriction on the constitutive law
$$\iiint_V B_i\,\delta u_i\,dv + \iint_S T_i^{(v)}\,\delta u_i\,ds$$ $$= \iiint_V \tau_{ij}\,\delta\epsilon_{ij}\,dv$$	$$\iiint_V u_i\,\delta B_i\,dv + \iint_S u_i\,\delta T_i^{(v)}\,ds$$ $$= \iiint_V \epsilon_{ij}\,\delta\tau_{ij}\,dv$$	Infinitesimal deformation
Dummy displacement method for trusses	Dummy load method for trusses	
Principle of Total Potential Energy	*Principle of Total Complementary Energy*	For elastic bodies only
$\delta^{(1)}(\pi) = 0$	$\delta^{(1)}(\pi^*) = 0$	
Form kinematically compatible deformation fields for variation process. Move pins for trusses.	Form statically compatible stress fields for variation process by cutting members and replacing cuts by forces which may be varied.	
First Castigliano Theorem	*Second Castigliano Theorem*	
$Q_i = \dfrac{\partial U}{\partial \Delta_i}$	$\Delta_i = \dfrac{\partial U^*}{\partial Q_i}$	Point loads and displacement components in the direction of loads
Stiffness Methods	Flexibility Methods	

[1] See "Introduction to Matrix Methods of Structural Analysis," H. C. Martin, McGraw-Hill Book Co.

Part C
QUADRATIC FUNCTIONALS

3.11   SYMMETRIC AND POSITIVE DEFINITE OPERATORS

In previous sections we have presented functionals whose Euler–Lagrange equations were important equations of solid mechanics. In this section we shall begin with a differential equation which is one of a certain class of differential equations and we shall show how under certain conditions we can reach the appropriate functional for which the aforesaid equation is the Euler–Lagrange equation. That is, we shall be moving in a direction *opposite* to that presented earlier in this chapter and in Chap. 2. And our results, although valid for many aspects of solid mechanics, have validity transcending this field. We have the opportunity here thus of opening the discussion toward other areas of mathematical physics.[1]   In order to achieve this greater generality we shall need now to introduce several definitions. (These definitions will also be needed for other purposes in Chaps. 7 and 9.)

We will represent a partial differential equation in the discussion as follows

$$Lu = f \qquad (3.46)$$

where L is an operator acting on the dependent variable u, (or variables $u, v \ldots$) and where f, the so-called driving (or forcing) function, represents a function of the independent variables. Examples of linear operators are the harmonic operator ∇^2 and the biharmonic operator ∇^4. A nonlinear operator is

$$L = \frac{d^2}{dx^2} + \sin$$

Next we define the *inner product* of two functions g and h over the domain of the problem V as follows:

$$(g,h) \equiv \text{INNER PRODUCT OF } g \text{ AND } h \equiv \iiint_V gh \, dv \qquad (a) \quad (3.47)$$

The inner product of two vector fields $\mathbf{A}$ and $\mathbf{B}$ over the domain of the problem V is defined as:

$$(\mathbf{A}, \mathbf{B}) \equiv \iiint_V \mathbf{A} \cdot \mathbf{B} \, dv \qquad (b) \quad (3.47)$$

We will limit the discussion now to operators L which are *self-adjoint* (or *symmetric*). Such operators have the following property

$$(Lu,v) = (u,Lv) \qquad (3.48)$$

---

[1] For an excellent exposition of this approach in detail see Mikhlin, "Variational Methods in Mathematical Physics," The Macmillan Co., N.Y., 1964.

where u and v are any two functions satisfying appropriate boundary conditions. To illustrate the establishment of the self-adjoint property for an operator L along with the appropriate boundary conditions, we shall consider two cases now for operators involved respectively in the theory of beams (Chap. 4) and the theory of plates (Chap. 6).

Case 1.

$$L = \frac{d^2}{dx^2}\left[g(x)\frac{d^2}{dx^2} \right]$$

We proceed to investigate the conditions under which Eq. (3.48) is valid. Thus we first examine the left side of this equation. Considering the domain of the problem to extend from $x = 0$ to $x = L$, we have:

$$(Lu,v) = \int_0^L \left(\frac{d^2}{dx^2} g(x)\frac{d^2 u}{dx^2} \right) v \, dx \qquad (3.49)$$

Integrating by parts twice on the right side of the equation we get:

$$\int_0^L \left(\frac{d^2}{dx^2} g(x)\frac{d^2 u}{dx^2} \right) v \, dx = \frac{d}{dx}\left[g(x)\frac{d^2 u}{dx^2} \right] v \Big|_0^L$$
$$- g(x)\frac{d^2 u}{dx^2}\frac{dv}{dx}\Big|_0^L + \int_0^L g(x)\frac{d^2 u}{dx^2}\frac{d^2 v}{dx^2} \, dx \qquad (3.50)$$

Similarly considering the right side of Eq. (3.48) we get:

$$\int_0^L u\left(\frac{d^2}{dx^2} g(x)\frac{d^2 v}{dx^2} \right) dx = \frac{d}{dx}\left[g(x)\frac{d^2 v}{dx^2} \right] u \Big|_0^L$$
$$- g(x)\frac{d^2 v}{dx^2}\frac{du}{dx}\Big|_0^L + \int_0^L g(x)\frac{d^2 u}{dx^2}\frac{d^2 v}{dx^2} \, dx \qquad (3.51)$$

Equating the right sides of Eqs. (3.50) and (3.51) we get, as a requirement of the self-adjoint property:

$$\left\{ \frac{d}{dx}\left[g(x)\frac{d^2 u}{dx^2} \right] v - g(x)\left(\frac{d^2 u}{dx^2} \right)\left(\frac{dv}{dx} \right) - \frac{d}{dx}\left[g(x)\frac{d^2 v}{dx^2} \right] u + g(x)\left(\frac{d^2 v}{dx^2} \right)\left(\frac{du}{dx} \right) \right\}_0^L = 0$$

Using primes for derivatives and rearranging we get the following requirement:

$$[v(gu'')' - u(gv'')' + g(v''u' - u''v')]\Big|_0^L = 0 \qquad (3.52)$$

Thus the above equation contains the appropriate boundary conditions for the operator

$$\frac{d^2}{dx^2}\left[g\frac{d^2}{dx^2} \right]$$

to be self-adjoint. It should be immediately clear that beams that are pin-ended, fixed at the ends,[1] or free at the ends satisfy the above requirements.

---

[1] We will see in the next chapter that the following are the appropriate boundary conditions:

Simple supports	$u = u'' = 0$
Fixed end	$u = u' = 0$
Free end	$u'' = (gu'')' = 0$

Case 2.

$$L = \nabla^2[g(x,y)\nabla^2] \quad \text{where} \quad \nabla^2 = \frac{\partial^2}{\partial x^2} + \frac{\partial^2}{\partial y^2}.$$

We proceed as before by considering the left side of Eq. (3.48) as follows

$$\iint_S [\nabla^2(g\nabla^2 u)]v \, dA = \iint_S \{\nabla \cdot [v\nabla(g\nabla^2 u)] - \nabla v \cdot \nabla(g\nabla^2 u)\} \, dA \tag{3.53}$$

where the right side may be verified by carrying out the divergence operator for the first term using the chain rule. Now using a well-known vector identity involving the dot product of two gradients we have for the last term in the above equation:

$$\nabla v \cdot \nabla(g\nabla^2 u) = \nabla \cdot [(g\nabla^2 u)(\nabla v)] - g\nabla^2 u \nabla^2 v$$

Substituting the above result into Eq. (3.53) we get:

$$\iint_S [\nabla^2(g\nabla^2 u)]v \, dA = \iint_S [\nabla \cdot \{v\nabla(g\nabla^2 u) - (g\nabla^2 u)(\nabla v)\} + g\nabla^2 u \nabla^2 v] \, dA \tag{3.54}$$

Similarly for the right side of Eq. (3.48) we get:

$$\iint_S u[\nabla^2(g\nabla^2 v)] \, dA = \iint_S [\nabla \cdot \{u\nabla(g\nabla^2 v) - (g\nabla^2 v)(\nabla u)\} + g\nabla^2 u \nabla^2 v] \, dA$$

Equating the right sides of the two equations above we get the following result:

$$\iint_S \nabla \cdot [v\nabla(g\nabla^2 u) - u\nabla(g\nabla^2 v) + (g\nabla^2 v)\nabla u - (g\nabla^2 u)\nabla v] \, dA = 0$$

Finally employing the divergence theorem we have

$$\oint_C [v\nabla(g\nabla^2 u) - u\nabla(g\nabla^2 v) + (g\nabla^2 v)\nabla u - (g\nabla^2 u)\nabla v] \cdot \hat{\mathbf{n}} ds = 0$$

where $\hat{\mathbf{n}}$ is the outward normal from the boundary. Since $\nabla(\ ) \cdot \hat{\mathbf{n}} = \partial(\ )/\partial n$ we can say for the above:

$$\oint_C \left[v\frac{\partial}{\partial n}(g\nabla^2 u) - u\frac{\partial}{\partial n}(g\nabla^2 v) + g\nabla^2 v\frac{\partial u}{\partial n} - g\nabla^2 u\frac{\partial v}{\partial n} \right] ds = 0 \tag{3.55}$$

This equation is the appropriate condition for the self-adjoint property of the operator $\nabla^2(g\nabla^2)$. For a clamped plate $w = \partial w/\partial n = 0$ (w is the deflection) on the boundary and for a simply-supported rectangular plate we will later see that

$$w = \frac{\partial^2 w}{\partial x^2} = \frac{\partial^2 w}{\partial y^2} = 0$$

on the boundaries. Accordingly the above condition is satisfied for such supports.

We now impose one more requirement on the operator L. We now require that L be *positive definite*. This means that

$$(Lu,u) \geq 0 \tag{3.56}$$

for all functions u satisfying *homogeneous* boundary conditions dependent for their form on L, and that the equality holds *only* when $u = 0$ everywhere in the domain.

We now go back to *Case 1* to consider the operator

$$L \equiv \frac{d^2}{dx^2}\left(g(x)\frac{d^2}{dx^2}\right)$$

for this property. It will now be assumed here that $g(x) > 0$ everywhere in the domain—a fact consistent with beam theory for which the operator can be applied. Then using Eq. (3.50) with v replaced by u we may say:

$$(Lu,u) = \int_0^L g\left(\frac{d^2u}{dx^2}\right)^2 dx + [(gu'')'u - gu''u']_0^L$$

The bracketed quantity is zero clearly for beams with simple, fixed, or free supports. Since g is never negative we can conclude immediately from the above that

$$(Lu,u) \geq 0$$

Now assume the equality is imposed in the above equation. To show that L is positive definite we must show that $u = 0$ everywhere for such a situation. Accordingly we consider:

$$\int_0^L g\left(\frac{d^2u}{dx^2}\right)^2 dx = 0$$

This means that

$$\frac{d^2u}{dx^2} = 0$$

therefore

$$u = C_1 x + C_2$$

in the domain. Imposing the simple and fixed end boundary conditions we see that $C_1 = C_2 = 0$ and indeed $u = 0$ everywhere. Clearly for such a function u the operator

$$\frac{d^2}{dx^2}g(x)\frac{d^2}{dx^2}$$

is positive-definite. Those functions u satisfying the boundary conditions for which L is positive-definite are called the *field of definition* of L.

As for *Case 2*, where the operator $L \equiv \nabla^2[g\nabla^2]$ is vital for the analysis of plates, we can consider Eq. (3.54) with v replaced by u. We get:

$$(u,\nabla^2(g\nabla^2u)) = \iint_S [\nabla \cdot \{u\nabla(g\nabla^2u) - (g\nabla^2u)\nabla u\} + g(\nabla^2u)^2]\,dA$$

Now using the divergence theorem in the above equation we get:

$$(u, \nabla^2(g\nabla^2u)) = \iint_S g(\nabla^2u)^2\,dA + \oint_C \left[u\frac{\partial}{\partial n}(g\nabla^2u) - (g\nabla^2u)\frac{\partial u}{\partial n}\right]ds$$

It is clear that for a clamped plate $u = \partial u/\partial n = 0$ on the boundary and the line integral vanishes. We will see that this integral vanishes also for rectangular, simply-supported plates. Accordingly for a positive function $g(x,y)$ as appears in plate theory we can conclude that

$$(u,\nabla^2(g\nabla^2u)) = \iint_S g(\nabla^2u)^2\,dA \geq 0$$

If we examine the case where the equality holds then we conclude that

$$\nabla^2 u = 0$$

in the domain—i.e., u is harmonic[1] for this case. We will now use a well-known theorem from potential theory[2] which says that at any point x, y in the domain having Γ as the boundary, the value of a harmonic function u is zero inside the boundary if u is zero on the boundary. We can then conclude for the clamped or simply-supported plate that since $u(x,y) = 0$ on Γ, hence $u = 0$ inside Γ. We thus have proven that the operator $\nabla^2(g\nabla^2)$ is positive-definite for such cases.

We have shown that two operators which are vital in the study of structural mechanics are self-adjoint and positive-definite for meaningful boundary conditions. Actually one can show[3] that the operator L for the Navier equation is also symmetric and positive-definite for homogeneous boundary conditions employed in the theory of elasticity. We can then conclude that the key equations of structural mechanics and elasticity have symmetric and positive-definite operators for many physically meaningful boundary conditions.

Other operators such as the Laplacian operator and the operator

$$\sum_{k=0}^{m} (-1)^k \frac{d^k}{dx^k}\left[p_k(x)\frac{d^k}{dx^k} \right]$$

are similarly symmetric and positive-definite for certain boundary conditions and we can conclude that this class of operators is rather large, encompassing many areas of study outside solid mechanics.

3.12   QUADRATIC FUNCTIONALS

At this time we introduce a function $I(u)$, which we denote as a *quadratic functional*, related to the equation $Lu = f$ for *homogeneous boundary* conditions. We define this functional as follows:

$$I(u) = (Lu,u) - 2(u,f) \qquad (3.57)$$

The significance of $I(u)$ stems from the following theorem.

Theorem   If the equation $Lu = f$ has a solution and if L is a self-adjoint, positive-definite operator, then the function u, from the field of definition of operator L, that minimizes $I(u)$ is the function that is the solution of the differential equation $Lu = f$. Conversely, if there is a function u, from the field of

---

[1] Harmonic functions are those satisfying Laplace's equation, $\nabla^2 \phi = 0$. The theory of such and related functions is often called potential theory.
[2] See Kellogg: "Potential Theory," Dover Publications.
[3] See Prob. 3.19.

definition of L, that is the solution of the equation $Lu = f$ then that function minimizes $I(u)$ with respect to the functions of the field of definition of L.

PROOF   Let us say that u_0 minimizes $I(u)$. Hence consider a one-parameter family of admissible functions $u_0 + \epsilon\eta$ where, for $u_0 + \epsilon\eta$ to belong to the field of definition of L and thus satisfy homogeneous boundary conditions, η must be zero at the boundary. We then have:

$$I(u_0 + \epsilon\eta) - I(u_0) \geq 0 \qquad (3.58)$$

Now using the definition of the functional I in the above expression:

$$(L(u_0 + \epsilon\eta), u_0 + \epsilon\eta) - 2(u_0 + \epsilon\eta, f) - (Lu_0, u_0) + 2(u_0, f) \geq 0$$

From the definition of the inner product this becomes:

$$(Lu_0, u_0) + (Lu_0, \epsilon\eta) + (\epsilon L\eta, u_0) + \epsilon^2(L\eta, \eta) - 2(u_0, f)$$
$$- 2\epsilon(\eta, f) - (Lu_0, u_0) + 2(u_0, f) \geq 0$$

On canceling terms and regrouping, this may be written as:

$$[(Lu_0, \eta) + (L\eta, u_0) - 2(\eta, f)]\epsilon + (L\eta, \eta)\epsilon^2 \geq 0$$

Now using the self-adjoint property of the operator L, we note that $(Lu_0, \eta) = (L\eta, u_0)$, and so we have:

$$2[(Lu_0, \eta) - (\eta, f)]\epsilon + (L\eta, \eta)\epsilon^2 \geq 0$$

The left side of the inequality which we denote as $g(\epsilon)$ is quadratic in ϵ and must have a plot such as "a" in Fig. 3.21 where there are no roots, or curve "b" where there is only a single root. Two real roots are not possible. Hence the discriminant of this quadratic must be negative (for no real roots) or be zero (for a single real root). Thus we can say for the discriminant, since $(Lu_0, \eta) - (\eta, f) = (Lu_0 - f, \eta)$ that:

$$4[(Lu_0 - f, \eta)^2 - (L\eta, \eta)(0)] \leq 0$$

Since the expressions Lu_0, f, and η are real this means that:

$$(Lu_0 - f, \eta) = \iiint_V (Lu_0 - f)\eta \, dv = 0$$

Using the fundamental lemma of the calculus of variations we conclude that

$$Lu_0 - f = 0$$

and so we see that u_0 satisfies the differential equation.

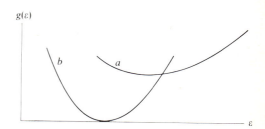

FIGURE 3.21

Now we turn to the *converse* of the theorem. That is, we take u_0 to be the solution of the differential equation, and we are to prove that u_0 minimizes $I(u)$. Thus we express $I(u)$, using Lu_0 to replace f, as follows:

$$I(u) = (Lu,u) - 2(Lu_0,u)$$

We add and subtract the quantity (Lu_0,u_0) in order to rearrange the above equation. Thus:

$$I(u) = (Lu,u) - 2(Lu_0,u) + (Lu_0,u_0) - (Lu_0,u_0)$$
$$= (L(u - u_0), u - u_0) - (Lu_0,u_0)$$

You may readily verify the last step making use of the self-adjoint property of the operator L. Since L is positive definite, *it is clear from the above that when $u = u_0$ the first expression on the right reaches its smallest value, zero, and the right side becomes a minimum.* We thus have proven that the solution to the equation $Lu = f$ minimizes the corresponding quadratic functional $I(u)$.   ////

We now can conclude that, for the conditions specified on L, the extremal function for the functional $I(u)$ satisfies the differential equation $Lu = f$ and so the latter is the Euler–Lagrange equation for the particular functional. It is to be noted that in both Chap. 2 and in this chapter we started with the functional and showed how to get the appropriate differential equation and boundary conditions. Now under certain conditions we can produce the appropriate functional for a given differential equation and boundary conditions. We have thus presented the reverse process. Why would we want the functional for a given differential equation? We will soon see that for producing *approximate* solutions to the differential equation it is sometimes more feasible to work with the corresponding quadratic functional with the view toward minimizing its value with respect to an admissible class of functions.

We now show that the total potential energy is actually one half the quadratic functional for Navier's equation for homogeneous boundary conditions. We rewrite this equation (Eq. 1.106) in vector notation. One may show that this equation may be expressed as follows with the aid of vector identities.

$$-[(\lambda + 2G)\nabla(\nabla \cdot \mathbf{u}) - G\nabla \times (\nabla \times \mathbf{u})] = \mathbf{B} \qquad (3.59)$$

We have here a vector operator acting on a vector, and we denote this as $\mathbf{Lu}$ for the left side of the equation. Thus we have for the above

$$\mathbf{Lu} = \mathbf{B} \qquad (3.60)$$

where

$$\mathbf{L} = -[(\lambda + 2G)\nabla(\nabla \cdot \quad) - G\nabla \times (\nabla \times \quad)]$$

In the ensuing development, we shall make use of Eq. (3.3) which is seen from its development to be an identity and may equally well be written with u_i replacing δu_i and with ϵ_{ij} replacing $\delta \epsilon_{ij}$ as follows:

$$\iiint_V \mathbf{u} \cdot \mathbf{B} \, dv = \iiint_V \tau_{ij}\epsilon_{ij} \, dv - \iint_S \mathbf{T}^{(v)} \cdot \mathbf{u} \, dS$$

Noting from earlier work that the first integral on the right-hand side is twice the strain energy for linear elastic bodies and replacing $\mathbf{B}$ by $\mathbf{Lu}$ from Eq. (3.60) we get:

$$\iiint_V \mathbf{u} \cdot \mathbf{Lu} \, dv = 2\iiint_V U_0 \, dv - \iint_S \mathbf{u} \cdot \mathbf{T}^{(v)} \, dS \qquad (3.61)$$

We now give the quadratic functional for the Navier equation making use of the identity (3.61) and replacing $\mathbf{f}$ by $\mathbf{B}$:

$$I(\mathbf{u}) = (\mathbf{Lu},\mathbf{u}) - 2(\mathbf{f},\mathbf{u}) = \left\{ 2\iiint_V U_0 \, dv - \iint_S \mathbf{u} \cdot \mathbf{T}^{(v)} \, dS \right\} - 2\iiint_V \mathbf{B} \cdot \mathbf{u} \, dv$$

With $\mathbf{T} = \mathbf{0}$ on S_1 and $\mathbf{u} = \mathbf{0}$ on S_2 (homogeneous boundary conditions) the surface integral in the above equation then vanishes and we have:

$$(\mathbf{Lu},\mathbf{u}) - 2(\mathbf{f},\mathbf{u}) = 2\left[\iiint_V U_0 \, dv - \iiint_V \mathbf{B} \cdot \mathbf{u} \, dv \right]$$

We thus get an expression proportional to the total potential energy for this case. For non-homogeneous boundary conditions, one has to extremize a more complex quadratic functional in that additional terms are needed. Since this discussion is beyond the level of this text, we merely point out that for more general conditions we get as the functional $I(u)$ for the case at hand the expression:

$$I(u) = 2\left[\iiint_V U_0 \, dv - \iiint_V B_i u_i \, dv - \iint_S T_i^{(v)} u_i \, dS \right]$$

which again is just twice the total potential energy.

As a final comment, we wish to point out that there can be an infinity of functionals mutually related in a certain way for a given Euler–Lagrange equation. Thus the functional

$$I = \int_{x_1}^{x_2} F(x, y, y', \ldots, y^n) \, dx$$

can be altered, without changing the Euler–Lagrange equations, by adding to the integrand an expression of the form:

$$(\text{Const.}) \frac{dG(x, y, y', \ldots, y^{(n-1)})}{dx}$$

Also the functional:

$$I = \iiint_V F(x, y, z, w, w_x, w_y, w_z) \, dx \, dy \, dz$$

has the same Euler–Lagrange equation if the following "divergence like" expression is added to the integrand:

$$\alpha \frac{\partial A}{\partial x} + \beta \frac{\partial B}{\partial y} + \gamma \frac{\partial C}{\partial z}$$

where α, β, and γ are constants and A, B, and C are functions of the variables x, y, z, and w. Note that these functions have only w and not w_x, w_y, and w_z as in the functional I. For functionals I with higher-order partial derivatives than unity, it is not in general possible to choose A, B, and C with the derivatives appearing in these functions having a lower order than in the original functional.[1]

Part D
APPROXIMATE METHODS

3.13   INTRODUCTORY COMMENT

We have presented by various arguments a number of functionals whose Euler–Lagrange equations were of particular interest in solid mechanics in Part A, and in Part C we showed how for a certain rather broad class of ordinary and partial differential equations we could formulate the appropriate functional for which the equation was the corresponding Euler–Lagrange equation. Thus in either case we have available the functional and the Euler–Lagrange equations. The principal approach in Part D of this chapter will be to work with the functional for the purpose of finding approximate solutions to the corresponding differential equation.

We shall first consider the Ritz method as it is applied to the total potential energy functional. This method may also be used, however, for other functionals such as the total complementary energy and the quadratic functionals discussed in the previous section. We will then briefly consider the Galerkin method which, while not being a variational process, is nevertheless so closely related to the Ritz method

---

[1] See Courant and Hilbert, "Methods of Mathematical Physics," Vol. I, Interscience, pp. 194, 195.

and so useful as to warrant discussion at this time. We shall cover other approximation methods, such as the method of Trefftz and the method of Kantorovich, later in the text. They will be presented at a time when there will be suitable motivation and background to appreciate their particular characteristics.

3.14 THE RITZ METHOD

One of the most useful approximate methods stemming from variational considerations is the Ritz method wherein for the present we employ the following approximate displacement field components for expressing the total potential energy:

$$u_n = \phi_0(x,y,z) + \sum_{i=1}^{n} a_i\phi_i(x,y,z)$$

$$v_n = \psi_0(x,y,z) + \sum_{i=1}^{n} b_i\psi_i(x,y,z)$$

$$w_n = \gamma_0(x,y,z) + \sum_{i=1}^{n} c_i\gamma_i(x,y,z) \tag{3.62}$$

The functions with the subscript zero satisfy the *kinematic* boundary conditions on S_2 while the remaining $3n$ functions are zero there. The coefficients a_i, b_i, c_i are undetermined. The scheme for the Ritz method is to choose the values of these coefficients so as to minimize the total potential energy. When the values of the coefficients are thus determined they are called *Ritz coefficients*. It can be shown[1] that if $n \to \infty$ in this process then u_n, v_n, and w_n converge in energy[2] to the exact solution for the problem provided the functions ϕ_i, ψ_i, and γ_i are complete.[3] When n is small, we can still reach very good approximations to the exact solution if a judicious selection of the functions $\phi_0, \phi_1, \cdots, \phi_n, \psi_0, \psi_1, \cdots, \psi_n$, and $\gamma_0, \gamma_1, \cdots, \gamma_n$ is made.

[1] See Trefftz: "Handbuch der Physik," 1928 edition, Springer-Verlag, Berlin.
[2] A sequence of functions u_n may be said to converge in many different senses to a function u. Thus there may be *uniform convergence* where in the interval the quantity $[(u_n)_{max} - (u)_{max}]$ can be made arbitrarily small by increasing n. There is convergence in *the mean* whereby $\iiint [u_n - u]^2 \, dv$ can be made arbitrarily small by increasing n. The convergence referred to here, namely that of *energy*, requires that $(L(u - u_n), (u - u_n))$ which is in unabridged notation $\iiint [L(u - u_n)][u - u_n] \, dv$, can be made arbitrarily small by increasing n. We will see in Chap. 7 that (Lu,u) actually does relate to energy of the system and so this convergence does have physical significance.
[3] A system of functions can be *complete* in several ways. A series of functions ϕ_i may be complete if, for any function f, a set of constants c_i can be found so that $\sum c_i\phi_i$ converges (a) uniformly, (b) in the mean, or (c) in energy to this function. The nature of the completeness must of course be stated in a discussion.

The procedure is to use u_n, v_n, and w_n to formulate an approximate total potential energy $\tilde{\pi}$ which then becomes a function of the $3n$ undetermined constants. We minimize $\tilde{\pi}$ by imposing the $3n$ requirements:

$$\frac{\partial \tilde{\pi}}{\partial a_i} = 0$$

$$\frac{\partial \tilde{\pi}}{\partial b_i} = 0$$

$$\frac{\partial \tilde{\pi}}{\partial c_i} = 0 \qquad i = 1, 2, \cdots, n$$

This yields $3n$ equations for the unknown coefficients. The solution of these equations for these coefficients then yields the aforementioned Ritz coefficients.

Suppose next that the kinematic boundary conditions are homogeneous, as for example the condition

$$\alpha u_i + \beta \frac{\partial u_i}{\partial n} = 0$$

with α and β as constants and n as a normal direction to the boundary. Then if a function $\phi_0(x,y,z)$ satisfies the boundary conditions of the problem so will $K\phi_0(x,y,z)$ where K is an arbitrary constant. If we go back to Eq. (3.62) we can set forth the following possible arrangement for approximate displacement field components u_n, v_n, and w_n to be used in the Ritz method:

$$u_n = \left(\sum_1^n a_i\right)\phi_0(x,y,z) + \sum_1^n a_i\phi_i(x,y,z)$$

$$v_n = \left(\sum_1^n b_i\right)\psi_0(x,y,z) + \sum_1^n b_i\psi_i(x,y,z)$$

$$w_n = \left(\sum_1^n c_i\right)\gamma_0(x,y,z) + \sum_1^n c_i\gamma_i(x,y,z) \qquad (3.63)$$

This arrangement has the same properties as that given by Eq. (3.62). Now rearranging we get:

$$u_n = \sum_1^n a_i(\phi_0 + \phi_i) = \sum_1^n a_i\Phi_i$$

$$v_n = \sum_1^n b_i(\psi_0 + \psi_i) = \sum_1^n b_i\Psi_i$$

$$w_n = \sum_1^n c_i(\gamma_0 + \gamma_i) = \sum_1^n c_i\Gamma_i \qquad (3.64)$$

The functions Φ_i, Ψ_i, and Γ_i clearly satisfy the kinematic boundary conditions of the problem. For homogeneous boundary conditions we know, making use of the convergence of the form given by Eq. (3.62), that this sequence, with the undetermined coefficients chosen to minimize $\tilde{\pi}$, will converge in energy to the exact solution u_i in the limit as $n \to \infty$, provided the functions Φ_i, Ψ_i, and Γ_i are complete sets with respect to energy.

We shall have ample opportunity to employ the Ritz method in succeeding chapters. At that time we will, by example, discuss how best to choose the functions (usually called coordinate functions) for expeditious use in the Ritz method for small values of n. For such problems we will have restricted the admissible virtual displacements so that only one function, w, appears in π rather than the three functions u_1, u_2, and u_3. (This was discussed, you will recall, in Sec. 3.3.) The Ritz method in such cases will be given as

$$w = \sum_1^n a_i \Phi_i$$

where Φ_i satisfies the homogeneous boundary conditions for w.

We now turn to the remarks of Sec. 3.11 concerning the quadratic functionals. We may employ the Ritz method for minimizing the quadratic functional and thereby approximate the solution of the corresponding Euler–Lagrange equation. The proof of convergence of the process in this general case again requires that the coordinate function Φ_i be complete in energy. The proof of such completeness (this places further limitations on the operators L) is a difficult one.[1]

EXAMPLE 3.10   Consider the following ordinary differential equation:

$$x^2 \frac{d^2 y}{dx^2} + 2x \frac{dy}{dx} = 6x \qquad (a)$$

We wish to solve this equation approximately using the Ritz method for the following end conditions:

$$y(1) = y(2) = 0 \qquad (b)$$

Since an exact solution here is readily available we can then compare results.

We must first find the functional associated with this boundary-value problem. Note in this regard that the equation can be written as follows:

$$-\frac{d}{dx}\left(x^2 \frac{dy}{dx}\right) = -6x \qquad (c)$$

---

[1] See Mikhlin, S. G.: "Variational Methods in Mathematical Physics," The Macmillan Co., N.Y., 1964.

We leave it for you to show that the operator

$$L = -\left(\frac{d}{dx}x^2\frac{d}{dx}\right)$$

is a positive-definite, symmetric operator for the given boundary conditions. And so we have for the functional I:

$$I = -\int_1^2 \left[\frac{d}{dx}\left[(x^2)\frac{dy}{dx}\right]y - 12yx\right]dx$$

$$= -\int_1^2 \left[x^2y\frac{d^2y}{dx^2} + 2xy\frac{dy}{dx} - 12xy\right]dx \qquad (d)$$

We now choose a single coordinate function satisfying the boundary condition such that:

$$y_1 = C_1\Phi_1 = C_1(x - 1)(x - 2) = C_1(x^2 - 3x + 2) \qquad (e)$$

Substituting y_1 into the functional we get:

$$\tilde{I} = -\int_1^2 \left\{x^2\Phi_1\left(\frac{d^2\Phi_1}{dx^2}\right)C_1^2 + 2x\Phi_1\frac{d\Phi_1}{dx}C_1^2 - 12x\Phi_1C_1\right\}dx$$

Extremizing $\tilde{I}$ with respect to C_1 we get on canceling -2 from the result:

$$\int_1^2 \left\{x^2\Phi_1\frac{d^2\Phi_1}{dx^2}C_1 + 2x\Phi_1\frac{d\Phi_1}{dx}C_1 - 6x\Phi_1\right\}dx = 0 \qquad (f)$$

Substituting for Φ we have:

$$\int_1^2 [(3x^4 - 12x^3 + 15x^2 - 6x)C_1 - 3(x^3 - 3x^2 + 2x)]dx = 0$$

Integrating we get:

$$[(\tfrac{3}{5}x^5 - 3x^4 + 5x^3 - 3x^2)C_1 - (\tfrac{3}{4}x^4 - 3x^3 + 3x^2)]_1^2 = 0$$

therefore

$$C_1 = 1.875$$

The approximate solution from the Ritz method is then:

$$y_1 = 1.875(x^2 - 3x + 2)$$

We may get an exact solution readily for this equation by noting that it is an Euler–Cauchy type.[1] By making the following change of variable

$$x = e^t, \qquad t = \ln x$$

[1] See Golomb and Shanks: "Elements of Ordinary Differential Equations," McGraw-Hill Book Co., Chap. 7.

we can get an equation having constant coefficients. Thus for t as the independent variable we have:

$$\frac{d^2 y}{dt^2} + \frac{dy}{dt} = 6 e^t$$

The complementary solution is readily found to be

$$y_c = A e^{-t} + B$$

A particular solution is seen by inspection to be

$$y_p = 3 e^t$$

Hence the exact solution is

$$y_{ex} = A e^{-t} + B + 3 e^t = \frac{A}{x} + B + 3x$$

To satisfy the boundary conditions we require that:

$$0 = A + B + 3$$

$$0 = \frac{A}{2} + B + 6$$

Hence $A = 6$ and $B = -9$. The exact solution for the problem is then:

$$y_{ex} = \frac{6}{x} + 3x - 9 \tag{g}$$

To compare methods we examine results at position $x = 1.5$. We get for the approximate and exact solutions:

$$y_1(1.5) = -0.469, \qquad y_{ex}(1.5) = -0.50$$

To achieve greater accuracy we may form a more general coordinate function according to the following pattern:

$$y_2 = (x - 1)(x - 2)(C_1 + C_2 x)$$

The following results give the values of $y_2(1.5)$ showing an improvement in accuracy.

$$y_2(1.5) = -0.50895 \qquad y_{ex}(1.5) = -0.50$$

We will have much opportunity to employ the Ritz method in succeeding chapters.   ////

3.15   GALERKIN'S METHOD

We now examine another method for finding an approximate solution to a differential equation—namely the Galerkin method. This method involves direct use of the differential equation; it does not require the existence of a functional. For this reason the method has a broader range of application than does the Ritz method. Yet we will soon see that the methods are closely related in the area of solid mechanics.

Consider the linear equation

$$Lu = f \qquad (3.65)$$

where the boundary conditions are *homogeneous*. The right side may be thought of as a forcing function of some sort and we may formulate a "virtual work" expression for this function as follows:

$$(\delta W)_1 = \iiint_V f \, \delta u \, dv \qquad (3.66)$$

where δu is a "virtual displacement" consistent with the constraints. It must also be true from Eq. (3.65) that

$$\iiint_V (Lu) \, \delta u \, dv = \iiint_V f \, \delta u \, dv \qquad (3.67)$$

for any "virtual displacement field" δu consistent with the constraints. However, if we use an approximate $\tilde{u}$ for u in expressing Lu in Eq. (3.67), then the two expressions for "virtual work" will no longer be equal. And this lack of equality may be in some way a measure of the departure of $\tilde{u}$ from the exact solution u. We now express $\tilde{u}$ as follows

$$\tilde{u} = \sum_1^n a_i \Phi_i$$

where the functions Φ_i again called the coordinate functions, satisfy all the boundary conditions of the problem. Then we can say generally

$$\iiint_V (L\tilde{u}) \, \delta u \, dv \neq \iiint_V f \, \delta u \, dv$$

However, we can, for any given "virtual displacement," force an equality in the above statement by properly adjusting the constants a_i. With n constants a_i we can actually force an equality for n different "virtual displacements." Forcing such an equality for these virtual displacement fields brings $\tilde{u}$ closer to the correct function u. Indeed, if we had an infinite set of properly selected coordinate functions we might conceivably force an equality for *any* admissible displacement field and thus with the left sides of Eqs. (3.66) and (3.67) becoming equal presumably we would expect $\tilde{u}$ to converge in some way in the limit to u, the exact solution. We shall not involve ourselves here in considerations of convergence criteria for the Galerkin method; they are quite complicated and beyond the level of this text. In practice a judiciously chosen finite number of functions Φ_i to represent $\tilde{u}$ can lead to good approximations to the exact solution of the problem. For this calculation we choose for the n "virtual displacements" also the n coordinate functions Φ_i and we require that:

$$(L\tilde{u})\Phi_i \, dv = \iiint_V f \Phi_i \, dv \iiint_V \quad i = 1, 2, \cdots, n$$

therefore

$$\iiint_V (L\tilde{u} - f)\Phi_i \, dv = 0 \qquad (3.68)$$

(Another viewpoint is to say that we have required that $(L\tilde{u} - f)$ be orthogonal to the ith coordinate function Φ_i.) This gives us n equations to solve for the undetermined constants a_i. In Problem 3.23 we have asked the reader to show that for *the theory of elasticity the Ritz coefficients are identical to the coefficients found by the Galerkin method for the same system of coordinate functions Φ_i. Thus these methods are equivalent to each other in this area of physics.* There is some advantage in using the Galerkin method for problems of linear elasticity over the Ritz method in that the equations for the coefficients a_i are reached more directly via Eq. (3.68) than in the Ritz method where one deals first with the functional.

We now demonstrate the use of Galerkin's method.

EXAMPLE 3.11   We shall reconsider the previous example with the view toward employing the Galerkin method.

For this purpose we again choose a single coordinate function Φ_1 to be the same function of Example 3.10. That is:

$$\Phi_1 = (x - 1)(x - 2) = (x^2 - 3x + 2) \qquad (a)$$

Now we can give $\tilde{y}$ as follows:

$$\tilde{y} = C_1\Phi_1 = C_1(x^2 - 3x + 2) \qquad (b)$$

Employing next Eq. (3.68) we get:

$$\int_1^2 \left(x^2 \frac{d^2\tilde{y}}{dx^2} + 2x\frac{d\tilde{y}}{dx} - 6x \right)\Phi_1 \, dx = 0 \qquad (c)$$

Hence:

$$\int_1^2 \left(x^2 \Phi_1 \frac{d^2\Phi_1}{dx^2}C_1 + 2x\Phi_1\frac{d\Phi_1}{dx}C_1 - 6x\Phi_1 \right) dx = 0 \qquad (d)$$

But the above formulation is identical to Eq. (f) of Example 3.10. Hence the result for C_1 is identical for both the Ritz and the Galerkin methods.   ////

One can show[1] that the Ritz and Galerkin methods are identical for self-adjoint, second-order ordinary differential equations with homogeneous and even non-homogeneous boundary conditions on the dependent variable. (It is clear from the Examples 3.10 and 3.11 that the Galerkin method provides a more direct way of setting up the equations for determining the coefficients.) Also whereas the Ritz method requires the differential equation to have a functional, the Galerkin method poses no such restriction.

3.16   CLOSURE

In this key chapter we have presented variational and related principles that form the basis of this text. And in doing so, we have inserted a self-contained discussion of trusses forming the first of a continuing study of structures. Finally we have set forth certain mathematical notions and a theorem concerning functionals that leads us smoothly into the area of approximation techniques, forming one of the chief benefits of variational considerations.

In the following chapter we will be able to apply many of the principles and techniques presented in this chapter. In so doing, we will then be able to continue the development of structural mechanics when we consider beams, frames and rings.

READING

ARTHURS, A. M.: "Complementary Variational Methods," Oxford University Press.

BIEZENO, AND GRAMMEL: "Engineering Dynamics," Vol. I, Blackie and Son, London, 1956.

FORRAY, M.: "Variational Calculus in Science and Engineering," McGraw-Hill Book Co., 1968.

HOFF, N. J.: "The Analysis of Structures," John Wiley and Sons, Inc., N.Y., 1956.

KANTOROVICH, AND KRYLOV, "Approximate Methods in Higher Analysis," Interscience Publishers, 1964.

LANCZOS, C.: "The Variational Principles of Mechanics," University of Toronto Press.

LANGHAAR, H. L.: "Energy Methods in Applied Mechanics," John Wiley & Sons Inc., N.Y.

MIKHLIN, S. G.: "Variational Methods in Mathematical Physics," Macmillan Co., N.Y., 1964.

MOISEIWITCH, B. L.: "Variational Principles," Interscience Publishers Inc.

ODEN, J. T.: "Mechanics of Elastic Structures," McGraw-Hill Book Co., N.Y.

SHAMES, I. H.: "Mechanics of Deformable Solids," Prentice-Hall Inc, 1964.

SOKOLNIKOFF, I. S.: "The Mathematical Theory of Elasticity," McGraw-Hill Book Co., 1956.

TEMPLE, G., AND BICKLEY, W. G.: "Rayleigh's Principle," Dover.

WASHIZU, K.: "Variational Methods in Elasticity and Plasticity," Pergamon Press, 1968.

---

[1] See Kantorovich and Krylov: "Approximate Methods of Higher Analysis," Interscience Publishers, Chap. 4.

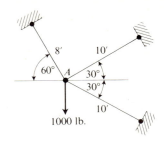

FIGURE 3.22

PROBLEMS

3.1   Use the method of virtual work to find the forces in the members shown in Fig. 3.22. What is the horizontal movement of pin A? Assume linear elastic behavior with each member having the same modulus of elasticity and the same cross-sectional area.

3.2   Do the previous problem for the case of nonlinear elastic behavior where for all members:

$$\tau = \epsilon^{0.8} \times 15 \times 10^6$$

All members are identical in cross-sectional area.

3.3   Show by extremizing π as given by Eq. (3.11) that we can arrive at the equations of equilibrium and Cauchy's formula.

3.4   Do Problem 3.1 by the method of total potential energy.

3.5   Consider a perfectly flexible membrane simply-supported at the boundary (see Fig. 3.23) on the xy plane and with a tensile force T per unit length which is everywhere a constant in the membrane. If a normal pressure distribution $q(x,y)$ is applied to cause small deflection of the membrane show that the static deflection $w(x,y)$ of this membrane is.

$$\frac{\partial^2 w}{\partial x^2} + \frac{\partial^2 w}{\partial y^2} = -\frac{q}{T} \qquad (a)$$

Hint: The area of the surface of the membrane in its deformed configuration is given (from differential geometry) as:

$$A = \iint_S \left[1 + \left(\frac{\partial w}{\partial x}\right)^2 + \left(\frac{\partial w}{\partial y}\right)^2 \right]^{1/2} dx\, dy$$

Show first that the strain energy is $T\Delta A$ where ΔA is the change in area of the membrane from deformation. Now show that:

$$V = \frac{1}{2} T \iint_S \left[\left(\frac{\partial w}{\partial x}\right)^2 + \left(\frac{\partial w}{\partial y}\right)^2 \right] dx\, dy$$

3.6   We have shown in this text that with $\delta^{(1)}\pi = 0$ we have:

$$\pi(\epsilon_{ij} + \delta\epsilon_{ij}) - \pi(\epsilon_{ij}) = \frac{1}{2!} \iiint_V \frac{\partial^2 U_0}{\partial \epsilon_{ij} \partial \epsilon_{kl}} \delta\epsilon_{ij}\, \delta\epsilon_{kl}\, dv + \cdots$$

FIGURE 3.23

Consider the stress-strain law in conjunction with the variation of τ_{ij}. Show that for any ϵ_{ij} the first term on the right side of the above equation is equal to:

$$\frac{1}{2!} \iiint_V \delta\tau_{ij} \, \delta\epsilon_{ij} \, dv$$

For linear elastic material explain why the above expression is positive-definite. Thus the minimum nature of the total potential energy functional is proved for all values of ϵ_{ij} for linear elastic materials.

3.7   For constant tension, T, we have derived the governing equation of equilibrium of a transversely loaded string (Example 3.3). Derive the analogous equation for the case where the tension is not fixed, i.e., $T = T(x)$. Is this a linear problem, mathematically speaking?

3.8   Use the dummy load method to compute the vertical movement of joint C of the truss in Fig. 3.24 from the external loads. Assume linear elastic behavior with all members identical in cross-section with the same modulus of elasticity.

3.9   Using Castigliano's theorem, find the deflection in the vertical direction for pin A of the truss in Fig. 3.25. Each member has a cross-sectional area "A" and is nonlinear-elastic with a stress–strain law given as:

$$\tau = 10 \times 10^6 \epsilon^{0.8}$$

Also solve the problem by method of virtual work.

3.10   Find the forces in the truss shown in Fig. 3.26. Assume linear elastic behavior with all members identical in cross section and with the same modulus of elasticity.

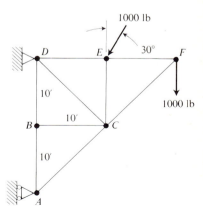

FIGURE 3.24

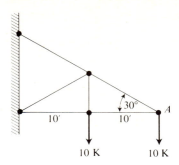

FIGURE 3.25

3.11 Find the horizontal and vertical movement of each joint in the truss of Problem 3.10. What are the forces in the members?

3.12 What is the movement of pin D in the horizontal direction in the truss shown in Fig. 3.27? The members are identical in cross-section having the same modulus of elasticity.

3.13 Show that $\delta^{(T)}I_R$ can be given as follows (see Eq. (3.27) and Eq. (3.29)):

$$\delta^{(T)}(I_R) = \delta^{(1)}I_R + \iiint_V \left[\delta\tau_{ij}\,\delta\epsilon_{ij} - \frac{1}{2!}\frac{\partial^2 U_0^*}{\partial\tau_{ij}\,\partial\tau_{kl}}\,\delta\tau_{ij}\,\delta\tau_{kl} + \cdots \right] dv$$

Take $\delta^{(1)}I_R = 0$. We argue, as we did in Problem 3.6, that:

$$\delta^{(T)}(I_R) = \tfrac{1}{2} \iiint_V \delta\tau_{ij}\,\delta\epsilon_{ij}\,dv + \cdots$$

Even for linear elastic materials why can't we argue that the expression on the right side of the equation is positive definite? Accordingly we *cannot show* that the Reissner principle is a *minimum* principle.

3.14 Show that the operator

$$-\frac{d}{dx}\left(x^2\frac{d}{dx}\right)$$

is symmetric and positive definite for certain boundary conditions at $x = a, b$. What are those boundary conditions?

3.15 Do Problem 3.14 for the operator $-\nabla^2$.

3.16 The one dimensional steady state heat conduction equation has the following form:

$$\frac{d^2 T}{dx^2} + \psi(x) = 0$$

where T is the temperature and ψ is a specified function representing a heat source distribution. If $T = 0$ at the end points show that the quadratic functional can be found directly to be

$$I = \int_a^b [\tfrac{1}{2}T_x^2 - T\psi]\,dx$$

where $T_x = dT/dx$.

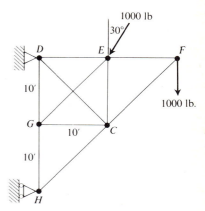

FIGURE 3.26

3.17 Consider the following functionals

$$\iiint_V H(x,y,z,w,w_x,w_y)\, dx\, dy\, dz \qquad (a)$$

$$\iiint_V \left[H(x,y,z,w,w_x,w_y) + \alpha \frac{\partial A(x,y,z,w)}{\partial x} + \beta \frac{\partial B(x,y,z,w)}{\partial y} + \gamma \frac{\partial C(x,y,z,w)}{\partial z} \right] dx\, dy\, dz \qquad (b)$$

where α, β, and γ are constants. Show that you get the same Euler–Lagrange equation for both functionals but that the boundary conditions are different.

3.18 Starting with the partial differential equation for the deflection of the flexible membrane of Problem 3.5 in the form

$$-\nabla^2 w = -\frac{q}{T}$$

formulate the corresponding quadratic functional. Compare the quadratic functional with the total potential energy functional for this problem. Need they be the same?

*3.19 (a) Show that the Navier equation for linear elasticity can be put in the form (see Eqs. (1.106) and (1.98)):

$$-G\left\{ \frac{1}{1-2v} u_{i,ij} + u_{j,ii} \right\} = B_j \qquad (a)$$

Consider the traction field $T_i^{(v)}$ associated with displacement field u_i to be denoted as $\overset{u}{T_i}^{(v)}$. From Cauchy's formula and Hooke's law (see Eqs. (1.96) and (1.98)) we can say that

$$\overset{u}{T_i}^{(v)} = \tau_{ij}n_j = G\left\{ \frac{2v}{1-2v} u_{m,m} \delta_{ij} + (u_{i,j} + u_{j,i}) \right\} n_j \qquad (b)$$

Let $\overset{u}{e} = u_{i,i}$ and $2\overset{u}{\epsilon}_{ij} = (u_{i,j} + u_{j,i})$. Show for the operator of Eq. (a) that by adding and subtracting $\overset{u}{e}_{,j}$ we can arrive at the following result on employing the divergence

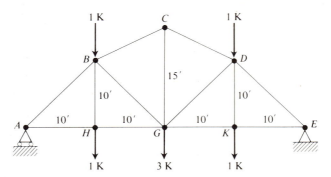

FIGURE 3.27

theorem

$$(\mathbf{Lu},\mathbf{v}) = G \iiint_V \left[\frac{2v}{1-2v} \overset{u}{e}\overset{v}{e} + 2\overset{u}{\epsilon_{ij}}\overset{v}{\epsilon_{ij}} \right] dv$$

$$- G \iint_S \left[\frac{2v}{1-2v} \overset{u}{e} v_j \, \delta_{ij} + 2\overset{u}{\epsilon_{ij}} v_j \right] n_j \, ds \qquad (c)$$

where n_j are components of the unit normal to the bounding surface s. Show next that

$$(\mathbf{Lu},\mathbf{v}) - (\mathbf{Lv},\mathbf{u}) = \iint_S \overset{v}{T_i^{(v)}} u_i \, ds - \iint_S \overset{u}{T_i^{(v)}} v_i \, ds \qquad (d)$$

The well-known *Maxwell–Betti-reciprocal theorem* states that for *linear elastic solids* the work done by a system of forces A, through displacements caused by a second system of forces B, equals the work done by the second system of forces B through displacements caused by the first system of forces A. What are the implications for the adjointness of the operator L from Navier's equation in light of the Maxwell–Betti reciprocity theorem? Can the latter be obtained from the result (d)?

(b) From the above result show that:

$$(\mathbf{Lu},\mathbf{u}) = - \iint_S \overset{u}{T_i^{(v)}} u_i \, ds + 2G \iiint_V \left[\frac{v}{1-2v}(u_{m,m})^2 + \epsilon_{ij}\epsilon_{ij} \right] dv$$

Discuss the implications of this result for the positive-definiteness of the operator L. Are there situations where $(\mathbf{Lu},\mathbf{u}) = 0$ while $\mathbf{u} \neq \mathbf{0}$? What do they mean? What occurs if rigid-body motions $u_i^R = u_i^0 + \omega_{ij}x_j$ are substituted above?

3.20 Show for the differential equation $Lu = f$, that if L is symmetric, then the Galerkin solution equation is obtainable as the *first variation* of the quadratic functional $I(u)$.

3.21 Consider the differential equation:

$$x^3\frac{d^4y}{dx^4} + 6x^2\frac{d^3y}{dx^3} + 6x\frac{d^2y}{dx^2} = 10x$$

If $y' = 0$ and $y = 0$ at $x = 1$ and $x = 3$ solve this boundary value problem by a one-parameter approximation using the Ritz method.

3.22 Do problem 3.21 using the Galerkin method.

*3.23 Starting from the Navier equations show that for the Galerkin method the required equations to determine the undetermined coefficients a_n are

$$\iiint_V \left(\nabla^2 \tilde{u}_1 + \frac{1}{1-2v} \frac{\partial \epsilon}{\partial x} + \frac{B_x}{G} \right) \Phi_i \, dv = 0 \qquad (a)$$

$$\iiint_V \left(\nabla^2 \tilde{u}_2 + \frac{1}{1-2v} \frac{\partial \epsilon}{\partial y} + \frac{B_y}{G} \right) \Psi_i \, dv = 0 \qquad (b)$$

$$\iiint_V \left(\nabla^2 \tilde{u}_3 + \frac{1}{1-2v} \frac{\partial \epsilon}{\partial z} + \frac{B_z}{G} \right) \Gamma_i \, dv = 0 \qquad (c)$$

where $\epsilon = \epsilon_{xx} + \epsilon_{yy} + \epsilon_{zz} = \epsilon_{ii}$. Now express the strain energy of a body as follows:

$$\tilde{U}_0 = G \left[\tilde{\epsilon}_{xx}{}^2 + \tilde{\epsilon}_{yy}{}^2 + \tilde{\epsilon}_{zz}{}^2 + \frac{\tilde{\epsilon}^2}{1-2v} + 2(\tilde{\epsilon}_{yz}{}^2 + \tilde{\epsilon}_{zx}{}^2 + \tilde{\epsilon}_{xy}{}^2) \right]$$

Formulate π, the total potential energy of the system. Noting that

$$\tilde{u}_1 = \sum a_i \Phi_i$$
$$\tilde{u}_2 = \sum b_i \Psi_i$$
$$\tilde{u}_3 = \sum c_i \Gamma_i$$

set $\partial \tilde{\pi}/\partial a_i = 0$ and reach the following result:

$$-2G \iiint_V \left[\left(\tilde{\epsilon}_{xx} + \frac{\tilde{\epsilon}}{v-2} \right) \frac{\partial \Phi_i}{\partial x} + \tilde{\epsilon}_{yx} \frac{\partial \Phi_i}{\partial y} + \tilde{\epsilon}_{zx} \frac{\partial \Phi_i}{\partial z} \right] dv$$

$$- \iiint_V B_x \Phi_i \, dv - \iint T_x^{(v)} \Phi_i \, ds = 0$$

Integrate by parts for the first integral (i.e. Eq. (I.51), Appendix I) to get the result;

$$2G \iint_S \left[\left(\tilde{\epsilon}_{xx} + \frac{\tilde{\epsilon}}{v-2} \right) a_{nx} + \tilde{\epsilon}_{yx} a_{ny} + \tilde{\epsilon}_{zx} a_{nz} - \frac{T_x^{(v)}}{2G} \right] \Phi_i \, ds$$

$$-2G \iiint_V \left[\frac{\partial[\tilde{\epsilon}_{xx} + \tilde{\epsilon}/(v-2)]}{\partial x} + \frac{\partial \tilde{\epsilon}_{yx}}{\partial y} + \frac{\partial \tilde{\epsilon}_{zx}}{\partial z} + \frac{B_x}{2G} \right] \Phi_i \, dv = 0$$

Get integrand of first expression in terms of stresses and show that it must be zero if the boundary conditions are maintained everywhere. Finally show that the second integral may be written in terms of $\tilde{u}$ and $\tilde{\epsilon}$ so that we have

$$\iiint_V \left(\nabla^2 \tilde{u} + \frac{1}{1-2v} \frac{\partial \tilde{\epsilon}}{\partial x} + \frac{B_x}{G} \right) \Phi_i \, dv = 0$$

This equation is identical to Eq. (a). By performing the preceding steps for $\partial \tilde{\pi}/\partial b_i$ and $\partial \tilde{\pi}/\partial c_i$ we can reach Eqs. (b) and (c). We thus show that the Galerkin method and the Ritz method give the *same set of simultaneous equations for the undetermined constants* when used with the Navier equation.

4

BEAMS, FRAMES AND RINGS

4.1 INTRODUCTION

In the previous chapter, while developing as the primary effort certain variational principles of mechanics, we entered into a discussion of trusses in order both to illustrate certain aspects of the theory and to present a discussion of the most simple class of structures. We could take on this dual task at this stage because the stress and deformation of any one single member of a truss is a very simple affair. That is, the only stress on any section (away from the ends)[1] is a uniform normal stress given

---

[1] At the ends, in reality, due to friction of the supports and complicated boundaries of the member it is unlikely that simple uniaxial tension or compression will exist.

as F/A while the strain at any section is that of normal strain directly available from F/A through an appropriate constitutive law.

We shall now consider a more complex structural member, the beam, wherein we will first develop approximate equations for determining stress and deformation of the beam via the method of total potential energy. The theory that we shall propose is called the *technical* or *engineering theory of beams*. We shall then consider approximate solutions to the deformation of beams by using the Ritz method and a method stemming from the Reissner principle. Next we shall use Castigliano's second theorem to consider statically indeterminate supporting force systems for beams.

With the technical theory of beams established, we will be ready to consider both open and closed frames. We are at this point in a position analogous to the previous chapter where, knowing the individual behavior of truss members, we could then use energy methods (either directly involving or closely related to the variational process) to study trusses themselves. Indeed we will here find computations for frames that are highly analogous to those for trusses.

We shall not present a development of curved beams and accordingly we will only be able to set forth a limited discussion of rings.

Part A
BEAMS

4.2 TECHNICAL THEORY OF BEAMS

In Chap. 1 we first presented the full theory of linear elasticity, and we then made simplifying assumptions as to stress distributions (plane stress) by stating that certain stresses were to be taken as zero or constant in the domain of interest. This decreased the number of dependent variables to be dealt with and permitted the solution of several problems of interest. We pointed out in Sec. 3.3 that in the structural studies that follow we make the assumptions primarily about displacement fields. Also, we usually make assumptions as to certain aspects of the constitutive law to be employed. The variational process, as it relates to the total potential energy, will then give us the proper equations of equilibrium and the proper boundary conditions for a problem like the one we start with, except that it now has certain internal constraints (as a result of the aforementioned displacement assumptions) and it now behaves according to our assumed constitutive law. Most importantly, by making appropriate assumptions for the displacement field we will be able to reduce the number of variables significantly and thus facilitate actual computations. For the technical theory of beams, we shall reduce the problem to a single dependent variable, $w(x)$, which represents the deflection of the centerline of the beam—thus

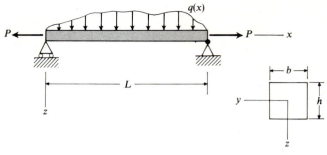

FIGURE 4.1

we are considering a curve in space to represent the beam. Similarly, in a later chapter, we shall be able to employ the considerations of a surface in space to represent a plate. How do we know when the equations we thus derive from variational considerations reasonably depict the actual problem? We can answer this by checking results for simple cases with those which stem from the full theory of elasticity, and also we can check results from our theory with those from experiments.

Consider then a prismatic rod of rectangular cross section as shown in Fig. 4.1. We assume that the dimensions h and b are both much smaller than the length L. The loading $q(x)$ is assumed in the plane of symmetry of the cross section, i.e., in the xz plane, and is directed transversely to the beam. (Such loadings might entail point forces or couples.) Axial tensile loading is also possible along the centroidal axis (taken as the x axis) of the beam.

We now present a simplified view of the deformation to be expected from these loads. In the x direction, we imagine a *stretching* action from the axial load wherein each section undergoes a displacement in the x direction, which is a function of the position x of the section. Thus using the notation $\mathbf{u} = u_1\hat{\mathbf{i}} + u_2\hat{\mathbf{j}} + u_3\hat{\mathbf{k}}$, the strain at any section due to the axial tensile loads is $d[u_1(x)]_s/dx$, where the subscript s refers to the axial stretching action. In addition we superpose a *bending* contribution from the transverse loads. To assess this contribution consider the line in the beam in the undeformed geometry coinciding with the x axis. As a result of deformation from the transverse loads, this line deforms into a curve whose position relative to the reference is denoted as $w(x)$ as shown in Fig. 4.2. For small deformation, a line element dx in the undeformed geometry is assumed to remain at coordinate x but translates in the z direction an amount $w(x)$ and rotates in the xz plane an amount given by the slope dw/dx of the deflection curve at coordinate x. We now assume that plane sections such as ab in Fig. 4.3 originally normal to the centerline of the beam in the undeformed geometry remain plane and normal to the centerline (see line $a'b'$) in the deformed geometry as the beam bends from the loads. Furthermore, in the

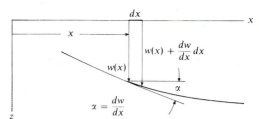

FIGURE 4.2 z

planes of these sections we assume there is no stretching or shortening whatsoever; they are assumed to act like rigid surfaces. Thus as a result of bending, sections of the beam translate vertically an amount $w(x)$ to new positions and, in addition, rotate by the amount dw/dx. This has been shown in Fig. 4.3. Accordingly in considering the bending displacement in the x direction of a point in a cross section of a beam at position z below the centroidal axis in the undeformed geometry, we get for small deflections:

$$(u_1)_{\text{bend}} = -z\left(\frac{dw(x)}{dx}\right)$$

The total displacement u_1 (i.e., the total displacement in the x direction) for the aforementioned point is then:

$$u_1 = (u_1)_s - z\left(\frac{dw}{dx}\right)$$

while the corresponding strain is then

$$\epsilon_{xx} = \frac{d[u_1(x)]_s}{dx} - z\frac{d^2w(x)}{dx^2} \tag{4.1}$$

There is no relative motion in the y direction at any time of points in the cross section of the beam. Accordingly we can say:

$$\epsilon_{yy} = 0 \tag{4.2}$$

We may conclude that $u_2 = 0$. Finally, we have pointed out that the translatory displacement of the section in the z direction is simply $w(x)$. Accordingly we may say, neglecting vertical movement stemming from rotations of the sections for small deformation:

$$u_3 = w(x) \tag{4.3}$$

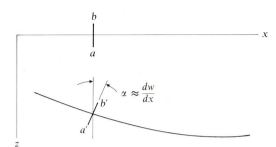

FIGURE 4.3 z

We now summarize the results presented for the proposed displacement field:

$$
\begin{aligned}
u_1 &= [u_1(x)]_s - z\frac{dw(x)}{dx} \\
u_2 &= 0 \\
u_3 &= w(x)
\end{aligned}
\qquad (4.4)
$$

It is clear that for the above displacement field all strains except ϵ_{xx} are zero. Also note that the bending deformation is now given in terms of the deformation of the centerline of the beam.

The assumed deformation field has obvious errors even for small deformation. For instance, in the case of a thin beam ($b \ll L$) and with no loading on the sides of the beam, we would expect intuitively that the stresses τ_{yy} would be very close to zero. However, it is clear for a Hookean material that the stress τ_{yy} as a result of our assumptions must be, in accordance with Eqs. (1.96) and (1.98):

$$
\tau_{yy} = \frac{vE}{(1 + v)(1 - 2v)}\epsilon_{xx} \qquad (4.5(a))
$$

Furthermore, in considering τ_{zz} we note that there is zero traction on the bottom surface of the beam and that we can consider the direct stress τ_{zz} at the surface of contact for q to be negligibly small compared to bending and stretching stress τ_{xx}. Accordingly for $h \ll L$, the stress τ_{zz} should be very small. Hooke's law, however, gives τ_{zz} as follows:

$$
\tau_{zz} = \frac{vE}{(1 + v)(1 - 2v)}\epsilon_{xx} \qquad (4.5(b))
$$

If we use a constitutive law wherein $v = 0$, we will get the desired zero value for the stresses τ_{yy} and τ_{zz}. We shall do this as we explained in Sec. 4.1 so as to facilitate the development of the theory, but we will be left with the burden later of showing that the results are at least partially valid for real materials having non-zero Poisson ratios.

It is also to be noted that the theory presented thus far yields a zero value for τ_{xz}. For beams with non-zero lateral loading, we know from earlier work in strength of materials that there will be a non-zero stress τ_{xz} over much of the beam. However, this shear stress and the accompanying shear deformation will be small for the class of problems we are studying, and despite the incorrect result from our theory for τ_{xz}, we still nevertheless get a good evaluation of the far more significant stress τ_{xx} and the quantity $w(x)$.[1]

The theory presented for evaluating τ_{xx} and $w(x)$ is called the *technical theory of beams*. We shall in succeeding sections make efforts to assess the accuracy of the technical theory of beams. It gives us effectively a one-dimensional stress problem to consider, and we shall study the variational aspects of this problem in the next section.

4.3 DEFLECTION EQUATIONS FOR THE TECHNICAL THEORY OF BEAMS

As pointed out in the previous section, we can formulate a technical theory of beams as a one-dimensional problem. In computing the strain energy U for the beam we will have for linear elastic behavior:

$$U = \tfrac{1}{2} \int \int \int_V \tau_{ij}\epsilon_{ij}dv$$

$$= \tfrac{1}{2}b \int_0^L \int_{-h/2}^{h/2} \tau_{xx}\epsilon_{xx}\, dz\, dx \qquad (4.6)$$

We shall wish to make use of the method of minimum total potential energy and so we next examine the potential of the applied loads to do work. At position $x = L$ the applied axial load P (see Fig. 4.1) is in the same direction as the positive displacement $[u_1(L)]_s$ while at $x = 0$ the applied load P is in the opposite direction as the positive displacement $[u_1(0)]_s$. Accordingly we have for loads P the following result:

$$V_p = -\{P[u_1(L)]_s - P[u_1(0)]_s\} = -P\int_0^L \frac{d}{dx}(u_1)_s\, dx$$

In considering the work of loading $q(x)$, we assume that this loading acts at the centroidal axis of the beam and not on the top surface. This means that $w(x)$ char-

---

[1] Note because we are neglecting the shear deformation associated with τ_{xz}, $w(x)$ represents deflection due only to bending.

acterizes the movement of the points of loading for $q(x)$, a conclusion consistent with the deformation assumption. Accordingly, since q and w are both positive when directed downward we have:

$$V_q = - \int_0^L q(x)w(x)\,dx \qquad (4.7)$$

The total potential energy for the problem then becomes:

$$\pi = \tfrac{1}{2}b \int_0^L \int_{-h/2}^{h/2} \tau_{xx}\epsilon_{xx}\,dx\,dz - P \int_0^L \frac{d}{dx}[u_1]_s\,dx - \int_0^L q(x)w(x)\,dx$$

In computing the total potential energy we shall use Hooke's law to replace τ_{xx}. Maintaining the earlier assumption that $v = 0$, we can then give τ_{xx} simply as $E\epsilon_{xx}$. Now replace ϵ_{xx} in the above equation in terms of the displacement field given by Eq. (4.4). Thus:

$$\pi = \tfrac{1}{2} \int_0^L \int_{-h/2}^{h/2} bE\left[\frac{d(u_1)_s}{dx} - z\frac{d^2w}{dx^2}\right]^2 dz\,dx - P \int_0^L \frac{d(u_1)_s}{dx}\,dx - \int_0^L qw\,dx$$

If we carry out the indicated squaring operation and note the following integrals

$$\int_{-h/2}^{h/2} \{1, z, z^2\}\,dz = \left\{h, 0, \frac{h^3}{12}\right\}$$

we easily obtain the following result:

$$\pi = \int_0^L \left\{\frac{EA}{2}\left(\frac{d(u_1)_s}{dx}\right)^2 + \frac{EI}{2}\left(\frac{d^2w}{dx^2}\right)^2 - P\frac{d(u_1)_s}{dx} - qw\right\} dx \qquad (4.8)$$

We have then a functional with one independent variable, x, and involving function $(u_1)_s$, with a derivative of order unity, and function w, with a derivative of order two. The Euler–Lagrange equations for finding the extremal functions are readily established with the aid of Eq. (2.64).[1] Thus for the stretching displacement $(u_1)_s$ we get:

$$\frac{d}{dx}\left[EA\frac{d(u_1)_s}{dx}\right] = 0 \qquad (4.9)$$

while for the bending displacement w we get:

$$\frac{d^2}{dx^2}\left[EI\frac{d^2w}{dx^2}\right] = q \qquad (4.10)$$

As for the boundary conditions (see Eq. (2.63)), we may specify the kinematic conditions at the ends of the beam. This means that $(u_1)_s$ is specified at an end as are w and w'. If we wish to employ the natural boundary conditions to establish an

---

[1] You will be asked as an exercise (Problem 4.1) to find these results using the δ operator.

extremum we see from Eq. (2.63) that for the function $(u_1)_s$ we get:

$$\frac{\partial F}{\partial (u_1)'_s} = EA\frac{d(u_1)_s}{dx} - P = 0 \qquad \text{at } x = 0, L \qquad (4.11)$$

and in place of specifying w itself we have the natural boundary conditions which in accordance with Eqs. (2.63) and (4.8) become:

$$-\frac{d}{dx}\left(\frac{\partial F}{\partial w''}\right) + \frac{\partial F}{\partial w'} = -\frac{d}{dx}\left(EI\frac{d^3 w}{dx^3}\right) = 0 \qquad \text{at } x = 0, x = L \qquad (4.12)$$

Finally in place of specifying w' at the ends we see that:

$$-\frac{\partial F}{\partial w''} = EI\frac{d^2 w}{dx^2} = 0 \qquad \text{at } x = 0, x = L \qquad (4.13)$$

We now summarize the boundary conditions as follows:

Kinematic Boundary Conditions		*Natural Boundary Conditions*
$(u_1)_s$ prescribed	or	$EA\dfrac{d(u_1)_s}{dx} = P$
$\dfrac{dw}{dx}$ prescribed	or	$EI\dfrac{d^2 w}{dx^2} = 0$
w prescribed	or	$\dfrac{d}{dx}\left(EI\dfrac{d^2 w}{dx^2}\right) = 0$ $\qquad$ (4.14)

We shall now give physical interpretations for the natural boundary conditions. Let us first define an axial stress resultant N and a moment M as:

$$N = \int_{-h/2}^{h/2} \tau_{xx} b\, dz \qquad (a)$$

$$M = \int_{-h/2}^{h/2} \tau_{xx} zb\, dz \quad (b) \quad (4.15)$$

Now expressing τ_{xx} as $E\epsilon_{xx}$ and using Eq. (4.1) for the strain we find after simple calculations that:

$$N = EA\frac{d(u_1)_s}{dx} \quad (a) \qquad M = -EI\frac{d^2 w}{dx^2} \quad (b) \quad (4.16)$$

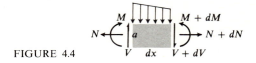

FIGURE 4.4

Now consider an element of the beam as shown in Fig. 4.4. The shear force V is the stress resultant due to τ_{xz}, that is:

$$V = \int_{-h/2}^{h/2} \tau_{xz} b \, dz$$

One way of computing the relationship between V and M as well as the distribution of shear stress in the cross-section is to take moments about the point "a" in Fig. 4.4, and then resort to the classical $\tau = VQ/Ib$ formula of strength of materials. We shall determine the shear stress distribution from the first equation of equilibrium of elasticity theory, and then we will evaluate the shear force resultant.

The equation of equilibrium in the x direction in the absence of body forces and with $\tau_{xy} = 0$ is

$$\frac{\partial \tau_{xx}}{\partial x} + \frac{\partial \tau_{xz}}{\partial z} = 0$$

Let us now evaluate the bending stress τ_{xx} due to the bending deflection $w(x)$. From Eq. (4.1) and the one-dimensional constitutive law,

$$\tau_{xx} = E\epsilon_{xx} = -Ez\frac{d^2w}{dx^2}$$

The rate of change of curvature, d^2w/dx^2, can be eliminated from the above equation using Eq. (4.16(b)) to yield the classical flexure formula:

$$\tau_{xx} = \frac{Mz}{I} \qquad (4.17)$$

Substituting this bending stress distribution into the elasticity equilibrium equation and integrating with respect to z yields

$$\tau_{xz} = \int \frac{dM(x)}{dx}\frac{z}{I} dz + g(x) = \frac{dM(x)}{dx}\left(\frac{z^2}{2I}\right) + g(x)$$

where $g(x)$ is an arbitrary function of x. Since the shear stress vanishes at $z = \pm h/2$, as we are applying only normal loading to the beam, we can determine $g(x)$ and find that

$$\tau_{xz} = \frac{dM}{dx}\left(\frac{1}{2I}\right)(z^2 - h^2/4) \qquad (4.18)$$

The shear resultant V can then be found by evaluating the integral

$$V = \int_{-h/2}^{h/2} \frac{dM}{dx} \frac{b}{2I}(z^2 - h^2/4)\,dz = \frac{dM}{dx} \qquad (4.19)$$

Hence we have derived another classical formula between the shear and moment resultants. We can combine Eqs. (4.16(b)) and (4.19) to relate the shear to the bending deflection, i.e.,

$$V = -\frac{d}{dx}\left(EI\frac{d^2w}{dx^2}\right) \qquad (4.20)$$

Using Eqs. (4.16) and (4.20) we can now go back to the natural boundary conditions as given by Eq. (4.14) and restate them in terms of forces and moments. We now give the boundary conditions as follows:

At $x = 0, x = L$:

	Kinematic Boundary Conditions		Natural Boundary Conditions
	$(u_1)_s$ prescribed	or	$N = P$
	$\dfrac{dw}{dx}$ prescribed	or	$M = 0$
	w prescribed	or	$V = 0$ (4.21)

We can now state the boundary conditions as follows:

(a) Either the longitudinal displacement or the total axial force is prescribed at the ends;

(b) Either the slope of the deflected centerline is prescribed at the ends or the total bending moment is zero there; and finally

(c) Either the vertical displacement is prescribed at the ends or the shear force is zero there.

Now we will note an important characteristic of the equations we have just derived. If we examine the differential equations (Eqs. (4.9) and (4.10)) and the boundary conditions (Eqs. (4.14) or (4.21)) we see that the *stretching problem* (involving the displacement $(u_1)_s$ and the axial stress resultant N) is completely solvable independently of the bending problem. That is, the stretching and bending problems are *uncoupled*. This is a direct consequence of the kinematic assumptions we have

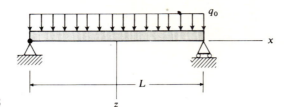

FIGURE 4.5

made (Eq. (4.4)). Because the stretching problem is very simple, we shall only concern ourselves in the sequel with the bending problem.

4.4   SOME JUSTIFICATIONS FOR THE TECHNICAL THEORY OF BEAMS

For the purpose of demonstrating the validity of the technical theory of beams, we now consider a simple problem that has already been investigated (Problem 1.26) in Chap. 1 as an example of plane stress. Thus we show in Fig. 4.5 a simply-supported beam loaded uniformly in the transverse direction. We shall place the origin of the reference at the center of the beam as shown in the diagram. The obvious boundary conditions that we shall use for this problem at $x = \pm(L/2)$ are that the deflection w is zero and that the bending moment M is zero. The latter is accomplished by requiring that $d^2w/dx^2 = 0$ at the ends (see Eq. (4.16)).

The equation for the deflection of the centerline (see Eq. (4.10)) may then be written as follows for uniform properties of the beam and uniform shape of the cross-section:

$$\frac{d^4w}{dx^4} = \frac{q_0}{EI}$$

Integrating four times we get:

$$w = \frac{q_0 x^4}{24EI} + \frac{C_1 x^3}{6} + \frac{C_2 x^2}{2} + C_3 x + C_4$$

We may determine the constants of integration from the boundary conditions. Note first that:

$$\frac{d^2w}{dx^2} = \frac{q_0 x^2}{2EI} + C_1 x + C_2$$

Hence at $x = \pm L/2$ we have:

$$0 = \frac{q_0 L^2}{8EI} \pm C_1 \frac{L}{2} + C_2$$

Clearly C_1 must be zero for the above to be correct and $C_2 = -qL^2/8EI$. Also, setting $w = 0$ at $x = \pm L/2$ we get:

$$0 = \frac{q_0 L^4}{384EI} \pm \frac{C_1 L^3}{48} - \frac{q_0 L^2}{8EI}\left(\frac{L^2}{4}\right) + C_4$$

Again it is immediately apparent from above that $C_1 = 0$ and that $C_4 = 5q_0L^4/(384EI)$. We then have for the deflection curve:

$$w = \frac{q_0 x^4}{24EI} - \frac{q_0 L^2}{8EI}\frac{x^2}{2} + \frac{5q_0 L^4}{384EI}$$

$$= \frac{q_0 L^4}{24EI}\left[\left(\frac{x}{L}\right)^4 - \frac{3}{2}\left(\frac{x}{L}\right)^2 + \frac{5}{16}\right] \tag{4.22}$$

As a first step in evaluating the validity of the above result, we show that there is inner consistency in the theory. Specifically, we have neglected the shear strain energy in the development of the theory, and we shall now examine the shear strain energy that stems from the resulting formulation of the theory. If there is consistency, shear strain energy from these formulations must also be small. Denoting shear strain energy as U_{shear} we have:

$$U_{\text{shear}} = \frac{1}{2}\iiint_V \tau_{xz}\gamma_{xz}\,dv = \frac{1}{2G}\iiint_V \tau_{xz}^2\,dv$$

Using Eq. (4.18) and noting for the problem that $dM/dx = q_0 x$

$$U_{\text{shear}} = \frac{1}{2G}\int_{-L/2}^{L/2}\int_{-h/2}^{h/2}\left\{-\frac{q_0 x}{2I}\left[\frac{h^2}{4} - z^2\right]\right\}^2 b\,dz\,dx$$

$$= \frac{q_0^2 L^5}{240EI}\left[2(1+v)\left(\frac{h}{L}\right)^2\right] \tag{4.23}$$

where we have replaced G by the expression $E/[2(1+v)]$.

Now compute the strain energy due to bending. We have, using the portion of Eq. (4.8) that is applicable and substituting from Eq. (4.22):

$$U_{\text{bend}} = \frac{EI}{2}\int_{-L/2}^{L/2}\left(\frac{d^2w}{dx^2}\right)^2 dx = \frac{q_0^2 L^5}{240EI} \tag{4.24}$$

Taking the ratio of Eqs. (4.23) and (4.24) for comparison purposes we have:

$$\frac{U_{\text{shear}}}{U_{\text{bend}}} = 2(1+v)\left(\frac{h}{L}\right)^2$$

Thus we see that the ratio of the shear strain energy to the bending strain energy is proportional to (h/L) squared. Thus for long thin beams the shear strain energy is very small compared to the bending strain energy. This result is entirely consistent with the assumption made in developing the technical theory.

We may also readily show here that the ratio of the maximum magnitudes of the shear stress τ_{xz} to that of the normal stress τ_{xx} is proportional to (h/L). To do this we return to Eq. (4.18) and, introducing dimensionless variables

$$ \xi = \frac{x}{L} \qquad \eta = \frac{z}{h} $$

we get:

$$ \tau_{xz} = -\frac{q_0 L^2 h}{2I}\left(\frac{h}{L}\right)\left(\frac{1}{4} - \eta^2\right)\xi $$

Now expressing $\tau_{xx} = E\epsilon_{xx} = -Ez(d^2w/dx^2)$ as per Eq. (4.1), and using Eq. (4.22) for w, we get:

$$ \tau_{xx} = -\frac{q_0 L^2 h}{2I}\left(\xi^2 - \frac{1}{4}\right)\eta \qquad (4.25) $$

Note that $|\tau_{xz}|_{max}$ occurs at $\xi = \frac{1}{2}$ and $\eta = 0$ while $|\tau_{xx}|_{max}$ occurs at $\xi = 0$ and $\eta = \frac{1}{2}$. We thus have for these quantities:

$$ |\tau_{xz}|_{max} = \frac{qL^2 h}{2I}\left(\frac{h}{L}\right)\left(\frac{1}{4} - 0\right)\left(\frac{1}{2}\right) $$

$$ |\tau_{xx}|_{max} = -\frac{qL^2 h}{2I}\left(0 - \frac{1}{4}\right)\left(\frac{1}{2}\right) $$

Clearly the ratio $|\tau_{xx}|_{max}/|\tau_{xz}|_{max}$ is (h/L) and we see that τ_{xx} is the important stress for long thin beams. Hence considering τ_{xz} small in the development of the theory is consistent with the results stemming from the theory.

The test presented in the example is an indirect one for our purposes. It shows no internal inconsistency. However, we need further evidence that it is correct, since the absence of internal inconsistencies is not sufficient for justification that the result is correct.

For this purpose we now shall compare the results given here for the simply-supported, uniformly loaded beam with the results presented in Problem 1.26 for the same problem as developed from the theory of elasticity. Thus we now rewrite the plane-stress solution of the aforementioned problem as given in Problem 1.26:

$$ \tau_{xx} = -\frac{q_0}{2I}\left[x^2 - \left(\frac{L}{2}\right)^2\right]z + \frac{q_0}{2I}\left(\frac{2}{3}z^3 - \frac{h^2}{10}z\right) $$

$$\tau_{zz} = -\frac{q_0}{2I}\left(\frac{1}{3}z^3 - \frac{h^2}{4}z + \frac{h^3}{12}\right)$$

$$\tau_{xz} = -\frac{q_0}{2I}\left(\frac{h^2}{4} - z^2\right)x \tag{4.26}$$

Introducing the dimensionless variables ξ and η we get for the above equations:

$$\tau_{xx} = -\frac{q_0 L^2 h}{2I}\left[\left(\xi^2 - \frac{1}{4}\right)\eta + \left(\frac{h}{L}\right)^2\left(\frac{2}{3}\eta^3 - \frac{1}{10}\eta\right)\right] \tag{a}$$

$$\tau_{zz} = -\frac{q_0 L^2 h}{2I}\left(\frac{h}{L}\right)^2\left(\frac{1}{3}\eta^3 - \frac{1}{4}\eta + \frac{1}{12}\right) \tag{b}$$

$$\tau_{xz} = -\frac{q_0 L^2 h}{2I}\left(\frac{h}{L}\right)\left(\frac{1}{4} - \eta^2\right)\xi \tag{c} \tag{4.27}$$

Now considering τ_{xx} we see that it consists of the stress given by the technical theory (see Eq. (4.25)) plus a corrective term containing the multiplicative parameter $(h/L)^2$. Thus the technical theory gives good results for τ_{xx} when compared to the exact theory for long slender beams where $(h/L)^2$ is very small.

Again from the results of the plane-stress solution (Problem (1.27)), we have for the deflection of the centerline of the simply-supported, uniformly-loaded beam:

$$w(x,0) = \frac{q_0 L^4}{24EI}\left[\frac{5}{16} - \frac{3}{2}\left(\frac{x}{L}\right)^2 + \left(\frac{x}{L}\right)^4\right] + \frac{q_0 L^4}{24EI}\left(\frac{h}{L}\right)^2\left[\frac{12}{5} + \frac{3v}{2}\right]\left[\frac{1}{4} - \left(\frac{x}{L}\right)^2\right] \tag{4.28}$$

We see that Eq. (4.28) contains the elementary solutions (see Eq. (4.22)) plus a "corrective" term multiplied by $(h/L)^2$. Again we see that for long slender beams the results of elasticity theory reduce to those of elementary beam theory.

Note from Eq. (4.27) that for long slender beams the stress τ_{zz}, because of the presence of term $(h/L)^2$, can be expected to be considerably smaller than τ_{xz} which has in its expression the term (h/L). In the following section, in which we present a more accurate theory, we accordingly take into account shear effects while still setting $\tau_{zz} = 0$.

We have thus demonstrated via this problem both the inner consistency of beam theory as well as the correlation of certain of its results with appropriate limiting cases of the elasticity solution.

4.5 TIMOSHENKO BEAM THEORY

The technical theory of beams presented in the previous section does not include the effects of shear deformation. For short stubby beams this contribution clearly cannot

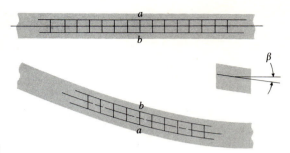

FIGURE 4.6

be neglected, and for this reason we now present the Timoshenko theory of beams as a means of accounting for the effects of shear in a simple manner.

Let us accordingly consider shear deformation *alone*. We will assume that line elements such as *ab* in Fig. 4.6 *normal* to the centerline of the beam in the undeformed state move only in the vertical direction and also *remain vertical* during deformation. Line elements *tangent* to the centerline meanwhile undergo a rotation $\beta(x)$ as has been shown in an exaggerated manner in the diagram. ($\beta(x)$ accordingly gives the shear angle γ_{xz} at points along the centerline.) The total slope dw/dx of the centerline stemming respectively from shear deformation and bending deformation can then be given as the sum of two parts in the following way:

$$\frac{dw}{dx} = \psi(x) + \beta(x) \qquad (4.29)$$

where $\psi(x)$ is the rotation of line elements along the centerline due to bending only. We now make a further assumption that will easily be recognized as being incorrect but which we shall later adjust. We assume that the shear strain is the same at *all* points over a given cross section of the beam. That is, the angle $\beta(x)$, used heretofore for rotation of elements along the centerline, is considered to measure the shear angle at all points in the cross section of the beam at position x. This has been shown in Fig. 4.7. Such an assumed action means that the shear stress is uniform through the thickness—a result that is not possible when only normal loads act on the upper surface and no traction whatever exists on the lower surface.[1] Such an assumed behavior nevertheless will greatly facilitate computations, and the error incurred can and will be reasonably taken into account later. The displacement field can now be considered the superposition of the bending action set forth in Sec. 4.2 and the aforestated shearing action. Clearly the axial displacement u_1 is then due only to

---

[1] The shear stress must then be zero at the upper and lower boundary surfaces.

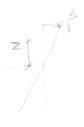

FIGURE 4.7

bending action. Employing Eq. (4.29) we can thus state:

$$u_1(x,y,z) = -z\psi(x) = -z\left[\frac{dw}{dx} - \beta(x)\right]$$

$$u_2(x,y,z) = 0$$

$$u_3(x,y,z) = w(x) \tag{4.30}$$

It then follows from strain-displacement relations that:

$$\epsilon_{xx} = -z\frac{d\psi}{dx} \quad (a)$$

$$\epsilon_{xz} = \tfrac{1}{2}\beta(x) \quad (b) \quad (4.31)$$

Now making use of the approximation $\tau_{xx} = E\epsilon_{xx}$ (i.e., using $v = 0$) and using Eq. (4.31) we find, employing Fig. 4.1:

$$M = \int_{-h/2}^{h/2} \tau_{xx} z b \, dz = -EI\frac{d\psi}{dx} \tag{a}$$

$$V = \int_{-h/2}^{h/2} \tau_{xz} b \, dz = \tau_{xz} \int_{-h/2}^{h/2} b \, dz = \tau_{xz} A = GA\beta \quad (b) \quad (4.32)$$

The assumption of a uniform shear stress through the thickness leads, from the above formulation, to the following result for such a stress:

$$\tau_{xz} = \frac{V}{A} = G\beta(x) \tag{4.33}$$

Actually a more proper statement for τ_{xz} would appear as follows:

$$\tau_{xz} = G\beta(x,z) \tag{4.34}$$

where a variation with z is taken into account. However, a procedure such as one taken in Eq. (4.34), would complicate the problem and remove the simplicity inherent in a one-dimensional beam approach. In order to recognize the non-uniform shear stress distribution at a section while still retaining the one-dimensional approach, we modify Eq. (4.34) by introducing a factor k as follows:

$$\tau_{xz} = kG\beta(x) \qquad (4.35)$$

Hence we shall give Eq. (4.32(b)) as follows:

$$V(x) = kGA\beta(x) \qquad (4.36)$$

There are virtually as many definitions[1] of k as there are published papers on the Timoshenko beam. A recent one by G. R. Cowper[2] gives an interesting review (as well as a new definition). Suffice it here to mention that it is a function of the cross section and, depending on the interpretation, may also be a function of the Poisson ratio v. In Table 4.1, we have shown a listing of k taken from Cowper's paper.

Table 4.1

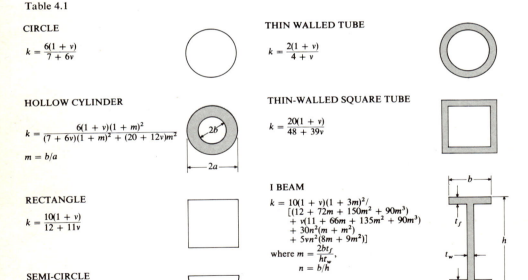

CIRCLE

$$k = \frac{6(1 + v)}{7 + 6v}$$

THIN WALLED TUBE

$$k = \frac{2(1 + v)}{4 + v}$$

HOLLOW CYLINDER

$$k = \frac{6(1 + v)(1 + m)^2}{(7 + 6v)(1 + m)^2 + (20 + 12v)m^2}$$

$$m = b/a$$

THIN-WALLED SQUARE TUBE

$$k = \frac{20(1 + v)}{48 + 39v}$$

RECTANGLE

$$k = \frac{10(1 + v)}{12 + 11v}$$

I BEAM

$$k = 10(1 + v)(1 + 3m)^2 / \\ [(12 + 72m + 150m^2 + 90m^3) \\ + v(11 + 66m + 135m^2 + 90m^3) \\ + 30n^2(m + m^2) \\ + 5vn^2(8m + 9m^2)]$$

$$\text{where } m = \frac{2bt_f}{ht_w},$$

$$n = b/h$$

SEMI-CIRCLE

$$k = \frac{1 + v}{1.305 + 1.273v}$$

---

[1] Among the many definitions are those stemming from relating the maximum shear stress through the thickness, as developed from a more exact solution, to the approximation (4.35), or those from matching of certain wave speeds from the dynamics of a Timoshenko beam to more accurate results of elasticity theory.

[2] *Journal of Applied Mechanics*, June 1966, p. 335.

We now find the total potential energy for the assumed deformation field of the Timoshenko beam. From the work of the previous section, it is clear that:

$$\pi = \frac{1}{2} \int_0^L \int_{-h/2}^{h/2} \tau_{xx} \epsilon_{xx} \, dx \, b \, dz + \int_0^L \int_{-h/2}^{h/2} \tau_{xz} \epsilon_{xz} \, dx \, b \, dz - \int_0^L qw \, dx$$

Using $\tau_{xx} = E\epsilon_{xx}$ and using Eq. (4.31(a)) in the first integral while employing $\tau_{xz} = kG\beta$ and Eq. (4.31(b)) in the second integral we get:

$$\pi = \frac{EI}{2} \int_0^L \left(\frac{d\psi}{dx}\right)^2 dx + \frac{GA}{2} \int_0^L k\beta^2 \, dx - \int_0^L q \, w \, dx$$

$$= \frac{EI}{2} \int_0^L \left(\frac{d\psi}{dx}\right)^2 dx + \frac{kGA}{2} \int_0^L \left(\frac{dw}{dx} - \psi\right)^2 dx - \int_0^L q \, w \, dx$$

wherein we have employed Eq. (4.29) to eliminate β in the second integral. Collecting terms, we have for the total potential energy:

$$\pi = \int_0^L \left[\frac{EI}{2}\left(\frac{d\psi}{dx}\right)^2 + \frac{kGA}{2}\left(\frac{dw}{dx} - \psi\right)^2 - qw \right] dx \qquad (4.37)$$

Thus the functional has two functions, ψ and w. The Euler–Lagrange equations for this case are then easily deduced from Eq. (2.64).[1] Thus we have

$$\frac{d}{dx}\left[EI\frac{d\psi}{dx}\right] + kGA\left(\frac{dw}{dx} - \psi\right) = 0 \qquad (a)$$

where for constant E and I we get:

$$EI\left[\frac{d^2\psi}{dx^2}\right] + kGA\left(\frac{dw}{dx} - \psi\right) = 0 \qquad (b) \quad (4.38)$$

Also:

$$\frac{d}{dx}\left[kGA\left(\frac{dw}{dx} - \psi\right)\right] + q = 0 \qquad (a)$$

where for constant G and A we get:

$$kGA\left(\frac{d\psi}{dx} - \frac{d^2w}{dx^2}\right) = q \qquad (b) \quad (4.39)$$

---

[1] As an exercise you will be asked to verify the following results by carrying out the extremization process.

Also from Eq. (2.63) adjusted for several dependent variables it becomes clear that for the end conditions at $x = 0$ and $x = L$:

(a) Either ψ is specified or $EI\dfrac{d\psi}{dx} = 0$

(b) Either w is specified or $kGA\left(\dfrac{dw}{dx} - \psi\right) = 0$ $(4.40(a))$

Considering Eqs. $(4.32(a))$ and $(4.32(b))$, with β replaced in the latter by $(dw/dx - \psi)$ according to Eq. (4.29) and using the factor k, we may give the above results as follows:

(a) Either ψ is specified or $M = 0$ at $x = 0, L$.

(b) Either w is specified or $V = 0$ at $x = 0, L$. $(4.40(b))$

Similarly making the same substitutions, it is seen that the Euler–Lagrange equations reduce to relations that are no doubt familiar to you from earlier course work in strength of materials. Thus:

$$\frac{dM}{dx} = V \quad (a)$$

$$\frac{dV}{dx} = -q \quad (b) \quad (4.41)$$

We may *uncouple* the Euler–Lagrange equations by differentiating Eq. $(4.38(b))$ with respect to x and then substituting from Eq. $(4.39(b))$. We get for this calculation:

$$EI\frac{d^3\psi}{dx^3} = q \quad (4.42)$$

To get the uncoupled differential equation for w first differentiate Eq. $(4.39(b))$ twice with respect to x and solve for ψ''' to get:

$$\psi''' = \frac{q''}{kGA} + w^{IV} \quad (4.43)$$

Now differentiate Eq. $(4.38(b))$ with respect to x and replace ψ''' using the above equation. Also replace the expression $kGA(\psi' - w'')$ by using Eq. $(4.39(b))$. We then get, on rearranging terms, the desired equation for w:

$$EIw^{IV} = q - \frac{EI}{kGA}q'' \quad (4.44)$$

Notice the resemblance of the above equation to the corresponding deflection equation from the technical theory of beams.

To compare the Timoshenko beam theory with the technical theory and the plane stress theory, we examine again the simply-supported beam carrying a uniform load q_0. The Eqs. (4.42) and (4.44) become for this case:[1]

$$EI\frac{d^3\psi}{dx^3} = q_0 \quad (a)$$

$$EI\frac{d^4w}{dx^4} = q_0 \quad (b) \quad (4.45)$$

We require that $w = 0$ and that $M = 0$ at the ends. The latter condition in turn means (see Eq. (4.32(a)) that $d\psi/dx = 0$ at the ends. We shall proceed by carrying out the quadratures for ψ as follows:

$$\psi = \frac{1}{EI}\left[\frac{q_0 x^3}{6} + \frac{C_1 x^2}{2} + C_2 x + C_3\right] \quad (4.46)$$

If we place the origin of the reference at the center of the beam, it should be clear from the symmetry of the problem that ψ is an *odd* function and so we set $C_1 = C_3 = 0$. Furthermore the end condition $d\psi/dx = 0$ at $x = \pm L/2$ gives us C_2 as

$$0 = \frac{1}{EI}\left[\frac{q_0 x^2}{2} + C_2\right]_{x = \pm L/2}$$

therefore

$$C_2 = -\frac{q_0 L^2}{8}$$

We then have for ψ:

$$\psi = \frac{1}{EI}\left(\frac{q_0 x^3}{6} - \frac{q_0 L^2}{8}x\right)$$

To get w most simply we will substitute the above result in Eq. (4.38(b)) as follows:

$$-qx + kGA\left[\frac{1}{EI}\left(\frac{q_0 x^3}{6} - \frac{q_0 L^2}{8}x\right) - \frac{dw}{dx}\right] = 0$$

Rearranging we get

$$\frac{dw}{dx} = -\frac{q_0 x}{kGA} + \frac{1}{EI}\left(\frac{q_0 x^3}{6} - \frac{q_0 L^2}{8}x\right)$$

Integrating we have:

$$w = -\frac{q_0 x^2}{2kGA} + \frac{1}{EI}\left[\frac{q_0 x^4}{24} - \frac{q_0 L^2 x^2}{16}\right] + C$$

Setting $w = 0$ at $x = \pm L/2$ we get:

$$-\frac{q_0 L^2}{8kGA} + \frac{1}{EI}\left[\frac{q_0 L^4}{384} - \frac{q_0 L^4}{64}\right] + C = 0$$

---

[1] Note before proceeding further that solutions of the uncoupled equations must be checked in the coupled form since the uncoupled equations form a higher-order set of equations having thus as an outcome, extraneous solutions.

therefore

$$C = \frac{q_0 L^2}{8kGA} + \frac{5q_0 L^4}{384EI}$$

We thus have for w:

$$w = \frac{q_0}{2kGA}\left[\frac{L^2}{4} - x^2\right] + \frac{q_0}{EI}\left[\frac{x^4}{24} - \frac{L^2 x^2}{16} + \frac{5L^4}{384}\right]$$

Rearranging we have:

$$w = \frac{q_0 L^4}{24EI}\left[\left(\frac{x}{L}\right)^4 - \frac{3}{2}\left(\frac{x}{L}\right)^2 + \frac{5}{16}\right] - \frac{q_0 L^2}{2k\left[\frac{E}{2(1+v)}\right]bh}\left[\left(\frac{x}{L}\right)^2 - \frac{1}{4}\right]$$

where we have replaced G by $E/(2(1+v))$ and A by bh in the last expression. Finally we give w as follows, on introducing $I = \frac{1}{12}bh^3$ into the last expression:

$$w = \frac{q_0 L^4}{24EI}\left\{\left[\left(\frac{x}{L}\right)^4 - \frac{3}{2}\left(\frac{x}{L}\right)^2 + \frac{5}{16}\right] - \left(\frac{h}{L}\right)^2\frac{2(1+v)}{k}\left[\left(\frac{x}{L}\right)^2 - \frac{1}{4}\right]\right\} \tag{4.47}$$

Comparing this result with that of the technical theory of beams (Eq. (4.22)), we see that the first part above is identical to the latter. Clearly the second part of the above result corresponds to the contribution of shear deformation. Now compare the above result with that of plane stress (Eq. (4.28)). We see the first expression in the bracket is identical to the dominant expression in the plane stress solution, while the second expression above (shear deformation) is of the same order of magnitude as the remaining part of the plane stress solution. In fact, if we choose Cowper's value of k for the rectangular cross section from Table 4.1, namely,

$$k = \frac{10(1+v)}{12+11v}$$

we find the ratio of these second-order contributions to be:

$$\frac{w_{\text{shear-Timo}}}{w_{\text{shear-plane stress}}} = \frac{24 + 22v}{24 + 15v}$$

This number ranges from 1.00 to 1.11 as the Poisson ratio ranges from 0 to $\frac{1}{2}$.

Thus we have a rather simple theory that includes shear deformation contribution with what would appear to be reasonably good accuracy. And so we may now be able to ascertain the stress τ_{xx} and the deflection curve $w(x)$ for short as well as long beams.

You will note that we have used the variational approach up to this time to formulate from the total potential energy functional the differential equations of interest and the natural boundary conditions. Now we will work directly with the functional in order to present approximate solutions to the technical beam equations.

4.6 COMMENTS ON THE RITZ METHOD

In Chapter 3 we pointed out that a limiting sequence of linearly independent functions u_n which is complete in energy and which minimizes the quadratic functional I

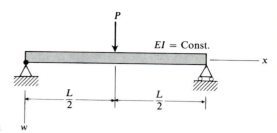

FIGURE 4.8

converges in energy to a function u_0, the solution to the differential equation associated with I. However, we indicated that by a judicious choice of functions, only a very few of them could yield, via the Ritz method, a very good approximation to the solution to the differential equation. In this section we shall demonstrate by example how this may be accomplished.

Accordingly we consider a simply-supported beam with a concentrated load P as has been shown in Fig. 4.8. The solution of the differential equation derived from the technical theory of beams (Eq. (4.10)) is easily found in this case to be, for $0 \le x \le L/2$:

$$w_{\text{ex}} = \frac{PL^3}{48EI}\left[3\left(\frac{x}{L}\right) - 4\left(\frac{x}{L}\right)^3\right] \qquad (4.48)$$

where the subscript "ex" indicates that the solution is exact for the given differential equation. And the total potential energy for the problem is:

$$\pi = \frac{EI}{2}\int_0^L \left(\frac{d^2 w}{dx^2}\right)^2 dx - Pw(x)|_{x=L/2} \qquad (4.49)$$

Now we will employ one function for the Ritz approach. Consider the following:

$$w_1 = A \sin \frac{\pi x}{L} \qquad (4.50)$$

where A is an undetermined constant. Notice from Eq. (4.14) that the function $\sin \pi x/L$ satisfies a kinematic boundary condition ($w = 0$ at the ends) and a natural boundary condition ($w'' = 0$ at the ends) and thus fully satisfies the requirements imposed on the extremizing functions by the variational process. Furthermore the assumed deflection curve has a shape that is "reasonable" in that it is symmetric about the center of the beam. Now substitute w_1 into Eq. (4.49). We get:

$$\pi = \frac{EI}{2}A^2\left(\frac{\pi}{L}\right)^4\frac{L}{2} - PA \qquad (4.51)$$

To establish the proper value of A for the Ritz method we differentiate with respect to A and set the result equal to zero. Thus:

$$\frac{EIL}{2}A\left(\frac{\pi}{L}\right)^4 - P = 0$$

therefore

$$A = \frac{2PL^3}{\pi^4 EI}$$

We then have for the corresponding approximate function w_{app}:

$$w_{app} = \frac{2PL^3}{\pi^4 EI}\sin\frac{\pi x}{L} \tag{4.52}$$

To compare the results from Eqs. (4.52) and (4.48) we examine the deflection at points $x = L/4$ and $x = L/2$ as has been given below:

	$x = L/4$	$x = L/2$
$\dfrac{EIw_{ex}}{PL^3}$	0.01432	0.02083
$\dfrac{EIw_{app}}{PL^3}$	0.01452	0.02053

We see that the results are very close indeed.

Now consider as the function w_1 the following form:

$$w_1 = B(x^3 - L^2 x) \tag{4.53}$$

Note that the kinematic boundary condition $w = 0$ is satisfied at both ends, but only at the left end is the natural boundary condition $w'' = 0$ satisfied. A full set of boundary conditions as required by the variational process has not accordingly been satisfied (see Eq. (4.14)). Then by substituting the above result into the total potential energy and extremizing the result with respect to B we get:

$$w_{app} = \frac{PL^3}{32EI}\left[\frac{x}{L} - \left(\frac{x}{L}\right)^3\right] \tag{4.54}$$

It is easy to see from the data below that the above result does not compare well with the exact solution.

	$x = L/4$	$x = L/2$
$\dfrac{EIw_{ex}}{PL^3}$	0.01432	0.02083
$\dfrac{EIw_{app}}{PL^3}$	0.00732	0.01172

We shall ascribe the reason for this poor result to the fact that one of the needed boundary conditions is not satisfied and the fact that the function $x^3 - L^2 x$ is not symmetric about $x = L/2$, a condition one would expect as a result of the loading and geometry of the problem. We will remedy these two conditions now and we will see a remarkable improvement in the accuracy. To do this, we express w_1 as follows:[1]

$$w_1 = B(3L^2 x - 4x^3) \qquad 0 \le x \le L/2 \qquad (4.55)$$

with the understanding that the function for $L/2 \le x \le L$ is given so as to form an even function w_1 over the interval. Now, substituting into the expression for the total potential energy and extremizing with respect to B, we get the following result:

$$w_1 = \frac{PL^3}{48EI}\left[3\left(\frac{x}{L}\right) - 4\left(\frac{x}{L}\right)^3\right] \qquad (4.56)$$

Indeed by comparing the above result with Eq. (4.48) we see that we have arrived at the exact solution as a result of the changes made above.

We next conclude from these calculations that for good results with few functions, it is important to satisfy a *full set* of boundary conditions set forth by the extremization process, and in addition take into account consideration of symmetry in choosing the functions.

To further bring out the importance of satisfying a full set of boundary conditions, we use next a third function $[1 - \cos(2\pi x/L)]$ so that:

$$w_1 = A\left(1 - \cos\frac{2\pi x}{L}\right) \qquad (4.57)$$

Inspection of this function shows clearly that at the ends the kinematic boundary condition $w = 0$ is the only one satisfied. We have accordingly not satisfied a complete set of conditions. Now substituting into the total potential energy and extremizing with respect to A we get:

$$w_{\text{app}} = \frac{PL^3}{4\pi^4 EI}\left(1 - \cos\frac{2\pi x}{L}\right) \qquad (4.58)$$

The deflection at $x = L/4$ and $x = L/2$ for the above approximation and for the exact solution of the differential equation are given below:

	$x = L/4$	$x = L/2$	
$EI\dfrac{w_{\text{ex}}}{PL^3}$	0.01432	0.02083	(a)
$EI\dfrac{w_{\text{app}}}{PL^3}$	0.00257	0.00513	(b) (4.59)

[1] The coefficients 3 and 4 have been chosen so as to give w_1 a zero slope at $x = L/2$ as is required of a beam with a continuous slope and symmetry about $x = L/2$.

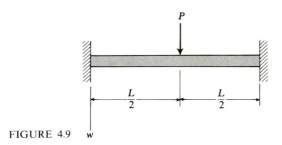

FIGURE 4.9

There is obviously very poor correlation here. If now one were to have used Eq. (4.57) for the *fixed-ended beam* in Fig. 4.9, then we see that the function satisfies a *complete set* of boundary conditions for this problem (two kinematic boundary conditions in this case) since $w_1 = w'_1 = 0$ at the ends. The results for an exact solution of the fixed-ended beam problem are given below for comparison:

	$x = L/4$	$x = L/2$
$EI \dfrac{w_{\text{ex}}}{PL^3}$	0.00261	0.00522

$$(4.60)$$

We see by comparing Eq. (4.59(b)) and the above result that, with a complete set of boundary conditions satisfied, we again have excellent correlation between the results from the exact solution of the differential equation and the approximate solution.

4.7   THE RITZ METHOD FOR A SERIES SOLUTION

In previous sections we investigated how one might best proceed via the Ritz method to produce good approximate solutions by employing one or several coordinate functions. Now we consider the Ritz method once again for the purpose of generating infinite series solutions to certain beam problems. We shall accordingly now consider beams supported by a continuous elastic foundation.

We examine the case of a beam simply-supported at the ends and having a uniform elastic foundation between the ends. This has been shown in Fig. 4.10 where you will note a uniform load q_0 has been indicated as acting on the beam. The elastic foundation is indicated graphically as a series of springs. The force developed by the elastic foundation per unit length of the foundation and per unit deflection at a point is called the *foundation modulus* and is denoted as k. The force distribution on the beam from the elastic foundation is then given as kw. The potential energy of the

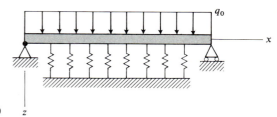

FIGURE 4.10

foundation for a given deflection w is computed as follows:

$$V_{\text{found}} = \int_0^L \left[\int_0^w kw \, dw \right] dx = \int_0^L \frac{1}{2} kw^2 \, dx \qquad (4.61)$$

We can then give the total potential energy for bending of the beam, using results from previous sections:

$$
\pi = \frac{EI}{2} \int_0^L [w''(x)]^2 \, dx + \frac{1}{2} k \int_0^L w^2 \, dx - q_0 \int_0^L w \, dx
$$

$$
= \int_0^L \left[\frac{EI}{2} (w'')^2 + \frac{1}{2} kw^2 - q_0 w \right] dx \qquad (4.62)
$$

It will be left to the reader to show (see Eq. (2.62)) that the following equation is the proper Euler–Lagrange equation for this functional:

$$EI \frac{d^4 w}{dx^4} + kw = q_0 \qquad (4.63)$$

We can get an exact solution to this differential equation satisfying a kinematic boundary condition at each end ($w = 0$) and a natural boundary condition at each end ($w'' = 0$). Using as trial solutions e^{px}, you may readily deduce that the complementary solution is:

$$
w_c = A \sinh \lambda x \sin \lambda x + B \cosh \lambda x \sin \lambda x +
$$
$$
C \sinh \lambda x \cos \lambda x + D \cosh \lambda x \cos \lambda x
$$

where

$$\lambda = \sqrt[4]{\frac{k}{4EI}}$$

and is the reciprocal of the so-called characteristic length. The particular solution is, by inspection,

$$w_p = \frac{q_0}{k}$$

Submitting the combined solution to the aforementioned boundary conditions leads us to the following result:

$$w = \frac{q_0}{k}\left[1 - \frac{\cosh \lambda x \cos \lambda(x - L) + \cosh \lambda(x - L) \cos \lambda x}{\cosh \lambda L + \cos \lambda L}\right] \quad (4.64)$$

The exact bending moment is then easily computed as (see Eq. (4.16)):

$$M = -EIw'' = -\frac{q_0}{2\lambda^2} \frac{\sinh \lambda x \sin \lambda(x - L) + \sinh \lambda(x - L) \sin \lambda x}{\cosh \lambda L + \cos \lambda L} \quad (4.65)$$

By way of comparison we now use the Ritz method for the functional given by Eq. (4.62). We employ a single function which satisfies a complete set of boundary conditions:

$$w_1 = a_1 \sin \frac{\pi x}{L}$$

A straightforward extremization of the functional with respect to a_1 then leads to the following approximation:

$$w_{\text{app}} = \frac{4q_0 L^4/(\pi^5 EI)}{1 + kL^4/(\pi^4 EI)} \sin \frac{\pi x}{L}$$

$$= \left\{\frac{16}{\pi^5}\left(\frac{q_0}{k}\right)(L\lambda)^4 \middle/ \left[1 + \frac{4}{\pi^4}(\lambda L)^4\right]\right\} \sin(\pi x/L) \quad (4.66)$$

with the corresponding moment distribution:

$$M_{\text{app}} = \frac{4q_0 L^2/\pi^3}{1 + kL^4/(\pi^4 EI)} \sin \frac{\pi x}{L}$$

$$= \left\{\frac{4q_0 L^2}{\pi^3} \middle/ \left[1 + \frac{4}{\pi^4}(\lambda L)^4\right]\right\} \sin \frac{\pi x}{L} \quad (4.67)$$

Now let us increase the number of terms in the series as follows:

$$w_p = \sum_{i=1}^{p} a_n \sin \frac{n\pi x}{L} \quad (4.68)$$

The functions $\phi_n = \sin(n\pi x/L)$ are eigenfunctions[1] of the operator $[d^4/dx^4 + k/EI]$ for the boundary conditions of the problem and are linearly independent. As such, one can show that they are complete in energy. We know from our discussion in Chap. 2 that the partial sums w_p form a minimizing sequence for this functional when the coefficients are chosen so as to extremize the value of the functional with respect to the a's (that is, the a's are the Ritz coefficients). Hence, with the a's so

---

[1] We shall consider eigenfunctions in Chap. 7.

chosen we can say that:

$$\lim_{n \to \infty} w_n = \sum_{n=1}^{\infty} (a_n)_{\text{Ritz}} \sin \frac{n\pi x}{L} \equiv \text{sol. to diff. eq.} \qquad (4.69)$$

To get the Ritz coefficients first substitute Eq. (4.69) into Eq. (4.62). We then get the following result:

$$\pi = \sum_{n=1,3,5\ldots}^{\infty} \left\{ a_n{}^2 \left[\left(\frac{EI}{2}\right)\left(\frac{L}{2}\right)\left(\frac{n\pi}{L}\right)^4 + \frac{kL}{4} \right] - a_n \frac{2q_0 L}{n\pi} \right\}$$

$$+ \sum_{n=2,4,\ldots}^{\infty} \left\{ a_n{}^2 \left[\left(\frac{EI}{2}\right)\left(\frac{L}{2}\right)\left(\frac{n\pi}{L}\right)^4 + \frac{kL}{4} \right] \right\}$$

wherein we have used the orthogonal properties of the eigenfunctions in carrying out the integrations. Noting that

$$\frac{\partial}{\partial a_n}(\pi) = 0$$

we get the Ritz coefficients to be for *odd* values of n:

$$a_n = \frac{4q_0 L^4/(n^5\pi^5 EI)}{1 + kL^4/(n^4\pi^4 EI)}$$

All other a_n's are zero. Accordingly the correct solution to the differential equation is given as:

$$w = \sum_{n=1,3,5,\ldots}^{\infty} \frac{4q_0 L^4/(n^5\pi^5 EI)}{1 + kL^4/(n^4\pi^4 EI)} \sin \frac{n\pi x}{L} \qquad (4.70)$$

Let us suppose that we can differentiate the above series term by term. We get:

$$M = \sum_{n=1,3,5,\ldots}^{\infty} \frac{4q_0 L^2/(n\pi)^3}{1 + kL^4/(n^4\pi^4 EI)} \sin \frac{n\pi x}{L} \qquad (4.71)$$

It is immediately clear from inspection of the coefficients that both series converge very rapidly. We present a table of results showing the exact results for w and M at center of beam along with approximate results, for $n = 1$ and $n = 1, 3$ etc. and for a given value of $kL^4/EI = 4$.

	Exact	$n = 1$	$n = 1, 3$	$n = 1, 3, 5$	$n = 1, 3, 5, 7$
$\dfrac{EIw(L/2)}{q_0 L^4}$	0.012505	0.12556	0.012502	0.012506	0.012505
$\dfrac{M(L/2)}{q_0 L^2}$	0.119914	0.123918	0.119142	0.120174	0.119798

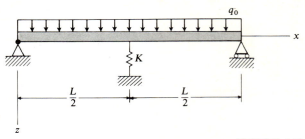

FIGURE 4.11

You will note in the previous calculations that we were able to compute the Ritz coefficients *individually* in a direct manner. That is, the algebraic equations for a_n were independent of each other. We next consider a problem where this is not the case. We have shown in Fig. 4.11 a simply-supported beam having an elastic support at the center. This support is represented by a linear spring having a spring constant K. The total potential energy for this system, assuming that the spring support is undeformed when there is no load on the beam, is given as follows:

$$\pi = \frac{EI}{2} \int_0^L [w''(x)]^2 \, dx + \tfrac{1}{2}K[(w)|_{x=L/2}]^2 - q_0 \int_0^L w(x) \, dx$$

We will now demonstrate that the spring introduces into calculations for the Ritz method a *coupling* between the coefficients. Thus suppose we assume for w the following:

$$w = a_1 \sin \frac{\pi x}{L} + a_3 \sin \frac{3\pi x}{L}$$

Substitution of the above into the expression for the total potential energy then gives us:

$$\pi = \frac{EIL}{4}\left[a_1{}^2 \left(\frac{\pi}{L}\right)^4 + a_3{}^2 \left(\frac{3\pi}{L}\right)^4\right] + \frac{1}{2}K(a_1 - a_3)^2 + \frac{2q_0L}{\pi}\left(a_1 + \frac{a_3}{3}\right)$$

Extremizing the total potential energy with respect to a_1 and a_3 then gives us the following pair of equations:

$$a_1\left(1 + \frac{2KL^3}{EI\pi^4}\right) - \frac{2KL^3}{EI\pi^4}a_3 = \frac{4q_0L^4}{EI\pi^5}$$

$$-\frac{2KL^3}{EI\pi^4}a_1 + \left(3^4 + \frac{2KL^3}{EI\pi^4}\right)a_3 = -\frac{4q_0L^4}{3EI\pi^5}$$

We see that for the determination of a_1 and a_3, the equations are now *simultaneous* equations. An interesting question that we shall now pursue centers around the calculation of the a's in the above problem when an infinite series is used to describe the deflection. That is, let

$$w = \sum_{n=1,3,5,\ldots}^{\infty} a_n \sin \frac{n\pi x}{L} \qquad (4.72)$$

where we use odd values of "n" to take into account the symmetry of the problem about the center of the beam.[1] The total potential energy then becomes:

$$\pi = \sum_{n=1,3,5,\ldots}^{\infty}\left[a_n{}^2 \frac{EIL}{4}\left(\frac{n\pi}{L}\right)^4 - a_n \frac{2q_0L}{n\pi}\right] + \frac{K}{2}\left[\sum_{m=1,3,5,\ldots}^{\infty} a_n \sin \frac{n\pi}{2}\right]^2 \qquad (4.73)$$

---

[1] Note that origin of coordinates is at left end of beam (see Fig. 4.11).

Now extremizing with respect to the coefficients we get for a_n the following result:

$$a_n = \frac{4q_0L^4}{\pi^5EI}\frac{1}{n^5} - \frac{2KL^3}{\pi^4EI}\frac{1}{n^4}\sin\frac{n\pi}{2}\left(\sum_{s=1,3,5,...}^{\infty} a_s \sin\frac{s\pi}{2}\right) \quad (4.74)$$

We see here that the value of a_n depends on the values of all the other a's. That is, the above represents one of an infinite set of linear simultaneous algebraic equations. To solve this set we introduce the following notation:

$$S = \sum_{s=1,3,5,...}^{\infty} a_s \sin\frac{s\pi}{2} \quad (4.75)$$

Now multiply Eq. (4.74) by $\sin(n\pi/2)$ and sum to infinity over odd n. We then have, using the above notation,

$$S = \frac{4q_0L^4}{\pi^5EI}\sum_{n=1,3,5,...}^{\infty}\frac{\sin(n\pi/2)}{n^5} - \frac{2KL^3}{\pi^4EI}\left(\sum_{n=1,3,5,...}^{\infty}\frac{1}{n^4}\right)S \quad (4.76)$$

wherein we have used the fact that, for odd "n", $\sin^2(n\pi/2) = 1$. The infinite sums in the above equation are readily obtained from tables. Thus:

$$\sum_{n=1,3,5,...}^{\infty}\frac{\sin(n\pi/2)}{n^5} = \frac{5\pi^5}{(16)(19)}$$

$$\sum_{n=1,3,5,...}^{\infty}\frac{1}{n^4} = \frac{\pi^4}{96}$$

Solving for S in Eq. (4.76), using the preceding sums, we get:

$$S = \frac{5q_0L^4}{384EI}\left/\left(1 + \frac{KL^3}{48EI}\right)\right.$$

We can now go back to Eq. (4.74) and, using the above result to replace the infinite sum on the right side of the equation, we can solve for a_n. The solution for w can then be given as:

$$w = \frac{4q_0L^4}{\pi^5EI}\sum_{n=1,3,5,...}^{\infty}\frac{\sin(n\pi x/L)}{n^5} - \frac{2KSL^3}{\pi^4EI}\sum_{n=1,3,5,...}^{\infty}\left(\frac{\sin(n\pi/2)\sin(n\pi x/L)}{n^4}\right)$$

4.8   USE OF THE REISSNER PRINCIPLE

We have been considering the Ritz method for estimating the dependent variable w of the differential equation. Let us go back to the simply-supported beam with the single point load P with the use of very few functions. We shall this time be concerned with finding the bending moment M at any section. Since from the flexure formula the stress at any point is proportional to the bending moment, we will in effect also be making inquiries as to the bending stress. The bending moment M was shown (Eq. 4.16(b)) to be equal to $-EIw''$ and so we have, using the approximate solution found in Sec. 4.6 for a single coordinate function, $\sin(\pi x/L)$ (see Eq. (4.52))

$$M_{\text{app}} = \frac{2PL}{\pi^2}\sin\frac{\pi x}{L} \quad (4.77)$$

The exact result from rigid-body considerations is:

$$M_{\text{ex}} = \frac{Px}{2} \qquad (4.78)$$

In the following table we compare these results at positions $x = L/4$ and $x = L/2$:

	$x = L/4$	$x = L/2$
$\dfrac{M_{\text{ex}}}{PL}$	0.125	0.250
$\dfrac{M_{\text{app}}}{PL}$	0.143	0.203

Note that unlike the earlier comparison of the deflection that showed excellent agreement, we now get poor agreement for moments, and hence stresses. The reason for this poor correlation is that in order to reach the approximate stresses we carry out a two-fold differentiation on an approximate deflection function. The differentiation process is not a smoothing operation as is integration, and small differences in the shape of the curve w_{app} as compared to w_{ex} result in large differences in the values of w''_{app} as compared to w''_{ex}. It is clear that the Ritz method with few functions generally cannot be expected to give good results for stress.

We may ameliorate this difficulty by employing the Reissner functional wherein we have several dependent variables to adjust independently of each other. We shall show for this problem that we can then pick moments and displacements independently. Accordingly we now restate the Reissner functional as given earlier

$$I_R = \iiint_V [\tau_{ij}\epsilon_{ij} - U_0^*(\tau_{ij})]\, dv - \iiint_V \bar{B}_i u_i\, dv - \iint_{S_1} \bar{T}_i^{(v)} u_i\, ds \qquad (4.79)$$

You will recall that the barred quantities are prescribed. For the one-dimensional technical beam theory that we have presented, with a zero body force distribution, the first integral may be reduced to:

$$\iiint_V [\tau_{ij}\epsilon_{ij} - U_0^*(\tau_{ij})]\, dv = \iiint_V \left[\tau_{xx}\epsilon_{xx} - \frac{1}{2E}\tau_{xx}^2 \right] dv \qquad (4.80)$$

where we take all other $\tau_{ij} = 0$ in the calculation, as explained earlier. The only traction forces are P at the center of the beam and the supporting forces at the ends of the beam. The traction everywhere else is zero. Since u_i in the direction of the supporting forces is zero at the ends, we accordingly get for the third integral in Eq. (4.79) the expression $Pw(x)|_{x=L/2}$.

Consequently the Reissner functional becomes:

$$I_R = \iiint_V \left[\tau_{xx}\epsilon_{xx} - \frac{1}{2}\frac{\tau_{xx}^2}{E} \right] dv - Pw(x)|_{x=L/2} \qquad (4.81)$$

Now we employ for ϵ_{xx} the following relation (see Eq. (4.1) applied to bending)

$$\epsilon_{xx} = -z\frac{d^2w}{dx^2} \qquad (4.82)$$

and for τ_{xx} the well known flexure formula (as developed in Sec. 4.3):

$$\tau_{xx} = \frac{Mz}{I} \qquad (4.83)$$

We then get:

$$
\begin{aligned}
I_R &= -\iiint_V \left\{ \frac{1}{2E}\left(\frac{Mz}{I}\right)^2 + \frac{Mz}{I}\left[z\frac{d^2w}{dx^2}\right] \right\} dv - Pw(x)|_{x=L/2} \\
&= -\iiint_V \left[\frac{M^2}{2EI^2} + \frac{M}{I}\frac{d^2w}{dx^2} \right] z^2 \, dv - Pw(x)|_{x=L/2} \\
&= -2\int_0^{L/2}\int_{-h/2}^{h/2} \left[\frac{M^2}{2EI^2} + \frac{M}{I}w'' \right] z^2 b \, dz \, dx - Pw(x)|_{x=L/2}
\end{aligned}
$$

where b is the width of the beam. Note we have used $2\int_0^{L/2}$ to replace $\int_0^L$ in the last equation because of the symmetry of deformation about position $x = L/2$. Now integrating with respect to z we get:

$$
\begin{aligned}
I_R &= -2\int_0^{L/2} \left(\frac{M^2}{2EI^2} + \frac{M}{I}w'' \right)\frac{bh^3}{12} \, dx - Pw(x)|_{x=L/2} \\
&= -2\int_0^{L/2} \left(\frac{M^2}{2EI} + Mw'' \right) dx - Pw(x)|_{x=L/2} \qquad (4.84)
\end{aligned}
$$

The functional has M and w as the variables. We shall use the Ritz method for extremizing this functional by employing the following functions:

$$w = A\sin\frac{\pi x}{L} \qquad (a)$$

$$M = Bx \quad \left(0 \le x \le \frac{L}{2}\right) \qquad (b) \quad (4.85)$$

Note the proposed equation for M satisfies the end conditions for M (see Eq. (4.21)) and is to be used as a symmetric function. Substituting these functions into the functional then gives us:

$$I_R = 2AB - \frac{B^2 L^3}{24EI} - PA$$

Extremizing I_R with respect to both A and B, we get immediately the results:

$$A = \frac{PL^3}{48EI}$$

$$B = \frac{P}{2}$$

The equations for moment and deflection become, for $0 \leq x \leq L/2$:

$$M = \frac{Px}{2}$$

$$w = \frac{PL^3}{48EI} \sin \frac{\pi x}{L} \tag{4.86}$$

The moment distribution is now exact, while the deflection approximation is now even better than the calculation made at the outset of the previous section wherein the total potential energy was used for the same deflection function.

4.9   ADDITIONAL PROBLEMS IN BENDING OF BEAMS

We have used the total potential energy principle in conjunction with certain assumptions as to the displacement field, to formulate the deflection equations of the technical theory of beams. After some justifications for the theory, we then used the total potential energy in conjunction with the Ritz method to generate approximate solutions to beam problems. In this regard the use of the Reissner functional was also illustrated. We shall now utilize certain of the results from the technical theory of beams to illustrate the determination of deflections of loading points on beams via the second Castigliano theorem. Also we shall determine shear force and bending moment distributions in beams supported in a statically indeterminate manner by using the total complementary energy principle.

Consider first beams supported in a statically determinate manner with zero axial force and with discrete vertical loads P_i. The strain energy of the beam in the absence of axial forces is easily found from Eq. (4.8) by setting $[(u)_1]_s = 0$ and dropping the loading term q. We then get:

$$U = \int_0^L \frac{EI}{2} \left(\frac{d^2 w}{dx^2} \right)^2 dx = \frac{1}{2} \int_0^L \frac{M^2}{EI} dx \tag{4.87}$$

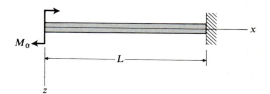

FIGURE 4.12

where we have again used Eq. (4.16(*b*)) to reach the last expression. The complementary strain energy for the beam is also given as above because we have assumed linear elastic behavior. The deflection of a loading point of force P_i in the direction of force P_i is thus given as follows according to the second Castigliano theorem:

$$\Delta_i = \frac{\partial U}{\partial P_i} = \int_0^L \frac{M(\partial M/\partial P_i)}{EI}\, dx \qquad (4.88)$$

We illustrate the use of this formulation in the following example.

EXAMPLE 4.1   A cantilever beam is shown in Fig. 4.12 loaded with a point couple M_0 at the end. What is the deflection at position $x = L/2$ and the slope of the deflection curve at this position?

To get the deflection, we first place a force P at $x = L/2$, later to be set equal to zero. This has been shown in Fig. 4.13. The bending moment along the beam is, from simple equilibrium considerations:

$$M = -M_0 + P\left(x - \frac{L}{2}\right)\left[u\left(x - \frac{L}{2}\right)\right] \qquad (a)$$

where $[u(x - L/2)]$ is a step function starting at $x = L/2$. We then get for $\partial M/\partial P$ from above:

$$\frac{\partial M}{\partial P} = \left(x - \frac{L}{2}\right)\left[u\left(x - \frac{L}{2}\right)\right] \qquad (b)$$

Substituting into Eq. (4.88) we get for the deflection Δ_P in the z direction of the point $x = L/2$:

$$\Delta_P = \frac{1}{EI}\int_0^L \left\{-M_0 + P\left(x - \frac{L}{2}\right)\left[u\left(x - \frac{L}{2}\right)\right]\right\}\left\{\left(x - \frac{L}{2}\right)\left[u\left(x - \frac{L}{2}\right)\right]\right\} dx$$

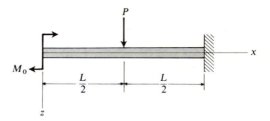

FIGURE 4.13

Setting $P = 0$ now we get for the above:

$$\Delta_P = \frac{1}{EI} \int_0^L (-M_0)\left(x - \frac{L}{2}\right)\left[u\left(x - \frac{L}{2}\right)\right] dx$$

$$= \frac{1}{EI} \int_{L/2}^L (-M_0)\left(x - \frac{L}{2}\right) dx$$

Integrating and inserting limits:

$$\Delta_P = -\frac{L^2 M_0}{8EI}$$

The minus sign indicates, for the z axis pointing downward, that the point moves upwards.

To get the slope at $x = L/2$ we next add to the loading of Fig. 4.12 a point couple C as shown in Fig. 4.14. We then have for M:

$$M = -M_0 - C\left[u\left(x - \frac{L}{2}\right)\right]$$

therefore

$$\frac{\partial M}{\partial C} = -\left[u\left(x - \frac{L}{2}\right)\right]$$

Going to Eq. (4.88) we get the result:

$$\Delta_C = \frac{1}{EI} \int_0^L \left\{-M_0 - C\left[u\left(x - \frac{L}{2}\right)\right]\right\}\left\{-\left[u\left(x - \frac{L}{2}\right)\right]\right\} dx$$

Setting $C = 0$ now we get:

$$\Delta_C = \frac{1}{EI} \int_{L/2}^L [-(-M_0)] dx$$

$$= \frac{LM_0}{2EI}$$

The above result must of course be in radians.   ////

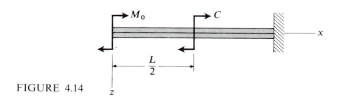

FIGURE 4.14

As the next step we consider a beam that is supported in a statically indeterminate manner such as the beam shown in Fig. 4.15.

To determine the supporting force system for such a beam and thereby to pave the way for determination of deflections of the beam, we may proceed as follows. Consider as unknown external loads as many of the supporting forces and torques as are necessary to render the beam statically determinate in the remaining supporting forces and torques. (That is, we choose as unknown forces and torques a set of *redundant constraints.*[1]) We next compute the total complementary energy in terms of these redundant constraints. Since the redundant constraints have been chosen so as to result in a statically determinate system, it is clear that, for any set of values we wish to assign to these constraints, there exists for the given external loads a stress field in the body that is statically compatible. Hence by varying the redundant constraints we may generate an infinity of statically compatible stress fields in the body. Thus we may appropriately vary the total complementary energy by varying the redundant constraints. By then extremizing the total complementary energy with respect to the redundant constraints we single out the *particular* stress field that is linked through the constitutive law employed, to a *kinematically compatible* strain field. We thus reach the appropriate values of the redundant constraints.

Note that the process is valid for non-linear elastic behavior and, unlike computations employing the Castigliano theorems, is not restricted to point loads. We next illustrate the aforestated procedure in the following example.

EXAMPLE 4.2 We have shown in Fig. 4.15 a cantilever beam uniformly loaded and having two additional supports beyond the base. This is clearly a statically indeterminate support system with two redundant constraints. Assume linear elastic behavior.

To solve the supporting force system we choose the two roller supports as the redundant constraints and show these as unknown external forces R_1 and R_2 in

---

[1] Constraints not needed for equilibrium.

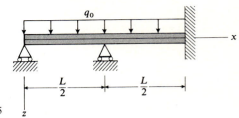

FIGURE 4.15

Fig. 4.16. We next compute the total complementary strain energy π^*. We note that:

$$\pi^* = U^* + V^*$$

where

$$V^* = -\iiint_V u_i B_i \, dv - \iint_{S_2} u_i T_i^{(v)} \, dS$$

Note that for the problem at hand $B_i = 0$ and that the surface integral above is zero since for the region S_2 (i.e., at the supports) $u_i = 0$. Accordingly $V^* = 0$. And since we have linear elastic behavior, we can use U in place of U^* to compute π^*. Thus we have for this case using Eq. (4.87).

$$\pi^* = \frac{1}{2} \int_0^L \frac{M^2}{EI} \, dx$$

The moment M at any section of the beam in terms of R_1 and R_2 is:

$$M(x) = -R_1 x + \frac{q_0 x^2}{2} - R_2\left(x - \frac{L}{2}\right)\left[u\left(x - \frac{L}{2}\right)\right]$$

The total complementary energy then is:

$$\pi^* = \frac{1}{2EI} \int_0^L \left\{ -R_1 x + \frac{q_0 x^2}{2} - R_2\left(x - \frac{L}{2}\right)\left[u\left(x - \frac{L}{2}\right)\right] \right\}^2 dx \qquad (a)$$

To extremize π^* we require that:

$$\frac{\partial \pi^*}{\partial R_1} = 0 \qquad (b)$$

$$\frac{\partial \pi^*}{\partial R_2} = 0 \qquad (c)$$

Employing Eq. (a) for π^* in Eq. (b) we get:

$$\frac{1}{2EI} \int_0^L 2\left\{ -R_1 x + \frac{q_0 x^2}{2} - R_2\left(x - \frac{L}{2}\right)\left[u\left(x - \frac{L}{2}\right)\right] \right\}(-x) \, dx = 0$$

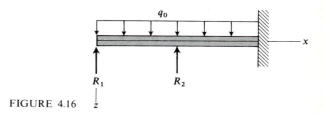

FIGURE 4.16

This becomes:

$$\int_0^L \left(R_1 x^2 - \frac{q_0 x^3}{2} \right) dx + \int_{L/2}^L R_2 \left(x - \frac{L}{2} \right) x \, dx = 0$$

Integrating we get:

$$\frac{R_1 L^3}{3} - \frac{q_0 L^4}{8} + \frac{5}{48} L^3 R_2 = 0$$

therefore

$$16 R_1 + 5 R_2 = 6 q_0 L \qquad (d)$$

Now doing the same for Eq. (c) we get:

$$5 R_1 + 2 R_2 = \frac{17}{8} q_0 L \qquad (e)$$

Solving Eqs. (d) and (e) simultaneously we have:

$$R_1 = \tfrac{11}{56} q_0 L; \qquad R_2 = \tfrac{4}{7} q_0 L \qquad (f)$$

We thus have determined the supporting force system and can then proceed to evaluate moments, stresses and deflections using the results of the technical theory of beams.   ////

The approach presented may be used for *nonlinear elastic behavior* and for *prescribed movement of the supports* corresponding to the redundant constraints. The total complementary energy would then be expressed as follows:

$$\pi^* = \int\int\int_V \int_{\tau_{ij}} \epsilon_{ij} \, d\tau_{ij} \, dv - \sum_i R^{(i)} \bar{u}^{(i)} \qquad (4.89)$$

where $R^{(i)}$ is the ith redundant constraint and $\bar{u}^{(i)}$ is the prescribed displacement in the direction of the ith redundant constraint. In the case of a beam of rectangular cross section the above equation becomes:

$$\pi^* = \int_{-h/2}^{h/2} \int_0^L \int_{\tau_{xx}} \epsilon_{xx} \, d\tau_{xx} \, b \, dz \, dx - \sum_i R^{(i)} \bar{u}^{(i)}$$

Now using the appropriate constitutive law for the problem we may get τ_{xx} as a function of ϵ_{xx} so that the above equation may be written as:

$$\pi^* = \int_{-h/2}^{h/2} \int_0^L G(\epsilon_{xx}) b \, dz \, dx - \sum_i R^{(i)} \bar{u}^{(i)} \qquad (4.90)$$

where $G(\epsilon_{xx})$ is the function resulting from integration with respect to strain. Assuming the deformation model set forth in Sec. 4.2 (plane sections remain plane) we may give ϵ_{xx} as:

$$\epsilon_{xx} = -z \frac{d^2 w}{dx^2}$$

Next integrating with respect to z we express π^* as follows:

$$\pi^* = \int_0^L H\left(h, b, \frac{d^2 w}{dz^2}\right) dx - \sum_i R^{(i)} \bar{u}^{(i)} \qquad (4.91)$$

where H is some function of h, b, and dw^2/dx^2.

We turn at this point to the computation of M at a section. Thus we have:

$$M = \int_{-h/2}^{h/2} \tau_{xx} z b \, dz$$

Employing the constitutive law to replace τ_{xx} by a function of ϵ_{xx} and then using the plane sections assumption to replace ϵ_{xx} in terms of $-z \, dw^2/dx^2$, we then have on carrying out integration with respect to z:

$$M = K\left(h, b, \frac{d^2 w}{dx^2}\right) \qquad (4.92)$$

The idea is first to solve for $d^2 w/dx^2$ in terms of M, h, and b in the above equation and to substitute this result into the function H in the complementary energy (Eq. (4.91)). Next we express M in terms of the external loads and redundant constraints $R^{(i)}$ from simple equilibrium considerations. Finally, by replacing M in Eq. (4.91), using this result, we have π^* as a function of the redundant constraints and we are ready to carry out the extremization process.

We have presented several problems as assignments wherein you may work out the details of the above outlined procedure for specific situations.

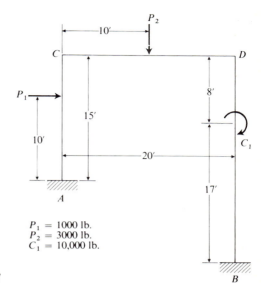

P_1 = 1000 lb.
P_2 = 3000 lb.
C_1 = 10,000 lb.

FIGURE 4.17

Part B
FRAMES AND RINGS

4.10 OPEN FRAMES

We shall now briefly consider a class of structural systems called frames. Such a system is composed of a group of members firmly fastened together at the ends, like a truss, but unlike a truss these connections are designed so that the members can support loads at places other than at the ends (i.e., the joints). This means that members of a frame are not generally two-force members but may transmit bending couples, twisting couples, and shearing forces in addition to axial forces. This in turn means that the joints must also be designed to transmit such force systems. There is a key assumption regarding joints that we now make, and that is—all the members at a joint must undergo the same rotation—i.e., *the joint is considered perfectly rigid* in this regard.

In this section we shall consider *open* frames—i.e., frames, some of whose members are attached to foundations (see Fig. 4.17 as an example) as opposed to closed frames whose members form a closed circuit (see Fig. 4.24). We shall set forth two procedures for open frames here. First we consider minimization of the total complementary energy for a frame supported in a statically indeterminate manner such as the one shown in Fig. 4.17 (wherein we have a degree of indeterminacy which

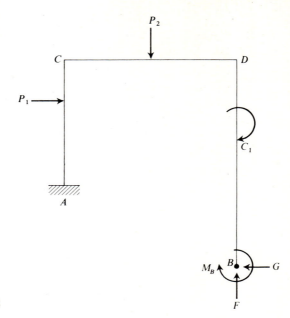

FIGURE 4.18

is three-fold). As in the discussion of beams we choose enough supporting forces and torques to render the problem statically determinate in terms of these forces and torques. We then compute the total complementary energy π^* in terms of these parameters. In doing so, we may consider that we have a system of beams and the procedure is as given in the previous section. We then extremize π^* with respect to these parameters for the reasons set forth in the previous section. This gives us enough equations for determining these parameters and we thus can readily establish all the supporting forces and torques for the frame. Furthermore, we can then easily find the shear force, the axial force and the bending moments along the frame by simple equilibrium considerations.

In this approach as well as the aforementioned second approach, to be explained later, we shall use the convention that a bending moment causing the outer fibers of the frame to be in tension is a positive bending moment. We now consider a simple frame problem.

EXAMPLE 4.3   We wish to determine the supporting force system for the frame shown in Fig. 4.17. All members are linearly elastic; all have the same modulus of elasticity E; and all have the same moment of inertia I.

We shall consider as the redundant supports, the forces F, G and the torque M_B as shown in Fig. 4.18. We can then give the bending moments in the members of the

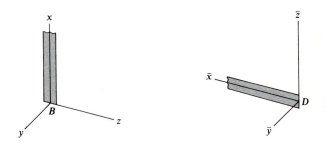

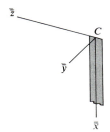

frame as follows noting the above diagrams:

From _B to D_ (measuring x from B)

$$M = M_B + Gx + C_1[u(x - 17)]$$

From _D to C_ (measuring $\bar{x}$ from D)

$$M = M_B + 25G + C_1 + P_2(\bar{x} - 10)[u(\bar{x} - 10)] - F\bar{x}$$

From _C to A_ (measuring $\bar{\bar{x}}$ from C)

$$M = M_B + G(25 - \bar{\bar{x}}) + C_1 + 10P_2 - 20F + P_1(\bar{\bar{x}} - 5)[u(\bar{\bar{x}} - 5)]$$

The total complementary energy can then be given as in Example 4.2:

$$
\begin{aligned}
\pi^* = \frac{1}{2EI}\Bigg[& \int_0^{24} \{M_B + Gx + C_1[u(x - 17)]\}^2 \, dx \\
& + \int_0^{20} \{M_B + 25G + C_1 + P_2(\bar{x} - 10)[u(\bar{x} - 10)] - F\bar{x}\}^2 \, dx \\
& + \int_0^{15} \{M_B + G(25 - \bar{\bar{x}}) + C_1 + 10P_2 - 20F \\
& + P_1(\bar{\bar{x}} - 5)[u(\bar{\bar{x}} - 5)]\}^2 \, dx \Bigg]
\end{aligned}
$$

(a)

We next set forth the following derivatives:

$$\frac{\partial \pi^*}{\partial F} = 0 \qquad \frac{\partial \pi^*}{\partial G} = 0 \qquad \frac{\partial \pi^*}{\partial M_B} = 0 \qquad (b)$$

We get the following simultaneous equations as a result of the above computations:

$$M_B - 17.333F + 20.500G = -1.000C_1 - 2.000P_1 - 7.667P_2$$
$$M_B - 8.333F + 17.917G = -0.717C_1 - 0.833P_1 - 3.333P_2$$
$$M_B - 9.535F + 21.008G = -0.866C_1 - 0.620P_1 - 3.605P_2$$

Solving for F, G, and M_B we get:

$$M_B = 0.174C_1 - 1.839P_1 - 1.058P_2$$
$$F = 0.020C_1 + 0.168P_1 + 0.514P_2$$
$$G = -0.040C_1 + 0.134P_1 + 0.112P_2 \qquad (c)$$

It is now an easy matter to determine shear and bending moments at any section of the truss.

The above procedure can be generalized to encompass non-linear elastic behavior in a manner analogous to that discussed for the corresponding case for beams. ////

In the second phase of our work in this section we shall set forth the so-called *slope-deflection method* for finding the deflection of the entire frame (supporting forces can also be deduced from this procedure). The procedure involves the use of Castigliano's first theorem and is completely analogous to the treatment of the truss problem in Example 3.7. (It is thus a stiffness method approach.) Accordingly we shall be restricted here to point force and point couple loadings and the material will have to be linear-elastic (the latter is to allow for the use of the superposition principle).

In the case of a truss the strain energy could be determined in terms of displacements of movable pins. For frames the comparable computation is more complex. For this reason we first consider an arbitrary member of a frame shown in a deformed state in Fig. 4.19 (diagram exaggerated). A load P and a point couple C are shown acting on the member. Note that we have indicated at each end a bending moment M, forces R and H, an angle θ of the end slope, and finally a transverse end deflection Δ. We shall assume small deformations and so we will neglect the contribution of the axial loads to the bending moments in the ensuing computations. Also we shall neglect the shear strain energy and the axial strain energy in comparison to the bending strain energy. We accordingly will not be interested in the axial forces H shown in the

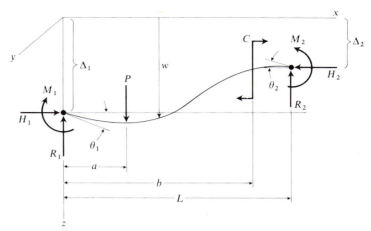

FIGURE 4.19

diagram. Employing equilibrium (take moments about the ends) we shall now give R_1 and R_2 in terms of M_1, M_2 and the external loads as follows:

$$R_1 = \frac{P}{L}(L - a) - \frac{C}{L} - \frac{M_1}{L} + \frac{M_2}{L} \quad (a)$$

$$R_2 = P\frac{a}{L} + \frac{C}{L} + \frac{M_1}{L} + \frac{M_2}{L} \quad (b) \quad (4.93)$$

The bending moment equation for the beam may be given as follows:

$$M = R_1 x + M_1 - [u(x - a)]P(x - a) + C[u(x - b)]$$

Employing Eq. (4.16(b)) we have for (d^2w/dx^2) the following result:

$$\frac{d^2w}{dx^2} = -\frac{M}{EI} = -\frac{1}{EI}\{R_1 x + M_1 - [u(x - a)]P(x - a) + C[u(x - b)]\} \quad (4.94)$$

Integrating twice we get:

$$w = -\frac{1}{EI}\left\{\frac{R_1 x^3}{6} + \frac{M_1 x^2}{2} - [u(x - a)]\frac{P(x - a)^3}{6} + \frac{Cx^2}{2}[u(x - b)]\right\}$$
$$+ C_1 x + C_2$$

Employing the condition $w(0) = \Delta_1$ and $w'(0) = \theta_1$ as shown in Fig. 4.19, we get $C_1 = \theta_1$ and $C_2 = \Delta_1$ so that the above equation becomes:

$$w = -\frac{1}{EI}\left\{\frac{R_1 x^3}{6} + \frac{M_1 x^2}{2} - [u(x - a)]\frac{P(x - a)^3}{6} + \frac{Cx^2}{2}[u(x - b)]\right\}$$
$$+ \theta_1 x + \Delta_1 \quad (4.95)$$

Next note in Fig. 4.19 that $w(L) = \Delta_2$ and $w'(L) = \theta_2$. Submitting Eq. (4.95) to these conditions and replacing R_1 using Eq. (4.93) we get the following equations:

$$\Delta_2 = -\frac{1}{EI}\left\{ \frac{P(L-a)L^2}{6} - \frac{CL^2}{6} - \frac{M_1 L^2}{6} + \frac{M_2 L^2}{6} + \frac{M_1 L^2}{2} \right.$$
$$\left. -\frac{P(L-a)^3}{6} + \frac{CL^2}{2} \right\} + \theta_1 L + \Delta_1$$

$$\theta_2 = -\frac{1}{EI}\left\{ \frac{P(L-a)L}{2} - \frac{CL}{2} - \frac{M_1 L}{2} + \frac{M_2 L}{2} + M_1 L \right.$$
$$\left. -\frac{P}{2}(L-a)^2 + CL \right\} + \theta_1$$

Solving for M_1 and M_2 from the preceding equations we get:

$$M_1 = \frac{6EI}{L^2}(\Delta_1 - \Delta_2) + \frac{2EI}{L}(\theta_2 + 2\theta_1) - C - \frac{Pa(L-a)^2}{L^2}$$

$$M_2 = -\frac{6EI}{L^2}(\Delta_1 - \Delta_2) - \frac{2EI}{L}(\theta_1 - 2\theta_2) - \frac{Pa^2(L-a)}{L^2} \tag{4.96}$$

Now going back to Eq. (4.94) we may give the following result, replacing R_1 and M_1 in terms of θ_1, θ_2, Δ_1, and Δ_2, as well as the external loads from Eqs. (4.93) and (4.96). Thus:

$$\frac{d^2 w}{dx^2} = -\frac{1}{EI}\left\{ \frac{P(L-a)^2}{L^2}\left[x\left(1 + \frac{2a}{L}\right) - a \right] - P(x-a)[u(x-a)] \right.$$
$$\left. + C\left([u(x-b)] - 1 \right) \right\} + \frac{2(\theta_2 + \theta_1)}{L^2}(3x - L)$$
$$-\frac{2\theta_1}{L} + \frac{6(\Delta_1 - \Delta_2)}{L^3}(2x - L) \tag{4.97}$$

And making the same substitutions in Eq. (4.95) we get:

$$w(x) = -\frac{1}{EI}\left\{ \frac{Px^2(L-a)^2}{2L^2}\left[\frac{x}{3}\left(1 + \frac{2a}{L}\right) - a \right] - \frac{P(x-a)^3}{6}[u(x-a)] \right.$$
$$\left. + C\frac{x^2}{2}\left([u(x-b)] - 1 \right) \right\} + \frac{\theta_1 + \theta_2}{L}x^2\left(\frac{x}{L} - 1\right)$$
$$+ \theta_1 x\left(1 - \frac{x}{L}\right) + \frac{\Delta_1 - \Delta_2}{L^2}x^2\left(\frac{2x}{L} - 3\right) + \Delta_1 \tag{4.98}$$

A comparable expression can be written for any system of point loads by simply adding contributions of the type $[u(x - a)]P(x - a)$ and $C[u(x - b)]$ for the additional loads. The strain energy for the beam can be given by this formula

$$U = \frac{1}{2}EI \int_0^L \left(\frac{d^2 w}{dx^2}\right)^2 dx$$

upon substitution of the appropriate $(d^2 w/dx^2)$ expression from Eq. (4.97).

Now the frame consists of a number of these members. We can compute the strain energies for each member in terms of the appropriate Δ_i and θ_i at each joint. However, simplifications can often be made at this time as to the number of Δ's and θ's needed. Thus we have shown the frame of Fig. 4.17 in a deflected form in Fig. 4.20. Note that for the upper member the transverse deflection (in the upward direction) may be taken as zero for small axial deformation of the vertical members. Furthermore the transverse deflections of the vertical members (horizontal movements) are zero at the bottom, and at the upper ends are taken as equal for both members (and measured by Δ). The latter reflects the small axial deformation assumption for the horizontal member. The left joint has been shown to rotate an angle α, while the right joint has been shown to rotate an angle β. We can then find from Δ, α, and β the appropriate Δ_1, Δ_2, θ_1, and θ_2 to be used in Eq. (4.97) for *each* member of the frame and we can then compute the strain energy for the frame in terms of Δ, α, and β. For this reason, we call Δ, α, and β the *frame parameters* of the problem. Hence we have:

$$U = U(\Delta, \alpha, \beta)$$

Since the deformation of the frame is now determined by the frame parameters, they may also be considered as *generalized coordinates*. And if we use the first Castigliano theorem, we can equate the partials of U with respect to Δ, α, and β to the generalized forces corresponding to these generalized coordinates. Thus we have

$$\frac{\partial U}{\partial \Delta} = Q_\Delta \quad (a)$$

$$\frac{\partial U}{\partial \alpha} = Q_\alpha \quad (b)$$

$$\frac{\partial U}{\partial \beta} = Q_\beta \quad (c) \quad (4.99)$$

where the Q's are the generalized forces. We may find the generalized forces by first computing the work done by the *actual external loads*, using Eq. (4.98) or its derivative, on the appropriate member of the frame to express the movement of forces or the rotation of torques. The work done is thus determined in terms of the actual loads and the generalized coordinates. We then equate this work to the sum $\frac{1}{2}(Q_\Delta \Delta + Q_\alpha \alpha +$

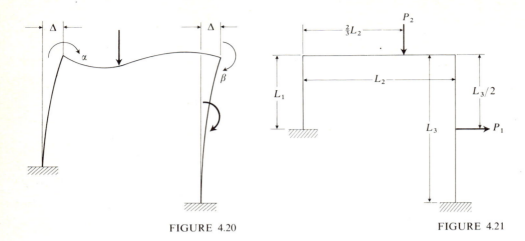

FIGURE 4.20   FIGURE 4.21

$Q_\beta \beta$), which is by definition the work input in terms of generalized forces. We can then solve for Q_Δ, Q_α, and Q_β in terms of the generalized coordinates for use in Eq. (4.99). We thus obtain three equations to determine α, β, and Δ for the problem. Using these results in Eq. (4.98), we can then get the deflection curve for each beam in the frame.

Although we have centered our discussion on the frame of Fig. 4.17 with three generalized coordinates, the reader should have no trouble in extending the comments made here to more complex frames with more than three generalized coordinates.

We now illustrate the details of the approach for a specific frame in the following example.

EXAMPLE 4.4 Shown in Fig. 4.21 is a frame with three members on which two forces P_1 and P_2 act. The members are linearly elastic, have uniform cross sections, and have the same modulus of elasticity, E. Find the deflection equations for the three members of the frame.

In Fig. 4.22 we have shown the frame in a highly exaggerated deformed state. The generalized coordinates to be used, are again Δ, α, and β. The frame members have been numbered, and in Fig. 4.23 we have shown each member separately. For convenience we have associated a reference for each member of the frame and, referring to Fig. 4.19 as a guide, we have established Δ_1, Δ_2, θ_1, and θ_2 in terms of the generalized coordinates. We can then apply Eq. 4.97 to each member. Thus:

Member 1

$$\frac{d^2 w_1}{dx_1^2} = \frac{2\alpha}{L_1^2}(3x_1 - L_1) - \frac{6}{L_1^3}(2x_1 - L_1) \qquad (a)$$

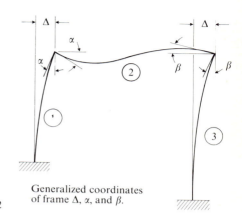

FIGURE 4.22

Generalized coordinates
of frame Δ, α, and β.

Member 1

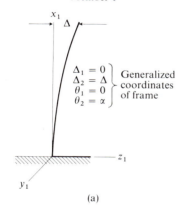

$\left.\begin{array}{l} \Delta_1 = 0 \\ \Delta_2 = \Delta \\ \theta_1 = 0 \\ \theta_2 = \alpha \end{array}\right\}$ Generalized
coordinates
of frame

(a)

Member 2

$\begin{array}{l} \Delta_1 = 0 \\ \Delta_2 = 0 \\ \theta_1 = \alpha \\ \theta_2 = \beta \end{array}$

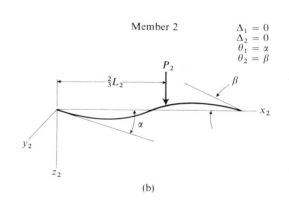

(b)

Member 3

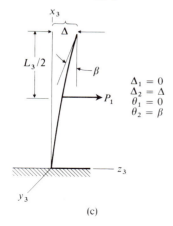

$\begin{array}{l} \Delta_1 = 0 \\ \Delta_2 = \Delta \\ \theta_1 = 0 \\ \theta_2 = \beta \end{array}$

(c)

FIGURE 4.23

Member 2

$$\frac{d^2 w_2}{dx_2{}^2} = -\frac{1}{EI}\left\{\frac{P_2(L_2 - 5L_2/3)^2}{L_2{}^2}\left[x_2\left(1 + \frac{2(2L_2/3)}{L_2}\right) - \frac{2L_2}{3}\right]\right.$$

$$\left. - P_2\left(x_2 - \frac{2L_2}{3}\right)\left[u\left(x_2 - \frac{2L_2}{3}\right)\right]\right\} + \frac{2(\alpha + \beta)}{L_2{}^2}(3x_2 - L_2)$$

$$- \frac{2\alpha}{L_2} + 0$$

therefore

$$\frac{d^2 w_2}{dx_2{}^2} = -\frac{P_2}{EI}\left\{\frac{1}{27}(7x_2 - 2L_2) - \frac{1}{3}(3x_2 - 2L_2)\left[u\left(x_2 - \frac{2L_2}{3}\right)\right]\right\}$$

$$+ \frac{2\alpha}{L_2{}^2}(3x_2 - 2L_2) + \frac{2\beta}{L_2{}^2}(3x_2 - L_2) \qquad (b)$$

Member 3

$$\frac{d^2 w_3}{dx_3{}^2} = -\frac{1}{EI}\left\{-\frac{P_1(L_3 - L_3/2)^2}{L_3{}^2}\left[x_3\left(1 + \frac{2L_3}{2L_3}\right) - \frac{L_3}{2}\right]\right.$$

$$\left. + P_1\left(x_3 - \frac{L_3}{2}\right)\left[u\left(x_3 - \frac{L_3}{2}\right)\right]\right\} + \frac{2\beta}{L_3{}^2}(3x_3 - L_3)$$

$$+ \frac{6(-\Delta)}{L_3{}^3}(2x_3 - L_3)$$

therefore

$$\frac{d^2 w_3}{dx_3{}^2} = \frac{P_1}{EI}\left\{-\frac{4x_3 - L_3}{8} - \frac{1}{2}(2x_3 - L_3)\left[u\left(x_3 - \frac{L_3}{2}\right)\right]\right\}$$

$$+ \frac{2\beta}{L_3{}^2}(3x_3 - L_3) - \frac{6\Delta}{L_3{}^3}(2x_3 - L_3) \qquad (c) \ (4.100)$$

To get the strain energy for the frame using the above results we have:

$$U = \frac{1}{2}EI \sum_{p=1}^{3} \int_0^{L_p} \left(\frac{d^2 w_p}{dx_p{}^2}\right)^2 dx_p \qquad (4.101)$$

Now using the first Castigliano theorem with respect to the generalized coordinates we have:

$$\frac{\partial U}{\partial \Delta} = Q_\Delta$$

$$\frac{\partial U}{\partial \alpha} = Q_\alpha$$

$$\frac{\partial U}{\partial \beta} = Q_\beta \qquad (4.102)$$

Employing Eq. (4.101) in (4.102) we see that:

$$\frac{\partial U}{\partial \Delta} = EI \sum_{p=1}^{3} \int_0^{L_p} \left(\frac{d^2 w_p}{dx_p^2}\right) \frac{\partial(d^2 w_p/dx_p^2)}{\partial \Delta} dx_p = Q_\Delta \quad (a)$$

$$\frac{\partial U}{\partial \alpha} = EI \sum_{p=1}^{3} \int_0^{L_p} \left(\frac{d^2 w_p}{dx_p^2}\right) \frac{\partial(d^2 w_p/dx_p^2)}{\partial \alpha} dx_p = Q_\alpha \quad (b)$$

$$\frac{\partial U}{\partial \beta} = EI \sum_{p=1}^{3} \int_0^{L_p} \left(\frac{d^2 w_p}{dx_p^2}\right) \frac{\partial(d^2 w_p/dx_p^2)}{\partial \beta} dx_p = Q_\beta \quad (c) \quad (4.103)$$

Now in computing the integrals of the above equations using Eq. (4.100) you will find that the coefficient of the forces P_1 and P_2 always turns out to be zero. We can accordingly delete in the above considerations the expressions in Eq. (4.100) with the brackets { }. They are constant expressions with respect to the generalized coordinates and give the curvature of the beam from the loads when the beam is fixed at each end. (That is, we get a curvature equal to $-1/(EI)$ times the bracketed expression if we set the Δ, α, and β equal to zero.) Accordingly we get the following results for Eq. (4.103).

$$EI\left[12\left(\frac{1}{L_1^3} + \frac{1}{L_3^3}\right)\Delta - \frac{6}{L_1^2}\alpha - \frac{6}{L_3^2}\beta \right] = Q_\Delta \quad (a)$$

$$EI\left[-\frac{6}{L_1^2}\Delta + 4\left(\frac{1}{L_1} + \frac{1}{L_2}\right)\alpha + \frac{2}{L_2}\beta \right] = Q_\alpha \quad (b)$$

$$EI\left[-\frac{6}{L_3^2}\Delta + \frac{2}{L_2}\alpha + 4\left(\frac{1}{L_2} + \frac{1}{L_3}\right)\beta \right] = Q_\beta \quad (c) \quad (4.104)$$

We must now find the generalized forces. For this we use the basic definition of these quantities when we equate $\frac{1}{2}(Q_\Delta\Delta + Q_\alpha\alpha + Q_\beta\beta)$ to the work W_k done by the external forces. Thus:

$$W_k = \tfrac{1}{2}(Q_\Delta\Delta + Q_\alpha\alpha + Q_\beta\beta)$$

Hence we see from above that:

$$\frac{\partial W_k}{\partial \Delta} = \tfrac{1}{2}Q_\Delta$$

$$\frac{\partial W_k}{\partial \alpha} = \tfrac{1}{2}Q_\alpha$$

$$\frac{\partial W_k}{\partial \beta} = \tfrac{1}{2}Q_\beta \quad (4.105)$$

To get W_k we compute the work done by forces P_1 and P_2 employing Eq. (4.98) taken for the appropriate member. Clearly the bracketed part { } of this equation will contribute nothing to the above equation since it is constant with respect to the generalized coordinates. We shall now delete this so that for members 2 and 3 we have for Eq. (4.98) using Figs. 4.23(b) and 4.23(c):

$$w_2(x) = \frac{\alpha + \beta}{L^2} x_2{}^2 \left(\frac{x_2}{L_2} - 1\right) + \alpha x_2 \left(1 - \frac{x_2}{L}\right)$$

$$w_3(x) = \frac{\beta}{L_3} x_3{}^2 \left(\frac{x_3}{L_3} - 1\right) + \beta x_3 \left(1 - \frac{x_3}{L_3}\right) - \frac{\Delta}{L_3{}^2} x_3{}^2 \left(\frac{2x_3}{L} - 3\right) - \Delta$$

Hence W_k is given as:

$$W_k = \frac{1}{2} P_2 \left[\frac{\alpha + \beta}{L_2}\left(\frac{2}{3}L_2\right)^2 \left(-\frac{1}{3}\right) + \alpha\left(\frac{2}{3}L_2\right)\left(\frac{1}{3}\right)\right]$$

$$+ \frac{(-P_1)}{2}\left[\frac{\beta}{L_3}\left(\frac{1}{2}L_3\right)^2 \left(-\frac{1}{2}\right) + \beta\left(\frac{1}{2}L_3\right)\left(\frac{1}{2}\right) + \frac{\Delta}{L_3{}^2}\left(\frac{1}{2}L_3\right)^2 (2) - \Delta\right]$$

$$= \frac{P_2 L_2}{27}[\alpha - 2\beta] - \frac{P_1 L_3}{16}\beta + \frac{P_1}{4}\Delta$$

Now substituting into Eq. (4.105) we get:

$$Q_\Delta = \frac{P_1}{2}$$

$$Q_\alpha = \frac{2}{27}P_2 L_2$$

$$Q_\beta = -\frac{4}{27}P_2 L_2 - \frac{P_1 L_3}{8} \qquad (4.106)$$

Finally substitute the above results for the Q's into Eq. (4.104). Multiply Eq. (a) of this set by $L_3{}^3/6$, Eq. (b) by $L_1{}^2/2$ and Eq. (c) by $L_3{}^2/2$. We then get the following system of equations:

$$
\begin{bmatrix}
2\left[1 + \left(\dfrac{L_3}{L_1}\right)^3\right] & -L_3\left(\dfrac{L_3}{L_1}\right)^2 & -L_3 \\[3mm]
-3 & 2L_1\left[1 + \dfrac{L_1}{L_2}\right] & L_1\left(\dfrac{L_1}{L_2}\right) \\[3mm]
-3 & L_3\left(\dfrac{L_3}{L_2}\right) & 2L_3\left[1 + \dfrac{L_3}{L_2}\right]
\end{bmatrix}
\begin{Bmatrix} \Delta \\ \alpha \\ \beta \end{Bmatrix}
$$

$$
= \frac{1}{EI}
\begin{Bmatrix}
\dfrac{P_1 L_3{}^3}{12} \\[4mm]
\dfrac{P_2 L_2 L_1{}^2}{27} \\[4mm]
\dfrac{-2P_2 L_2 L_3{}^2}{27} - \dfrac{P_1 L_3{}^3}{16}
\end{Bmatrix}
$$

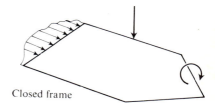

FIGURE 4.24   Closed frame

For the data

$$P_1 = 1000\ \text{lb}$$
$$P_2 = 3000\ \text{lb}$$
$$L_1 = 10\ \text{ft}$$
$$L_2 = 20\ \text{ft}$$
$$L_3 = 30\ \text{ft}$$

we solve the equations simultaneously for $\varDelta$, α, and β to get:

$$\begin{bmatrix} \varDelta \\ \alpha \\ \beta \end{bmatrix} = \frac{10^6}{EI} \begin{bmatrix} 0.169 \\ 0.032 \\ -0.044 \end{bmatrix}$$

The entire deformed geometry of the frame has thus been established.   ////

4.11 CLOSED FRAMES AND RINGS

In the previous section we considered open frames. At this time we examine closed frames and rings (see Figs. 4.24 and 4.25) with the view toward establishing the shear forces and bending moments in these members from external loads.

Up to now we have been able to make direct use of the results from the technical theory of beams presented earlier. In the study of closed frames and rings we shall continue to use the basic assumptions from this study and the concommittant results. For instance, we shall neglect deformation due to axial force and shear force. We shall also use the flexure formula for stress with the result that the strain energy of deformation of closed frames and rings, neglecting strain energy due to shear and axial forces, will be given as follows for linear elastic behavior

$$U = \oint \frac{M^2}{2EI}\,dS \qquad (4.107)$$

where S is taken along the centerline of the frame or ring. We will continue to use the convention that M is positive if the outer fibers of the frame or ring are in tension.

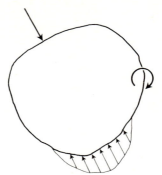

FIGURE 4.25

The procedure to be followed here is to cut the member at a convenient section 0 and to expose thereby a shear force V_0, an axial force T_0, and a bending moment M_0 (see Fig. 4.26). These, of course, are unknowns. Now the total complementary energy π^* is computed in terms of the external loads and the parameters M_0, V_0, and T_0. We then extremize π^* with respect to these parameters for reasons set forth in Sec. 4.9. This gives us three equations for the three unknowns. Knowing M_0, V_0, and T_0 we can then get M at any section of the frame and thus determine from the flexure formula the normal stress distribution at the section. It is to be pointed out that with zero body force and with no region S_2 of the boundary present for which u_i is prescribed, the total complementary energy becomes simply U^* and can be given by Eq. (4.107) for linear elastic behavior wherein we neglect shear deformation energy as well as axial deformation energy. Also, if only point loads P and point couples C are applied, we can determine the deflections and rotations at the points of application of these loads using Castigliano's second theorem. It will be simplest to illustrate this procedure directly with an example.

EXAMPLE 4.5   What are the bending moment, shear force, and axial force distributions for the closed frame shown in Fig. 4.27? A cut has been made at a corner as has been shown in Fig. 4.28 exposing bending moment M_0, shear force V_0, and axial force T_0. We will use the following variables in the ensuing discussion:

$$
\begin{aligned}
&\text{From } A \text{ to } B \quad x\\
&\text{From } B \text{ to } C \quad \bar{x}\\
&\text{From } C \text{ to } D \quad \bar{\bar{x}}\\
&\text{From } D \text{ to } A \quad \tilde{x}
\end{aligned}
\qquad (a)
$$

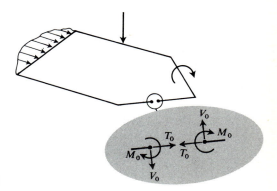

FIGURE 4.26

We can then give M as follows by considering free bodies starting from the cut at A and extending counterclockwise around the frame:

From A to B:

$$M(x) = M_0 + V_0 x$$

From B to C:

$$M(\bar{x}) = M_0 + V_0 b + T_0 \bar{x}$$

From C to D:

$$M(\bar{\bar{x}}) = M_0 + V_0(b - \bar{\bar{x}}) + T_0 a + P_1 \bar{\bar{x}} - C\left[u\left(\bar{\bar{x}} - \frac{b}{2}\right)\right]$$

From D to A:

$$M(\tilde{x}) = M_0 + T_0(a - \tilde{x}) + P_1 b - C \qquad (b)$$

The potential energy can then be given as follows:

$$U = \frac{1}{2EI}\left[\int_0^b (M_0 + V_0 x)^2\, dx + \int_0^a (M_0 + V_0 b + T_0 \bar{x})^2\, d\bar{x} \right.$$

$$+ \int_0^b \left\{ M_0 + V_0(b - \bar{\bar{x}}) + T_0 a + P_1 \bar{\bar{x}} - C\left[u\left(\bar{\bar{x}} - \frac{b}{2}\right)\right]\right\}^2 d\bar{\bar{x}}$$

$$\left. + \int_0^a \left\{ M_0 + T_0(a - \tilde{x}) + P_1 b - C\right\}^2 d\tilde{x} \right] \qquad (c)$$

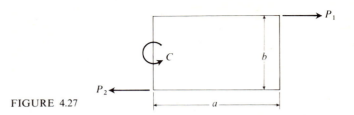

FIGURE 4.27

We next extremize U with respect to the parameters M_0, V_0, and T_0 as follows:

$$EI\frac{\partial U}{\partial M_0} = 0 = \int_0^b (M_0 + V_0 x)\, dx + \int_0^a (M_0 + V_0 b + T_0 \bar{x})\, d\bar{x}$$

$$+ \int_0^b \left\{ M_0 + V_0(b - \bar{\bar{x}}) + T_0 a + P_1 \bar{\bar{x}} - C\left[u\left(\bar{\bar{x}} - \frac{b}{2}\right) \right] \right\} d\bar{\bar{x}}$$

$$+ \int_0^a \left\{ M_0 + T_0(a - \tilde{x}) + P_1 b - C \right\} d\tilde{x}$$

$$EI\frac{\partial U}{\partial V_0} = 0 = \int_0^b (M_0 + V_0 x)x\, dx + \int_0^a (M_0 + V_0 b + T_0 \bar{x})b\, d\bar{x}$$

$$+ \int_0^b \left\{ M_0 + V_0(b - \bar{\bar{x}}) + T_0 a + P_1 \bar{\bar{x}} - C\left[u\left(\bar{\bar{x}} - \frac{b}{2}\right) \right] \right\}(b - \bar{\bar{x}})\, d\bar{\bar{x}}$$

$$EI\frac{\partial U}{\partial T_0} = 0 = \int_0^a (M_0 + V_0 b + T_0 \bar{x})\bar{x}\, d\bar{x}$$

$$+ \int_0^b \left\{ M_0 + V_0(b - \bar{\bar{x}}) + T_0 a - P_1 \bar{\bar{x}} - C\left[u\left(\bar{\bar{x}} - \frac{b}{2}\right) \right] \right\} a\, d\bar{\bar{x}}$$

$$+ \int_0^a \left\{ M_0 + T_0(a - \tilde{x}) + P_1 b - C \right\}(a - \tilde{x})\, d\tilde{x} \qquad (d)$$

We then arrive at the following set of simultaneous equations for M_0, V_0, and T_0:

$$4M_0 + 2bV_0 + 2aT_0 = \frac{2a + b}{a + b}(C - bP_1)$$

$$2M_0 + \frac{2b(3a + 2b)}{3(a + b)}V_0 + aT_0 = \frac{b}{a + b}\left(\frac{C}{4} - \frac{bP_1}{3}\right)$$

$$2M_0 + bV_0 + \frac{2a(3b + 2a)}{3(a + b)}T_0 = C - bP_1 \qquad (e)$$

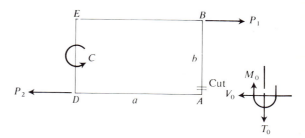

FIGURE 4.28

Solving simultaneously we get:

$$V_0 = \frac{6a + b}{2(3a + b)}P_1 - \frac{3(4a + b)}{4b(3a + b)}C$$

$$M_0 = \frac{2a + b}{4(a + b)}(C - bP_1) + \frac{3c(4a^2 + 7ab + b^2) - 4abP_1(3a + 5b)}{8(3a + b)(a + 3b)}$$

$$+ \frac{bc}{2(3a + b)}$$

$$T_0 = \frac{3b}{2a(a + 3b)}(C - bP_1) \qquad\qquad (f)$$

The bending moment M, shear force V, and axial force T are now available at any position of the beam. For instance to get M we simply use the appropriate equation from the set (b) inserting the values of V_0, M_0, and T_0 established above. As for V or T at any section we would merely employ statical considerations for a free body starting at the cut at A (Fig. 4.28) and ending at the section of interest. Again using the values for V_0, M_0, and T_0 as determined above, we can establish the desired result.

////

We now consider the case of a *ring* (see Fig. 4.25) which differs from a closed frame in that the centerline is a curve rather than a system of straight intersecting lines. If we take the radius of curvature of the ring at all points to be large compared to the cross sectional dimensions, we may use for linearly elastic behavior the results for normal stress on the cross section developed from the technical theory of beams— namely the flexure formula $\tau = My/I$ where y is the distance from the centerline of the ring measured normal to this centerline in the plane of the ring, and I is the moment of inertia of the section about an axis through the centerline normal to the plane of the ring.

As in the case of the closed frame we shall be concerned with determining the bending moment, shear force, and axial force in a ring of given geometry acted on by

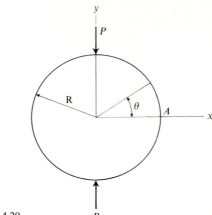

FIGURE 4.29

external forces. The procedure again will be to expose a section of the ring by a cut, thereby introducing M_0, V_0, and T_0 at the section thus exposed. We then find these quantities by extremizing the total complementary energy computed with the assumption that shear strain energy and axial strain energy are negligible compared with bending strain energy.

The following example illustrates the procedure.

EXAMPLE 4.6   A circular ring loaded by point forces P is shown in Fig. 4.29. It has a radius R and for linear elastic behavior has an elastic modulus E uniform around the ring. The moment of inertia I of the cross section is constant for the ring.

We will institute a cut at A (see Fig. 4.30) thus introducing the quantities M_0, V_0, and T_0. We will use θ measured counterclockwise from A (as shown in the diagram) as the independent variable. Hence to get M we have at any position θ:

$$M = M_0 + V_0 R \sin \theta + T_0 R(1 - \cos \theta) + P\left[u\left(\theta - \frac{\pi}{2}\right)\right](R)(-\cos \theta)$$

$$+ P\left[u\left(\theta - \frac{3\pi}{2}\right)\right] R \cos \theta \qquad (a)$$

As in the discussion of closed frames the total complementary energy equals the strain energy. We shall again neglect strain energy due to axial deformation and strain energy due to shear deformation. Furthermore, the theory of curved beams, which we shall not discuss here, shows that for cases where the cross sectional dimensions are small compared to the undeformed radius of curvature, in this case R for all values of θ, we can use the results from the technical theory of straight beams as to both stress and strain energy. Accordingly, we may employ Eq. (4.107) in

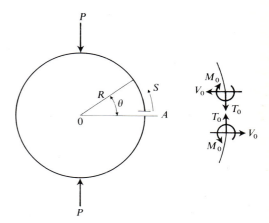

FIGURE 4.30

conjunction with Eq. (*a*) for this problem. Thus

$$U = \frac{1}{2EI} \int_0^{2\pi} \left\{ M_0 + V_0 R \sin \theta + T_0 R(1 - \cos \theta) \right.$$

$$\left. + P\left[u\left(\theta - \frac{\pi}{2}\right) \right] R(-\cos \theta) + P\left[u\left(\theta - \frac{3\pi}{2}\right) \right] R \cos \theta \right\}^2 R \, d\theta \qquad (b)$$

We now extremize U with respect to M_0, V_0, and T_0 in accordance with the principle of total complementary energy. Thus:

$$\frac{\partial U}{\partial M_0} = 0 = \int_0^{2\pi} 2\left\{ M_0 + V_0 R \sin \theta + T_0 R(1 - \cos \theta) \right.$$

$$\left. + P\left[u\left(\theta - \frac{\pi}{2}\right) \right] R(-\cos \theta) + P\left[u\left(\theta - \frac{3\pi}{2}\right) \right] R \cos \theta \right\} R \, d\theta$$

$$\frac{\partial U}{\partial T_0} = 0 = \int_0^{2\pi} 2\left\{ M_0 + V_0 r \sin \theta + T_0 R(1 - \cos \theta) \right.$$

$$\left. + P\left[u\left(\theta - \frac{\pi}{2}\right) \right] R(1 - \cos \theta) + P\left[u\left(\theta - \frac{3\pi}{2}\right) \right] R \right\} R^2(1 - \cos \theta) \, d\theta$$

$$\frac{\partial U}{\partial V_0} = 0 = \int_0^{2\pi} 2\left\{ M_0 + V_0 R \sin \theta + T_0 R(1 - \cos \theta) \right.$$

$$\left. + P\left[u\left(\theta - \frac{\pi}{2}\right) \right] R(-\cos \theta) + P\left[u\left(\theta - \frac{3\pi}{2}\right) \right] R \cos \theta \right\} R^2 \sin \theta \, d\theta \qquad (c)$$

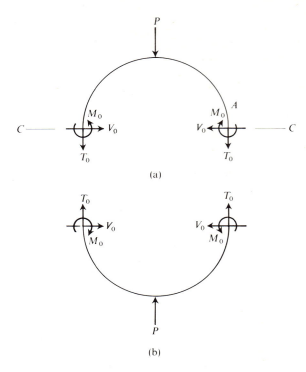

(a)

(b)

FIGURE 4.31

Carrying out the integrations and solving for M_0, V_0, and T_0 simultaneously we get:

$$M_0 = \frac{RP}{2\pi}(\pi - 2)$$

$$V_0 = 0$$

$$T_0 = -\frac{P}{2} \qquad (d)$$

We could have easily foreseen here that V_0 had to be zero by making use of the *symmetry* that is present in this problem. Accordingly consider Fig. 4.31(a) wherein we have shown a semi-circular portion of the ring. Now since the lower half of the ring is a mirror image in every respect with the upper half of the ring, we may rotate the free body shown in Fig. 4.31(a) 180° about the axis CC to get the proper free body diagram for the lower half of the ring. This has been done in Fig. 4.31(b). The important thing to note is that V_0 for each cut if not zero has the same direction for both free bodies—thus violating Newton's third law. We must conclude from the above observation that $V_0 = 0$.

Knowing M_0, V_0, and T_0 we may readily find the corresponding quantities at any section of the ring using simple rigid-body mechanics. Hence we have:

$$M(\theta) = M_0 + V_0 R \sin \theta + T_0 R (1 - \cos \theta)$$
$$+ P\left[u\left(\theta - \frac{\pi}{2}\right)\right] R(-\cos \theta) + P\left[u\left(\theta - \frac{3\pi}{2}\right)\right] R \cos \theta$$

$$V(\theta) = -\frac{P}{2} \sin \theta$$

$$T(\theta) = -\frac{P}{2} \cos \theta$$

Since we have not developed the theory of curved beams, we shall stop at this point and not get into deformation considerations of the ring. You are referred to other texts for such considerations.

4.12 CLOSURE

In this chapter we have dealt with only one independent variable. The problems were *one-dimensional* from this point of view. The results of Chaps. 2 and 3 sufficed our needs in this chapter. We now turn our attention in Chap. 5 to the torsion of shafts. Here we must employ two independent variables. As a consequence we will introduce several new approximation procedures stemming from variational considerations. We thus once again combine developments of structural mechanics with the continued development of variational methods.   ////

READING

Hoff, N. J.: "The Analysis of Structures," John Wiley and Sons, Inc., N.Y., 1956.
Kinney, J. S.: "Indeterminate Structural Analysis," Addison-Wesley Co., Reading, Mass., 1957.
Langhaar, H. L.: "Energy Methods in Applied Mechanics," John Wiley and Sons, Inc., N.Y., 1962.
Oden, J. T.: "Mechanics of Elastic Structures," McGraw-Hill Book Co., N.Y., 1967.
Rivello, R. M.: "Theory and Analysis of Flight Structures," McGraw-Hill Book Co., N.Y., 1969.
Vankatramen, B., and Patel, S. A.: "Structural Mechanics with Introductions to Elasticity and Plasticity," McGraw-Hill Book Co., N.Y., 1970.

PROBLEMS

4.1 Carry out the extremization of π for the beam (see Eq. (4.8)) using the delta operator and verify the results (Eqs. (4.9), (4.10), and (4.14)) that were simply taken from the formulations of Chap. 2.

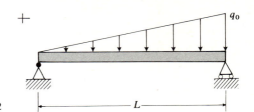

FIGURE 4.32

4.2 What is the deflection curve for a simply-supported beam (Fig. 4.32) carrying a triangular loading distribution according to the Timoshenko theory of beams? The cross section of the beam is rectangular. What is the ratio of the deflection at the endpoint according to the Timoshenko theory of beams to that of the technical theory of beams? Take $v = 0.3$.

4.3 Using the flexure formula $\tau_{xx} = Mz/I$ and the simple result $\tau_{xz} = V/A$ show that the Reissner functional for the Timoshenko beam is given as:

$$I_R = \int_0^L \left\{ -M(w'' - \beta) + \beta V - \frac{M^2}{2EI} - \frac{V^2}{2kAG} - qw \right\} dx$$

where we use the shear coefficient, k, in the complementary energy term. Extremizing this functional, show that we arrive at Eqs. (4.41(a)) and (4.41(b)), the differential equations of equilibrium for the Timoshenko beam, as well as Eqs. (4.32(a)) and (4.33), which are effectively stress–strain relations. Also show that the boundary conditions (4.40(b)) are satisfied. We have thus derived Timoshenko beam theory from the Reissner functional.

4.4 For the beam shown in Fig. 4.33, form one-term and two term approximate solutions. Compute the deflection under the load for each solution and compare these results with those from the technical theory of beams.

4.5 Devise a function that satisfies a complete set of boundary conditions for the cantilever beam shown in Fig. 4.34. Calculate the maximum deflection via the Ritz method and compare this with the result from the technical theory of beams. Take $P = 1000$ lb., $q_0 = 100$ lb/ft., and $L = 20$ ft.

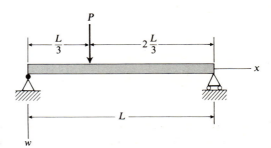

FIGURE 4.33

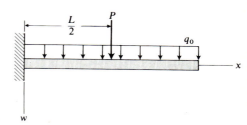

FIGURE 4.34   w

4.6   Obtain an approximate solution for a *clamped-clamped* beam, under a concentrated load
at $x = \xi$. Show how this may be used to obtain the deflection of a uniformly loaded beam,
and of a beam loaded by: $q(x) = 0$, $0 \leq x \leq \epsilon$; $q(x) = q_0$, $\epsilon \leq x \leq (L - \epsilon)$; $q(x) = 0$,
$(L - \epsilon) \leq x \leq L$. Show how the last solution can be made to yield the deflection of a
centrally loaded beam.

4.7   We wish to show that the flexure formula from the technical theory of beams is exact provided the bending moment M is a constant or a linear function of x along the beam.

Using equilibrium in the x direction, the flexure formula for τ_{xx}, and Hooke's law for
the shear stresses show that:

$$\frac{y}{I}\frac{\partial M}{\partial x} + G\left\{\left(\frac{\partial^2 u_x}{\partial y^2} + \frac{\partial^2 u_x}{\partial z^2} - \frac{\partial^2 u_x}{\partial x^2}\right) + \frac{\partial}{\partial x}(\text{div } \mathbf{u})\right\} = 0 \qquad (a)$$

Show from Hooke's law, with the aid of the flexure formula, that for beams:

$$\text{div } \mathbf{u} = \frac{My}{I\lambda} - \frac{2G}{\lambda}\frac{\partial u_x}{\partial x} \qquad (b)$$

Accordingly combine Eqs. (a) and (b) to reach:

$$\left(\frac{y}{GI} + \frac{y}{I\lambda}\right)\frac{\partial M}{\partial x} - \left(1 + \frac{2G}{\lambda}\right)\frac{\partial^2 u_x}{\partial x^2} = -\left(\frac{\partial^2 u_x}{\partial y^2} + \frac{\partial^2 u_x}{\partial z^2}\right) \qquad (c)$$

Using the flexure formula again show that:

$$u_x = \frac{y}{EI}\int M\, dx + \chi(y,z) \qquad (d)$$

where χ is an arbitrary function of y and z. Using Eq. (d) in Eq. (c) show that the following
equation holds:

$$D\frac{\partial M}{\partial x} = -\frac{1}{y}\left(\frac{\partial^2 \chi}{\partial y^2} + \frac{\partial^2 \chi}{\partial z^2}\right)$$

where D is a constant. Why can you conclude from this equation that:

$$M = C_1 x + C_2$$

where C_1 and C_2 are constants? We have thus shown that for the flexure formula to give
exact results it is necessary that the bending moment be a linear function of x.

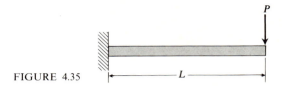

FIGURE 4.35

4.8 Give an approximation for the bending moment distribution and the deflection curve for the cantilever beam in Fig. 4.35 using the Reissner functional approach.

4.9 Express the Reissner functional for the cantilever beam shown in Fig. 4.36 having a triangular axial force distribution, and a uniform normal loading distribution. Get an approximate solution to the bending moment, the vertical deflection curve w and the horizontal stretching curve u_s using the Reissner functional. Compare with results from the technical theory of beams at $x = L$. Assume linear elastic behavior.

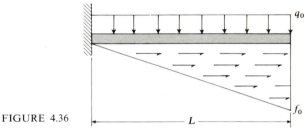

FIGURE 4.36

4.10 For the continuous beam shown in Fig. 4.37, find the magnitude and direction of the centerline deflection of the right half of the structure. Assume a uniform EI.

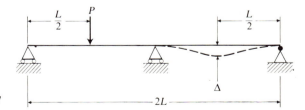

FIGURE 4.37

4.11 Determine the force of the spring for the beam shown in Fig. 4.38 using energy methods. First consider the spring force as the redundant quantity and then consider the support force as the redundant quantity. What are the limiting cases that occur when $\alpha \to 0$ or $\alpha \to 1$ and when $K \to 0$ or $K \to \infty$. Assume uniform EI.

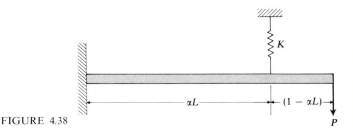

FIGURE 4.38

4.12 Find the deflection and slope at point *a* of the simply-supported beam shown in Fig. 4.39 using the second Castigliano theorem. Assume linear elastic behavior.

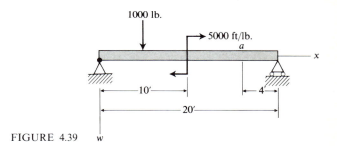

FIGURE 4.39   w

4.13 Find the supporting forces in the beam of Fig. 4.40 if support *B* sags 0.1 in. as a result of the loading. Assume linear elastic behavior.

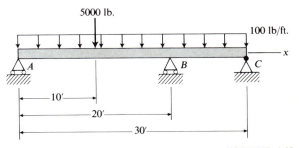

FIGURE 4.40

4.14 It is desired to limit the centerline deflection of the spandrel beam (Fig. 4.41) to $(1/288)L$, where *L* is the length. Expressed in terms of the usual beam parameters (E,I,L,P), how small must the load *P* be so as not to violate the constraint?

4.15 How large a moment is required at the free end of a cantilever beam to provide a slope there of $\theta(x = L) = 0.036$ in/in? Let the beam have the properties $E = 30 \times 10^6$ psi, $L = 90$ in., $I = 1$ in^4. How does the slope vary along the beam?

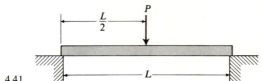

FIGURE 4.41

4.16 Starting with the reciprocal theorem (Problem 1.22) and using the basic equations of beam theory (strain-displacement, equilibrium) develop the reciprocal theorem for beams in the form

$$\int_0^L q^{(1)}(x)w^{(2)}(x)\,dx = \int_0^L q^{(2)}(x)w^{(1)}(x)\,dx.$$

4.17 Using the result of Problem 4.16, and the Dirac delta function, show that the deflection at a point at $x = \xi_2$ produced by a load P at $x = \xi_1$, is identically equal to the deflection at $x = \xi_1$ produced by the load P at ξ_2.

4.18 The beam shown in Fig. 4.42 is made of material that behaves non-linearly. In tension the stress-strain relation is given as

$$\tau_{xx} = 30 \times 10^6(\epsilon_{xx})^{3/4}$$

while in compression the stress-strain law is given as:

$$\tau_{xx} = -30 \times 10^6(-\epsilon_{xx})^{3/4}$$

Formulate an equation for finding one of the supporting forces. If time allows solve for this supporting force numerically with the aid of a computer for $b = 2$ in. and $d = 4$ in.

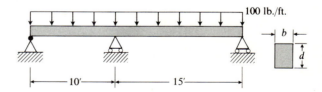

FIGURE 4.42

4.19 The material in the beam shown in Fig. 4.43 behaves according to the following stress-strain law:

$$\tau_{xx} = C_1 \sinh (C_2 \epsilon_{xx})$$

Find π^* as a function of d^2w/dx^2. Now using the left support R as the redundant constraint find an equation involving d^2w/dx^2, x, and R stemming from the bending moment at a

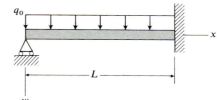

FIGURE 4.43 w

section. Explain how with the aid of a computer you could then find an approximation of the proper value of R.

4.20   Find the supporting forces for the frame shown in Fig. 4.44. Assume linear elastic behavior.

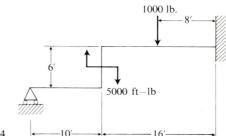

FIGURE 4.44

4.21   For the given frame, find the horizontal deflection at A (Fig. 4.45). Would the deflection there vanish if $Q = 0$? Explain. All members of same length L and uniform stiffness EI.

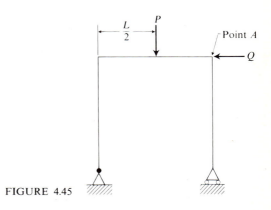

FIGURE 4.45

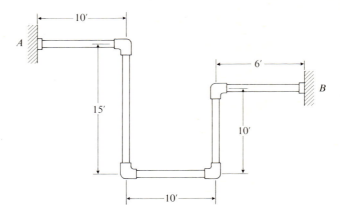

FIGURE 4.46

4.22 A pipe (see Fig. 4.46) in the vertical plane is fixed at *A* and *B*. What are the supporting forces and moments at *A* and *B* if the pipes and their contents weigh 50 lb/ft.? Assume linear elastic behavior.

4.23 What are the deflection equations for the frame shown in Fig. 4.47? Assume linear elastic behavior.

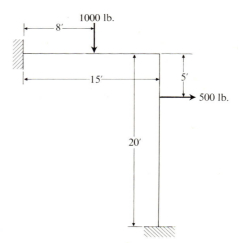

FIGURE 4.47

4.24 If the bottom support in the preceding problem were pin-connected, what would then be the deflection equations for the frame?

4.25 Give the deflection curves for the frame shown in Fig. 4.48. Assume linear elastic behavior.

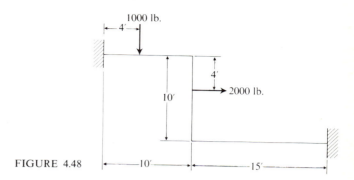

FIGURE 4.48

4.26   Find the bending moment along the portion *AB* of the rectangular frame shown in Fig. 4.49. One corner *A* is pinned; the other corner *B* is supported by rollers. Assume linear elastic behavior.

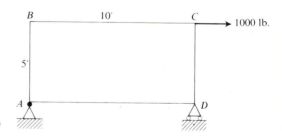

FIGURE 4.49

4.27   What is the bending moment distribution around the closed ring shown in Fig. 4.50? Assume linear elastic behavior.

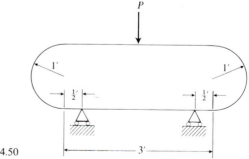

FIGURE 4.50

5

TORSION

5.1 INTRODUCTION

In this chapter we shall consider the Saint Venant theory of torsion for uniform prismatic elastic rods loaded by twisting couples at the ends of the rod (see Fig. 5.1). You will recall that for the special case of a circular cross section, it is assumed in strength of materials (and shown valid in the theory of elasticity) that cross sections of the rod merely rotate as rigid surfaces under the action of the twisting couples. Thus if θ is the angle of rotation for a cross section at a position z, then we have the

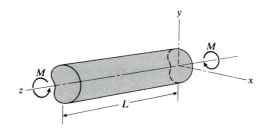

FIGURE 5.1

following result for small deformations (see Fig. 5.2):

$$u = -r\theta \sin \beta = -y\theta$$
$$v = r\theta \cos \beta = x\theta$$
$$w = 0$$

We now introduce the *rate of twist*, $\alpha = d\theta/dz$, which is a constant for any given problem with end couples. Taking $\theta = 0$ when $z = 0$, we conclude that $\theta = \alpha z$ and we can give the above equations as follows:

$$u = -\alpha yz \quad = -\frac{d\theta}{dz} yz$$
$$v = \alpha xz$$
$$w = 0 \qquad (5.1)$$

For an *arbitrary cross section* we must abandon the assumption that plane sections remain plane (i.e., $w = 0$) and introduce a function $\kappa(x,y)$ to account for "warping" in the z direction. Thus:

$$u = -\alpha yz$$
$$v = \alpha xz$$
$$w = \kappa(x,y) \qquad (5.2)$$

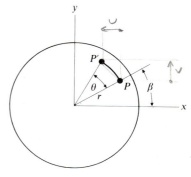

FIGURE 5.2

Note that the warping shape of a cross section is not a function of z; all sections have the same deformed shape. We shall use the above model for torsional deformation. The corresponding strain field is:

$$\epsilon_{xx} = 0 \qquad \epsilon_{xy} = 0$$

$$\epsilon_{yy} = 0 \qquad \epsilon_{xz} = \frac{1}{2}\left(-\alpha y + \frac{\partial \kappa}{\partial x}\right)$$

$$\epsilon_{zz} = 0 \qquad \epsilon_{yz} = \frac{1}{2}\left(\alpha x + \frac{\partial \kappa}{\partial y}\right) \qquad (5.3)$$

We now examine the principle of total potential energy for this assumed deformation field.

5.2   TOTAL POTENTIAL ENERGY; EQUATION FOR TORSION

The total potential energy for torsion can readily be given for linear elastic behavior:

$$\pi = \iiint_V (\tau_{xz}\epsilon_{xz} + \tau_{yz}\epsilon_{yz})\, dv - ML\alpha \qquad (5.4)$$

Hence from the principle of total potential energy:

$$\delta^{(1)}\pi = \delta^{(1)}\left\{\iiint_V (\tau_{xz}\epsilon_{xz} + \tau_{yz}\epsilon_{yz})\, dv\right\} - ML\,\delta\alpha = 0$$

Using Hooke's law directly, $\tau_{xz} = 2G\epsilon_{xz}$ and $\tau_{yz} = 2G\epsilon_{yz}$, the above equation becomes:

$$\delta^{(1)}\pi = 2\delta^{(1)}\left\{\iiint_V G(\epsilon_{xz}^2 + \epsilon_{yz}^2)\, dv\right\} - ML\,\delta\alpha = 0$$

Substituting from Eq. (5.3) we get:

$$\delta^{(1)}\pi = \delta^{(1)}\left\{\frac{1}{2}G \iiint_V\left[\left(\frac{\partial \kappa}{\partial x} - y\alpha\right)^2 + \left(\frac{\partial \kappa}{\partial y} + x\alpha\right)^2\right] dv\right\} - ML\,\delta\alpha = 0$$

$$= \iiint_V G\left[\left(\frac{\partial \kappa}{\partial x} - \alpha y\right)\left(\frac{\partial(\delta\kappa)}{\partial x} - y\,\delta\alpha\right)\right.$$

$$+ \left.\left(\frac{\partial \kappa}{\partial y} + x\alpha\right)\left(\frac{\partial(\delta\kappa)}{\partial y} + x\,\delta\alpha\right)\right] dv - ML\,\delta\alpha = 0$$

Collecting terms we obtain:

$$\left\{\iiint_V G\left[\left(\frac{\partial \kappa}{\partial x} - y\alpha\right)(-y) + \left(\frac{\partial \kappa}{\partial y} + x\alpha\right)(x)\right] dv - ML\right\}\delta\alpha$$

$$+ \iiint_V G\left\{\left(\frac{\partial \kappa}{\partial x} - y\alpha\right)\frac{\partial}{\partial x}(\delta\kappa) + \left(\frac{\partial \kappa}{\partial y} + x\alpha\right)\frac{\partial}{\partial y}(\delta\kappa)\right\} dv = 0$$

Integrating over z from 0 to L for both integrals and then using Green's theorem in the second integral (in order to integrate by parts) we find that:

$$\left\{ \iint_R LG\left[\left(\frac{\partial \kappa}{\partial x} - y\alpha\right)(-y) + \left(\frac{\partial \kappa}{\partial y} + \alpha x\right)(x) \right] dA - ML \right\} \delta\alpha$$

$$+ GL \iint_R \left[\frac{\partial}{\partial x}\left(\frac{\partial \kappa}{\partial x} - y\alpha\right) + \frac{\partial}{\partial y}\left(\frac{\partial \kappa}{\partial y} + x\alpha\right) \right] \delta\kappa \, dA$$

$$- GL \oint_C \left[\left(\frac{\partial \kappa}{\partial x} - y\alpha\right)a_{nx} + \left(\frac{\partial \kappa}{\partial y} + \alpha x\right)a_{ny} \right] \delta\kappa \, dS = 0$$

Since $\delta\alpha$ and $\delta\kappa$ are independent it is clear that:

$$\iint_R LG\left[\left(\frac{\partial \kappa}{\partial x} - y\alpha\right)(-y) + \left(\frac{\partial \kappa}{\partial y} + \alpha x\right)(x) \right] dA - ML = 0$$

therefore

$$M = G \int \int_R \left[x\frac{\partial \kappa}{\partial y} - y\frac{\partial \kappa}{\partial x} + \alpha(x^2 + y^2) \right] dA \quad (a)$$

Also:

$$GL\left[\frac{\partial}{\partial x}\left(\frac{\partial \kappa}{\partial x} - y\alpha\right) + \frac{\partial}{\partial y}\left(\frac{\partial \kappa}{\partial y} + x\alpha\right) \right] = 0$$

therefore

$$\frac{\partial^2 \kappa}{\partial x^2} + \frac{\partial^2 \kappa}{\partial y^2} = 0 \quad (b)$$

Finally we have the following natural boundary condition:

$$\left[\left(\frac{\partial \kappa}{\partial x} - y\alpha\right)a_{nx} + \left(\frac{\partial \kappa}{\partial y} + x\alpha\right)a_{ny} \right] = 0 \qquad \text{on boundary} \quad (c) \quad (5.5)$$

Now let κ be given as $\alpha\phi$ where ϕ is called the *warping function*. Thus:

$$\kappa = \alpha\phi \qquad (5.6)$$

Then on substitution into Eqs. (5.5) (b), (a), and (c) we get the following equations:

$$\nabla^2 \phi = 0 \qquad (a)$$

$$M = G\alpha \int \int_R \left[x\frac{\partial \phi}{\partial y} - y\frac{\partial \phi}{\partial x} + (x^2 + y^2) \right] dA \qquad (b)$$

$$\left(\frac{\partial \phi}{\partial x} - y\right)a_{nx} + \left(\frac{\partial \phi}{\partial y} + x\right)a_{ny} = 0 \qquad \text{(on boundary)} \qquad (c) \quad (5.7)$$

At this stage of the discussion we have used the total potential energy associated with a proposed deformation which involves the unknown warping function ϕ as well as the rate of twist α. The extremization of π has led to equations of equilibrium for the proposed system in terms of ϕ, and has given us as well the appropriate natural boundary conditions. It is well known from experimental and exact analyses that the equations, when solved for the aforementioned system, give results that are very close to the actual case. The next step in our study would be to consider methods arising from the variational approach to reach approximate solutions to the torsion problem. We shall find that it is often desirable in this regard to use the total complementary energy functional for torsion. We accordingly now consider this functional and the associated boundary value problem.

5.3   THE TOTAL COMPLEMENTARY ENERGY FUNCTIONAL

For linear elastic behavior there are only two non-zero stresses associated with the deformation proposed in Sec. 5.1. They are the shear stresses τ_{xz} and τ_{yz}. Accordingly the equation of equilibrium for such a situation is (in the absence of body forces):

$$\frac{\partial \tau_{xz}}{\partial x} + \frac{\partial \tau_{yz}}{\partial y} = 0 \qquad (5.8)$$

As in the case of plane stress we may satisfy this equation exactly if we express the stresses in terms of a *stress function*, Ψ, as follows:

$$\tau_{xz} = G\alpha \frac{\partial \Psi}{\partial y} \qquad \tau_{yz} = -G\alpha \frac{\partial \Psi}{\partial x} \qquad (5.9)$$

Thus working with Ψ permits the employment at all times of a *statically compatible* stress field. On the lateral boundary we have $T^{(v)} = 0$ and the following condition of equilibrium must be satisfied (see Eq. (1.5))

$$\tau_{zx} a_{nx} + \tau_{zy} a_{ny} = 0$$

therefore

$$G\alpha \left(\frac{\partial \Psi}{\partial y} a_{nx} - \frac{\partial \Psi}{\partial x} a_{ny} \right) = 0 \qquad (5.10)$$

From Fig. 5.3 we see that:

$$a_{nx} = \frac{dy}{ds} \qquad a_{ny} = -\frac{dx}{ds} \qquad (5.11)$$

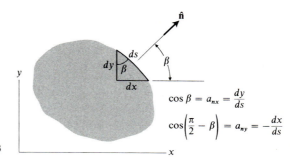

FIGURE 5.3

Hence on substitution into Eq. (5.10) we get:

$$\frac{\partial \Psi}{\partial x}\frac{dx}{ds} + \frac{\partial \Psi}{\partial y}\frac{dy}{ds} = 0$$

therefore

$$\frac{d\Psi}{ds} = 0 \qquad \text{on boundary} \qquad (5.12)$$

We conclude that Ψ must be a constant on the boundary. For a simply-connected domain, there is no loss in generality in taking $\Psi = 0$ on the boundary.

We are now ready to express the complementary energy functional in terms of Ψ. Note that the tractions have been specified as zero on the lateral periphery of the body, but only resultants M have been specified at the ends. We shall for convenience consider that the end $z = 0$ is fixed so that $u = v = 0$ there. Furthermore we shall later show that specifying M at the ends is equivalent to specifying the rate of twist α and is thus equivalent to specifying the displacement field at the end $z = L$. Hence region S_2 consists of the end faces of the shaft, and with $u_i = 0$ at $z = 0$, actually only the end at $z = L$ need be considered. Thus we have for the total complementary energy:

$$\pi^* = \iiint_V (\tau_{xz}\epsilon_{xz} + \tau_{yz}\epsilon_{yz})\, dv - \left[\iint_R T_i^{(v)} u_i\, dA\right]_{\text{at } z = L} \qquad (5.13)$$

We now employ Hooke's law to replace the strain terms in the first integral and, using $L = 1$, integrate with respect to z for this integral. In the second integral carry out the inner product using stress components for $T_i^{(v)}$ and displacement components given by Eqs. (5.2) for u_i. We then get:

$$\pi^* = \frac{1}{2G}\iint_R (\tau_{xz}{}^2 + \tau_{yz}{}^2)\, dA - \alpha\left[\iint_R (-\tau_{xz}y + \tau_{yz}x)\, dA\right]_{z = 1} \qquad (5.14)$$

Since the integrand of the second integral, like that of the first, is not a function of z it applies for all values of z and so we may delete the subscript notation $z = 1$. Now introducing the stress function Ψ, we obtain the following result for π^*:

$$\pi^* = \frac{G\alpha^2}{2} \int\int_R \left[\left(\frac{\partial \Psi}{\partial x}\right)^2 + \left(\frac{\partial \Psi}{\partial y}\right)^2 \right] dA - \int\int_R G\alpha^2 \left(-\frac{\partial \Psi}{\partial y} y - \frac{\partial \Psi}{\partial x} x \right) dA$$

Also, integrate the last surface integral by parts. We find:

$$\pi^* = \frac{G\alpha^2}{2} \int\int_R \left[\left(\frac{\partial \Psi}{\partial x}\right)^2 + \left(\frac{\partial \Psi}{\partial y}\right)^2 \right] dA - \int\int_R 2G\alpha^2 \Psi \, dA$$
$$+ \oint_C G\alpha^2 \Psi (ya_{ny} + xa_{nx}) \, dl$$

Noting that Ψ is zero on the boundary, we now collect terms:

$$\pi^* = \frac{G\alpha^2}{2} \int\int_R \left[\left(\frac{\partial \Psi}{\partial x}\right)^2 + \left(\frac{\partial \Psi}{\partial y}\right)^2 - 4\Psi \right] dA \quad (a)$$
$$= \frac{G\alpha^2}{2} \int\int_R [(\nabla \Psi)^2 - 4\Psi] \, dA \quad\quad (b) \quad (5.15)$$

Now we are in a position to carry out the variational process for π^*, according to the principle of total complementary energy:

$$\delta^{(1)}\pi^* = 0 = \frac{G\alpha^2}{2} \int\int_R \left[2\frac{\partial \Psi}{\partial x}\frac{\partial \delta\Psi}{\partial x} + 2\frac{\partial \Psi}{\partial y}\frac{\partial \delta\Psi}{\partial y} - 4\,\delta\Psi \right] dA$$

After integrating by parts and dropping $G\alpha^2$ we see that:

$$\int\int_R \left[-\frac{\partial^2 \Psi}{\partial x^2} - \frac{\partial^2 \Psi}{\partial y^2} - 2 \right] \delta\Psi \, dA + \oint_C \left(\frac{\partial \Psi}{\partial x}a_{nx} + \frac{\partial \Psi}{\partial y}a_{ny} \right) \delta\Psi \, dl = 0$$

Since Ψ is specified on the boundary then $\delta\Psi$ must be zero there and so we have for the above equation:

$$\int\int_R [\nabla^2 \Psi + 2] \delta\Psi \, dA = 0$$

Now using the fundamental lemma we have the result:

$$\nabla^2 \Psi = -2 \quad\quad (5.16)$$

We would expect from considerations of the total complementary energy that Eq. (5.16) is a kinematic compatibility equation of some sort. To investigate this[1] examine the differential of the

---

[1] Recall we have done this for the case of plane stress in Sec. 3.4.

displacement component w as follows:

$$dw = \frac{\partial w}{\partial x} dx + \frac{\partial w}{\partial y} dy \qquad (5.17)$$

Note that

$$\epsilon_{xz} = \frac{1}{2}\left(\frac{\partial w}{\partial x} + \frac{\partial u}{\partial z}\right) = \frac{1}{2}\left(\frac{\partial w}{\partial x} - \alpha y\right)$$

$$\epsilon_{yz} = \frac{1}{2}\left(\frac{\partial w}{\partial y} + \frac{\partial v}{\partial z}\right) = \frac{1}{2}\left(\frac{\partial w}{\partial y} + \alpha x\right)$$

where we have used Eq. (5.2) for u and v. Hence we have from above:

$$\frac{\partial w}{\partial x} = 2\epsilon_{xz} + \alpha y = \frac{\tau_{xz}}{G} + \alpha y = \alpha\left(\frac{\partial \Psi}{\partial y} + y\right)$$

$$\frac{\partial w}{\partial y} = 2\epsilon_{yz} - \alpha x = \frac{\tau_{yz}}{G} - \alpha x = \alpha\left(-\frac{\partial \Psi}{\partial x} - x\right)$$

Substituting the above results into Eq. (5.17) we get:

$$dw = \alpha\left(\frac{\partial \Psi}{\partial y} + y\right) dx - \alpha\left(\frac{\partial \Psi}{\partial x} + x\right) dy$$

Now integrate over any closed path in the domain including the boundary. We get:

$$\oint_C dw = \oint_C \left\{ \alpha\left(\frac{\partial \Psi}{\partial y} + y\right) dx - \alpha\left(\frac{\partial \Psi}{\partial x} + x\right) dy \right\}$$

$$= \oint_C \left\{ \alpha\left(\frac{\partial \Psi}{\partial y} + y\right) \frac{dx}{ds} - \alpha\left(\frac{\partial \Psi}{\partial x} + x\right) \frac{dy}{ds} \right\} ds$$

Employing Eqs. (5.11) we may restate the above result as follows:

$$\oint_C dw = -\oint_C \alpha\left\{ \left(\frac{\partial \Psi}{\partial y} + y\right) a_{ny} + \left(\frac{\partial \Psi}{\partial x} + x\right) a_{nx} \right\} ds$$

Now employ the integration by parts formula (Eq. (I.52) Appendix I) on the right side of the equation to get:

$$\oint_C dw = -\alpha \iint_R \left[\left(\frac{\partial^2 \Psi}{\partial y^2} + 1\right) + \left(\frac{\partial^2 \Psi}{\partial x^2} + 1\right) \right] dA$$

$$= -\alpha \iint_R (\nabla^2 \Psi + 2)\, dA$$

If we are to have kinematic compatibility, then $\oint dw = 0$ to prevent a dislocation in w. This then means that the integrand in the above surface integral must be zero. But this requirement is simply Eq. (5.16), showing that the latter equation is in fact a kinematic compatibility requirement.

There is a constraint that we must impose on Ψ to give the resultant torque on the end surface its proper value M. That is:

$$M = \iint_R (-y\tau_{zx} + x\tau_{zy})\, dA$$

$$= -\iint_R G\alpha\left(y\frac{\partial \Psi}{\partial y} + x\frac{\partial \Psi}{\partial x}\right) dA$$

Integrating by parts we get:

$$M = 2G\alpha \iint_R \Psi\, dA - \oint_C G\alpha(ya_{ny} + xa_{nx})\Psi\, dl$$

Since $\Psi = 0$ on the boundary we obtain the following constraining equation to go with Eq. (5.16)[1]:

$$M = 2G\alpha \iint_R \Psi\, dA$$

To summarize, we have from the principle of total complementary energy the following boundary value problem:

$$\nabla^2 \Psi = -2 \qquad\qquad (a)$$

$$\Psi = 0 \qquad \text{on boundary} \quad (b)$$

$$M = 2G\alpha \iint_R \Psi\, dA \qquad (c) \ \ (5.18)$$

The idea is to find a solution Ψ from Eq. 5.18(a) satisfying Eq. 5.18(b) in terms of the parameter α. Now going to Eq. (5.18(c)) we can determine α in terms of M.

We have thus presented two systems of equations for torsion—one deals with the *warping function*, ϕ, (see Eq. (5.7)) and the present case where we employ the *stress function*, Ψ. We will now set forth yet a third statement of the boundary value problem using the *conjugate function* ψ. Since ϕ is a harmonic function we know that there is a conjugate function ψ, also harmonic and related to ϕ through the Cauchy-Riemann equations. That is:

$$\frac{\partial \phi}{\partial x} = \frac{\partial \psi}{\partial y}$$

$$\frac{\partial \phi}{\partial y} = -\frac{\partial \psi}{\partial x} \qquad (5.19)$$

---

[1] We can now see from the following equation that specifying the torques M at the ends of the shaft is equivalent to specifying the rate of twist α. That is, M and α play equivalent roles in the mathematical aspects of the problem.

We leave it for you as an exercise to show that in terms of ψ we may represent the torsion problem as follows:

$$\nabla^2 \psi = 0 \qquad\qquad\qquad\qquad (a)$$

$$\psi = \tfrac{1}{2}(x^2 + y^2) \qquad \text{on boundary} \qquad\qquad (b)$$

$$M = G\alpha \iint_R \left(-x\frac{\partial\psi}{\partial x} - y\frac{\partial\psi}{\partial y} + x^2 + y^2\right) dA \quad (c) \;\; (5.20)$$

Also, we can show readily that the following equation

$$\psi = \Psi + \tfrac{1}{2}(x^2 + y^2) \qquad\qquad (5.21)$$

relates the conjugate function ψ to the stress function Ψ.

5.4 APPROXIMATE SOLUTIONS FOR LINEAR ELASTIC BEHAVIOR VIA THE RITZ METHOD

We shall now use the total complementary energy coupled with the Ritz method to formulate approximate solutions to the torsion problem, first for a linear elastic body and then for a non-linear elastic body.

We consider first a rectangular cross-section, having dimensions $2a$ and $2b$ as shown in Fig. 5.4. As a first choice for a coordinate function we consider

$$\Psi_1 = C_1(a^2 - x^2)(b^2 - y^2) \qquad\qquad (5.22)$$

Note that the boundary condition (Eq. (5.18(b))) is satisfied and the function is even with respect to x and y as would be expected of the solution. Now substitute into the total complementary energy functional (Eq. (5.15)). Extremizing with respect to C_1, we get for this constant:

$$C_1 = \frac{5}{4}\left(\frac{1}{b^2 + a^2}\right)$$

The *torsional rigidity* D defined as M/α is then given by the following formula (see Eq. (5.18(c))):

$$D = 2G \iint_R \Psi \, dA = \frac{40}{9}\frac{a^3 b^3}{b^2 + a^2}G \qquad\qquad (5.23)$$

For the maximum stress (occurring at the middle of the widest section) we get for $b > a$:

$$\tau_{\max} = \frac{5}{2}\frac{b^2 a}{b^2 + a^2}\alpha G \qquad\qquad (5.24)$$

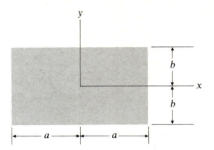

FIGURE 5.4

We may compare results from the above procedure with those from an exact solution from the theory of elasticity in the form of an infinite series.[1] Taking $a = 1$ and computing (D/G) for various values b/a we have shown the results in Table 5.1. Notice that the obtained approximation is *from below*. (We shall soon present a way of formulating an approximation for D *from above* by the Trefftz method so as to provide a means of *bracketing* the desired result.) To improve the result we may use a more complex function as follows:

$$\Psi = [x^2 - a^2][y^2 - b^2]\{C_1 + C_2 x^2 y^2 + C_3 x^4 y^4 + C_4 x^6 y^6 + \cdots\} \qquad (5.25)$$

Employing this expression with only three unknown coefficients and extremizing the complementary potential energy expression (Eq. (5.15)) with respect to C_i we obtain the following equations for the C's for the rectangular section:

$$
\begin{bmatrix}
5.69 & 0.162 & 0.0232 \\
0.162 & 0.0851 & 0.0246 \\
0.0232 & 0.0246 & 0.0112
\end{bmatrix}
\begin{Bmatrix}
C_1 \\
a^2 b^2 C_2 \\
a^4 b^4 C_3
\end{Bmatrix}
= \frac{1}{a^2 + b^2}
\begin{Bmatrix}
7.111 \\
0.284 \\
0.0522
\end{Bmatrix}
$$

Table 5.1

	$(D/G)_{app}$	$(D/G)_{exact}$
$b/a = 1$	2.22	2.25
$b/a = 2$	7.12	7.32
$b/a = 3$	12.00	12.60
$b/a = 4$	16.73	18.60
$b/a = 5$	21.37	23.30

[1] See Timoshenko and Goodier: "Theory of Elasticity," McGraw-Hill Book Co., Chap. 11.

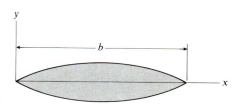

FIGURE 5.5

Improvement of the solution is illustrated in Table 5.2.

We now consider a class of problems for a section having a chord of width b as has been shown in Fig. 5.5 where the upper and lower curves are given respectively as follows:

$$y = K_1\left(\frac{x}{b}\right)^n\left[1 - \left(\frac{x}{b}\right)^m\right]^p = K_1 H_1\left(\frac{x}{b}\right)$$

$$y = -K_2\left(\frac{x}{b}\right)^q\left[1 - \left(\frac{x}{b}\right)^r\right]^s = -K_2 H_2\left(\frac{x}{b}\right) \qquad (5.26)$$

K_1 and K_2 are positive constants and n,m,p,q,r, and s are positive integers. Now we form the following function:

$$\Psi = C_1(y - K_1 H_1)(y + K_2 H_2) \qquad (5.27)$$

Clearly such a function is zero on the entire boundary of the region. The total complementary energy then becomes (see Eq. (5.15(a))) using the variable ξ for x/b and dropping $G\alpha^2/2$:

$$\pi^* = \int_0^1 \int_{y=-K_2 H_2}^{y=K_1 H_1} \left\{ \frac{C_1^2}{b^2}[(y - K_1 H_1)^2(K_2 H_2')^2 + (y + K_2 H_2)^2(K_1 H_1')^2 \right.$$
$$+ 2(y - K_1 H_1)(y + K_2 H_2)K_1 H_1' K_2 H_2'] + C_1^2[(y - K_1 H_1)^2$$
$$+ (y + K_2 H_2)^2 + 2(y + K_2 H_2)(y - K_1 H_1)]$$
$$\left. - 4C_1(y - K_1 H_1)(y + K_2 H_2) \right\} dy\, b\, d\xi$$

Table 5.2

	Ψ	$\dfrac{(a^2+b^2)D}{a^3 b^3 G}$
1 Parameter C_1		4.4444
2 Parameters C_1, C_2		4.4854
3 Parameters C_1, C_2, C_3		4.4856
Exact		4.5000

Hence

$$
\begin{aligned}
\frac{\partial \pi^*}{\partial C_1} = 0 = \int_0^1 \int_{y=-K_2H_2}^{y=K_1H_1} 2 \Big\{ & \frac{C_1}{b^2}[(y - K_1H_1)^2(K_2H_2')^2 + (y + K_2H_2)^2(K_1H_1')^2 \\
& + 2(y - K_1H_1)(y + K_2H_2)K_1K_2H_1'H_2'] + 2C_1[(y - K_1H_1)^2 \\
& + (y + K_2H_2)^2 + 2(y + K_2H_2)(y - K_1H_1)] \\
& - 4(y - K_1H_1)(y + K_2H_2) \Big\} dy \, b \, d\xi
\end{aligned}
$$

Now after carrying out the integration with respect to the variable y, we end up with a series of integrals of the form:

$$
I(a,g,h) = \int_0^1 \xi^a[1 - \xi^g]^h \, d\xi \tag{5.28}
$$

We have listed results for various values of a, g, and h for the above integration in Appendix III and so we can readily get C_1 and thus find an approximate solution for the torsional rigidity.

As an example we note that for the case where $n = m = r = s = 2$, $p = q = 1$, and $K_1 = K_2 = b = 1$ we have a shape as shown in Fig. 5.5. The torsional rigidity then can be computed to be:

$$
D = 0.01122 \, G
$$

5.5 APPROXIMATE SOLUTIONS FOR A NONLINEAR ELASTIC TORSION PROBLEM

One of the virtues of the methods of total potential energy and total complementary energy is that these methods are valid for *nonlinear* elastic behavior. In Chap. 3 we considered a simple nonlinear elastic truss using the second Castigliano theorem (which you will recall was derived from the principle of total complementary energy). And in Chap. 4 we outlined a procedure for solving statically indeterminate non-linearly elastic beam problems via the method of total complementary energy. We now consider a nonlinearly elastic torsion problem.

An approximation for the one-dimensional test of certain materials such as copper can be achieved by using a stress-strain relation of the following form:

$$
\epsilon = \epsilon_0 \sinh \frac{\tau}{\tau_0} \tag{5.29}
$$

where ϵ_0 and τ_0 are constants dependent on the material to be represented. An example of such a stress-strain curve is shown in Fig. 5.6. For more general states of stress we replace ϵ and τ by quantities proportional to the *octahedral shear strain* and the *octahedral shear stress* respectively.[1] Thus:

$$
C_1 \epsilon_{\text{oct}} = \epsilon_0 \sinh \frac{C_2 \tau_{\text{oct}}}{\tau_0} \tag{5.30}
$$

[1] See Problems 1.7 and 1.8 of Chap. 1.

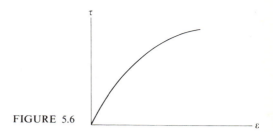

FIGURE 5.6

We choose the values of C_1 and C_2 so that when Cartesian stress and strain components are respectively used for the octahedral stress and the octahedral strain, the formulation reduces to that of Eq. (5.29) when applied to a one-dimensional test case for a Poisson ratio ν of $1/2$. You will be asked to show that $C_1 = \sqrt{2}$ and $C_2 = 3/\sqrt{2}$ for this requirement.[1] Accordingly we have:

$$\sqrt{2}\epsilon_{\text{oct}} = \epsilon_0 \sinh\left(\frac{3/\sqrt{2}}{\tau_0}\tau_{\text{oct}}\right) \qquad (5.31)$$

We now consider the case of torsion where we have, in accordance with the model of Sec. 5.2, only strains γ_{xz} and γ_{yz} as nonzero and where we assume only stresses τ_{xz} and τ_{yz} as nonzero. The result from Eq. (5.31) is then (see footnote):

$$\sqrt{\gamma_{yz}^2 + \gamma_{xz}^2} = \sqrt{3}\epsilon_0 \sinh\left[\frac{\sqrt{3}}{\tau_0}\sqrt{\tau_{yz}^2 + \tau_{xz}^2}\right] \qquad (5.32)$$

Within the framework of the above formulation we now may make another stipulation. We will assume that the *Hencky total strain theory*, which is used in the theory of plasticity, is valid for the problem at hand.[2] The theory stipulates that:

$$\frac{\gamma_{xz}}{2\tau_{xz}} = \frac{\gamma_{yz}}{2\tau_{yz}} = \frac{\gamma_{xy}}{2\tau_{xy}} \qquad (5.33)$$

---

[1] Recall from Problem 1.8 that:

$$\tau_{\text{oct}}^2 = \tfrac{1}{9}[2(I_t)^2 + 6(II_t)^2]$$

From this formulation we see that:

$$\tau_{\text{oct}} = \tfrac{1}{3}\sqrt{(\tau_{xx} - \tau_{yy})^2 + (\tau_{xx} - \tau_{zz})^2 + (\tau_{yy} - \tau_{zz})^2 + 6(\tau_{xy}^2 + \tau_{xz}^2 + \tau_{yz}^2)}$$

Similarly ϵ_{oct} is given as:

$$\epsilon_{\text{oct}} = \tfrac{1}{3}\sqrt{(\epsilon_{xx} - \epsilon_{yy})^2 + (\epsilon_{xx} - \epsilon_{zz})^2 + (\epsilon_{yy} - \epsilon_{zz})^2 + \tfrac{3}{2}(\gamma_{xy}^2 + \gamma_{yz}^2 + \gamma_{xz}^2)}$$

[2] For a more detailed explanation of this step as well as the problem of inelastic torsion under discussion see Smith and Sidebottom: "Inelastic Behavior of Load Carrying Members," John Wiley and Sons, Chaps. 2–8.

For our case we can say that:

$$\frac{\gamma_{xz}}{\tau_{xz}} = \frac{\gamma_{yz}}{\tau_{yz}} \tag{5.34}$$

Replace γ_{yz} in Eq. (5.32) using the above result and replace the stresses with appropriate derivatives of the stress function Ψ. Finally let $G\alpha\Psi$ be replaced by the function $\mathscr{H}$. Denoting $(\partial/\partial x)\mathscr{H}$ as $\mathscr{H}_x$, etc. we then get for γ_{xz}:

$$\gamma_{xz} = \frac{\sqrt{3}\epsilon_0 \mathscr{H}_y}{\sqrt{\mathscr{H}_x^2 + \mathscr{H}_y^2}} \sinh\left[\frac{\sqrt{3}}{\tau_0}\sqrt{\mathscr{H}_x^2 + \mathscr{H}_y^2}\right] \tag{5.35}$$

Similarly we may get the following result for γ_{yz}:

$$\gamma_{yz} = -\frac{\sqrt{3}\epsilon_0 \mathscr{H}_x}{\sqrt{\mathscr{H}_x^2 + \mathscr{H}_y^2}} \sinh\left[\frac{\sqrt{3}}{\tau_0}\sqrt{\mathscr{H}_x^2 + \mathscr{H}_y^2}\right] \tag{5.36}$$

Noting that

$$\tau_{yz} = -G\alpha\frac{\partial \Psi}{\partial x} = -\mathscr{H}_x$$

$$\tau_{xz} = G\alpha\frac{\partial \Psi}{\partial y} = \mathscr{H}_y \tag{5.37}$$

we can give the relation between the complementary strain energy, U_0^*, and the strain field as follows (see Eq. (1.83)):

$$\gamma_{yz} = \frac{\partial U_0^*}{\partial \tau_{yz}} = -\frac{\partial U_0^*}{\partial \mathscr{H}_x}$$

$$\gamma_{xz} = \frac{\partial U_0^*}{\partial \tau_{xz}} = \frac{\partial U_0^*}{\partial \mathscr{H}_y}$$

For U_0^* to satisfy the above basic requirements and at the same time yield the constitutive law given by Eqs. (5.35) and (5.36) it must be of the form

$$U_0^* = \tau_0\epsilon_0 \cosh\left(\frac{\sqrt{3}}{\tau_0}\sqrt{\mathscr{H}_x^2 + \mathscr{H}_y^2}\right) \tag{5.38}$$

a result you may yourself verify.

Next consider the shaft to be of unit length and take the end at $z = 0$ to be fixed. The end surface at $z = 1$ then becomes region S_2 in the evaluation of π^* for reasons set forth earlier for the linear elastic torsion. Here using Eqs. (5.2) for the displacement components u and v we get for V^*

$$V^* = -\iint_{S_2} T_i^{(v)}u_i \, dA = -\iint_R (-\tau_{zx}\alpha y + \tau_{zy}\alpha x) \, dA$$

The total complementary energy functional becomes, on integrating U_0^* with respect to z:

$$\pi^* = \iint_R \tau_0 \epsilon_0 \cosh\left(\frac{\sqrt{3}}{\tau_0}\sqrt{\mathscr{H}_x^2 + \mathscr{H}_y^2}\right) dA - \alpha \iint_R (-\tau_{zx}y + \tau_{zy}x)\, dA$$

Now replace τ_{zx} and τ_{zy} by $\mathscr{H}_y$ and $-\mathscr{H}_x$ respectively and integrate the second integral by parts. Since $\mathscr{H} = 0$ on the boundary the resulting line integral vanishes. The total complementary energy then is given as:

$$\pi^* = \iint_R \tau_0 \epsilon_0 \cosh\left(\frac{\sqrt{3}}{\tau_0}\sqrt{\mathscr{H}_x^2 + \mathscr{H}_y^2}\right) dA - 2\iint_R \alpha\mathscr{H}\, dA$$

At this point it is easiest to express the hyperbolic function as a power series as follows retaining only the first three terms:

$$\cosh\eta = 1 + 0.472\eta^2 + 0.0574\eta^4$$

We then have for π^*:

$$\pi^* = \iint_R \tau_0\epsilon_0\left[1 + \frac{1.416}{\tau_0^2}(\mathscr{H}_x^2 + \mathscr{H}_y^2) + \frac{0.5166}{\tau_0^4}(\mathscr{H}_x^2 + \mathscr{H}_y^2)^2\right] dA$$
$$- 2\iint_R \alpha\mathscr{H}\, dA$$

Since additive constants in π^* as well as multiplicative constants over the entire function do not affect the results of a variational process we will alter π^*, with no loss in generality, by deleting the number 1 in the first integral and by multiplying through by $1/(\tau_0\epsilon_0)$. We shall accordingly henceforth use the following form for π^*:

$$\pi^* = \iint_R \left\{ 1.416\left[\left(\frac{\mathscr{H}_x}{\tau_0}\right)^2 + \left(\frac{\mathscr{H}_y}{\tau_0}\right)^2\right] + 0.5166\left[\left(\frac{\mathscr{H}_x}{\tau_0}\right)^2 + \left(\frac{\mathscr{H}_y}{\tau_0}\right)^2\right]^2 \right. $$
$$\left. - \frac{2\alpha\mathscr{H}}{\tau_0\epsilon_0} \right\} dA \tag{5.39}$$

We may now use the Ritz procedure to find approximations for the stress function for particular cross sections. As an example we consider the case of an equilateral triangle shown in Fig. 5.7. Due to arguments of symmetry we can assume that the axis of rotation for the prism goes through the centroid of the section and for this reason we have set up the reference with the origin at this point. For the approximate coordinate function we use the following expression:

$$\tilde{\mathscr{H}} = C_1\left[x + \frac{a}{3}\right]\left[y - \frac{1}{\sqrt{3}}\left(\frac{2a}{3} - x\right)\right]\left[y + \frac{1}{\sqrt{3}}\left(\frac{2a}{3} - x\right)\right] \tag{5.40}$$

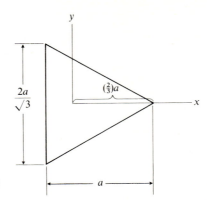

FIGURE 5.7

Clearly this function vanishes on the boundary of the triangle. Now we substitute the function into Eq. (5.39) and extremize with respect to the parameter C_1 to get:

$$1.88a\left(\frac{C_1}{\tau_0}\right) + 0.070a^5\left(\frac{C_1}{\tau_0}\right)^3 = -\frac{\alpha}{\epsilon_0} \qquad (5.41)$$

When the values of ϵ_0, τ_0, and a are chosen we can solve for C_1 in terms of the rate of twist α. Now Eq. (5.18(c)) relates M with α and the constant C_1. For a given value of M (or α) we can solve Eq. (5.41) and (5.18(c)) (with $Ga\Psi$ replaced by $\mathcal{H}$) simultaneously to establish both C_1 and α (or M). The torsional rigidity is then established for a given value of M (or a given value of α).

The choice of ϵ_0 and τ_0 is made so that the hyperbolic sine curve fits as well as possible, over the range of the problem, the actual stress-strain diagram. In particular τ_0/ϵ_0 corresponds to the slope E_0 of the stress-strain curve at the beginning of loading (see Fig. 5.8); one then chooses either ϵ_0 or τ_0.[1]

To exemplify the procedure we consider a stress-strain curve for the data $\tau_0 = 18 \times 10^3$ psi and $\epsilon_0 = 0.000900$. The result we get for the torsional stiffness is plotted against α as curve (1) in Fig. 5.9. Now if we had linear elastic behavior of the rod with a modulus of elasticity $E = \tau_0/\epsilon_0 = 20 \times 10^6$ psi we could use the function as given by Eq. (5.40) in conjunction with the total complementary energy as given by Eq. (5.15) to get, via the Ritz procedure, a torsional stiffness of:

$$\frac{M}{\alpha} = 5.5 \times 10^5$$

For comparison this value is shown as curve (2) in Fig. 5.9 for copper ($v = 0.3$). Note that as the α becomes smaller both curves come together since the stress-strain curves approach each other for the linear and nonlinear cases. This correlation serves to give

---

[1] A procedure for accomplishing this effectively is developed in Smith and Sidebottom: "Inelastic Behavior of Load Carrying Members," John Wiley and Sons, Chap. 1.

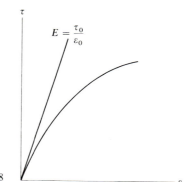

FIGURE 5.8

credence to the many assumptions made in developing the nonlinear elastic theory.

For a more careful study of the nonlinear torsion problem we can use for $\tilde{\mathscr{H}}$ a function with two constants C_1 and C_2 of the form:

$$\tilde{\mathscr{H}} = (C_1 + C_2 x^2)\left[x + \frac{a}{3}\right]\left[y - \frac{1}{\sqrt{3}}\left(\frac{2a}{3} - x\right)\right]\left[y + \frac{1}{\sqrt{3}}\left(\frac{2a}{3} - x\right)\right]$$

Extremization of π^* then gives the following equations which may be used to determine C_1 and C_2 in terms of α.

$$1.88a\left(\frac{C_1}{\tau_0}\right) + 0.060a^3\left(\frac{C_2}{\tau_0}\right) + 0.070a^5\left(\frac{C_1}{\tau_0}\right)^3 = -\frac{\alpha}{\epsilon_0}$$

$$8.58a^3\left(\frac{C_2}{\tau_0}\right) + 0.731a^5\left(\frac{C_1}{\tau_0}\right)^3 + 1.0a^7\left(\frac{C_1}{\tau_0}\right)^2\left(\frac{C_2}{\tau_0}\right) = 0$$

The procedure is then as discussed earlier.

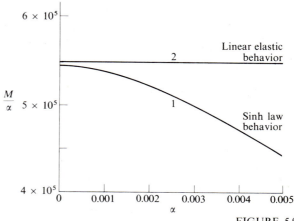

FIGURE 5.9

We may point out that we can readily consider the rectangular cross section for the kind of nonlinear behavior presented in this section by using the function of Eq. (5.22) or more complex functions such as are given by Eq. (5.25).

5.6   THE METHOD OF TREFFTZ; UPPER BOUND FOR TORSIONAL RIGIDITY

In the linear elastic problems considered in Sec. 5.4 we used the Ritz method and illustrated via the problem of the rectangular cross section that we could then approximate the torsional rigidity *from below*. We now set forth a method whereby the torsional rigidity is approximated *from above*. This is the method of Trefftz. By this means we can then *bracket* the correct value of the torsional rigidity.

For this purpose we now consider again the conjugate function ψ for which the boundary-value problem in torsion is given by Eqs. (5.20). (You may recognize this to be the so-called *Dirichlet* problem wherein the function is harmonic and is specified on the boundary.) In Problem 3.15 we showed for homogeneous boundary conditions that $-\nabla^2$ is both symmetric and positive-definite so that the quadratic functional I for this boundary value problem is:

$$I(\psi) = -(\psi, \nabla^2 \psi) = -\iint_R \left(\psi \frac{\partial^2 \psi}{\partial x^2} + \psi \frac{\partial^2 \psi}{\partial y^2} \right) dA$$

Integrating by parts and using the homogeneous boundary conditions we get:

$$I(\psi) = \iint_R \left[\left(\frac{\partial \psi}{\partial x} \right)^2 + \left(\frac{\partial \psi}{\partial y} \right)^2 \right] dA \qquad (5.42)$$

One can show[1] that for non-homogeneous boundary conditions the quadratic functional again has the above form. Thus we may state the boundary-value problem for the conjugate function ψ as follows:

$$-\nabla^2 \psi = 0 \qquad\qquad \text{in } R \qquad (a)$$

$$\psi = \tfrac{1}{2}(x^2 + y^2) \qquad\qquad \text{on } C \qquad (b)$$

$$I(\psi) = \iint_R [\psi_x^2 + \psi_y^2] \, dA$$

$$= \iint_R (\nabla \psi)^2 \, dA \qquad\qquad (c) \quad (5.43)$$

We now prove the following theorem by Weinstein which is vital for the Trefftz method.

---

[1] See Mikhlin: "Variational Methods in Mathematical Physics," The Macmillan Co., 1964.

Theorem If Θ is a harmonic function in R, and ψ is the solution to a Dirichlet problem for this domain then

$$I(\Theta) \leq I(\psi) \qquad (5.44)$$

provided that on the boundary

$$\oint_C [\psi - \Theta] \frac{d\Theta}{dv}\, ds = 0 \qquad (5.45)$$

The *proof* is quite straightforward. Define η as follows:

$$\eta = \psi - \Theta \qquad (a)$$

therefore

$$\psi = \Theta + \eta \qquad (b)\ (5.46)$$

Clearly η is a harmonic function. Now examine $I(\psi)$ as follows, using Eq. (5.43(c)) and Eq. (5.46(b)):

$$I(\psi) = \iint_R \{[\Theta_x + \eta_x]^2 + [\Theta_y + \eta_y]^2\}\, dA$$

$$= \iint_R (\Theta_x{}^2 + \Theta_y{}^2)\, dA + \iint_R (\eta_x{}^2 + \eta_y{}^2)\, dA$$

$$+ 2\iint_R (\eta_x \Theta_x + \eta_y \Theta_y)\, dA$$

$$= I(\Theta) + I(\eta) + 2\iint_R (\eta_x \Theta_x + \eta_y \Theta_y)\, dA$$

Now using Green's theorem in two dimensions (see Eq. (I.55) Appendix I) for the last integration we get:

$$\iint_R (\Theta_x \eta_x + \Theta_y \eta_y)\, dA = -\iint_R \eta \nabla^2 \Theta\, dA + \oint_C \eta \frac{d\Theta}{dv}\, ds$$

But $\nabla^2 \Theta = 0$, and by virtue of Eqs. (5.45) and (5.46(a)) the line integral is zero. Hence we get:

$$I(\psi) = I(\Theta) + I(\eta)$$

Since $I(\eta)$, $I(\Theta)$, and $I(\psi)$ are positive-definite functionals we can conclude from above that

$$I(\Theta) \leq I(\psi)$$

thus proving the theorem.

Suppose next we form a parameter-laden sum of linearly independent functions to represent Θ in the following way

$$\Theta = \sum_i^n a_i v_i \qquad (5.47)$$

where functions v_i must be harmonic. To ensure that $I(\Theta) \leq I(\psi)$ we select the constants a_i so as to satisfy the condition given by Eq. (5.45) for the problem at hand. That is, noting Eq. (5.43(b)) we may state:

$$\oint_C \left[\psi - \sum_{i=1}^n a_i v_i \right] \frac{d}{dv} \left(\sum_{j=1}^n a_j v_j \right) ds = 0$$

therefore

$$\oint_C \left[\tfrac{1}{2}(x^2 + y^2) - \sum_{i=1}^n a_i v_i \right] \left(\sum_{j=1}^n a_j \frac{dv_j}{dv} \right) ds = 0$$

To satisfy the above equation it is sufficient to require that the coefficients of each a_j be zero. That is:

$$\oint_C \left[\tfrac{1}{2}(x^2 + y^2) - \sum_{i=1}^n a_i v_i \right] \frac{dv_j}{dv} ds = 0 \qquad j = 1, 2, \ldots, n \qquad (5.48)$$

We have here a system of n equations with n unknowns a_i. One can prove that if the boundary is such that $\tfrac{1}{2}(x^2 + y^2) \neq 0$ on the boundary and if the functions v_i are linearly independent then there is a *unique* set of constants a_i that can be found from the above set of equations.[1] Now by letting $n \to \infty$ one would expect that the set of a_i determined from Eq. (5.48) to be such that $\sum_{i=1}^n a_i v_i$ approaches $\tfrac{1}{2}(x^2 + y^2)$ on the boundary in some manner. Indeed it can be shown (we shall as usual not get into the convergence question) that if functions v_i are complete in the domain except at isolated points[2] then $\sum_{i=1}^n a_i v_i$ converges to ψ uniformly in the region R. This means that $I(\Theta)$ can be made arbitrarily close to $I(\psi)$ everywhere by choosing n large enough, and Eq. (5.44) adds the fact that this approach must be *from below*.

There is a relation between an approximate torsional rigidity using $\sum a_i v_i$ as formulated above and the actual torsional rigidity. For this consideration examine $I(\psi)$ while replacing ψ by $[\Psi + \tfrac{1}{2}(x^2 + y^2)]$ in accordance with Eq. (5.21). Thus we

---

[1] See Sokolnikoff: "Mathematical Theory of Elasticity," McGraw-Hill Book Co., 1956, pp. 427, 428.

[2] That is, if we can approximate any harmonic function arbitrarily closely at all but isolated points by the sum

$$\sum_{i=1}^n a_i v_i$$

by properly choosing the a's and making n large enough.

get:

$$I(\psi) = \iint_R \{\nabla[\Psi + \tfrac{1}{2}(x^2 + y^2)]\}^2 \, dA$$

$$= \iint_R [(\nabla\Psi)^2 + \nabla\Psi \cdot \nabla(x^2 + y^2) + (x^2 + y^2)] \, dA \qquad (5.49)$$

Now from Green's theorem (Eq. (I.56) Appendix I) we have for the second expression of the integrand on the right side of the above equation:

$$\iint_R \nabla\Psi \cdot \nabla(x^2 + y^2) \, dA = -\iint_R \Psi\nabla^2(x^2 + y^2) \, dA + \oint_C \Psi\frac{\partial}{\partial \nu}(x^2 + y^2) \, ds$$

Since $\Psi = 0$ on the boundary the line integral vanishes. The surface integral on the right side of the equation is simply $\iint 4\Psi \, dA$. We then get on substitution into Eq. (5.49):

$$I(\psi) = \iint_R [(\nabla\Psi)^2 - 4\Psi] \, dA + \iint_R (x^2 + y^2) \, dA \qquad (5.50)$$

But the last integral is simply the *polar moment of inertia* of the cross section and we denote it simply as J. Thus we can say from Eq. (5.15):

$$I(\psi) = \frac{2}{G\alpha^2}\pi^* + J$$

therefore

$$\pi^* = \frac{G\alpha^2}{2}[I(\psi) - J] \qquad (5.51)$$

We shall now find another formulation of π^* in terms of the torsional stiffness D. The torsional stiffness has been shown to be (see Eq. 5.18(c)):

$$\frac{M}{\alpha} = D = 2G \iint_R \Psi \, dA \qquad (5.52)$$

But using Green's theorem (Eq. (I.56) Appendix I) in two dimensions with ψ and ϕ replaced by Ψ, while noting that $\Psi = 0$ on the boundary and that $\nabla^2\Psi = -2$ everywhere, the above result can be given as:

$$D = G \iint_R (\nabla\Psi)^2 \, dA \qquad (5.53)$$

We may now give π^* in terms of D. Thus employing Eqs. (5.53) and (5.52) in Eq. (5.15) we get another formulation of π^*:

$$\pi^* = \frac{G\alpha^2}{2}\left[\frac{D}{G} - 4\frac{D}{2G}\right] = -\frac{D\alpha^2}{2} \qquad (5.54)$$

Equating the right side of Eq. (5.54) and Eq. (5.51) and solving for D we get:

$$D = G[J - I(\psi)] \qquad (5.55)$$

Now consider a quantity which we denote as $\tilde{D}_{\text{Trefftz}}$ formed by using the right side of the above equation with ψ replaced by Θ in the functional I. That is:

$$\tilde{D}_{\text{Trefftz}} = G[J - I(\Theta)] \qquad (5.56)$$

We have already shown by the Weinstein theorem that for properly chosen a's $I(\Theta) \leq I(\psi)$ and so it is clear on comparing the right sides of the above equations for D and $\tilde{D}_{\text{Trefftz}}$ that:

$$\tilde{D}_{\text{Trefftz}} \geq D \qquad (5.57)$$

We see that $\tilde{D}_{\text{Trefftz}}$ *forms an approximation from above of the torsional rigidity.* Furthermore, because $I(\Theta)$ converges to $I(\psi)$, we can make $\tilde{D}_{\text{Trefftz}}$ arbitrarily close to D. Hence by a judicious selection of functions v_i and using the proper values of coefficients a_i we can compute $\tilde{D}_{\text{Trefftz}}$ using Eq. (5.56).

We have shown earlier (Sec. 5.4) that, for the example considered, the torsional rigidity approximation reached by the Ritz method was an approximation from below. We can now readily see that the torsional rigidity computed in this way must always be an approximation from below. Note from Eq. (5.54) that π^* must be a negative number for torsion since D must be positive.[1] Now an approximate total complementary energy $\tilde{\pi}^*_{\text{Ritz}}$ (via the Ritz method) must be greater than π^*. This means $\tilde{\pi}^*_{\text{Ritz}}$ has a smaller negative value than π^*. Hence the value of torsional stiffness D that you would get by using the aforementioned Ritz approximation of the total complementary energy will, according to Eq. (5.54), perforce be a *smaller* positive number than the exact value of D. Thus we conclude that the Ritz approximation for the torsional stiffness must be *from below.*

Thus using the Ritz approach and the Trefftz approach here we can *bracket* the exact result for the torsional rigidity. We illustrate this procedure now by going back to the rectangular shaft discussed in Sec 5.4. To generate suitable harmonic functions for the calculation we may consider the real and imaginary parts of the analytic function[2] $(x + iy)^n$. For various values of n we then select harmonic functions that are even functions with respect to coordinates x and y (see Fig. 5.4). Thus for $n = 2$ we have $(x^2 - y^2)$. Hence we may consider for Θ the function:

$$\Theta = a_1(x^2 - y^2) \qquad (5.58)$$

---

[1] Remember $D = M/\alpha$. It is clear that M and α must have the same sense and so M/α must be positive.

[2] Such a set of polynomials can be shown to be complete.

To get a_1 we employ Eq. (5.45) as follows:

$$\oint_C \left[\frac{1}{2}(x^2 + y^2) - a_1(x^2 - y^2) \right](a_1)\frac{d}{dv}(x^2 - y^2)\, ds = 0 \qquad (5.59)$$

This becomes, on considering the boundary:

$$\int_{-a}^{+a} [\tfrac{1}{2}(x^2 + b^2) - a_1(x^2 - b^2)](2y)|_{y=-b}\, dx$$

$$+ \int_{-b}^{+b} [\tfrac{1}{2}(a^2 + y^2) - a_1(a^2 - y^2)](2x)|_{x=a}\, dy$$

$$+ \int_{+a}^{-a} [\tfrac{1}{2}(x^2 + b^2) - a_1(x^2 - b^2)](-2y)|_{y=b}(-dx)$$

$$+ \int_{+b}^{-b} [\tfrac{1}{2}(a^2 + y^2) - a_1(a^2 - y^2)](-2x)|_{x=-a}(-dy) = 0$$

We thus have for a_1:

$$a_1 = \frac{a^2 - b^2}{2(a^2 + b^2)}$$

To get the torsional rigidity we have:

$$\tilde{D}_{\text{Trefftz}} = G[J - I(\Theta)]$$

$$= G\left\{ \int_{-b}^{+b}\int_{-a}^{+a}(x^2 + y^2)\, dx\, dy - \int_{-b}^{+b}\int_{-a}^{+a}\{\nabla[a_1(x^2 - y^2)]\}^2\, dx\, dy \right\}$$

$$= \frac{16Ga^3b^3}{3(a^2 + b^2)}$$

Table 5.3

	Ritz	Trefftz			Exact
a/b	$\Phi_1 = C_1(a^2 - x^2)(b^2 - y^2)$	$\Theta_1 = a_1(x^2 - y^2)$	$\Theta_2 = a_1(x^2 - y^2) + a_2(x^4 - 6x^2y^2 + y^4)$	$\Theta_3 = a_1(x^4 - 6x^2y^2 + y^4)$	Timoshenko
1.00	2.22	2.67	2.25	2.25	2.25
1.20	3.14	3.78	3.32	3.45	3.19
1.50	4.62	5.54	5.50	6.35	4.70
2.00	7.11	8.53	10.00	13.01	7.83
2.50	9.58	11.49	14.80	20.81	9.96
3.00	12.00	14.40	19.59	29.39	12.62
4.00	16.73	20.08	28.89	49.17	17.98
5.00	21.37	25.64	37.89	73·60	23.30
10.00	44.00	52.80	80.84	314.5	49.92

From the Ritz method meanwhile we have (see Eq. (5.23)):

$$\tilde{D}_{\text{Ritz}} = \frac{40a^3b^3}{9(a^2 + b^2)}G = \frac{13.3Ga^3b^3}{3(a^2 + b^2)}$$

We may conclude that the torsional stiffness is bracketed as follows:

$$\left(D_{\text{Ritz}} = \frac{13.3a^3b^3}{3(a^2 + b^2)}G\right) \le \left(D_{\text{Exact}} = \frac{13.5a^3b^3}{3(a^2 + b^2)}G\right)$$

$$\le \left(D_{\text{Trefftz}} = \frac{16a^3b^3}{3(a^2 + b^2)}G\right)$$

We have shown in Table 5.3 torsional rigidities computed via the Ritz method for a function Ψ_1 given as

$$\Psi_1 = C_1(a^2 - x^2)(b^2 - y^2)$$

as well as torsional rigidities via the Trefftz method employing the following functions:

$$\Theta_1 = a_1(x^2 - y^2)$$
$$\Theta_2 = a_1(x^2 - y^2) + a_2(x^4 - 6x^2y^2 + y^4)$$
$$\Theta_3 = a_1(x^4 - 6x^2y^2 + y^4)$$

Also a set of exact results are included for comparison. The torsional rigidity terms are given for $a = 1$ and as a function of the ratio a/b.

5.7   THE METHOD OF KANTOROVICH

A serious shortcoming of the Ritz method as well as the Galerkin method is that the obtained results have a strong dependence on the coordinate functions chosen. This was made most apparent in Chap. 4 on beams. The method of Kantorovich which we now consider will decrease this dependence of the results on the choice of the coordinate function thereby making the process more effective. However, this gain will not be reached without additional computational efforts. (In the following section we shall present recent work which is an extension of the Kantorovich method that actually eliminates the dependence of obtained results from the initial choice of the coordinate functions.)

We shall consider in this regard functionals of the form:

$$I(w) = \int\int_R F(x, y, w, w_x, w_y)\, dx\, dy \qquad (5.60)$$

Now in the Kantorovich method we shall again use coordinate functions, which we now denote as H_p, as we did in the Ritz method, but instead of using undetermined constants a_i, we shall now use undetermined coefficient *functions* $C_p(x)$. Thus for the dependent variable w we form the nth partial sum:

$$w_n = H_0(x,y) + \sum_{p=1}^{n} C_p(x)H_p(x,y) \qquad (5.61)$$

H_0 has the specified value of w on the boundary and functions $H_p(x,y)$ are zero on part or all of the boundary; where H_p is not zero the coefficient $C_p(x)$ must be zero on the boundary. We may substitute the above partial sum into Eq. (5.60) for the dependent variable w. Since F is specified and the H's are known, the functional $I(w_n)$ on carrying out the integration with respect to y becomes one involving x as the integration variable with the C's as functions. Thus we get:

$$I(w_n) = \int_{x_1}^{x_2} F[x, C_1(x), \ldots, C_p(x), \ldots, C_n(x), C_1'(x), \ldots, C_p'(x), \ldots, C_n'(x)]\, dx \qquad (5.62)$$

The procedure is now to invoke the variational process to establish the Euler–Lagrange equations for the functions $C_p(x)$ rendering them extremals for the functional $I(w_n)$. The resulting function w_n is then the desired approximation to the problem according to this method.

Let us in particular consider the *torsion problem* using the stress function Ψ as the dependent variable. The boundary value problem for this case was shown to be:

$$\nabla^2\Psi = -2$$

$$\Psi = 0 \qquad \text{on boundary} \qquad (5.63)$$

We have shown in Sec. 5.3 that the above boundary-value problem could be reached by extremizing the total complementary energy π^* as given by Eq. (5.15). Dropping the constant $G\alpha^2/2$, the functional I that we may consider for the Kantorovich method as applied to the above boundary value problem is:

$$I = \int\int_R [(\nabla\Psi)^2 - 4\Psi]\, dx\, dy \qquad (5.64)$$

In considering w_n (see Eq. (5.61)) to replace the dependent variable Ψ above we may delete the function H_0 in this case and employ the following partial sum, where we use the notation Ψ_n in place of w_n:

$$\Psi_n = \sum_{p=1}^{n} C_p(x)H_p(x,y) \qquad (5.65)$$

Now form varied functions $\tilde{C}_p(x)$ as follows

$$\tilde{C}_p(x) = C_p(x) + \epsilon_p\eta_p(x) \qquad (5.66)$$

with the condition that $\eta_p = 0$ on the boundary. Now employ $\tilde{\Psi}_n$, with $\tilde{C}_p$ as given above, in the functional (5.64). We then find $C_p(x)$ by carrying out the extremization

$$\left(\frac{\partial \tilde{I}}{\partial \epsilon_p}\right)_{\substack{\epsilon_1 = 0 \\ \vdots \\ \epsilon_n = 0}} = 0 \qquad p = 1, 2, \ldots, n$$

We have for $\tilde{I}$ from Eq. (5.64) the following result:

$$\tilde{I} = \int\int_R \tilde{F}\, dx\, dy = \int\int_R \left[\left(\frac{\partial \tilde{\Psi}_n}{\partial x}\right)^2 + \left(\frac{\partial \tilde{\Psi}_n}{\partial y}\right)^2 - 4\tilde{\Psi}_n\right] dx\, dy$$

where:

$$\tilde{\Psi}_n = \sum_{p=1}^{n} \tilde{C}_p(x)H_p(x, y)$$

$$= \sum_{p=1}^{n} [C_p(x) + \epsilon_p\eta_p(x)]H_p(x,y) \qquad (5.67)$$

Hence:

$$\tilde{F} = \left\{\sum_{p=1}^{n} [(\tilde{C}_p)(H_p)_x + (\tilde{C}_p)_x(H_p)]\right\}^2 + \left\{\sum_{p=1}^{n}(\tilde{C}_p)(H_p)_y\right\}^2 - 4\left[\sum_{p=1}^{n}(\tilde{C}_p)(H_p)\right] \qquad (5.68)$$

where $(H_p)_x = \partial(H_p)/\partial x$ etc. Now:

$$\left(\frac{\partial \tilde{I}}{\partial \epsilon_\alpha}\right)_{\substack{\epsilon_1 = 0 \\ \vdots \\ \epsilon_n = 0}} = 0 = \int\int_R \left[\frac{\partial \tilde{F}}{\partial \tilde{C}_\alpha}\eta_\alpha + \frac{\partial \tilde{F}}{\partial(\tilde{C}_\alpha)_x}(\eta_\alpha)_x\right]_{\substack{\epsilon_1 = 0 \\ \vdots \\ \epsilon_n = 0}} dx\, dy$$

$$= \int\int_R \left\{\frac{\partial \tilde{F}}{\partial \tilde{C}_\alpha} - \frac{\partial}{\partial x}\left[\frac{\partial \tilde{F}}{\partial(\tilde{C}_\alpha)_x}\right]\right\}_{\substack{\epsilon_1 = 0 \\ \vdots \\ \epsilon_n = 0}} \eta_\alpha\, dx\, dy$$

$$+ \text{ Line Integral} \qquad \alpha = 1, 2, \ldots, n$$

Substituting from Eq. (5.68) we get

$$\int\int_R \left\{2\sum_p [(C_p)(H_p)_x + (C_p)_x(H_p)](H_\alpha)_x + 2\left(\sum_p(C_p)(H_p)_y\right)(H_\alpha)_y\right.$$

$$\left. -4H_\alpha - 2\frac{\partial}{\partial x}\left(\sum_p [(C_p)(H_p)_x + (C_p)_x(H_p)][H_\alpha]\right)\right\}\eta_\alpha\, dx\, dy$$

$$+ \text{ Line Integral} = 0 \qquad \alpha = 1, 2, \ldots, n$$

where we have dropped the tildes and the conditions $\epsilon_1 = \epsilon_2 = \cdots \epsilon_n = 0$ from the notation. Next, carry out the differentiation of the indicated $\partial/\partial x$ above over the bracketed quantities using the product rule. We see by inspection that one of the resulting expressions is minus the first expression in the surface integral and accordingly we can cancel these expressions. We then have on dividing through by 2 and rearranging the terms:

$$\int\int_R \left\{ \left(-\frac{\partial}{\partial x} \sum_p [C_p(H_p)_x + (C_p)_x(H_p)] \right) H_\alpha + \sum_p C_p(H_p)_y(H_\alpha)_y - 2H_\alpha \right\} \eta_\alpha \, dx \, dy$$

$$+ \text{ Line Integral} = 0 \qquad \alpha = 1, 2 \cdots n$$

Note that the expression

$$\sum_p [C_p(H_p)_x + (C_p)_x H_p]$$

is simply

$$\frac{\partial}{\partial x}\left(\sum_p C_p H_p \right)$$

which is $(\Psi_n)_x$. We then have:

$$\int\int_R \left[-(\Psi_n)_{xx} H_\alpha + \left(\sum_p C_p(H_p)_y \right)(H_\alpha)_y - 2H_\alpha \right] \eta_\alpha \, dx \, dy + \text{ Line Integral} = 0$$

$$\alpha = 1, 2, \ldots, n \qquad (5.69)$$

Now integrate the second expression by parts using Green's theorem as follows:

$$\int\int_R \sum_p C_p(H_p)_y(H_\alpha)_y \eta_\alpha \, dx \, dy = -\int\int_R \frac{\partial}{\partial y}\left(\eta_\alpha \sum_p [C_p(H_p)_y] \right) H_\alpha \, dx \, dy$$

$$+ \text{ Line Integral} \qquad \alpha = 1, 2, \ldots, n$$

Recalling that C_p and η_α are functions only of x, we can express the above equation as:

$$\int\int_R \left(\sum_p C_p(H_p)_y(H_\alpha)_y \right) \eta_\alpha \, dx \, dy = -\int\int_R \left[\frac{\partial^2}{\partial y^2}\left(\sum_p C_p H_p \right) \right] \eta_\alpha H_\alpha \, dx \, dy$$

$$+ \text{ Line Integral}$$

$$= -\int\int_R (\Psi_n)_{yy} H_\alpha \eta_\alpha \, dx \, dy + \text{ Line Integral}$$

$$\alpha = 1, 2, \ldots, n$$

Substituting the above result into Eq. (5.69) we reach the following result upon setting the line integral equal to zero as required by the extremization process:

$$\int\int_R [\nabla^2 \Psi_n + 2] H_\alpha(x,y) \, \eta_\alpha(x) \, dx \, dy = 0 \qquad \alpha = 1, \ldots, n \qquad (5.70)$$

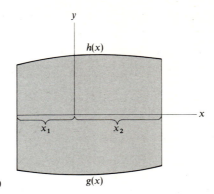

FIGURE 5.10

Now restricting ourselves to boundaries of the type shown in Fig. 5.10 we can rewrite the above equation as follows:

$$\int_{-x_1}^{x_2} \eta_\alpha(x)\, dx \int_{y=g(x)}^{y=h(x)} (\nabla^2 \Psi_n + 2) H_\alpha(x,y)\, dy = 0 \qquad \alpha = 1, 2, \ldots, n \qquad (5.71)$$

Since the η's are arbitrary we can conclude that[1]:

$$\int_{y=g(x)}^{y=h(x)} (\nabla^2 \Psi_n + 2) H_\alpha(x,y)\, dy = 0 \qquad \alpha = 1, 2, \ldots, n \qquad (5.72)$$

When we carry out the integration with respect to y in the above n equations, we obtain n ordinary differential equations for the desired functions $C_p(x)$. The constants of integration for the solution $C_p(x)$ are determined so as to maintain Ψ_n, the approximate stress function, equal to zero over the entire boundary.

We illustrate the procedure for using the Kantorovich method by considering the torsion of the shaft with a rectangular cross section (see Fig. 5.4) as presented by Kantorovich and Krylov.[2] We consider first a one-term approximating function of the form:

$$\Psi_1 = C_1(x)(b^2 - y^2) \qquad (a)$$

therefore

$$H_1 = (b^2 - y^2) \qquad (b) \quad (5.73)$$

---

[1] Note that by this process the problem has been altered such as to require the solution of a *Galerkin integral* (see Sec. 3.15). We shall make use of this transition in the Kantorovich method at later times in the text.

[2] L. V. Kantorovich and V. I. Krylov: "Approximate Methods of Higher Analysis," Interscience, N.Y., 1964.

Notice that H_1 satisfies the boundary conditions on two edges of the cross section; we will have accordingly to adjust $C_1(x)$ to satisfy the boundary condition on the other two edges. Substituting into Eq. 5.72 we get:

$$\int_{-b}^{+b} \left[C_1(-2) + (b^2 - y^2)\left(\frac{d^2 C_1}{dx^2}\right) + 2 \right] [(b^2 - y^2)] \, dy = 0$$

therefore

$$\int_{-b}^{+b} \left[2b^2(1 - C_1) - 2y^2(1 - C_1) + \frac{d^2 C_1}{dx^2}(b^4 - 2b^2 y^2 + y^4) \right] dy = 0$$

Integrating we get:

$$2b^2(1 - C_1)(2b) - 2(1 - C_1)\left(\frac{2b^3}{3}\right) + \frac{d^2 C_1}{dx^2}\left(2b^5 - \frac{4b^5}{3} + \frac{2b^5}{5}\right) = 0$$

therefore

$$\frac{d^2 C_1(x)}{dx^2} - \frac{5}{2b^2}C_1(x) = -\frac{5}{2b^2} \qquad (5.74)$$

The general solution to the above differential equation is

$$C_1(x) = A_1 \sinh \frac{\lambda_1 x}{a} + B_1 \cosh \frac{\lambda_1 x}{a} + 1 \qquad (5.75)$$

where:

$$\left(\frac{\lambda_1}{a}\right) = \sqrt{\frac{5}{2b^2}} \qquad (5.76)$$

Now we must determine A_1 and B_1 to satisfy the boundary conditions at $x = \pm a$. We thus find from Eq. (5.75) that:

$$C_1(\pm a) = 0 = A_1 \sinh(\pm \lambda_1) + B_1 \cosh(\pm \lambda_1) + 1$$

Since the hyperbolic sine function is an odd function and the hyperbolic cosine is an even function we set $A_1 = 0$ and solve for B_1 from the above to get:

$$B_1 = -\frac{1}{\cosh \lambda_1}$$

We can then give Ψ_1 from this calculation as follows, indicating it now as $(\Psi_1)_{\text{Kant}}$:

$$[\Psi_1]_{\text{Kant}} = \left[1 - \left(\frac{\cosh \frac{\lambda_1 x}{a}}{\cosh \lambda_1}\right) \right](b^2 - y^2) \qquad (5.77)$$

It will be of interest to compare the results of the Ritz method with the above results for the shaft with a rectangular cross section. For a one parameter approach we got for the Ritz method (see beginning of Sec. 5.4) the following result for Ψ_1:

$$[\Psi_1]_{Ritz} = \left[\frac{5}{4}\left(\frac{1}{b^2 + a^2}\right)(a^2 - x^2)\right](b^2 - y^2) \qquad (5.78)$$

Actually, the coefficients of $(b^2 - y^2)$ in Eqs. (5.77) and (5.78) are closely related in that the latter is essentially the first term in a power series expansion of the former.

We now compute the approximation for the torsional stiffness from the one-term Kantorovich calculation. Thus:

$$D_1 = 2G \int\int_R \Psi_1 \, dA = 2G \int_{-b}^{+b} \int_{-a}^{+a} \left[1 - \frac{\cosh\dfrac{\lambda_1 x}{a}}{\cosh \lambda_1}\right](b^2 - y^2) \, dx \, dy$$

Integrating we get:

$$D_1 = \frac{16}{3}Gb^3 a\left[1 - \frac{1}{\lambda_1}\tanh \lambda_1\right] \qquad (5.79)$$

In Table 5.4 we compare results from above with corresponding results from the Ritz method as well as the exact results from the theory of elasticity. This is done for various values of a and b.

Increased accuracy using the Kantorovich method can be achieved using for Ψ_n the following two-parameter approximation:

$$\Psi_2 = C_1(x)(b^2 - y^2) + C_2(x)(b^2 - y^2)y^2 \qquad (5.80)$$

Table 5.4

a	b	D/G (Ritz)	D/G (Kantorovich)	D/G (Exact)
1	1	2.22222222	2.23420152	2.24922043
2	1	7.11111111	7.30563645	7.31780011
3	1	12.00000000	12.62741545	12.63919901
4	1	16.73202614	17.96025882	17.97201565
5	1	21.36752137	23.29357141	23.30532665
6	1	25.94594595	28.62690387	28.63365902
7	1	30.48888889	33.96023716	33.97199231
8	1	35.00854701	39.29357050	39.30532565

5.8 EXTENDED KANTOROVICH METHOD

We shall now present some recent work by Kerr which may be described as an extension of the Kantorovich method.[1] As was pointed out in the previous section this method eliminates the dependence of the obtained results from the arbitrariness of the chosen coordinate functions and, as will be shown for the torsion problem, yields highly accurate results albeit with a certain amount of increased labor.

The procedure is as follows. We start exactly as in the Kantorovich method by selecting n functions H_p and expressing Ψ_n for a functional having r independent variables x_r in the form:

$$\Psi_n = \sum_{p=1}^{n} [C_p(x_1)][H_p(x_2, x_3, x_4, \cdots, x_r)] \qquad (5.81)$$

Note that the H's are *not* functions of variable x_1; this is an advisable step so as to have the resulting differential equations of the extremization process result in constant coefficients rather than variable coefficients. We find C_p exactly as described in the previous section. Taking the extremal functions C_p without the accompanying constant coefficients and denoting them as $a_{p1}(x_1)$, we now introduce a *new* set of *unknown* functions this time of x_2, namely $D_p(x_2)$, and a specified set of n functions $\bar{H}_p(x_3, x_4, \cdots, x_n)$ to express Ψ_n as follows:

$$\Psi_n = \sum_{p=1}^{n} [a_{p1}(x_1)][D_p(x_2)][\bar{H}_p(x_3, x_4, \ldots, x_r)] \qquad (5.82)$$

Again we substitute Ψ_n into the functional I and find the extremal functions $D_p(x_2)$ as described in the method of Kantorovich. We denote the resulting extremal functions without the accompanying constant coefficients as $a_{p2}(x_2)$. We continue further by using $a_{p1}(x_1)$ and $a_{p2}(x_2)$ and introducing unknown functions $E_p(x_3)$ and specified functions $\bar{\bar{H}}_p(x_4, x_5, \ldots, x_r)$. Proceeding in this way we may finally arrive at the following expression for Ψ_n, remembering to retain this time the constant coefficients A_p of the last expression found

$$\Psi_n^{(\mathrm{I})} = \sum_{p=1}^{n} [a_{p1}(x_1)][a_{p2}(x_2)] \cdots \{(A_p)[a_{pr}(x_r)]\} \qquad (5.83)$$

where the superscript I on Ψ_n indicates the first cycle of such a series of computations.

We may start a second cycle of such calculations by considering $G_p(x_1)$ to be a set of unknown functions and to associate each such function with the known function $[a_{p2}(x_2)a_{p3}(x_3) \cdots a_{pr}(x_r)]$. We can again determine $G_p(x_1)$ as in the Kantorovich method and, denoting the result without the accompanying coefficient as $a_{p1}^{(2)}(x_1)$, we now introduce a new set of undetermined functions $K_p(x_2)$ to be associated with $[a_{p1}^{(2)}(x_1)a_{p3}(x_3) \cdots a_{pr}(x_r)]$ as above. We thus may institute a second cycle of computations of the kind formulated in the preceding paragraph to yield $\Psi_n^{(\mathrm{II})}$, the superscript identifying the second cycle of such calculations.

$$\Psi_n^{(\mathrm{II})} = \sum_{p=1}^{n} [a_{p1}^{(2)}(x_1)][a_{p2}^{(2)}(x_2)] \cdots \{(A_p^{(2)})[a_{pr}^{(2)}(x_r)]\} \qquad (5.84)$$

Note that the last expression in the cycle retains its constant coefficient $A_p^{(2)}$. Proceeding in this way we may reach the ith cycle, to get:

$$\Psi_n^{(i)} = \sum_{p=1}^{n} [a_{p1}^{(i)}(x_1)][a_{p2}^{(i)}(x_2)]\{(A_p^{(i)})[a_{pr}^{(i)}(x_r)]\} \qquad (5.85)$$

---

[1] Arnold Kerr, "An Extension of the Kantorovich Method," *Quarterly of Applied Mathematics,* July, 1968.

The expectation is that $I(\Psi_n^{(i)})$ should approach an extreme value as "i" increases indefinitely, and that $\Psi_n^{(i)}$ should approach in some way the exact solution as "i" increases without limit.

We now demonstrate this process by examining again the torsion of a shaft with a rectangular cross-section (see Fig. 5.4). We shall employ a single-parameter family of functions for Ψ. Since we have only two variables x and y we can use a double subscript notation for Ψ which identifies the two functions used at any time. Thus at the outset we have

$$\Psi_{10} = [f_1(x)][g_0(y)] \qquad (5.86)$$

where $f_1(x)$ will be the first set of functions to be ascertained and where $g_0(y) = b^2[1 - (y/b)^2]$. From the previous section we have already found for $f_1(x)$, considering κ_1 to be the constant coefficient and $a_1(x)$ to be the functions that

$$f_1 = \kappa_1 a_1(x) = \left[\frac{1}{\cosh \lambda_1}\right]\left[\cosh \lambda_1 - \cosh \frac{\lambda_1 x}{a}\right] \qquad (5.87)$$

Thus we have for Ψ_{10}:

$$\Psi_{10} = \frac{b^2}{\cosh \lambda_1}\left[\cosh \lambda_1 - \cosh \frac{\lambda_1 x}{a}\right]\left[1 - \left(\frac{y}{b}\right)^2\right]$$

We next consider the function Ψ_{11} having the following form:

$$\Psi_{11} = [a_1(x)][g_1(y)] = \left[\cosh \lambda_1 - \cosh \frac{\lambda_1 x}{a}\right][g_1(y)]$$

To find $g_1(y)$ we may use Eq. (5.72) adjusted so that x is the integration variable.

$$\int_{-a}^{+a} (\nabla^2 \Psi_{11} + 2)a_1(x)\, dx = 0 \qquad (5.88)$$

You may verify yourself (see Problem 5.12) that after carrying out the above integration we may then solve the resulting differential equation to reach the following result

$$g_1(y) = B_{11} \cosh\left(\gamma_1 \frac{y}{b}\right) + B_{21} \sinh\left(\gamma_1 \frac{y}{b}\right) + C_1 \qquad (a)$$

where:

$$\left(\frac{\gamma_1}{b}\right)^2 = \frac{\lambda_1(\sinh 2\lambda_1 - 2\lambda_1)}{4a^2[\cosh^2 \lambda_1 - (3/(4\lambda_1))\sinh 2\lambda_1 + \frac{1}{2}]} \qquad (b)$$

and

$$C_1 = \frac{8a^2(\lambda_1 \cosh \lambda_1 - \sinh \lambda_1)}{\lambda_1^2(\sinh 2\lambda_1 - 2\lambda_2)} \qquad (c) \quad (5.89)$$

Now considering the boundary conditions along $y = \pm b$ we see that $B_{21} = 0$ and $B_{11} = -C_1/\cosh \gamma_1$. Hence using Γ_1 as the constant coefficient and $b_1(y)$ as the function for $g_1(y)$ we have:

$$g_1(y) = \Gamma_1 b_1(y) = \left[\frac{C_1}{\cosh \gamma_1}\right]\left[\cosh \gamma_1 - \cosh\left(\gamma_1 \frac{y}{b}\right)\right]$$

Hence, for the first cycle, we have:

$$\Psi^{(1)} = \frac{C_1}{\cosh \gamma_1}\left[\cosh \lambda_1 - \cosh\left(\lambda_1 \frac{x}{a}\right)\right]\left[\cosh \gamma_1 - \cosh\left(\gamma_1 \frac{y}{b}\right)\right]$$

For the beginning of the next cycle we form Ψ_{21} as follows

$$\Psi_{21} = [f_2(x)][b_1(y)] = [f_2(x)]\left[\cosh \gamma_1 - \cosh\left(\gamma_1 \frac{y}{b}\right)\right]$$

and we proceed as above to get $\Psi^{(\text{II})}$. Kerr has shown in his paper a recursion process in these calculations so that for $\Psi^{(N)}$ for the Nth cycle we have

$$\Psi^{(N)} = \frac{C_N}{\cosh \gamma_N}\left[\cosh \lambda_N - \cosh\left(\lambda_N \frac{x}{a}\right)\right]\left[\cosh \gamma_N - \cosh\left(\gamma_N \frac{y}{b}\right)\right] \tag{5.90}$$

where:

$$\gamma_N = \left\{\frac{\lambda_N \sinh(2\lambda_N) - 2\lambda_N}{4(a/b)^2[\cosh^2 \lambda_N - (3/(4\lambda_N))\sinh(2\lambda_N) + \frac{1}{2}]}\right\}^{1/2} \tag{5.91}$$

$$C_N = \frac{8a^2(\lambda_N \cosh \lambda_N - \sinh \lambda_N)}{\lambda_N^2(\sinh(2\lambda_N) - 2\lambda_N)} \tag{5.92}$$

$$\lambda_{N+1} = \left\{\frac{\gamma_N(\sinh(2\gamma_N) - 2\gamma_N)}{4(b/a)^2[\cosh^2 \gamma_N - (3/(4\gamma_N))\sinh(2\gamma_N) + \frac{1}{2}]}\right\}^{1/2} \tag{5.93}$$

To use the above results determine λ_1 from Eq. (5.76) (i.e., from the original Kantorovich calculation) as a first step. Next from Eq. (5.91) find γ_1 and going to Eq. (5.93) get λ_2. Now return to Eq. (5.91) to find γ_2. Hence going back and forth between Eqs. (5.91) and (5.93) we may eventually reach the desired value λ_N for the problem. Then $\Psi^{(N)}$ is readily available from the above formulas. Knowing $\Psi^{(N)}$ we can then approximate the torsional stiffness as $2G \iint \Psi^{(N)} dA$.

The parameters λ_N and C_N converge very rapidly to specific values for each ratio a/b when N is increased from unity. In most cases we can take three cycles and assume the parameters have converged, as demonstrated in Kerr's paper. We now compare the result for torsional stiffness with that stemming from exact solutions for $N = 3$. Thus:

	$a/b = 1$	$a/b = \frac{1}{2}$	$a/b = \frac{1}{4}$
$D_{\text{app.}}/Gab^3$	2.247	0.914	0.281
$D_{\text{ex.}}/Gab^3$	2.250	0.916	0.281

We see that the results from the extended Kantorovich method compare very closely indeed with those from the exact theory.

Knowing the torsional rigidity we may for a given torque M, determine the rate of twist α. And, using the approximate stress function, Eq. (5.90), we may then compute approximate stresses. In Fig. 5.11 we have shown from Kerr's paper shear stress along the axes of the cross section, i.e., along AB and AC, as well as shear stresses along the edge of the cross section, i.e., along BD and DC. Also shown are corresponding results for an exact analysis. Notice the close agreement.

5.9 CLOSURE

At the outset of this chapter we formulated various forms of the boundary value problem for torsion using the total potential energy and the total complementary energy functionals. We then employed the familiar Ritz method for finding approximate solutions to particular linear elastic and nonlinear elastic problems. Now the

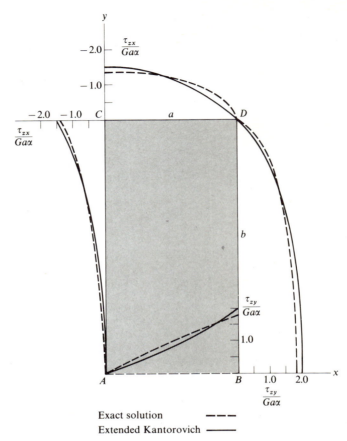

Exact solution — — —
Extended Kantorovich ———

From A. Kerr, An Extension of the Kantorovich Method, Quarterly of Applied Mathematics, July, 1968.

FIGURE 5.11

chief weakness of the Ritz method is that the obtained results depend heavily on the chosen coordinate function or functions. We have discussed this in Chap. 4 on beams. Because we are now dealing with functionals with two dependent variables we can compensate for this difficulty at least along one direction by using the Kantorovich method. Here the form of the function at least as far as one coordinate is concerned is altered by the variational process itself and hence the obtained results are less dependent, at least for the aforementioned coordinate, on the initial choice of the coordinate function. The extended Kantorovich method permits the variational process to alter the coordinate function for all coordinates and hence makes the obtained results not dependent on the initial functions. There is of course additional labor involved for these gains. We shall have occasion to further illustrate the use of the Kantorovich method and its extensions in later chapters when we consider plates and elastic stability.

A desirable goal in approximation calculations is to bracket the sought result—i.e., find upper and lower bounds for a desired result. In this chapter we have been able to use the Ritz method in conjunction with the total potential energy and the Trefftz method to formulate lower and upper bounds respectively for the torsional rigidity.

We now turn to the subject of plates, wherein we shall again formulate appropriate equations from the variational approach and will present approximate solutions to these equations, again employing variational and closely related methods.

READING

SOKOLNIKOFF, I. S.: "Mathematical Theory of Elasticity," McGraw-Hill Book Co., N.Y., 1956.
KANTOROVICH, L. V., AND KRYLOV, V. I.: "Approximate Methods of High Analysis," Interscience Publishers Inc., N.Y., 1964.
WASHIZU, K.: "Variational Methods in Elasticity and Plasticity," Pergamon Press, 1968.
TIMOSHENKO, S., AND GOODIER, J. N.: "Theory of Elasticity," McGraw-Hill Book Co., 1951.

PROBLEMS

5.1 Verify Eqs. (5.20) and (5.21).

5.2 Find the total potential energy in terms of ϕ. Now find the quadratic functional for the equation (5.7(a)). Show that they differ by the inclusion of a "divergence-like" expression in the integrand. Compare the boundary conditions for each case.

5.3 Estimate the torsional rigidity of a rectangular shaft $2a \times 2b$ using as a coordinate function $\cos(\pi x/2a)\cos(\pi y/2b)$. What is the maximum stress? Get D/G for $a/b = 1$ and $a = 1$.

5.4 Consider linear elastic torsion of a shaft with a triangular cross section as shown in Fig. 5.12. Show that the function

$$\Psi_1 = C_1\left[x + \frac{a}{3}\right]\left[y - \frac{1}{\sqrt{3}}\left(\frac{2a}{3} - x\right)\right]\left[y + \frac{1}{\sqrt{3}}\left(\frac{2a}{3} - x\right)\right]$$

satisfies the boundary conditions for the stress function. Using the Ritz method show that the torsional rigidity (for $v = 0.3$ and $E = 20 \times 10^6$ psi) is $2.96 \times 10^4 \, a^4$ in-lb/rad (for "a" in inches).

5.5 Form an approximation for the torsional rigidity of a shaft with a cross section shown in Fig. 5.13. The upper surface is described by a fourth degree polynomial.

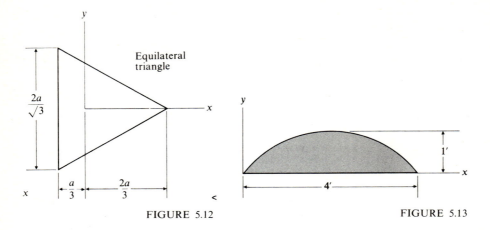

FIGURE 5.12   FIGURE 5.13

5.6   The cross section (Fig. 5.14) is parabolic on top and bottom. What are the equations of the upper and lower curves? Find an approximation of the torsional rigidity by the Ritz method as explained in Sec. 5.4.

5.7   Show that you get constants $C_1 = \sqrt{2}$ and $C_2 = 3/\sqrt{2}$ in the sinh law.

5.8   Find an approximation for the torsional rigidity (Fig. 5.15) when $\alpha = 0.002$, of a square shaft behaving according to the sinh law wherein $\tau_0 = 30 \times 10^6$ and ϵ_0 is 0.000800. Use a single parameter approach for $\mathcal{H}$.

5.9   Using a single-parameter approach for $\mathcal{H}$ find an approximation of the torsional rigidity when $\alpha = 0.003$ for a shaft having a circular cross section (Fig. 5.16) and composed of a material which behaves according to the sinh law. Take $E_0 = 20 \times 10^6$ and $\epsilon_0 = 0.000900$.

5.10 Show that by extremizing the functional

$$ I = \iint \left[\nabla \left(\psi - \sum_{i=1}^{n} a_i v_i \right) \right]^2 dA $$

with respect to a_i we can arrive at Eq. (5.48) (which is very important to the Trefftz method). Thus the important criterion for establishing the a's for the Trefftz method can be reached via a variational process.

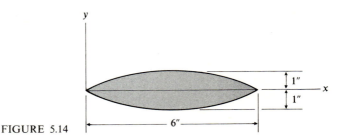

FIGURE 5.14

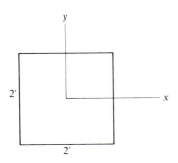

FIGURE 5.15

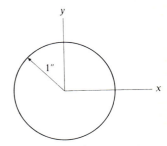

FIGURE 5.16

5.11   Using the method of Trefftz find an upper bound for the torsional rigidity of a shaft having the cross section shown in Fig. 5.17. Use a one-parameter approximation.

5.12   Justify the results developed from Eq. (5.88), i.e., get Eqs. (5.89).

5.13   For torsion of a rectangular shaft set up the differential equations for $C_1(x)$ and $C_2(x)$ using Ψ_2, as given by Eq. (5.80), for the Kantorovich method.

5.14   Using same coordinate functions for the rectangular shaft as was used in Sec. 5.4, find the same approximate solution via the Galerkin method.

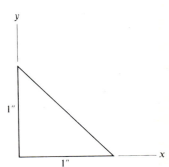

FIGURE 5.17

6

CLASSICAL THEORY OF PLATES

6.1 INTRODUCTION

In Chap. 4 on Beams, Frames, and Rings we were concerned with structures having the distinguishing feature wherein one of the geometric dimensions dominated the configuration. This feature permitted us to make vast simplifications in that we could replace the three-dimensional body by a curve and thus sharply reduce the number of variables of the problem while still yielding important information with considerable accuracy.

We shall now consider certain bodies where more than one geometric dimension dominates the configuration. Specifically we are interested in bodies bounded by surfaces whose lateral dimensions are large compared to the separation between these surfaces. When the bounding surfaces are flat the body is called a *plate*; when the surfaces are curved the body is called a *shell*. As examples of such bodies we have:

(*a*) the hull of a ship
(*b*) the cover of an airplane wing
(*c*) a sheet pile retaining wall
(*d*) reinforced concrete roofs used in modern architecture.

It is clear from this list alone that the subject of this chapter has considerable significance for structural technology. We will develop a theory of plates, consider its validity, and then solve a number of problems. You will note that whereas we ended up working with a curve in the earlier structural considerations of Chap. 4, we shall in the present undertaking end up working with a surface.

6.2 KINEMATICS OF THE DEFORMATION OF PLATES

We shall now propose a simple mode of deformation for the plate akin to that for deformation of beams which led to the technical theory of beams. As a result of such simplifications we will need only to consider the deformation of the midplane of the plate in order to find certain significant information for the structure as a whole. Such simplifications entail the inclusion of hidden constraints in the body as discussed in previous chapters, and the equations of equilibrium resulting from the variational process applied to the total potential energy are accordingly those for a "stiffer" system than the actual case.

We have shown a portion of a plate of thickness h in Fig. 6.1 wherein the xy coordinate plane corresponds to the *midplane* or *middle surface* in the undeformed geometry. On the top face of the plate we have indicated a normal load distribution $q(x,y)$ while at the edge we have shown a shear force distribution Q and a bending moment distribution M. These quantities will be discussed in the next section. Note that the bounding curve of the plate as seen in the z direction is denoted as Γ while the interior region is denoted as R.

As a first step we assume, in view of the small thickness of the plate, that the vertical movement of any point of the plate is identical to that of the point (below or above it) in the middle surface. Thus

$$u_3(x_1,x_2,x_3) = w(x,y) \qquad (6.1)$$

where w is the vertical displacement function of the midsurface.

As for movements parallel to the midsurface we consider two contributions. The first are *stretching actions* due to loads at the edge of the plate, the loads being

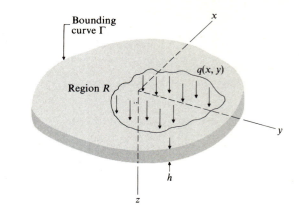

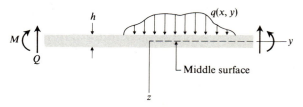

FIGURE 6.1

parallel to the midsurface of the plate. For these displacement components, denoted as $(u_1)_s$ and $(u_2)_s$, we assume that a point at position (x_1, x_2, x_3) has again identically the same displacement components as the corresponding points (above or below) in the midsurface of the plate. Thus we can say

$$[u_1(x_1, x_2, x_3)]_s = [u(x,y)]_s \qquad (a)$$

$$[u_2(x_1, x_2, x_3)]_s = [v(x,y)]_s \qquad (b) \quad (6.2)$$

where $(u)_s$ and $(v)_s$ refer to stretching action of the midsurface. Thus lines connecting the surfaces of the plate and normal to the xy plane in the undeformed geometry *translate horizontally* as a result of stretching action. The second contribution is attributed to *bending*. For this action, lines normal to the midsurface in the undeformed geometry remain normal to this surface in the deformed geometry. More specifically, lines, such as ab in Fig. 6.2 connecting the surfaces of the plate and normal to the xy plane in the undeformed geometry, translate vertically and in addition rotate as rigid elements as a result of bending (see $a'b'$ in Fig. 6.2). The displacement in

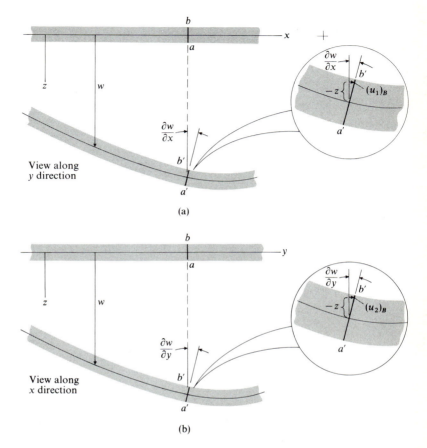

View along
y direction

(a)

View along
x direction

(b)

FIGURE 6.2

the x and y directions as a result of the bending action we denote respectively as $(u_1)_B$ and $(u_2)_B$. We have shown these quantities in the inscribed enlargement of Fig. 6.2. It should be clear that:

$$[u_1(x_1,x_2,x_3)]_B = -z\frac{\partial w(x,y)}{\partial x} \qquad (a)$$

$$[u_2(x_1,x_2,x_3)]_B = -z\frac{\partial w(x,y)}{\partial y} \qquad (b) \quad (6.3)$$

Note that in accordance with the initial assumption (leading to Eq. (6.1)), we may take z as the coordinate of the point in the *undeformed geometry* despite the fact that it is used above in the *deformed geometry* (see diagram) to identify the same point.

We now can give the proposed displacement field as follows combining stretching and bending actions

$$u_1 = u_s(x,y) - z\frac{\partial w(x,y)}{\partial x} \qquad (a)$$

$$u_2 = v_s(x,y) - z\frac{\partial w(x,y)}{\partial y} \qquad (b)$$

$$u_3 = w(x,y) \qquad (c) \quad (6.4)$$

Notice that the displacement field (u_1, u_2, u_3) is now fully described in terms of deformation of the midsurface $(u_s, v_s,$ and $w)$.[1]

Using the above displacement field it is a simple matter to compute the strain field. We get:

$$\epsilon_{xx} = \frac{\partial u_s}{\partial x} - z\frac{\partial^2 w}{\partial x^2}$$

$$\epsilon_{yy} = \frac{\partial v_s}{\partial y} - z\frac{\partial^2 w}{\partial y^2}$$

$$\epsilon_{xy} = \frac{1}{2}\left(\frac{\partial u_s}{\partial y} + \frac{\partial v_s}{\partial x}\right) - z\frac{\partial^2 w}{\partial x \, \partial y} \qquad (6.5)$$

All other strains are zero. We note immediately an obvious difficulty in that the transverse shear stresses (τ_{xz}, τ_{yz}) will be zero for the proposed displacement field. That these quantities cannot always be so is clear from simple equilibrium considerations. (Recall we had the same kind of difficulty in the technical theory of beams.) We shall accept this discrepancy for now but will give it proper attention later when discussing "improved theories" of plates.

6.3 STRESS RESULTANT INTENSITY FUNCTIONS AND THE EQUATIONS OF EQUILIBRIUM

In the study of beams we employed such resultant forces at a section as the shear force V, the bending moment M, and the axial force N. For the study of plates we find it useful to introduce *distributions* per unit length of these and other quantities.

---

[1] It is interesting to note that Eqs. (6.4(a)) and (6.4(b)) can be considered as the first two terms of a Taylor series in powers of the thickness coordinate z. In fact an alternate derivation of plate theory consists of expanding all quantities (stresses and displacements) into such Taylor series and then matching the coefficients of the corresponding powers of z. It is well to remember here that in retaining only the first two terms of the series we limit ourselves to plates of small thickness h compared to the other lateral dimensions of the plate surfaces.

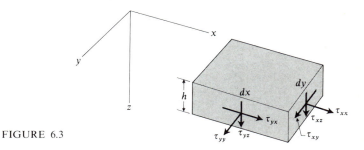

FIGURE 6.3

For this purpose we have shown in Fig. 6.3 an element of a plate with stresses at the midplane of the plate (these stresses vary in the z direction over the thickness h of the plate). We now define *shear force intensities*, Q_x and Q_y, as follows:

$$Q_x = \int_{-h/2}^{h/2} \tau_{xz}\, dz \qquad (a)$$

$$Q_y = \int_{-h/2}^{h/2} \tau_{yz}\, dz \qquad (b) \quad (6.6)$$

It is clear on considering Fig. 6.3 that Q_x is the shear force distribution on a face with a normal in the x direction given per unit length in the y direction, while Q_y is the shear force distribution on a face with a normal in the y direction given per unit length in the x direction. These force intensities per unit length have been shown in Fig. 6.4. We may next introduce *bending moment intensities* per unit length as follows:

$$M_x = \int_{-h/2}^{h/2} \tau_{xx} z\, dz \qquad (a)$$

$$M_y = \int_{-h/2}^{h/2} \tau_{yy} z\, dz \qquad (b) \quad (6.7)$$

These quantities are shown in Fig. 6.4. It is clear by considering the definition in conjunction with Fig. 6.3 that M_x is the bending moment distribution about the y axis on a section having a normal in the x direction per unit length in the y direction, etc. Finally we introduce two *twisting moment intensities* per unit length as follows:

$$M_{xy} = \int_{-h/2}^{h/2} \tau_{xy} z\, dz \qquad (a)$$

$$M_{yx} = \int_{-h/2}^{h/2} \tau_{yx} z\, dz \qquad (b) \quad (6.8)$$

Again considering the definition of M_{xy} we see that it represents the twisting moment distribution about the x axis at a section whose normal is the x axis, per unit length

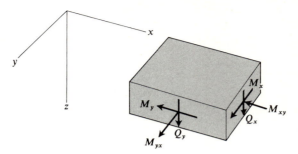

FIGURE 6.4

in the y direction, etc. It is immediately apparent from the complementary property of shear stress that:

$$M_{xy} = M_{yx} \qquad (6.9)$$

We may readily evaluate the various moment intensity functions in terms of w for the special case of linear elastic behavior by using Hooke's law for plane *stress*[1] to replace stresses in the defining equations. Thus we have from Hooke's law (see Eq. (1.117))

$$\tau_{xx} = \frac{E}{1 - v^2}(\epsilon_{xx} + v\epsilon_{yy}) \qquad (a)$$

$$\tau_{yy} = \frac{E}{1 - v^2}(\epsilon_{yy} + v\epsilon_{xx}) \qquad (b)$$

$$\tau_{xy} = 2G\epsilon_{xy} \qquad (c) \quad (6.10)$$

We then have:

$$M_x = \int_{-h/2}^{h/2} \frac{E}{1 - v^2}(\epsilon_{xx} + v\epsilon_{yy})z\, dz \qquad (a)$$

$$M_y = \int_{-h/2}^{h/2} \frac{E}{1 - v^2}(\epsilon_{yy} + v\epsilon_{xx})z\, dz \qquad (b)$$

$$M_{xy} = \int_{-h/2}^{h/2} 2G\epsilon_{xy}z\, dz \qquad (c) \quad (6.11)$$

[1] This simplification is made, as in the study of beams where we assumed $\tau_{zz} = 0$, despite the fact that the application of transverse loading $q(x,y)$ on the surface $z = -h/2$ leads to a nonzero stress τ_{zz}; and also in spite of the fact that the assumed displacement field with $\epsilon_{zz} = 0$ is not a case of plane stress.

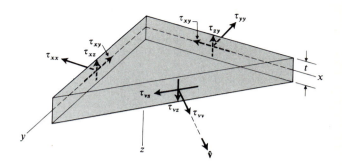

FIGURE 6.5

Now substituting for the strains from Eq. (6.5) we get the following results on carrying out the integration

$$M_x = -D\left(\frac{\partial^2 w}{\partial x^2} + v\frac{\partial^2 w}{\partial y^2}\right) \quad (a)$$

$$M_y = -D\left(\frac{\partial^2 w}{\partial y^2} + v\frac{\partial^2 w}{\partial x^2}\right) \quad (b)$$

$$M_{xy} = -(1 - v)D\frac{\partial^2 w}{\partial x\,\partial y} \quad (c) \quad (6.12)$$

where D, called the *bending rigidity*, is a constant given as:

$$D = \frac{Eh^3}{12(1 - v^2)} \quad (6.13)$$

As a next step we will formulate the stress resultant intensity functions along a section of the plane inclined arbitrarily in the $\hat{v}$ direction relative to the x, y directions. We shall accomplish this through the stress transformation equations for the stresses illustrated in Fig. 6.5. From a two-dimensional subset of the stress transformation equations (Eqs. (1.11)) it can be seen that

$$\tau_{vv} = a_{vx}^2\tau_{xx} + 2a_{vx}a_{vy}\tau_{xy} + a_{vy}^2\tau_{yy}$$

$$\tau_{vs} = a_{vx}a_{vy}(\tau_{yy} - \tau_{xx}) + (a_{vx}^2 - a_{vy}^2)\tau_{xy} \quad (6.14)$$

If moment resultants M_v and M_{vs} are defined in a manner analogous to Eqs. (6.7) and (6.8), Eqs. (6.14) may be multiplied by z and integrated over the plate thickness to yield the following moment transformation equations:

$$M_v = a_{vx}^2 M_x + 2a_{vx}a_{vy}M_{xy} + a_{vy}^2 M_y$$

$$M_{vs} = a_{vx}a_{vy}(M_y - M_x) + (a_{vx}^2 - a_{vy}^2)M_{xy} \quad (6.15)$$

From simple equilibrium considerations in the vertical direction for the element in Fig. 6.5

$$\tau_{vz}\,ds = \tau_{yz}\,dx + \tau_{xz}\,dy$$

We may note that[1]

$$a_{vx} = \frac{dy}{ds}, \qquad a_{vy} = \frac{dx}{ds}$$

Hence:

$$\tau_{vz} = a_{vy}\tau_{yz} + a_{vx}\tau_{xz}.$$

After defining a shear resultant Q_v as in Eqs. (6.6) and integrating over the thickness of the plate we find

$$Q_v = a_{vy}Q_y + a_{vx}Q_x. \qquad (6.16)$$

Quantities Q_x, Q_y, M_x, M_y and M_{xy} can be related by considering equilibrium of the plate element shown in Fig. 6.6, or by integrating the three-dimensional equilibrium equations of elasticity, as we did for the beam in Sec. 4.3. Thus for equilibrium in the x direction, in the absence of body forces,

$$\frac{\partial \tau_{xx}}{\partial x} + \frac{\partial \tau_{xy}}{\partial y} + \frac{\partial \tau_{xz}}{\partial z} = 0$$

If we multiply this equation by z and integrate over the plate thickness, noting that the operations $\partial/\partial x$ and $\partial/\partial y$ can be interchanged with the z integration, we find

$$\frac{\partial M_x}{\partial x} + \frac{\partial M_{xy}}{\partial y} + \int_{-h/2}^{h/2} z\frac{\partial \tau_{xz}}{\partial z}\,dz = 0$$

If the remaining thickness integral is integrated by parts,

$$\int_{-h/2}^{h/2} z\frac{\partial \tau_{xz}}{\partial z}\,dz = [z\tau_{xz}]_{-h/2}^{h/2} - \int_{-h/2}^{h/2} \tau_{xz}\,dz = -Q_x$$

We have noted here the definition (6.6a), and that no shear stresses are applied at the plate surfaces $z = \pm h/2$. Then we find that

$$Q_x = \frac{\partial M_x}{\partial x} + \frac{\partial M_{xy}}{\partial y} \qquad (6.17)$$

In exactly the same way we may integrate the equation of equilibrium in the y direction to obtain

$$Q_y = \frac{\partial M_{xy}}{\partial x} + \frac{\partial M_y}{\partial y} \qquad (6.18)$$

---

[1] Note that ds here is simply a distance increment. Later it will represent the differential of a coordinate s and then we will see that a_{vy} will be $-dx/ds$.

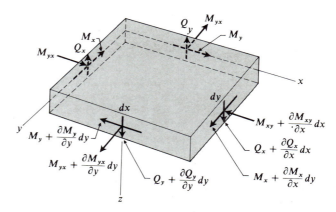

FIGURE 6.6

Finally we consider the integration over the thickness of the last equilibrium equation, i.e.,

$$\int_{-h/2}^{h/2} \left[\frac{\partial \tau_{xz}}{\partial x} + \frac{\partial \tau_{yz}}{\partial y} + \frac{\partial \tau_{zz}}{\partial z} \right] dz = 0$$

In view of the definition of the shear resultants, we find from above that

$$\frac{\partial Q_x}{\partial x} + \frac{\partial Q_y}{\partial y} = \tau_{zz}(z)_{z=-h/2} - \tau_{zz}(z)_{z=h/2}$$

Noting that $\tau_{zz}(z)_{z=h/2} = 0$ and that $\tau_{zz}(z)_{z=-h/2} = -q(x,y)$, we obtain finally

$$\frac{\partial Q_x}{\partial x} + \frac{\partial Q_y}{\partial y} + q(x,y) = 0 \qquad (6.19)$$

You may note that the operations carried out to obtain Eqs. (6.17)–(6.19) correspond to taking moments about the y and x axes, and considering force equilibrium in the z direction. In Sec. 6.6 we shall return to the equilibrium equations of elasticity and integrate them in a different fashion to obtain the details of the distribution of the shear stresses through the thickness.

We may now reduce the three equations of equilibrium (Eqs. (6.17)–(6.19)) to a single equation by eliminating Q_x and Q_y from the equation to form the following equation

$$\boxed{\frac{\partial^2 M_x}{\partial x^2} + 2\frac{\partial^2 M_{xy}}{\partial x\, \partial y} + \frac{\partial^2 M_y}{\partial y^2} + q = 0} \qquad (6.20)$$

where we have used the fact that $M_{xy} = M_{yx}$. Finally using Eqs. (6.12) we may introduce the function w into the above formulation as the dependent variable. We then get the following equation:

$$\boxed{\nabla^4 w = \frac{q}{D}} \qquad (6.21)$$

This is a non-homogeneous *biharmonic* equation[1] first obtained for plate theory by Sophie Germain in 1815.

It should be apparent from the above that those results involving the stress resultant intensity function alone (Eqs. (6.15) through Eq. (6.20)) are valid for all materials. Those results involving w are valid only for linear elastic materials.

Equation (6.21) will form one of the governing equations of classical plate theory. Needed in addition are specifications of appropriate boundary conditions (edge conditions) for classical plate theory. In the next section, using the variational approach we shall arrive at these boundary conditions in a straightforward manner. And Sophie Germain's equation will also be derived simultaneously.

6.4   MINIMUM TOTAL POTENTIAL ENERGY APPROACH

The strain energy U of the plate for linear elastic behavior is found by evaluating the following integral:

$$U = \frac{1}{2} \iint_R \int_{-h/2}^{h/2} \tau_{ij}\epsilon_{ij} \, dz \, dx \, dy \qquad (6.22)$$

In spite of our comments earlier concerning the transverse strains, we shall employ the state of plane stress in evaluating the above expression. Thus replacing τ_{ij} in terms of strains (see Eq. (6.10)) we get:

$$U = \frac{E}{2(1 - v^2)} \iint_R \int_{-h/2}^{h/2} \{\epsilon_{xx}^2 + 2v\epsilon_{xx}\epsilon_{yy} + \epsilon_{yy}^2 + 2(1 - v)\epsilon_{xy}^2\} \, dz \, dx \, dy \qquad (6.23)$$

The potential energy for the external loads meanwhile is given as follows

$$V = -\iint_R q(x,y)w(x,y) \, dx \, dy \qquad (6.24)$$

---

[1] The biharmonic operator ∇^4 is the same as two successive harmonic operators $\nabla^2(\nabla^2)$ and for rectangular coordinates is

$$\left(\frac{\partial^4}{\partial x^4} + 2\frac{\partial^2}{\partial x^2}\frac{\partial^2}{\partial y^2} + \frac{\partial^4}{\partial x^4} \right)$$

wherein the loads q are assumed to act on the midplane surface of the plate. Considering Eqs. (6.23) and (6.24) and expressing the strains in terms of the displacement field of the midsurface (Eqs. (6.5)) we get the following expression for the total potential energy:

$$
\pi = \frac{E}{2(1-v^2)} \int\!\!\int_R \int_{-h/2}^{h/2} \left\{ \left(\frac{\partial u_s}{\partial x} - z\frac{\partial^2 w}{\partial x^2} \right)^2 + \left(\frac{\partial v_s}{\partial y} - z\frac{\partial^2 w}{\partial y^2} \right)^2 \right.
$$
$$
+ 2v\left(\frac{\partial u_s}{\partial x} - z\frac{\partial^2 w}{\partial x^2} \right)\left(\frac{\partial v_s}{\partial y} - z\frac{\partial^2 w}{\partial y^2} \right)
$$
$$
\left. + 2(1-v)\left[\frac{1}{2}\left(\frac{\partial u_s}{\partial y} + \frac{\partial v_s}{\partial x} \right) - z\frac{\partial^2 w}{\partial x\,\partial y} \right]^2 \right\} dz\,dx\,dy
$$
$$
- \int\!\!\int_R q(x,y)w(x,y)\,dx\,dy \tag{6.25}
$$

Carrying out the squaring operation and integrating through the thickness while noting that

$$
\int_{-h/2}^{h/2} \{1,z,z^2\}\,dz = \left\{h, 0, \frac{h^3}{12}\right\} \tag{6.26}
$$

we may obtain the total potential energy in the form

$$
\pi = \frac{C}{2}\int\!\!\int_R \left[\left(\frac{\partial u_s}{\partial x} \right)^2 + \left(\frac{\partial v_s}{\partial y} \right)^2 + 2v\frac{\partial u_s}{\partial x}\frac{\partial v_s}{\partial y} \right.
$$
$$
\left. + \frac{1-v}{2}\left(\frac{\partial u_s}{\partial y} + \frac{\partial v_s}{\partial x} \right)^2 \right] dx\,dy
$$
$$
+ \frac{D}{2}\int\!\!\int_R \left[\left(\frac{\partial^2 w}{\partial x^2} \right)^2 + \left(\frac{\partial^2 w}{\partial y^2} \right)^2 + 2v\left(\frac{\partial^2 w}{\partial x^2} \right)\left(\frac{\partial^2 w}{\partial y^2} \right) \right.
$$
$$
\left. + 2(1-v)\left(\frac{\partial^2 w}{\partial x\,\partial y} \right)^2 \right] dx\,dy - \int\!\!\int_R qw\,dx\,dy \tag{6.27}
$$

where C is a new constant called the *extensional stiffness* and is given as:

$$
C = \frac{Eh}{1-v^2} \tag{6.28}
$$

The total potential energy functional has three dependent variables—the stretching components u_s and v_s and the vertical displacement variable w. We may use Eq. (2.70) limited to first-order derivatives and extended to apply to several functions. Thus we have for the Euler–Lagrange equations for u_s and v_s:

$$\frac{\partial F}{\partial u_s} - \frac{\partial}{\partial x}\frac{\partial F}{\partial(\partial u_s/\partial x)} - \frac{\partial}{\partial y}\frac{\partial F}{\partial(\partial u_s/\partial y)} = 0$$

$$\frac{\partial F}{\partial v_s} - \frac{\partial}{\partial x}\frac{\partial F}{\partial(\partial v_s/\partial x)} - \frac{\partial}{\partial y}\frac{\partial F}{\partial(\partial v_s/\partial y)} = 0$$

Substituting for F from the integrands of Eq. (6.27) we get:

$$\frac{\partial^2 u_s}{\partial x^2} + \frac{1-v}{2}\frac{\partial^2 u_s}{\partial y^2} + \frac{1+v}{2}\frac{\partial^2 v_s}{\partial x\,\partial y} = 0 \quad (a)$$

$$\frac{\partial^2 v_s}{\partial y^2} + \frac{1-v}{2}\frac{\partial^2 v_s}{\partial x^2} + \frac{1+v}{2}\frac{\partial^2 u_s}{\partial x\,\partial y} = 0 \quad (b) \quad (6.29)$$

Note the equations for u_s and v_s are *uncoupled* from the equation for w. This is exactly as was the case for bending of beams. We shall be concerned here with the stresses and moments in plates as a result of transverse loads. Since the strains due to this action will primarily stem from bending effects rather than stretching effects we shall only consider the former. Accordingly we return now to the total potential energy functional and deleting the u_s and v_s variables we rewrite the expression in the following form:

$$\pi = \frac{D}{2}\iint_R \left\{(\nabla^2 w)^2 + 2(1-v)\left[\left(\frac{\partial^2 w}{\partial x\,\partial y}\right)^2 - \left(\frac{\partial^2 w}{\partial x^2}\right)\left(\frac{\partial^2 w}{\partial y^2}\right)\right]\right\}dx\,dy$$

$$- \iint_R qw\,dx\,dy \tag{6.30}$$

We now extremize the above total potential energy functional as follows

$$\delta^{(1)}\pi = 0 = \frac{D}{2}\iint_R \left\{2(\nabla^2 w)\left[\frac{\partial^2(\delta w)}{\partial x^2} + \frac{\partial^2(\delta w)}{\partial y^2}\right]\right.$$

$$+ (1-v)\left[2\frac{\partial^2 w}{\partial x\,\partial y}\frac{\partial^2(\delta w)}{\partial x\,\partial y} + 2\frac{\partial^2 w}{\partial y\,\partial x}\frac{\partial^2(\delta w)}{\partial y\,\partial x}\right.$$

$$\left.\left. - 2\left(\frac{\partial^2 w}{\partial x^2}\right)\frac{\partial^2\,\delta w}{\partial y^2} - 2\left(\frac{\partial^2 w}{\partial y^2}\right)\left(\frac{\partial^2\,\delta w}{\partial x^2}\right)\right]\right\}dx\,dy$$

$$- \iint_R q\,\delta w\,dx\,dy \tag{6.31}$$

where we have split up the expression $2(\partial^2 w/\partial x\,\partial y)^2$ into

$$\left(\frac{\partial^2 w}{\partial x\,\partial y}\right)^2 + \left(\frac{\partial^2 w}{\partial y\,\partial x}\right)^2$$

in carrying out the above formulation. We now employ Green's theorem successively. You may wish to take the time to justify that the following result is reached:

$$\iint_R (D\nabla^4 w - q)\,\delta w\,dx\,dy + D \oint_\Gamma \left(\frac{\partial^2 w}{\partial x^2} + v\frac{\partial^2 w}{\partial y^2}\right)\frac{\partial\,\delta w}{\partial x}\,dy$$

$$- D \oint_\Gamma \left(\frac{\partial^2 w}{\partial y^2} + v\frac{\partial^2 w}{\partial x^2}\right)\frac{\partial\,\delta w}{\partial y}\,dx + D \oint_\Gamma (1-v)\frac{\partial^2 w}{\partial x\,\partial y}\frac{\partial\,\delta w}{\partial y}\,dy$$

$$- D \oint_\Gamma (1-v)\frac{\partial^2 w}{\partial x\,\partial y}\frac{\partial\,\delta w}{\partial x}\,dx + D \oint_\Gamma \left(\frac{\partial^3 w}{\partial y^3} + v\frac{\partial^3 w}{\partial x^2\,\partial y}\right)\delta w\,dx$$

$$- D \oint_\Gamma \left(\frac{\partial^3 w}{\partial x^3} + v\frac{\partial^3 w}{\partial x\,\partial y^2}\right)\delta w\,dy + D \oint_\Gamma (1-v)\frac{\partial^3 w}{\partial x^2\,\partial y}\delta w\,dx$$

$$- D \oint_\Gamma (1-v)\frac{\partial^3 w}{\partial x\,\partial y^2}\delta w\,dy = 0 \tag{6.32}$$

Now considering Eqs. (6.12) we see that the integrands of the line integrals can be expressed in terms of the stress resultant intensity functions. Thus we may rewrite the above equation as follows:

$$\iint_R (D\nabla^4 w - q)\,\delta w\,dx\,dy - \oint_\Gamma M_x \frac{\partial\,\delta w}{\partial x}\,dy + \oint_\Gamma M_y \frac{\partial\,\delta w}{\partial y}\,dx$$

$$- \oint_\Gamma M_{xy}\frac{\partial\,\delta w}{\partial y}\,dy + \oint_\Gamma M_{xy}\frac{\partial\,\delta w}{\partial x}\,dx$$

$$- \oint_\Gamma \left(\frac{\partial M_y}{\partial y} + \frac{\partial M_{xy}}{\partial x}\right)\delta w\,dx$$

$$+ \oint_\Gamma \left(\frac{\partial M_x}{\partial x} + \frac{\partial M_{xy}}{\partial y}\right)\delta w\,dy = 0 \tag{6.33}$$

Now using Eqs. (6.18) and (6.19) we may rewrite the last two integrals in terms of shear force intensities. We thus have:

$$\iint_R (D\nabla^4 w - q)\,\delta w\,dx\,dy - \oint_\Gamma M_x \frac{\partial\,\delta w}{\partial x}\,dy + \oint_\Gamma M_y \frac{\partial\,\delta w}{\partial y}\,dx$$

$$- \oint_\Gamma M_{xy}\frac{\partial\,\delta w}{\partial y}\,dy + \oint_\Gamma M_{xy}\frac{\partial\,\delta w}{\partial x}\,dx$$

$$- \oint_\Gamma Q_y\,\delta w\,dx + \oint_\Gamma Q_x\,\delta w\,dy = 0 \tag{6.34}$$

To simplify the functional further, examine a portion of the path Γ as shown in Fig. 6.7. Considering v and s to be a rectangular set of coordinates at a point on the boundary

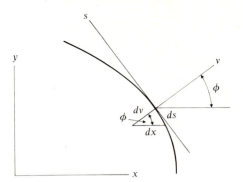

FIGURE 6.7

we may say

$$\cos \varphi = a_{vx} = \frac{dv}{dx}$$

$$\sin \varphi = a_{vy} = -\frac{ds}{dx} \qquad (6.35)$$

where the minus sign results from the fact that in going in the plus s direction on Γ we then move in the minus x direction. Also from Fig. 6.8 we can similarly conclude:

$$\cos \varphi = a_{vx} = \frac{ds}{dy}$$

$$\sin \varphi = a_{vy} = \frac{dv}{dy} \qquad (6.36)$$

Employing the results of Eqs. (6.35) and (6.36) we may next state that:

$$\frac{\partial}{\partial x} = \frac{\partial}{\partial v}\frac{dv}{dx} + \frac{\partial}{\partial s}\frac{ds}{dx} = a_{vx}\frac{\partial}{\partial v} - a_{vy}\frac{\partial}{\partial s}$$

$$\frac{\partial}{\partial y} = \frac{\partial}{\partial v}\frac{dv}{dy} + \frac{\partial}{\partial s}\frac{ds}{dy} = a_{vy}\frac{\partial}{\partial v} + a_{vx}\frac{\partial}{\partial s} \qquad (6.37)$$

Now substitute for $\partial/\partial x$ and $\partial/\partial y$ in Eq. (6.34) in accordance with the above equation. We get on collecting terms:

$$\iint_R (D\nabla^4 w - q)\,\delta w\,dx\,dy - \oint_\Gamma M_x\left(a_{vx}\frac{\partial}{\partial v}(\delta w) - a_{vy}\frac{\partial(\delta w)}{\partial s}\right)(dy)$$

$$+ \oint_\Gamma M_y\left(a_{vy}\frac{\partial(\delta w)}{\partial v} + a_{vx}\frac{\partial(\delta w)}{\partial s}\right)(dx)$$

$$- \oint_\Gamma M_{xy}\left(a_{vy}\frac{\partial(\delta w)}{\partial v} + a_{vx}\frac{\partial(\delta w)}{\partial s}\right)(dy) +$$

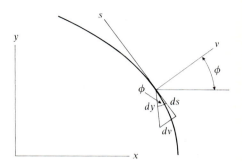

FIGURE 6.8

$$\oint_{\Gamma} M_{xy}\left(a_{vx}\frac{\partial(\delta w)}{\partial v} - a_{vy}\frac{\partial(\delta w)}{\partial s}\right)(dx)$$

$$-\oint_{\Gamma} Q_y\,\delta w(dx) + \oint_{\Gamma} Q_x\,\delta w(dy) = 0 \qquad (6.38)$$

We may introduce the differential along a boundary ds in place of dx and dy in the above formulation by noting from Fig. 6.9 that

$$dx = -ds\,a_{vy} \qquad (a)$$
$$dy = ds\,a_{vx} \qquad (b) \quad (6.39)$$

We thus have:

$$\iint_R (D\nabla^4 w - q)\,\delta w\,dx\,dy - \oint_{\Gamma} M_x\left[a_{vx}\frac{\partial\,\delta w}{\partial v} - a_{vy}\frac{\partial\,\delta w}{\partial s}\right](a_{vx}\,ds)$$

$$+ \oint_{\Gamma} M_y\left[a_{vy}\frac{\partial\,\delta w}{\partial v} + a_{vx}\frac{\partial\,\delta w}{\partial s}\right](-ds\,a_{vy})$$

$$- \oint_{\Gamma} M_{xy}\left[a_{vy}\frac{\partial\,\delta w}{\partial v} + a_{vx}\frac{\partial\,\delta w}{\partial s}\right](ds\,a_{vx})$$

$$+ \oint_{\Gamma} M_{xy}\left[a_{vx}\frac{\partial\,\delta w}{\partial v} - a_{vy}\frac{\partial\,\delta w}{\partial s}\right](-ds\,a_{vy})$$

$$- \oint_{\Gamma} Q_y\,\delta w(-dx\,a_{vy}) + \oint_{\Gamma} Q_x\,\delta w(ds\,a_{vx}) = 0$$

Now collecting terms we get:

$$\iint_R (D\nabla^4 w - q)\,\delta w\,dx\,dy + \oint_{\Gamma}[-M_x a_{vx}{}^2 - M_y a_{vy}{}^2 - 2M_{xy}a_{vx}a_{vy}]\left(\frac{\partial\,\delta w}{\partial v}\right)ds$$

$$+ \oint_{\Gamma}[M_x a_{vx}a_{vy} - M_y a_{vx}a_{vy} - M_{xy}a_{vx}{}^2 +$$

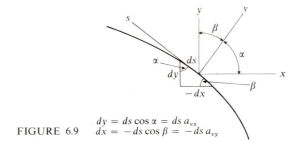

FIGURE 6.9 $\begin{aligned} dy &= ds \cos \alpha = ds\, a_{vx} \\ dx &= -ds \cos \beta = -ds\, a_{vy} \end{aligned}$

$$M_{xy}a_{vy}{}^2]\left(\frac{\partial\,\delta w}{\partial s}\right) ds + \oint_{\Gamma}[Q_y a_{vy} + Q_x a_{vx}]\,\delta w\, ds = 0$$

By employing Eqs. (6.15) and (6.16) to replace bracketed expressions in the integrands of the line integrals we may rewrite the above equation as follows:

$$\iint_{R}(D\nabla^4 w - q)\,\delta w\, dx\, dy - \oint_{\Gamma}M_v\frac{\partial\,\delta w}{\partial v}\,ds - \oint_{\Gamma}M_{vs}\frac{\partial\,\delta w}{\partial s}\,ds + \oint_{\Gamma}Q_v\,\delta w\, ds = 0 \qquad (6.40)$$

Let us examine the third integral. Consider it to be for a moment a line integral not closed. Then, if M_{vs} is continuous and the curve is smooth we may integrate by parts as follows:

$$\int_{1}^{2}M_{vs}\frac{\partial\,\delta w}{\partial s}\,ds = [M_{vs}\,\delta w]_1^2 - \int_{1}^{2}\frac{\partial M_{vs}}{\partial s}\,\delta w\, ds$$

For a closed smooth curve[1] the bracketed expression is clearly zero and so we have for this case:

$$\oint_{\Gamma}M_{vs}\frac{\partial\,\delta w}{\partial s}\,ds = -\oint_{\Gamma}\frac{\partial M_{vs}}{\partial s}\,\delta w\, ds \qquad (6.41)$$

We then get for the variation of the total potential energy:

$$\iint_{R}(D\nabla^4 w - q)\,\delta w\, dx\, dy - \oint_{\Gamma}M_v\,\delta\!\left(\frac{\partial w}{\partial v}\right)ds$$

$$+ \oint_{\Gamma}\left(Q_v + \frac{\partial M_{vs}}{\partial s}\right)\delta w\, ds = 0 \qquad (6.42)$$

[1] We shall examine later a rectangular plate which has four sections of smooth curves terminating in corners. In that case the integration may be carried out over each edge of the plate and so the closed integral would have contributions from the bracketed expression at each corner. These are the so-called corner conditions. The bracketed expression leads to these corner conditions whenever the bounding curve of the plate is only piecewise smooth, i.e., rectangular, triangular, polygonal, etc.

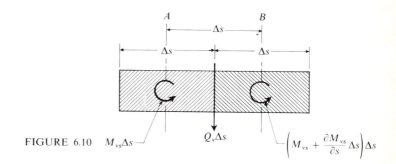

FIGURE 6.10 $M_{vs}\Delta s$ $Q_v\Delta s$ $\left(M_{vs} + \dfrac{\partial M_{vs}}{\partial s}\Delta s\right)\Delta s$

It is then apparent that the Euler–Lagrange equation for this problem is

$$\nabla^4 w = \frac{q}{D} \qquad (6.43)$$

the equation presented earlier. Now we get, as a result of the variational process two sets of boundary conditions, namely the natural and the kinematic boundary conditions. Thus on Γ we require that:

EITHER $\qquad\qquad M_v = 0 \quad$ OR $\quad \dfrac{\partial w}{\partial v} \quad$ IS PRESCRIBED

EITHER $\quad Q_v + \dfrac{\partial M_{vs}}{\partial s} = 0 \quad$ OR $\quad w \quad$ IS PRESCRIBED $\qquad (6.44)$

That there are only two conditions, in spite of the fact that there are three variables M, Q_v, and M_{vs} present, may come as a surprise. The first condition is acceptable by physical considerations and needs no further comment. We shall now examine the second condition with the view toward reaching some physical explanation. We present for this purpose the explanation set forth by Thomson and Tait in their classic treatise "Natural Philosophy." Accordingly we have shown in Fig. 6.10 part of the edge of a plate wherein two "panels" of length Δs have been identified. The twisting moment has been expressed in the second panel as a Taylor expansion with one term in terms of the twisting moment in the first panel. A third panel of length Δs may be imagined at the center of the aforementioned panels. This is shown in Fig. 6.10 as AB. The shear force for this panel is shown as $Q_v\Delta s$. Now we make use of the Saint Venant principle by replacing the twisting moment distribution in the original two panels by two couples (see Fig. 6.11) having forces of value M_{vs} and

$$\left(M_{vs} + \frac{\partial M_{vs}}{\partial s}\Delta s\right)$$

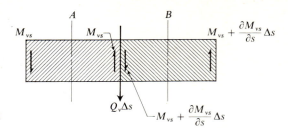

FIGURE 6.11

respectively with a separation of Δs between the forces. Clearly any conclusion arising from the new arrangement is valid "away" from the edge. Now we focus our attention on the center panel (between A and B). The "effective" shear force intensity, Q_{eff}, for this panel is then seen to be:

$$Q_{\text{eff}} = \frac{1}{\Delta s}\left[Q_v \Delta s + \left(M_{vs} + \frac{\partial M_{vs}}{\partial s}\Delta s\right) - M_{vs}\right]$$

$$= \left(Q_v + \frac{\partial M_{vs}}{\partial s}\right) \tag{6.45}$$

We may now conclude that the second of our natural boundary conditions renders the *effective shear force intensity equal to zero*.

Thus by the method of minimum total potential energy we have generated the entire boundary value problem for the classical thin plate theory. Because we employed a constitutive law (Hooke's law) in the formulations it might seem that the natural boundary conditions are restricted to Hookean materials. In the next section we shall present an alternative derivation of the boundary value problem by the method of virtual work. No constitutive law is used and we will generate the general moment intensity and shear force intensity equations presented in Sec. 6.3. Furthermore we shall verify the fact that the natural boundary conditions presented in this section are indeed valid for all structural materials.

6.5 PRINCIPLE OF VIRTUAL WORK; RECTANGULAR PLATES

In using the principle of virtual work we shall consider a rectangular plate. In particular we shall examine closely the corner conditions for such a problem.

We now apply the principle of virtual work, under the assumption of plane stress, to a rectangular plate (see Fig. 6.12) having dimensions $a \times b \times h$ and loaded normal to the centerplane of the plate by a loading intensity $q(x,y)$. Thus we have for zero body forces:

$$\int_0^a \int_0^b q(x,y)\,\delta[w(x,y)]\,dx\,dy = \int_0^a \int_0^b \int_{-h/2}^{h/2} (\tau_{xx}\,\delta\epsilon_{xx} + \tau_{yy}\,\delta\epsilon_{yy}$$
$$+ 2\tau_{xy}\,\delta\epsilon_{xy})\,dx\,dy\,dz \tag{6.46}$$

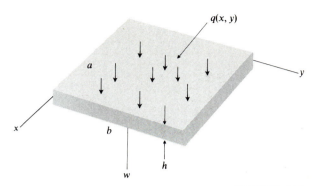

FIGURE 6.12

Now replace the strains using only the bending contributions of the displacement field (see Eq. (6.5)). We thus get, on rearranging the equation:

$$\int_0^a \int_0^b \int_{-h/2}^{h/2} z \left(\tau_{xx} \frac{\partial^2 \delta w}{\partial x^2} + \tau_{yy} \frac{\partial^2 \delta w}{\partial y^2} + 2\tau_{xy} \frac{\partial^2 \delta w}{\partial x \partial y} \right) dx\, dy\, dz$$

$$+ \int_0^a \int_0^b q\, \delta w\, dx\, dy = 0$$

Now integrate in the first integral with respect to z and use the stress resultant intensity functions presented in Sec. 6.3 to arrive at the following equation:

$$\int_0^a \int_0^b \left(M_x \frac{\partial^2 \delta w}{\partial x^2} + 2M_{xy} \frac{\partial^2 \delta w}{\partial x \partial y} + M_y \frac{\partial^2 \delta w}{\partial y^2} \right) dx\, dy$$

$$+ \int_0^a \int_0^b q(x,y)\, \delta w\, dx\, dy = 0 \qquad (6.47)$$

As in the previous section we decompose $2\partial^2\, \delta w / \partial x\, \partial y$ into $(\partial^2\, \delta w / \partial x\, \partial y + \partial^2\, \delta w / \partial y\, \partial x)$ and employ Green's theorem. We arrive at the following result:

$$\int_0^a \int_0^b \left(\frac{\partial^2 M_x}{\partial x^2} + 2\frac{\partial^2 M_{xy}}{\partial x \partial y} + \frac{\partial^2 M_y}{\partial y^2} + q \right) \delta w\, dx\, dy$$

$$+ \int_0^b \left(M_x \frac{\partial \delta w}{\partial x} \right)\Big|_{x=0}^{x=a} dy + \int_0^a \left(M_y \frac{\partial \delta w}{\partial y} \right)\Big|_{y=0}^{y=b} dx$$

$$- \int_0^b \left\{ \left[\frac{\partial M_x}{\partial x} + \frac{\partial M_{xy}}{\partial y} \right] \delta w \right\}\Big|_{x=0}^{x=a} dy - \int_0^a \left\{ \left[\frac{\partial M_y}{\partial y} + \frac{\partial M_{xy}}{\partial x} \right] \delta w \right\}\Big|_{y=0}^{y=b} dx$$

$$+ \int_0^a \left(M_{xy} \frac{\partial \delta w}{\partial x} \right)\Big|_{y=0}^{y=b} dx + \int_0^b \left(M_{xy} \frac{\partial \delta w}{\partial y} \right)\Big|_{x=0}^{x=a} dy = 0 \qquad (6.48)$$

Now integrate the last two expressions by parts. We get:

$$\int_0^a \left(M_{xy} \frac{\partial \delta w}{\partial x} \right)\Bigg|_{y=0}^{y=b} dx = \left[(M_{xy}\delta w) \Big|_{y=0}^{y=b} \right]_{x=0}^{x=a} - \int_0^a \left(\frac{\partial M_{xy}}{\partial x}\delta w \right)\Bigg|_{y=0}^{y=b} dx$$

$$= (M_{xy}\delta w)_{(a,b)} - (M_{xy}\delta w)_{(a,0)}$$

$$- (M_{xy}\delta w)_{(0,b)} + (M_{xy}\delta w)_{(0,0)}$$

$$- \int_0^a \left(\frac{\partial M_{xy}}{\partial x}\delta w \right)\Bigg|_{y=0}^{y=b} dx$$

$$\int_0^b \left(M_{xy} \frac{\partial \delta w}{\partial y} \right)\Bigg|_{x=0}^{x=a} dy = \left[(M_{xy}\,\delta w) \Big|_{x=0}^{x=a} \right]_{y=0}^{y=b} - \int_0^b \left(\frac{\partial M_{xy}}{\partial y}\delta w \right)\Bigg|_{x=0}^{x=a} dy$$

$$= (M_{xy}\,\delta w)_{(a,b)} - (M_{xy}\,\delta w)_{(a,0)}$$

$$- (M_{xy}\,\delta w)_{(0,b)} + (M_{xy}\,\delta w)_{(0,0)}$$

$$- \int_0^b \left(\frac{\partial M_{xy}}{\partial y}\delta w \right)\Bigg|_{x=0}^{x=a} dy$$

Inserting these results and noting Eqs. (6.17) and (6.18) we get:

$$\int_0^a \int_0^b \left[\frac{\partial^2 M_x}{\partial x^2} + 2\frac{\partial^2 M_{xy}}{\partial x \partial y} + \frac{\partial^2 M_y}{\partial y^2} + q \right] \delta w \, dx \, dy$$

$$+ \int_0^b \left[M_x \delta\left(\frac{\partial w}{\partial x} \right) \right]_{x=0}^{x=a} dy + \int_0^a \left[M_y \delta\left(\frac{\partial w}{\partial y} \right) \right]_{y=0}^{y=b} dx$$

$$- \int_0^b \left[\left(Q_x + \frac{\partial M_{xy}}{\partial y} \right) \delta w \right]_{x=0}^{x=a} dy - \int_0^a \left[\left(Q_y + \frac{\partial M_{xy}}{\partial x} \right) \delta w \right]_{y=0}^{y=b} dx$$

$$+ 2M_{xy}\delta w \bigg|_{(a,b)} - 2M_{xy}\,\delta w \bigg|_{(0,b)} - 2M_{xy}\,\delta w \bigg|_{(a,0)} + 2M_{xy}\,\delta w \bigg|_{(0,0)} = 0$$

$$(6.49)$$

Now we can see from the above equation that necessarily the following equation must be satisfied:

$$\frac{\partial^2 M_x}{\partial x^2} + 2\frac{\partial^2 M_{xy}}{\partial x \partial y} + \frac{\partial^2 M_y}{\partial y^2} + q = 0 \qquad (6.50)$$

This is identical to the equation (6.20) developed directly from equilibrium requirements and valid, consequently, for all materials. And on the boundaries of the plate we see that the following conditions must be met:

Along x = 0 and x = a

$$\text{EITHER } M_x = 0 \qquad \text{OR} \qquad \frac{\partial w}{\partial x} \text{ IS PRESCRIBED}$$

and

$$\text{EITHER } Q_x + \frac{\partial M_{xy}}{\partial y} = 0 \quad \text{OR} \quad w \text{ IS PRESCRIBED}$$

Along y = 0 and y = b

$$\text{EITHER } M_y = 0 \qquad \text{OR} \qquad \frac{\partial w}{\partial y} \text{ IS PRESCRIBED}$$

and

$$\text{EITHER } Q_y + \frac{\partial M_{xy}}{\partial x} = 0 \quad \text{OR} \quad w \text{ IS PRESCRIBED} \qquad (6.51)$$

Thus we see that the same edge conditions result as from the analysis using minimum total potential energy. We thus fully verify in the more general undertaking the results of the more limited analysis of the previous section.

Finally we have to consider the so-called *corner conditions*. Clearly we can either specify w at the corners, thus rendering $\delta w = 0$, or we can set $M_{xy} = 0$ at the corners. Beyond this, we can conclude from Eq. (6.49) that there actually exist corner forces given by the value $2M_{xy}$ at the corner.[1] We shall demonstrate this directly in Sec. 6.7. To get the proper direction of a corner force once M_{xy} has been determined, we may use the scheme of Thomson and Tait presented earlier to set forth the concept of the effective shear. Thus at a corner we use couples giving the proper directions of M_{xy} and M_{yx} on the orthogonal interfaces (see Fig. 6.13(a)). The forces coinciding at the corner edge then give the direction of the corner force. In Fig. 6.13(b)) we have shown corner forces for positive values of M_{xy} at all corners.

It is to be pointed out that all these conclusions for the corner conditions could have been reached from the formulations in the preceding section by not limiting ourselves as we did to smooth boundaries.

Now that the theory has been fully set forth we shall briefly examine some aspects of the validity of this theory in the following section.

---

[1] The expressions $2M_{xy} \delta w|_{a,b}$ etc., are just the virtual work expressions of these forces.

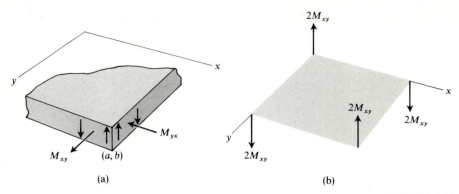

(a)

(b)

FIGURE 6.13

6.6 A NOTE ON THE VALIDITY OF CLASSICAL PLATE THEORY

As in the case of the technical theory of beams there are obvious discrepancies in the classical theory of plates that has been presented. Most notably, we have employed the assumption of plane stress in the plate thus rendering equal to zero transverse shears τ_{zx} and τ_{zy} as well as the transverse normal stress τ_{zz}. It is obvious from equilibrium considerations that such stresses will seldom be zero. Accordingly we need assurance that the classical plate theory can, under proper conditions, yield meaningful results.

In this regard we can follow one of two approaches. We might test our approximate results against corresponding results stemming from a more exact theory. (This was the procedure we took when we were at this juncture in the study of beams.) Or we may attempt to determine the bounds on the magnitudes of the stresses (given above) that have been deleted from the theory to show that these stresses, although not zero, are nevertheless small compared to the other stresses. In view of the scarcity of exact solutions for plate bending problems from the three-dimensional theory of elasticity we shall follow the latter procedure.

For this purpose we substitute the assumed form for the strain components as given by Eq. (6.5) into Hooke's law for plane stress as given by Eq. (6.10). The results are stated as follows in Cartesian coordinates:

$$\tau_{xx} = -\frac{Ez}{1 - v^2}\left(\frac{\partial^2 w}{\partial x^2} + v\frac{\partial^2 w}{\partial y^2}\right)$$

$$\tau_{yy} = -\frac{Ez}{1 - v^2}\left(\frac{\partial^2 w}{\partial y^2} + v\frac{\partial^2 w}{\partial x^2}\right)$$

$$\tau_{xy} = -\frac{Ez}{1 + v}\left(\frac{\partial^2 w}{\partial x\partial y}\right) \tag{6.52}$$

Now observing Eq. (6.12), we see that the bracketed expressions above may be replaced in favor of quantities involving the stress resultant intensity functions. Thus we have:

$$\tau_{xx} = \frac{M_x z}{h^3/12}$$

$$\tau_{yy} = \frac{M_y z}{h^3/12}$$

$$\tau_{xy} = \frac{M_{xy} z}{h^3/12} \qquad (6.53)$$

These results are the analogs to the flexure formula of beams and contain the key simplifications made in the classical theory of plates.

Now let us use these stresses in the equations of equilibrium with the view towards determining a τ_{zx}, a τ_{zy}, and a τ_{zz}. Thus in the x direction we have for the case of no body forces present:

$$\frac{\partial \tau_{xx}}{\partial x} + \frac{\partial \tau_{xy}}{\partial y} + \frac{\partial \tau_{xz}}{\partial z} = 0$$

therefore

$$\left(\frac{z}{h^3/12}\right)\frac{\partial M_x}{\partial x} + \left(\frac{z}{h^3/12}\right)\frac{\partial M_{xy}}{\partial y} + \frac{\partial \tau_{xz}}{\partial z} = 0$$

Collecting terms and then using Eq. (6.17) we have:

$$\frac{\partial \tau_{xz}}{\partial z} = -\frac{z}{h^3/12}\left[\frac{\partial M_x}{\partial x} + \frac{\partial M_{xy}}{\partial y}\right]$$

$$= -\left(\frac{z}{h^3/12}\right)(Q_x)$$

Now integrating with respect to z we get:

$$\tau_{xz} = -\frac{z^2/2}{h^3/12}Q_x + f(x,y)$$

To determine the arbitrary function $f(x,y)$ note that when $z = \pm h/2$, $\tau_{xz} = 0$.[1] Hence:

$$0 = -\frac{h^2/8}{h^3/12}Q_x + f$$

[1] We assume that no shears are applied at the top and bottom of the plate; only transverse loading $q(x,y)$ is given.

therefore

$$f = \frac{3}{2}hQ_x$$

We may thus give τ_{xz} as follows:

$$\tau_{xz} = \frac{3}{2}\frac{Q_x}{h}\left(1 - 4\frac{z^2}{h^2}\right) \qquad (6.54)$$

Next using the equilibrium equation in the y direction we may show in a similar fashion that:

$$\tau_{yz} = \frac{3}{2}\frac{Q_y}{h}\left(1 - 4\frac{z^2}{h^2}\right) \qquad (6.55)$$

Finally we consider the z direction and we note that:

$$\frac{\partial \tau_{zz}}{\partial z} = -\frac{\partial \tau_{xz}}{\partial x} - \frac{\partial \tau_{yz}}{\partial y}$$

$$= -\frac{3}{2h}\left(1 - \frac{4z^2}{h^2}\right)\left(\frac{\partial Q_x}{\partial x} + \frac{\partial Q_y}{\partial y}\right)$$

where we have used Eq. (6.54) and (6.55) to replace the stresses on the right side of the above equation. Now noting Eq. 6.19 we may replace the last bracketed quantity by $-q$ to get the following result:

$$\frac{\partial \tau_{zz}}{\partial z} = \frac{3q}{2h}\left(1 - 4\frac{z^2}{h^2}\right) \qquad (6.56)$$

Integrating we get for τ_{zz}:

$$\tau_{zz} = \frac{3q}{2h}\left(z - \frac{4z^3}{3h^2}\right) + g(x,y)$$

The boundary conditions for τ_{zz} are:

$$(a) \text{ when } z = -\frac{h}{2}$$

$$\tau_{zz} = -q$$

$$(b) \text{ when } z = \frac{h}{2}$$

$$\tau_{zz} = 0$$

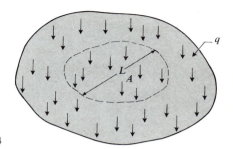

FIGURE 6.14

Hence we have from condition (a):

$$-q = \frac{3q}{2h}\left(-\frac{h}{2} + \frac{4h^3}{24h^2}\right) + g(x,y) \qquad (6.57)$$

therefore

$$g = -\tfrac{1}{2}q$$

Using this result for g, it is immediately clear that condition (b) is satisfied. We thus have for τ_{zz}:

$$\tau_{zz} = -q\left[\frac{1}{2} - \frac{3}{2}\frac{z}{h} + 2\frac{z^3}{h^3}\right] \qquad (6.58)$$

We thus have a set of stresses τ_{xz}, τ_{yz}, and τ_{zz} (Eqs. (6.54), (6.55), and (6.58)) which are computed from rigorous equilibrium considerations using results from a theory that neglected these very stresses. If it turns out that the above computed stresses are small, we have inner consistency in our theory. To do this, we shall next make an order-of-magnitude study of all six stresses for comparison purposes. For this purpose consider then a portion of a plate A (see Fig. 6.14) having some dimension L that characterizes lateral distances for this portion of the plate. Then we can say, taking q as some average loading intensity for the portion of the plate shown, that the order of magnitude of the total external force is:

$$O(F) = [O(q)][O(L^2)]$$

Considering equilibrium of the plate section A in the vertical direction we can equate the transverse load given above with the resultant of shear force distribution. Thus using the order of magnitude of the average value of Q_y we may write the following order-of-magnitude equation:

$$[O(Q_y)][O(L)] = [O(q)][O(L^2)]$$

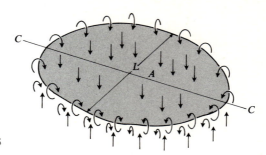

FIGURE 6.15

therefore

$$O(Q_v) = [O(q)][O(L)] = O(qL) \qquad (6.59)$$

We have thus established the order of magnitude of an average value of Q_v for A in terms of the order of magnitude of qL. Now consider an order of magnitude study of moments acting on the element. Equate moments about an axis C–C going through the line of action of the resultant of the force distribution q (see Fig. 6.15) with those of force distributions on the edge of the plate. We can conclude from equilibrium:

$$0 = \underbrace{[O(Q_v)][O(L)]}_{\text{shear force}} \underbrace{[O(L)]}_{\text{arm}} + [O(M_v)][O(L)] + [O(M_{vs})][O(L)] \qquad (6.60)$$

Replace $O(Q_v)$ using Eq. (6.59) in the above equation. Assume next that shear effects are *as significant* as either the bending or the twisting effects. We thus deliberately make shear effects significant at this point in the discourse. If the resulting shear stresses turn out *still* to be small in our order-of-magnitude studies, then we can rest assured that this did not arise from our diminishing prematurely their significance somewhere in the development. We then equate the order-of-magnitude of bending and twisting with that of the shear effects (using Eq. (6.59)) as they appear in the above equation:

$$O(M_v) = O(qL^2) \quad (a)$$
$$O(M_{vs}) = O(qL^2) \quad (b) \quad (6.61)$$

We thus have order-of-magnitude formulations for the stress resultant intensity functions Q_v, M_v, and M_{vs} which may now be used in conjunction with stresses for comparison purposes. First go back to Eqs. (6.53) to consider stresses τ_{xx}, τ_{yy}, and

τ_{xy}. Since z will be of the order of magnitude of h we can say, using Eq. (6.61):

$$O(\tau_{xx}) = O(qL^2/h^2)$$
$$O(\tau_{yy}) = O(qL^2/h^2)$$
$$O(\tau_{xy}) = O(qL^2/h^2) \qquad (6.62)$$

Now considering τ_{zx}, τ_{zy}, and τ_{zz} from Eqs. (6.54), (6.55), and (6.58) in a similar fashion, we conclude using Eqs. (6.59) and (6.61)

$$O(\tau_{zx}) = [O(qL)]\left[O\left(\frac{1}{h}\right)\right] = O\left(q\frac{L}{h}\right)$$

$$O(\tau_{zy}) = [O(qL)]\left[O\left(\frac{1}{h}\right)\right] = O\left(q\frac{L}{h}\right)$$

$$O(\tau_{zz}) = [O(q)] \qquad (6.63)$$

We can now make comparisons. The transverse shears τ_{zx} and τ_{zy} are smaller by an order of magnitude h/L than the midplane shear τ_{xy}. Also the transverse normal stress τ_{zz} is smaller by an order h^2/L^2 than the midplane normal stresses τ_{xx} and τ_{yy}. Thus for a very thin plate where $L/h \gg 10$ we have inner consistency and the classical theory may be expected to give good results. For a moderately thin plate, especially in the vicinity of concentrated loads it might be well to account for the effects of transverse shear deformation as we did earlier for beam theory. Improved theories derived by Mindlin and Reissner for this purpose will be presented in a later section of the chapter.

We now examine solutions of simple problems using the classical theory of plates.

6.7 EXAMPLES FROM CLASSICAL PLATE THEORY; SIMPLY-SUPPORTED RECTANGULAR PLATES

In this and the following two sections we shall consider examples illustrating certain aspects of the classical theory of plates. By no means are we presenting a comprehensive discussion of the work done in applying plate theory;[1] rather we shall examine only salient features of the classical applications.

In this section we consider rectangular plates having as domains $0 \leq x \leq a$, $0 \leq y \leq b$ (see Fig. 6.12). We consider first the case where the edges are simply-supported which means that $w = 0$ on all edges (this is the kinematic condition of Eq. (6.51)). Also from Eq. (6.51) we use the natural boundary condition $M_x = 0$

---

[1] For a more complete compendium of solutions to plate problems the reader is referred to the classic "Theory of Plates and Shells," by S. P. Timoshenko and S. Woinowsky-Krieger, McGraw-Hill Book Co., N.Y.

for $x = 0$ and $x = a$ and $M_y = 0$ for $y = 0$ and $y = b$. But from Eq. (6.12) this means that:

$$\left(\frac{\partial^2 w}{\partial x^2} + v\frac{\partial^2 w}{\partial y^2}\right) = 0 \qquad \text{for } x = 0 \text{ and } x = a \qquad (a)$$

$$\left(\frac{\partial^2 w}{\partial y^2} + v\frac{\partial^2 w}{\partial x^2}\right) = 0 \qquad \text{for } y = 0 \text{ and } y = b \qquad (b) \quad (6.64)$$

Since $w = 0$ along $x = 0$ and $x = a$ it means that $\partial^2 w/\partial y^2 = 0$ along these edges. Similarly $\partial^2 w/\partial x^2 = 0$ along edges $y = 0$ and $y = a$. The boundary condition for the plate can then be given as follows:

At $x = 0$, $x = a$:

$$w = \frac{\partial^2 w}{\partial x^2} = 0 \qquad (a)$$

At $y = 0$, $y = b$:

$$w = \frac{\partial^2 w}{\partial y^2} = 0 \qquad (b) \quad (6.65)$$

Now, using as coordinate functions in the Ritz method the eigenfunctions[1] for this problem, $\sin(m\pi x/a)\sin(n\pi y/b)$, we form the following infinite sequence for w

$$w = \sum_{m=1}^{\infty} \sum_{n=1}^{\infty} w_{mn} \sin\frac{m\pi x}{a} \sin\frac{n\pi y}{b} \qquad (6.66)$$

where w_{mn} are undetermined coefficients. Now we can determine w_{mn} by minimizing the total potential energy functions in accordance with the Ritz method (Chap. 3). Since the eigenfunctions are a complete set here, and because the eigenfunctions satisfy a full set of boundary conditions, we know from our earlier work that a solution so found must be an exact solution to the problem.

A simpler way to get the coefficients w_{mn}, rather than by formally using the approach of Ritz, is to proceed by expanding the loading $q(x,y)$ into a double Fourier series using the same eigenfunctions as in Eq. (6.66). Thus using i and j as dummy indices in place of m and n we have

$$q(x,y) = \sum_{i=1}^{\infty} \sum_{j=1}^{\infty} q_{ij} \sin\frac{i\pi x}{a} \sin\frac{j\pi y}{b} \qquad (6.67)$$

where the coefficients q_{mn} are determined by the familiar technique of multiplying both sides of the equation by $\sin(m\pi x/a)\sin(n\pi y/b)$ and integrating over the domain.

---

[1] We shall discuss eigenfunctions in Chap. 7.

Thus:

$$\int_0^a \int_0^b q(x,y) \sin \frac{m\pi x}{a} \sin \frac{n\pi y}{b} \, dx \, dy =$$

$$\int_0^a \int_0^b \sum_{i=1}^{\infty} \sum_{j=1}^{\infty} q_{ij} \sin \frac{i\pi x}{a} \sin \frac{m\pi x}{a} \sin \frac{j\pi y}{b} \sin \frac{n\pi y}{b} \, dx \, dy$$

Making use of the orthogonality properties of eigenfunctions[1] we can solve for q_{mn} from the above equations to get:

$$q_{mn} = \frac{4}{ab} \int_0^a \int_0^b q(x,y) \sin \frac{m\pi x}{a} \sin \frac{n\pi y}{b} \, dx \, dy \qquad (6.68)$$

Now substituting $q(x,y)$ as given by Eq. (6.67) and w as given by Eq. (6.66) into the basic equation (6.43), we get the following result, on equating the coefficients of $\sin(m\pi x/a) \sin(n\pi y/b)$,

$$w_{mn} = \frac{1}{\pi^4 D} \frac{q_{mn}}{[(m/a)^2 + (n/b)^2]^2} \qquad (6.69)$$

These are actually the Ritz coefficients as you may readily verify.

As a specific illustration, consider the rectangular plate of the previous discussion to have an applied loading distribution given as:

$$q(x,y) = q_{11} \sin \frac{\pi x}{a} \sin \frac{\pi y}{b}$$

We see from Eqs. (6.68) and (6.69) that there is only one non-zero Ritz coefficient, w_{11}, given as:

$$w_{11} = \frac{1}{\pi^4 D} \frac{q_{11}}{[(1/a)^2 + (1/b)^2]^2} \qquad (6.70)$$

and so the exact solution for this problem simply is:

$$w = \frac{q_{11}}{\pi^4 D [(1/a)^2 + (1/b)^2]^2} \sin \frac{\pi x}{a} \sin \frac{\pi y}{b} \qquad (6.71)$$

We shall now illustrate the existence of the corner forces discussed earlier. First note that the total downward force F_D from the loading is:

$$F_D = \int_0^a \int_0^b q(x,y) \, dx \, dy = \frac{4q_{11}ab}{\pi^2} \qquad (6.72)$$

---

[1] The functions $f_1, f_2, \ldots, f_n$ are orthogonal in a domain if $\iint f_i f_j \, dx \, dy = 0$ for $i \neq j$ for the domain. We shall discuss this property in more detail in Chap. 7.

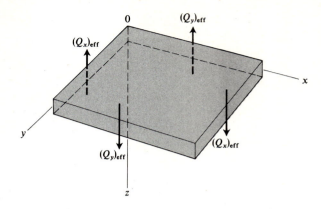

FIGURE 6.16

Next, consider the effective shear forces at the edges of the plate. These have been shown in Fig. 6.16. The effective shear forces $(Q_x)_{eff}$ and $(Q_y)_{eff}$ can be given as follows by first using Eqs. (6.17) and (6.18) and then using Eq. (6.12):

$$(Q_x)_{eff} = Q_x + \frac{\partial M_{xy}}{\partial y} = \frac{\partial M_x}{\partial x} + 2\frac{\partial M_{xy}}{\partial y}$$

$$= -D\left[\frac{\partial^3 w}{\partial x^3} + (2 - v)\frac{\partial^3 w}{\partial x \partial y^2}\right]$$

$$(Q_y)_{eff} = Q_y + \frac{\partial M_{xy}}{\partial x} = \frac{\partial M_y}{\partial y} + 2\frac{\partial M_{xy}}{\partial x}$$

$$= -D\left[\frac{\partial^3 w}{\partial y^3} + (2 - v)\frac{\partial^3 w}{\partial x^2 \partial y}\right]$$

Now employing Eq. (6.71) to replace w for the specific problem at hand, we get:

$$(Q_x)_{eff} = Dw_{11}\left[\left(\frac{\pi}{a}\right)^3 + (2 - v)\frac{\pi^3}{ab^2}\right]\cos\frac{\pi x}{a}\sin\frac{\pi y}{b}$$

$$(Q_y)_{eff} = Dw_{11}\left[\left(\frac{\pi}{b}\right)^3 + (2 - v)\frac{\pi^3}{a^2 b}\right]\sin\frac{\pi x}{a}\cos\frac{\pi y}{b} \qquad (6.73)$$

By integrating the above expressions along the edges of the plate we may determine a net vertical force F_u from the effective shear (actually the forces on all four edges turn out to be in the upward direction). Thus using Eq. (6.70) to replace w_{11} we get:

$$F_u = \frac{4q_{11}ab}{\pi^2} + \frac{8q_{11}(1 - v)}{\pi^2 ab(1/a^2 + 1/b^2)^2}$$

Comparing this upward force with the downward force from the loading we see that there is a net *upward* force of

$$8q_{11}(1 - v)\bigg/\left[\pi^2 ab\left(\frac{1}{a^2} + \frac{1}{b^2}\right)^2\right].$$

Clearly the corner forces must be taken into account. These you will recall are given as $2M_{xy}$ with a sign convention given by Fig. 6.13(b). Thus at the origin we have using Eqs. (6.12(c)) and (6.71):

$$[F_{(corner)}]_{(0,0)} = 2(M_{xy})_{\substack{x=0 \\ y=0}} = -2(1 - v)D\left[\frac{\partial^2 w}{\partial x \partial y}\right]_{\substack{x=0 \\ y=0}}$$

$$= \left\{-2(1 - v)(D)\frac{q_{11}}{\pi^4 D[(1/a)^2 + (1/b)^2]^2}\left(\frac{\pi}{a}\right)\left(\frac{\pi}{b}\right)\cos\frac{\pi x}{a}\cos\frac{\pi y}{b}\right\}_{(0,0)}$$

$$= -\frac{2q_{11}(1 - v)}{\pi^2 ab(1/a^2 + 1/b^2)^2}$$

Consulting Fig. 6.13(b) we see that the force at (0,0) is downward. We may similarly show that the forces at the other three corners are downward having the same value as given above. Thus there is a total *downward* force of

$$8q_{11}(1 - v)\bigg/\left[\pi^2 ab\left(\frac{1}{a^2} + \frac{1}{b^2}\right)^2\right]$$

from the corners. The net force on the plate is then clearly zero as required by equilibrium.

6.8 RECTANGULAR PLATES; LÉVY'S METHOD

We now examine a more general boundary condition for the rectangular plate, namely a plate which is simply-supported on two opposite edges (say at $x = 0$ and at $x = a$) while the other edges ($y = 0$ and $y = b$) are supported in an arbitrary manner. We follow the procedure set forth by the French mathematician Lévy. He assumed a solution of the form

$$w(x, y) = \sum_{m=1}^{\infty} [Y_m(y)] \sin\frac{m\pi x}{a} \qquad (6.74)$$

where $Y_m(y)$ is yet to be determined and must satisfy appropriate boundary conditions at $y = 0$ and $y = b$. We substitute for w, as given above, into the total potential energy as given by Eq. (6.30). Thus:

$$\pi = \frac{D}{2} \int_0^b \int_0^a \left\{ \left[\sum_{m=1}^{\infty} \left(-Y_m \left(\frac{m\pi}{a} \right)^2 \sin \frac{m\pi x}{a} + Y_m'' \sin \frac{m\pi x}{a} \right) \right]^2 \right.$$

$$+ 2(1-v) \left[\sum_{m=1}^{\infty} Y_m' \frac{m\pi}{a} \cos \frac{m\pi x}{a} \right]^2$$

$$\left. - 2(1-v) \left[\sum_{m=1}^{\infty} -Y_m \left(\frac{m\pi}{a} \right)^2 \sin \frac{m\pi x}{a} \right] \left[\sum_{p=1}^{\infty} Y_p'' \sin \frac{p\pi x}{a} \right] \right\} dx \, dy$$

$$- \int_0^b \int_0^a q \left(\sum_{m=1}^{\infty} Y_m \sin \frac{m\pi x}{a} \right) dx \, dy \qquad (6.75)$$

Note that in the first integration each term, after squaring operations are carried out, contains either $\sin^2 (m\pi x/a)$, $\cos^2 (m\pi x/a)$, or some product of $\sin (p\pi x/a) \sin (m\pi x/a)$ or $\cos (p\pi x/a) \cos (m\pi x/a)$ with $p \neq m$. The integration with respect to x yields $a/2$ for the squares of the sine and the cosine and yields zero for the sine and cosine products with different indices. The resulting equation can then be given as follows

$$\pi = \left(\frac{D}{2} \right) \left(\frac{a}{2} \right) \int_0^b \left\{ \sum_{m=1}^{\infty} \left[-Y_m \left(\frac{m\pi}{a} \right)^2 + (Y_m'') \right]^2 \right.$$

$$+ 2(1-v) \left[\sum_{m=1}^{\infty} (Y_m')^2 \left(\frac{m\pi}{a} \right)^2 \right] - 2(1-v) \left[\sum_{m=1}^{\infty} (-Y_m)(Y_m'') \left(\frac{m\pi}{a} \right)^2 \right]$$

$$\left. - \sum_{m=1}^{\infty} 2q_m Y_m \right\} dy \qquad (6.76)$$

where

$$q_m = \frac{2}{aD} \int_0^a q \sin \frac{m\pi x}{a} dx \qquad (6.77)$$

We have here a functional with one independent variable y and with infinitely many functions Y_m of this variable having as the highest order derivative the value two. From Eq. (2.64) we get as the appropriate Euler-Lagrange equations for extremizing the above functional

$$-\frac{d^2}{dy^2} \left(\frac{\partial F}{\partial Y_m''} \right) + \frac{d}{dy} \left(\frac{\partial F}{\partial Y_m'} \right) - \frac{\partial F}{\partial Y_m} = 0 \qquad m = 1, 2, \ldots$$

We then get for Y_m the following ordinary differential equation on substituting for F:

$$Y_m^{\text{IV}} - 2 \left(\frac{m\pi}{a} \right)^2 Y_m'' + \left(\frac{m\pi}{a} \right)^4 Y_m = q_m \qquad (6.78)$$

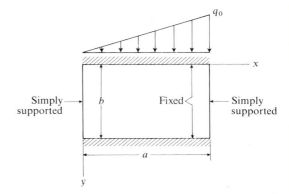

FIGURE 6.17

We also have from the variational process the following boundary conditions, as can readily be seen by consulting Eq. (2.63) and extrapolating the results to apply to a problem with many functions Y_m. Thus we have at the ends $y = 0$ and $y = b$:

$$\frac{\partial F}{\partial Y_m''} = 0 \quad \text{OR} \quad Y_m' \text{ PRESCRIBED}$$

$$-\frac{d}{dy}\left(\frac{\partial F}{\partial Y_m''}\right) + \frac{\partial F}{\partial Y_m'} = 0 \quad \text{OR} \quad Y_m \text{ PRESCRIBED} \qquad (6.79)$$

Using F from Eq. (6.76) we may write the above boundary conditions as follows:

$$Y_m'' - v\left(\frac{m\pi}{a}\right)^2 Y_m = 0 \quad \text{OR} \quad Y_m' \text{ PRESCRIBED}$$

$$Y_m''' - (2 - v)\left(\frac{m\pi}{a}\right)^2 Y_m' = 0 \quad \text{OR} \quad Y_m \text{ PRESCRIBED} \qquad (6.80)$$

The first of the above conditions represents the moment-slope duality while the second equation gives the effective shear-displacement pair.

We thus see that the problem is reduced to solving ordinary differential equations (Eqs. (6.78)). We illustrate this in the following example:

EXAMPLE 6.1 As a simple example to illustrate the method of Lévy, consider a rectangular plate of dimensions $a \times b$ simply-supported at edges $x = 0$ and $x = b$ and fixed at edges $y = 0$ and $y = b$. This has been shown in Fig. 6.17. A triangular load has been shown with a maximum intensity of q_0 at $x = a$.

For the problem at hand we can immediately give the boundary conditions for Y_m as follows (see Eq. (6.80)):

$$Y_m = Y'_m = 0 \quad \text{at } y = 0 \quad \text{and} \quad y = b \qquad (a)$$

The term q_m can next readily be determined. Thus employing $q = (x/a)q_0$ in Eq. (6.77) we get:

$$q_m = \frac{2}{aD} \int_0^a \frac{x}{a} q_0 \sin \frac{m\pi x}{a} \, dx$$

$$= \frac{2q_0}{a^2 D} \int_0^a x \sin \frac{m\pi x}{a} \, dx = -\frac{2q_0}{m\pi D} \cos m\pi$$

$$= (-1)^{m+1} \frac{2q_0}{m\pi D} \qquad (b)$$

The differential equation to be solved is then (see Eq. (6.78)):

$$Y_m^{\text{IV}} - 2\left(\frac{m\pi}{a}\right)^2 Y_m'' + \left(\frac{m\pi}{a}\right)^4 Y_m = (-1)^{m+1} \frac{2q_0}{m\pi D} \qquad (c)$$

The complementary solution is readily determined by employing the trial solution e^{py}. The characteristic equation becomes:

$$P^4 - 2\left(\frac{m\pi}{a}\right)^2 p^2 + \left(\frac{m\pi}{a}\right)^4 = 0$$

therefore

$$\left[p^2 - \left(\frac{m\pi}{a}\right)^2 \right]^2 = 0$$

Clearly $p = \pm m\pi/a$ has double roots. Hence we can give the complementary solution as follows:

$$(Y_m)_{\text{comp.}} = (C_1)_m \, e^{(m\pi/a)y} + (C_2)_m \, e^{(-m\pi/a)y}$$
$$+ (C_3)_m y \, e^{(m\pi/a)y} + (C_4)_m y \, e^{(-m\pi/a)y}$$

Noting by inspection that a particular solution is

$$(Y_m)_{\text{part}} = (-1)^{m+1} \frac{2q_0 a^4}{(m\pi)^5 D}$$

We may give the general solution as follows:

$$Y_m = (C_1)_m \, e^{(m\pi/a)y} + (C_2)_m \, e^{(-m\pi/a)y} + (C_3)_m y \, e^{(m\pi/a)y} + (C_4)_m y \, e^{(-m\pi/a)y}$$

$$+ (-1)^{m+1} \frac{2q_0 a^4}{(m\pi)^5 D} \qquad (d)$$

Now apply the boundary conditions (a) to the above solution:

$Y_m = 0$ *at* $y = 0$

$$(C_1)_m + (C_2)_m = (-1)^m \frac{2q_0 a^4}{(m\pi)^5 D}$$

$Y'_m = 0$ *at* $y = 0$

$$\left(\frac{m\pi}{a}\right)(C_1)_m - \left(\frac{m\pi}{a}\right)(C_2)_m + (C_3)_m + (C_4)_m = 0$$

$Y_m = 0$ *at* $y = b$

$$e^{(m\pi/a)b}(C_1)_m + e^{(-m\pi/a)b}(C_2)_m + b\,e^{(m\pi/a)b}(C_3)_m + b\,e^{(-m\pi/a)b}(C_4)_m$$
$$= \frac{(-1)^m 2q_0 a^4}{(m\pi)^5 D}$$

$Y'_m = 0$ *at* $y = b$

$$\frac{m\pi}{a} e^{(m\pi/a)b}(C_1)_m - \frac{m\pi}{a} e^{(-m\pi/a)b}(C_2)_m + e^{(m\pi/a)b}\left(1 + \frac{m\pi b}{a}\right)(C_3)_m$$
$$+ e^{(-m\pi/a)b}\left(1 - \frac{m\pi b}{a}\right)(C_4)_m = 0 \qquad (e)$$

Let us introduce the notation next

$$\beta_m = m\pi \frac{b}{a}$$

$$\alpha_m = \frac{(-1)^m 2q_0 a^4}{(m\pi)^5 D} \qquad (f)$$

In matrix form Eqs. (e) then become:

$$
\begin{bmatrix}
1 & 1 & 0 & 0 \\
\dfrac{\beta_m}{b} & -\dfrac{\beta_m}{b} & 1 & 1 \\
e^{\beta_m} & e^{-\beta_m} & b e^{\beta_m} & b e^{-\beta_m} \\
\dfrac{\beta_m}{b} e^{\beta_m} & -\dfrac{\beta_m}{b} e^{-\beta_m} & e^{\beta_m}(1 + \beta_m) & e^{-\beta_m}(1 - \beta_m)
\end{bmatrix}
\begin{bmatrix}
(C_1)_m \\
(C_2)_m \\
(C_3)_m \\
(C_4)_m
\end{bmatrix}
= \alpha_m
\begin{bmatrix}
1 \\
0 \\
1 \\
0
\end{bmatrix}
$$

We get the following results for the coefficients:

$$(C_1)_m = \frac{(\alpha_m)\{1 + 2\beta_m + 2\beta_m^2 - e^{\beta_m}(1 + \beta_m) + e^{-\beta_m}(1 - \beta_m) - e^{-2\beta_m}\}}{(2 + 4\beta_m^2 - e^{2\beta_m} - e^{-2\beta_m})}$$

$$(C_2)_m = \frac{(\alpha_m)\{-e^{-\beta_m}(1-\beta_m) + e^{\beta_m}(1+\beta_m) - e^{2\beta_m} + 1 - 2\beta_m + 2\beta_m^2\}}{(2 + 4\beta_m^2 - e^{2\beta_m} - e^{-2\beta_m})}$$

$$(C_3)_m = \frac{\dfrac{\alpha_m \beta_m}{b}\{e^{-\beta_m}(2\beta_m - 1) + e^{-2\beta_m} + e^{\beta_m} - 1 - 2\beta_m\}}{(2 + 4\beta_m^2 - e^{2\beta_m} - e^{-2\beta_m})}$$

$$(C_4)_m = \frac{\dfrac{\alpha_m \beta_m}{b}\{e^{\beta_m}(1 + 2\beta_m) - e^{-\beta_m} - e^{2\beta_m} + 1 - 2\beta_m\}}{(2 + 4\beta_m^2 - e^{2\beta_m} - e^{-2\beta_m})}$$

We can then give the solution as follows:

$$w(x,y) = \sum_{m=1}^{\infty} [(C_1)_m e^{(m\pi/a)y} + (C_2)_m e^{(-m\pi/a)y} + (C_3)_m y\, e^{(m\pi/a)y}$$

$$+ (C_4)_m y\, e^{(-m\pi/a)y} + \alpha_m] \sin \frac{m\pi x}{a} \qquad (g)$$

If we wish to compute M_x and M_y we may use Eqs. (6.12(a)) and (6.12(b)). Thus

$$M_x = -D\left[\frac{\partial^2 w}{\partial x^2} + v\frac{\partial^2 w}{\partial y^2}\right]$$

$$M_y = -D\left[\frac{\partial^2 w}{\partial y^2} + v\frac{\partial^2 w}{\partial x^2}\right] \qquad (h)$$

wherein we may use Eq. (g) for w. In the Table 6.1 we have given results from the above calculations for various values of m in the case of a square plate ($a/b = 1$). All results apply to the center of the plate.

Table 6.1

Highest value of m	$\dfrac{wD}{q_0 a^4}$	$\dfrac{M_x}{q_0 a^2}$	$\dfrac{M_y}{q_0 a^2}$
1	9.8093×10^{-4}	1.4002×10^{-2}	1.7307×10^{-2}
2	9.8093×10^{-4}	1.4002×10^{-2}	1.7307×10^{-2}
3	9.5679×10^{-4}	1.1810×10^{-2}	1.6505×10^{-2}
4	9.5679×10^{-4}	1.1810×10^{-2}	1.6505×10^{-2}
5	9.5887×10^{-4}	1.2324×10^{-2}	1.6661×10^{-2}
6	9.5887×10^{-4}	1.2324×10^{-2}	1.6661×10^{-2}
7	9.5848×10^{-4}	1.2136×10^{-2}	1.6605×10^{-2}
8	9.5848×10^{-4}	1.2136×10^{-2}	1.6605×10^{-2}
9	9.5859×10^{-4}	1.2224×10^{-2}	1.6632×10^{-2}
10	9.5859×10^{-4}	1.2224×10^{-2}	1.6632×10^{-2}
11	9.5855×10^{-4}	1.2176×10^{-2}	1.6617×10^{-2}

Timoshenko[1] has given an infinite series solution to the above problem by superposing the solution for a simply-supported rectangular plate under the triangular load, $(x/a)q_0$, with the solution to the same problem under an applied torque distribution along edges $y = 0$ and $y = a$ wherein the torque is of such a value as to render $\partial w/\partial y = 0$ at these edges. The results are identical to those found here since basically the same approach (sinusoidal expansions in the direction x) underlies both computations.

It may be of interest to compare deflection at the center of the plate results from the Lévy method with those stemming from a finite difference calculation for the same problem. Such results are shown in Table 6.2.   ////

6.9   THE CLAMPED RECTANGULAR PLATE; APPROXIMATE SOLUTIONS

In this section we shall examine the case of the clamped plate having dimensions $2a \times 2b$ and subject to a uniform load distribution q_0 (see Fig. 6.18). We shall present approximate solutions via several methods and will compare certain results of such computations with those from exact solutions.

A.   Ritz Method

As a first step we shall use the Ritz method in conjunction with the total potential energy π (see Eq. (6.30)) which we now rewrite as follows:

$$\pi = \frac{D}{2} \int \int_R \left\{ (\nabla^2 w)^2 + 2(1 - v)\left[\left(\frac{\partial^2 w}{\partial x \partial y} \right)^2 - \frac{\partial^2 w}{\partial x^2} \frac{\partial^2 w}{\partial y^2} \right] - \frac{2q_0 w}{D} \right\} dA \qquad (6.81)$$

We will first show that the above integral can be considerably simplified as a result of the boundary conditions of the problem, namely:

$$w = \frac{\partial w}{\partial x} = 0 \qquad x = \pm a$$

$$w = \frac{\partial w}{\partial y} = 0 \qquad y = \pm b \qquad (6.82)$$

Table 6.2

b/a	Lévy (m goes to 11)	Finite difference	
0.5	$8.1576 \times 10^{-5} q_0 a^4/D$	$10.813 \times 10^{-5} q_0 a^4/D$	(8×4 grid)
1	$9.5855 \times 10^{-4} q_0 a^4/D$	$11.014 \times 10^{-4} q_0 a^4/D$	(6×4 grid)
2	$4.2224 \times 10^{-3} q_0 a^4/D$	$4.4547 \times 10^{-3} q_0 a^4/D$	(4×8 grid)

[1] Timoshenko and Woinowsky-Krieger: "Theory of Plates and Shells," McGraw-Hill Book Co., p. 190.

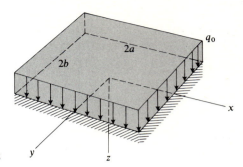

FIGURE 6.18

Thus consider the integral[1] from Eq. (6.81)

$$\iint_R \left[\left(\frac{\partial^2 w}{\partial x\, \partial y} \right)^2 - \frac{\partial^2 w}{\partial x^2} \frac{\partial^2 w}{\partial y^2} \right] dA$$

We shall first integrate by parts each expression in the integrand. Thus

$$\iint_R \left(\frac{\partial^2 w}{\partial x\, \partial y} \right)^2 dA = \iint_R \frac{\partial^2 w}{\partial x\, \partial y} \frac{\partial}{\partial x}\left(\frac{\partial w}{\partial y} \right) dA$$

$$= -\iint_R \frac{\partial^3 w}{\partial x^2\, \partial y} \frac{\partial w}{\partial y} dA + \oint \frac{\partial^2 w}{\partial x\, \partial y} \frac{\partial w}{\partial y} a_{vx}\, ds$$

$$= -\iint_R \frac{\partial^3 w}{\partial x^2\, \partial y} \frac{\partial w}{\partial y} dA + \oint \frac{\partial^2 w}{\partial x\, \partial y} \frac{\partial w}{\partial y}\, dy \qquad (6.83)$$

where we have used Eq. (6.39(b)) to replace $a_{vx}\, ds$. Also

$$\iint_R \frac{\partial^2 w}{\partial x^2} \frac{\partial^2 w}{\partial y^2} dA = \iint_R \frac{\partial^2 w}{\partial x^2} \frac{\partial}{\partial y}\left(\frac{\partial w}{\partial y} \right) dA$$

$$= -\iint_R \frac{\partial^3 w}{\partial y\, \partial x^2} \frac{\partial w}{\partial y} dA + \oint \frac{\partial^2 w}{\partial x^2} \frac{\partial w}{\partial y} a_{vy}\, ds$$

$$= -\iint_R \frac{\partial^3 w}{\partial y\, \partial x^2} \frac{\partial w}{\partial y} dA - \oint \frac{\partial^2 w}{\partial x^2} \frac{\partial w}{\partial y}\, dx \qquad (6.84)$$

[1] The expression

$$\left(\frac{\partial^2 w}{\partial x\, \partial y} \right)^2 - \frac{\partial^2 w}{\partial x^2} \frac{\partial^2 w}{\partial y^2}$$

is called the *Gaussian curvature*. It plays an important role in the differential geometry of curved surfaces.

where we have used Eq. (6.39(a)) to replace $a_{vy}\,ds$. We now may form the integral given by (6.83) in the following way:

$$\iint_R \left[\left(\frac{\partial^2 w}{\partial x\,\partial y}\right)^2 - \frac{\partial^2 w}{\partial x^2}\frac{\partial^2 w}{\partial y^2}\right]dA = \oint \frac{\partial w}{\partial y}\left[\frac{\partial}{\partial y}\left(\frac{\partial w}{\partial x}\right)dy + \frac{\partial}{\partial x}\left(\frac{\partial w}{\partial x}\right)dx\right]$$

$$= \oint \frac{\partial w}{\partial y}\,d\left(\frac{\partial w}{\partial x}\right) = \oint \frac{\partial w}{\partial y}\frac{d}{ds}\left(\frac{\partial w}{\partial x}\right)ds \qquad (6.85)$$

If the boundary is piecewise smooth, such as the rectangular plate we will be considering, then the line integral on the right side of the above equation is actually a sum of line integrals between the corners of the boundary. It is immediately apparent from above that for a clamped rectangular plate $\partial w/\partial y = 0$ on the boundaries $x = \pm a$ and $\partial w/\partial x = 0$ on the boundaries $y = \pm b$ (see Fig. 6.18). For the case at hand then we can conclude that the Gaussian curvature is zero and can be deleted from Eq. (6.81).

Let us consider the case of a plate with a *smooth boundary* such as an elliptic or circular plate. For this purpose we integrate the right side of Eq. (6.85) by parts

$$\iint_R \left[\left(\frac{\partial^2 w}{\partial x\,\partial y}\right)^2 - \frac{\partial^2 w}{\partial x^2}\frac{\partial^2 w}{\partial y^2}\right]dA = -\oint \frac{\partial w}{\partial x}\frac{d}{ds}\left(\frac{\partial w}{\partial y}\right)ds + \left[\frac{\partial w}{\partial y}\frac{\partial w}{\partial x}\right]_{\substack{\text{end}\\\text{pts.}}}$$

It is apparent that for the case of a continuous curve the last expression is zero. Now by expressing $(\partial^2 w/\partial x\,\partial y)^2$ in Eq. (6.83) and $(\partial^2 w/\partial x^2)(\partial^2 w/\partial y^2)$ in Eq. (6.84), respectively, as

$$\left[\frac{\partial^2 w}{\partial x\,\partial y}\frac{\partial}{\partial y}\left(\frac{\partial w}{\partial x}\right)\right] \qquad \text{and} \qquad \left[\frac{\partial^2 w}{\partial y^2}\frac{\partial}{\partial x}\left(\frac{\partial w}{\partial x}\right)\right]$$

rather than as we did, we may conclude by proceeding as above from Eq. (6.83) to Eq. (6.85) that:

$$\iint_R \left[\left(\frac{\partial^2 w}{\partial x\,\partial y}\right)^2 - \frac{\partial^2 w}{\partial x^2}\frac{\partial^2 w}{\partial y^2}\right]dA = -\oint \frac{\partial w}{\partial x}\frac{d}{ds}\left(\frac{\partial w}{\partial y}\right)ds \qquad (6.86)$$

We can then conclude from Eqs. (6.85) and (6.86) that:

$$\iint_R \left[\left(\frac{\partial^2 w}{\partial x\,\partial y}\right)^2 - \frac{\partial^2 w}{\partial x^2}\frac{\partial^2 w}{\partial y^2}\right]dA = \frac{1}{2}\oint \left[\frac{\partial w}{\partial y}\frac{d}{ds}\left(\frac{\partial w}{\partial x}\right) - \frac{\partial w}{\partial x}\frac{d}{ds}\left(\frac{\partial w}{\partial y}\right)\right]ds \qquad (6.87)$$

We see that if

$$\frac{\partial w}{\partial x} = \frac{\partial w}{\partial y} = 0$$

at all points along the boundary then again we can delete the integral of the Gaussian curvature in the total potential energy expression. But if the above conditions hold we can also conclude

$$\frac{\partial w}{\partial v} = \frac{\partial w}{\partial x}\frac{dx}{dv} + \frac{\partial w}{\partial y}\frac{dy}{dv} = 0 \qquad (a)$$

$$\frac{\partial w}{\partial s} = \frac{\partial w}{\partial x}\frac{dx}{ds} + \frac{\partial w}{\partial y}\frac{dy}{ds} = 0 \qquad (b)$$

where v and s are respectively normal and tangential to the boundary. The last condition (b) is satisfied if $w = 0$ on the boundary and so we can conclude from above that for a *clamped plate* $(\partial w/\partial v = w = 0)$ having a continuous boundary we can drop the integral of the Gaussian curvature from the total potential energy expression.[1]

Returning to the case of the clamped rectangular plate we then have for π:

$$\pi = \frac{D}{2} \int_{-a}^{+a} \int_{-b}^{+b} \left[(\nabla^2 w)^2 - \frac{2q_0 w}{D} \right] dx\, dy \qquad (6.88)$$

We may use for a one-parameter approach with the Ritz method the following function:

$$w = C_1 (x^2 - a^2)^2 (y^2 - b^2)^2 \qquad (6.89)$$

Note that the boundary conditions are satisfied and the function is even in the x and y coordinates as is to be expected of the solution. Substituting the above function into Eq. (6.88) and extremizing with respect to C_1 you may readily verify that we get:

$$C_1 = 0.383 \frac{q_0/D}{(7a^4 + 4a^2 b^2 + 7b^4)}$$

The approximate solution is then

$$w_1 = \frac{0.383(q_0/D)}{(7a^4 + 4a^2 b^2 + 7b^4)} (x^2 - a^2)^2 (y^2 - b^2)^2 \qquad (6.90)$$

The maximum deflection occurs at the center of the plate $x = 0$, $y = 0$ and for a square plate $a = b$ we get from Eq. (6.90)

$$w_1 = 0.02127 \frac{q_0 a^4}{D} \qquad (6.91)$$

A series solution[2] for the same problem gives the following result:

$$w_{max} = 0.0202 \left(\frac{q_0 a^4}{D} \right) \qquad (6.92)$$

---

[1] We shall make use of this in Sec. 6.10 when we consider clamped elliptic plates. Also in Problem 6.21 we will ask you to prove that for *polygonal plates* the first variation of the integral of the Gaussian curvature vanishes if simply $w = 0$ on the boundary. This means we can delete the Gaussian curvature expression in π for such situations (a simply-supported rectangular plate is an example) when we anticipate taking the first variation of the total potential energy.

[2] See Table 35, Timoshenko and Woinowsky-Krieger, "Theory of Plates and Shells," McGraw-Hill Book Co., p. 202. (Note we must adjust this result since our "a" is twice the "a" of Table 35.)

We can improve the result by employing three coordinate functions as follows:

$$w_3 = C_1(x^2 - a^2)^2(y^2 - b^2)^2 + C_2 x^2(x^2 - a^2)^2(y^2 - b^2)^2$$
$$+ C_3 y^2(x^2 - a^2)^2(y^2 - b^2)^2$$

The result for such a calculation gives

$$w_{max} = 0.02067 \frac{q_0 a^4}{D} \qquad (6.93)$$

which is closer to the series solution.

B.   Galerkin's Method

We now consider the same problem via Galerkin's method. We accordingly go to Eq. (3.68) and apply it to the plate equation

$$\nabla^4 w = \frac{q_0}{D} \qquad (6.94)$$

to get the following equation

$$\int_a^{+a} \int_b^{+b} \left(\nabla^4 \tilde{w} - \frac{q_0}{D} \right) \Phi_i \, dA = 0 \qquad i = 1, 2, \ldots \qquad (6.95)$$

The boundary conditions for the problem are given as follows

$$w = \frac{\partial w}{\partial x} = 0 \quad \text{at } x = -a \quad \text{and} \quad x = +a$$

$$w = \frac{\partial w}{\partial y} = 0 \quad \text{at } y = -b \quad \text{and} \quad y = +b \qquad (6.96)$$

We will now introduce the nondimensional variables η and ζ defined as follows:

$$\zeta = \frac{x}{a}$$

$$\eta = \frac{y}{b} \qquad (6.97)$$

This means that the boundary conditions are given as follows:

$$w = \frac{\partial w}{\partial \zeta} = 0 \quad \text{on} \quad \zeta = \pm 1$$

$$w = \frac{\partial w}{\partial \eta} = 0 \quad \text{on} \quad \eta = \pm 1 \qquad (6.98)$$

With this in mind we now set forth the following functions:

$$g_1 = (\zeta^2 - 1)^2$$
$$g_2 = (\zeta^3 - \zeta)^2$$
$$h_1 = (\eta^2 - 1)^2$$
$$h_2 = (\eta^3 - \eta)^2 \qquad (6.99)$$

We can then formulate a set of four coordinate functions Φ_i that satisfy the boundary conditions given above

$$\Phi_1 = g_1 h_1 = (\zeta^2 - 1)^2(\eta^2 - 1)^2$$
$$\Phi_2 = g_1 h_2 = (\zeta^2 - 1)^2(\eta^3 - \eta)^2$$
$$\Phi_3 = g_2 h_1 = (\zeta^3 - \zeta)^2(\eta^2 - 1)^2$$
$$\Phi_4 = g_2 h_2 = (\zeta^3 - \zeta)^2(\eta^3 - \eta)^2 \qquad (6.100)$$

We can then give $\tilde{w}$ as follows

$$\tilde{w} = C_{11}g_1 h_1 + C_{12}g_1 h_2 + C_{21}g_2 h_1 + C_{22}g_2 h_2 \qquad (6.101)$$

where the C's are to be determined by the Galerkin method.

Now going back to Eq. (6.95) we need only integrate over one quadrant as a result of symmetry. (The resulting multiplier 4 can then be discarded.) Also note that

$$\nabla^2 \equiv \frac{\partial^2}{\partial x^2} + \frac{\partial^2}{\partial y^2} = \frac{\partial^2}{a^2\,\partial\zeta^2} + \frac{\partial^2}{b^2\,\partial\eta^2}$$

We thus have the following four equations on multiplying through by b^4:

$$\int_0^1 \int_0^1 \left[\left(\frac{b^2}{a^2}\frac{\partial^2}{\partial\zeta^2} + \frac{\partial^2}{\partial\eta^2} \right)^2 (C_{11}g_1 h_1 + C_{12}g_1 h_2 + C_{21}g_2 h_1 \right.$$
$$\left. + C_{22}g_2 h_2) - \frac{b^4 q_0}{D} \right](g_1 h_1)\, d\zeta\, d\eta = 0$$

$$\int_0^1 \int_0^1 \left[\left(\frac{b^2}{a^2}\frac{\partial^2}{\partial\zeta^2} + \frac{\partial^2}{\partial\eta^2} \right)^2 (C_{11}g_1 h_1 + C_{12}g_1 h_2 + C_{21}g_2 h_1 \right.$$
$$\left. + C_{22}g_2 h_2) - \frac{b^4 q_0}{D} \right](g_1 h_2)\, d\zeta\, d\eta = 0$$

$$\int_0^1 \int_0^1 \left[\left(\frac{b^2}{a^2}\frac{\partial^2}{\partial\zeta^2} + \frac{\partial^2}{\partial\eta^2} \right)^2 (C_{11}g_1 h_1 + C_{12}g_1 h_2 + C_{21}g_2 h_1 \right.$$
$$\left. + C_{22}g_2 h_2) - \frac{b^4 q_0}{D} \right](g_2 h_1)\, d\zeta\, d\eta = 0$$

$$\int_0^1 \int_0^1 \left[\left(\frac{b^2}{a^2} \frac{\partial^2}{\partial \zeta^2} + \frac{\partial^2}{\partial \eta^2} \right)^2 (C_{11} g_1 h_1 + C_{12} g_1 h_2 + C_{21} g_2 h_1 \right.$$

$$\left. + C_{22} g_2 h_2) - \frac{b^4 q_0}{D} \right] (g_2 h_2) \, d\zeta \, d\eta = 0$$

For a given ratio b^2/a^2 we may solve the preceding equations for coefficients C_{ij} in terms of D, b, and q_0. In the case of a *square plate* we have $a = b$ and $b^2/a^2 = 1$. The above equations become

$$\begin{bmatrix} 13.3747 & 1.21588 & 1.21588 & 0.135098 \\ 1.21588 & 2.60843 & 0.135098 & 0.218235 \\ 1.21588 & 0.135098 & 2.60843 & 0.218235 \\ 0.135088 & 0.218235 & 0.218235 & 0.118993 \end{bmatrix} \begin{bmatrix} C_{11} \\ C_{12} \\ C_{21} \\ C_{22} \end{bmatrix} = \frac{a^4 q_0}{D} \begin{bmatrix} 0.284444 \\ 0.0406349 \\ 0.0406349 \\ 0.00580499 \end{bmatrix}$$

Solving with the aid of a computer we get

$$\begin{bmatrix} C_{11} \\ C_{12} \\ C_{21} \\ C_{22} \end{bmatrix} = \frac{a^4 q_0}{D} \begin{bmatrix} 0.0202319 \\ 0.00534806 \\ 0.00534806 \\ 0.00624451 \end{bmatrix}$$

The approximate solution to the problem is then given as follows:

$$\tilde{w} = \left\{ (0.0202319) \left[\left(\frac{x}{a} \right)^2 - 1 \right]^2 \left[\left(\frac{y}{b} \right)^2 - 1 \right]^2 \right.$$

$$+ (0.00534806) \left[\left(\frac{x}{a} \right)^2 - 1 \right]^2 \left[\left(\frac{y}{b} \right)^3 - \left(\frac{y}{b} \right) \right]^2$$

$$+ (0.00534806) \left[\left(\frac{x}{b} \right)^3 - \left(\frac{x}{b} \right) \right]^2 \left[\left(\frac{y}{b} \right)^2 - 1 \right]^2$$

$$\left. + (0.00624451) \left[\left(\frac{x}{b} \right)^3 - \left(\frac{x}{b} \right) \right]^2 \left[\left(\frac{y}{b} \right)^3 - \left(\frac{y}{b} \right) \right]^2 \right\} \frac{a^4 q_0}{D} \quad (6.102)$$

As a check on the results we consider the deflection at the center of the plate, $(x = y = 0)$. We get:

$$\tilde{w}_{\max} = 0.0202319 \frac{a^4 q_0}{D} \quad (6.103)$$

From an analytic series solution we have the following corresponding result:

$$w_{max} = 0.0202\left(\frac{a^4 q_0}{D}\right)$$

The results are clearly very close to each other.

C.   Kantorovich's Method

As a final step in this section we turn to the method of Kantorovich for the clamped plate under a uniform load q_0. As we showed in Sec. 5.7, the Kantorovich method[1] can be reformulated to be the solution of a Galerkin integral. Accordingly, we can use Eq. (6.95) presented for the Galerkin method in the previous section. We shall consider $(y^2 - b^2)^2$ to be the coordinate function $\Phi_1(y)$ and we shall use for w the function w_1 given by $C_1(x)(y^2 - b^2)^2$ with the unknown function as $C_1(x)$. Thus we have initially

$$\int_{-a}^{+a}\left\{\int_{-b}^{+b}\left[\nabla^4 w_1 - \frac{q_0}{D}\right]\Phi_1\,dy\right\}dx = 0$$

This equation will be satisfied if we set the bracketed expression equal to zero. Thus:

$$\int_{-b}^{+b}\left[\nabla^4 w_1 - \frac{q_0}{D}\right]\Phi_1(y)\,dy = 0$$

Substituting for w_1 and Φ_1 we have:

$$\int_{-b}^{+b}\left\{\nabla^4[C_1(x)(y^2 - b^2)^2] - \frac{q_0}{D}\right\}(b^2 - y^2)^2\,dy = 0$$

therefore

$$\int_{-b}^{+b}\left[24C_1 + 2(12y^2 - 4b^2)\frac{d^2 C_1}{dx^2} + (y^2 - b^2)^2\frac{d^4 C_1}{dx^4} - \frac{q_0}{D}\right](y^2 - b^2)^2\,dy = 0$$

Integrating, we then get on multiplying through by $315/(256b^5)$:

$$b^4\frac{d^4 C_1}{dx^4} - 6b^2\frac{d^2 C_1}{dx^2} + \frac{63}{2}C_1 = \frac{21}{16}\frac{q_0}{D}$$

The characteristic equation for the above differential equation is given as follows:

$$b^4 p^4 - 6b^2 p^2 + \tfrac{63}{2} = 0$$

---

[1] See Kantorovich and Krylov: "Approximate Methods of Higher Analysis," Interscience, 1958, p. 322.

Solving for p^2b^2 using the quadratic formula we get:

$$p^2b^2 = 3 \pm i(4.75)$$

Hence taking roots we get four values of p:

$$p_{1,2,3,4} = \frac{1}{b}(\pm\alpha \pm i\beta) \quad \text{where} \quad \begin{cases} \alpha = 2.975 \\ \beta = 1.143 \end{cases}$$

The complementary solution then is

$$(C_1)_c = B_1\, e^{(\alpha+i\beta)x/b} + B_2\, e^{(\alpha-i\beta)x/b} + B_3\, e^{(-\alpha-i\beta)x/b} + B_4\, e^{(-\alpha+i\beta)x/b}$$

This may be written as follows on expanding $e^{\pm\alpha}$ in terms of hyperbolic functions and expanding $e^{\pm i\beta}$ in terms of sines and cosines.

$$(C_1)_c = A_1 \cosh\alpha\frac{x}{b}\cos\beta\frac{x}{b} + A_2 \cosh\alpha\frac{x}{b}\sin\beta\frac{x}{b} + A_3 \sinh\alpha\frac{x}{b}\sin\beta\frac{x}{b}$$

$$+ A_4 \sinh\alpha\frac{x}{b}\cos\beta\frac{x}{b}$$

The particular solution is

$$(C_1)_p = \left(\frac{2}{63}\right)\left(\frac{21}{16}\right)\left(\frac{q_0}{D}\right) = \frac{1}{24}\frac{q_0}{D}$$

Since we expect that $C_1(x)$ will be an even function of x then we can immediately set $A_2 = A_4 = 0$ so that the general solution for $C_1(x)$ is as follows:

$$C_1(x) = A_1 \cosh\alpha\frac{x}{b}\cos\beta\frac{x}{b} + A_3 \sinh\alpha\frac{x}{b}\sin\beta\frac{x}{b} + \frac{1}{24}\frac{q_0}{D}$$

We next use the condition $C(a) = C'(a) = 0$ to satisfy the boundary condition for w_1 on edges $x = \pm a$. In this way we get the remaining constants A_1 and A_3. Thus:

$$A_1 \cosh\alpha\frac{a}{b}\cos\beta\frac{a}{b} + A_3 \sinh\alpha\frac{a}{b}\sin\beta\frac{a}{b} + \frac{1}{24}\frac{q_0}{D} = 0$$

$$A_1\left(\frac{\alpha}{b}\sinh\alpha\frac{a}{b}\cos\beta\frac{a}{b}\right) - A_1\left(\frac{\alpha}{b}\cosh\alpha\frac{a}{b}\sin\beta\frac{a}{b}\right)$$

$$+ A_3\left(-\frac{\beta}{b}\cosh\alpha\frac{a}{b}\sin\beta\frac{a}{b} + \frac{\beta}{b}\sinh\alpha\frac{a}{b}\cos\beta\frac{a}{b}\right) = 0$$

We get

$$A_1 = \frac{\gamma_1}{\gamma_0}\left(\frac{1}{24}\frac{q_0}{D}\right)$$

$$A_3 = \frac{\gamma_2}{\gamma_0}\left(\frac{1}{24}\frac{q_0}{D}\right)$$

where taking $a/b = \mu$ we have

$$\gamma_0 = \beta \sinh \alpha\mu \cosh \alpha\mu + \alpha \sin \beta\mu \cos \beta\mu$$

$$\gamma_1 = -(\alpha \cosh \alpha\mu \sin \beta\mu + \beta \sinh \alpha\mu \cos \beta\mu)$$

$$\gamma_2 = \alpha \sinh \alpha\mu \cos \beta\mu - \beta \cosh \alpha\mu \sin \beta\mu$$

The solution stemming from this analysis is then

$$w = \frac{1}{24}\frac{q_0}{D}\left\{\left[\frac{\gamma_1}{\gamma_0}\cosh \alpha\frac{x}{b}\cos \beta\frac{x}{b} + \frac{\gamma_2}{\gamma_0}\sinh \alpha\frac{x}{b}\sin \beta\frac{x}{b}\right] + 1\right\}(y^2 - b^2)^2$$

To find the maximum deflection set $x = y = 0$ above and find w. We get for a square plate where $a = b$:

$$w_{max} = \frac{1}{24}\frac{q_0}{D}\left[\frac{\gamma_1}{\gamma_0} + 1\right](a^4)$$

$$= \frac{1}{24}\left(\frac{q_0 a^4}{D}\right)(0.479) = 0.01992\left(\frac{q_0 a^4}{D}\right)$$

Note that using a single coordinate function we have achieved comparable accuracy using the Kantorovich method involving a variable coefficient as was achieved using four coordinate functions with constant coefficients.[1]

6.10 ELLIPTIC AND CIRCULAR PLATES

Consider an *elliptic* plate shown in Fig. 6.19 wherein the boundary is expressed as follows

$$\left(\frac{x}{a}\right)^2 + \left(\frac{y}{b}\right)^2 - 1 = 0$$

We shall examine the case where the edge is clamped and where a uniform load q_0 acts over the plate surface. The appropriate boundary conditions are then

$$w = \frac{\partial w}{\partial v} = 0 \qquad (6.104)$$

Because we have a smooth boundary which is clamped we know from the previous section that we can delete the Gaussian curvature expression from the total potential

---

[1] Far greater accuracy can be achieved to a point where even the stresses may be found with considerable accuracy by the *extended Kantorovich method* discussed in Chap. 5. The process is lengthy, however, and we refer you to the paper: "An Application of the Extended Kantorovich Method to the Stress Analysis of a Clamped Rectangular Plate" by A. D. Kerr and H. Alexander, *Acta Mechanica*, **6**, 1968.

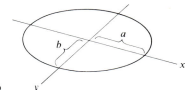

energy functional. We have then for π for this case:

$$\pi = \frac{D}{2} \int\!\!\int_R (\nabla^2 w)^2 \, dx \, dy - \int\!\!\int_R q_0 w \, dx \, dy \qquad (6.105)$$

By using Green's theorem and the simple vector identity

$$\nabla \cdot (\Phi \mathbf{F}) = \Phi \nabla \cdot \mathbf{F} + \mathbf{F} \cdot \nabla \Phi$$

we have asked you to demonstrate in Problem 6.13 that the above form of the total potential energy can be expressed as follows to simplify later calculations:

$$\pi = \frac{D}{2} \int\!\!\int_R w \nabla^4 w \, dx \, dy - \int\!\!\int_R q_0 w \, dx \, dy \qquad (6.106)$$

We shall consider as a coordinate function for the Ritz method the following function:

$$\Phi_1 = \left[\left(\frac{x}{a}\right)^2 + \left(\frac{y}{b}\right)^2 - 1 \right]^2$$

Clearly such a function satisfies the boundary conditions of the problem. We consider a one-term sequence given as follows:

$$w_1 = C_1 \left[\left(\frac{x}{a}\right)^2 + \left(\frac{y}{b}\right)^2 - 1 \right]^2 \qquad (6.107)$$

Substituting into Eq. (6.106) we get:

$$\pi = \frac{D}{2} \int\!\!\int_R C_1{}^2 [\Phi_1] \left[\frac{24}{a^4} + \frac{24}{b^4} + \frac{16}{a^2 b^2} \right] dA - q_0 \int\!\!\int_R C_1 \Phi_1 \, dA$$

Hence setting $\partial \pi / \partial C_1 = 0$ we have:

$$DC_1 \left[\frac{24}{a^4} + \frac{24}{b^4} + \frac{16}{a^2 b^2} \right] \left(\int\!\!\int_R \Phi_1 \, dA \right) - q_0 \left(\int\!\!\int_R \Phi_1 \, dA \right) = 0$$

therefore

$$C_1 = \frac{q_0}{D[24/a^4 + 24/b^4 + 16/a^2 b^2]}$$

Thus we get for w_1:

$$w_1 = \frac{q_0}{D(24/a^4 + 24/b^4 + 16/a^2 b^2)}\left[\left(\frac{x}{a}\right)^2 + \left(\frac{y}{b}\right)^2 - 1\right]^2 \qquad (6.108)$$

Actually the above solution is an exact solution to the problem as can readily be demonstrated by direct substitution into the differential equation:

$$\nabla^4 w = \frac{q_0}{D}$$

We can then compute strains and stresses from Eq. (6.108) by employing Eqs. (6.5) and (6.10).

To get the solution for a clamped *circular* plate uniformly loaded by q_0 is now a simple matter. Say the radius of the plate is a. We merely replace "b" by "a" in Eq. (6.108) and let $x^2 + y^2 = r^2$. The result becomes:

$$w = \frac{q_0}{D(64/a^4)}\left(1 - \frac{r^2}{a^2}\right)^2 \qquad (6.109)$$

Other coordinate functions that may be used in the Ritz method for loadings symmetric about the major and minor diameters of the elliptic plate are

$$\Phi_n = \left[\left(\frac{x}{a}\right)^2 + \left(\frac{y}{b}\right)^2 - 1\right]^n \qquad (6.110)$$

where $n > 2$. In the case of a clamped circular plate with loadings symmetric about the origin we have as possible coordinate functions

$$\Phi_n = \left(1 - \frac{r^2}{a^2}\right)^n \qquad (6.111)$$

where $n > 2$.

We now consider a circular plate that is *free* at the edges but is supported by an *elastic foundation* (see Fig. 6.20). A concentrated load P acts at the center of the plate. To account for the elastic support a potential energy term

$$U_{\text{found}} = \tfrac{1}{2}\iint_R kw^2\, dA \qquad (6.112)$$

must be added to the total potential energy where the term k is the modulus of the foundation. Accordingly the total potential energy functional is:

$$\pi = \frac{D}{2}\iint_R \left\{(\nabla^2 w)^2 - 2(1-v)\left[\left(\frac{\partial^2 w}{\partial x\, \partial y}\right)^2 - \frac{\partial^2 w}{\partial x^2}\frac{\partial^2 w}{\partial y^2}\right]\right\} dx\, dy$$
$$- \iint_R (qw - \tfrac{1}{2}kw^2)\, dA \qquad (6.113)$$

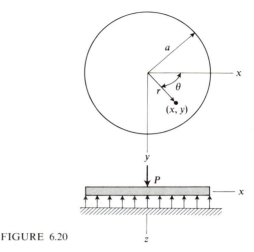

FIGURE 6.20

In the development of Sec. 6.4 the inclusion of the elastic support may be accomplished by replacing q in Eq. (6.31) by $(q - kw)$. The new Euler–Lagrange equation then becomes:

$$DV^4w + kw = q$$

Solutions to the above equation can be given in terms of so-called modified Kelvin functions.[1] These are rather difficult functions to use for numerical calculations. For this reason we shall consider obtaining an approximate solution by working with the total potential energy functional π. It is best to employ polar coordinates here and so we transform the coordinates from rectangular to polar in Eq. (6.113) to form the following functional

$$\pi = \int_0^{2\pi} \int_0^a \left[\frac{D}{2} \left\{ \left(\frac{\partial^2 w}{\partial r^2} + \frac{1}{r} \frac{\partial w}{\partial r} + \frac{1}{r^2} \frac{\partial^2 w}{\partial \theta^2} \right)^2 \right. \right.$$
$$- 2(1 - v) \left[\left(\frac{\partial^2 w}{\partial r^2} \right) \left(\frac{1}{r} \frac{\partial w}{\partial r} + \frac{1}{r^2} \frac{\partial^2 w}{\partial \theta^2} \right) \right.$$
$$\left. \left. \left. - \left(\frac{1}{r} \frac{\partial^2 w}{\partial r \partial \theta} - \frac{1}{r^2} \frac{\partial w}{\partial \theta} \right)^2 \right] \right\} - q(r,\theta)w + k \frac{w^2}{2} \right] r \, dr \, d\theta \qquad (6.114)$$

We shall consider an approximate solution of the form

$$w = A + B \left(\frac{r}{a} \right)^2 \qquad (6.115)$$

---

[1] See Timoshenko and Woinowsky-Krieger: "Theory of Plates and Shells," McGraw-Hill Book Co., Chap. 8.

where A and B are to be determined. Substituting into Eq. (6.114) we get:

$$\pi = \frac{4\pi D}{a^2}(1 + v)B^2 + \pi k(\tfrac{1}{2}A^2 a^2 + \tfrac{1}{2}ABa^2 + \tfrac{1}{6}B^2 a^2) - PA$$

Minimizing with respect to A and B yields the following pair of simultaneous differential equations

$$A + B\left[\frac{2}{3} + \frac{16D(1 + v)}{ka^4}\right] = 0 \qquad (a)$$

$$A + \tfrac{1}{2}B = \frac{P}{\pi ka^2} \qquad (b) \quad (6.116)$$

For a particular numerical example where

$$P = 10,000 \text{ lb} \qquad h = \tfrac{1}{2} \text{ in.}$$
$$k = 50 \text{ lb/in}^3 \qquad E = 30 \times 10^6 \text{ psi}$$
$$a = 5 \text{ ft} \qquad v = 0.3$$

we find at the center that

$$w_{r=0} = 2.552 \text{ in}$$

The exact solution is given by Timoshenko as

$$w_{\text{exact}} = 2.546 \text{ in}$$

The difference is less than 0.25%.

We might point out here that if we obtained moments (and stresses) from the above approximate deflection function we would be considerably in error. The reason is that for circular plates, concentrated loads introduce singularities in the moments and shears at the position of application of the concentrated loads. Such singularities require w to have the form $(r^2 \log r)$. Such a form appears in the aforementioned Kelvin function but not in our approximate function.

6.11 SKEWED PLATE PROBLEM

We now consider the problem of a skewed plate (see Fig. 6.21) fixed at the left side B. Since this plate might conceivably represent a swept wing of a high-speed aircraft, we have employed aeronautical nomenclature in the diagrams. Thus there is a parabolic line load with a maximum value Γ_0 lb/ft. at the tip chord while the root chord is clamped. We wish here to find the rotation of the tip chord relative to the root chord.

A Cartesian reference will be employed. However, it will be desirable to introduce dimensionless variables involving the chord length C and the span length L. Thus

$$\xi = \frac{x}{L} \qquad \eta = \frac{y}{C} \qquad (6.117)$$

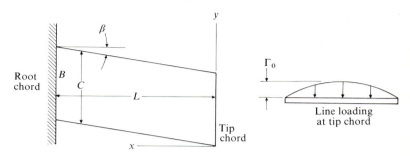

FIGURE 6.21

We shall now formulate a reasonable coordinate function to be used in the total potential energy functional in conjunction with the Ritz method. For this purpose recall from elementary strength of materials that the deflection curve for a simple straight cantilever beam is

$$w = \frac{PL^3}{6EI}\left[2 - 3\left(\frac{x}{L}\right) + \left(\frac{x}{L}\right)^3\right]$$

We now propose to use the bracketed expression in the above equation to give the variation in the x direction for the coordinate function. As for the y direction, we assume that the variation is linear with y—that is, we assume the deformation is chordwise linear. The coordinate function having these properties is then

$$w_1 = (A + B\eta)(2 - 3\xi + \xi^3) \qquad (6.118)$$

where we have used dimensionless variables and where A and B are undetermined coefficients.

We now express the strain energy of the skewed plate in terms of the dimensionless variables as follows:

$$U = \frac{LCD}{2}\int_0^1 d\xi \int_{\frac{L}{C}\xi\tan\beta}^{1+\frac{L}{C}\xi\tan\beta}\left\{\left[\frac{1}{L^2}\frac{\partial^2 w}{\partial\xi^2} + \frac{1}{C^2}\frac{\partial^2 w}{\partial\eta^2}\right]^2\right.$$

$$\left. + \frac{2(1-v)}{L^2C^2}\left[\left(\frac{\partial^2 w}{\partial\xi\,\partial\eta}\right)^2 - \left(\frac{\partial^2 w}{\partial\xi^2}\right)\left(\frac{\partial^2 w}{\partial\eta^2}\right)\right]\right\}d\eta \qquad (6.119)$$

Note that variable limits must be used for the first integration because of the skewed geometry. As for the loading we may say:

$$q = -\frac{4\Gamma_0}{C^2}(y^2 - Cy)[\delta(x - 0)]$$

where the last bracket is the familiar delta function. In terms of dimensionless variables we have

$$q = -4\Gamma_0(\eta^2 - \eta)[\delta(\xi - 0)] \qquad (6.120)$$

Now we employ Eqs. (6.119) and (6.120) to give the total potential energy as follows:

$$
\pi = \frac{LCD}{2} \int_0^1 d\xi \int_{\frac{L}{C}\xi\tan\beta}^{1+\frac{L}{C}\xi\tan\beta} \left\{ \left[\frac{1}{L^2}\frac{\partial^2 w}{\partial\xi^2} + \frac{1}{C^2}\frac{\partial^2 w}{\partial\eta^2} \right]^2 \right.
$$
$$
\left. + \frac{2(1-v)}{L^2 C^2}\left[\left(\frac{\partial^2 w}{\partial\xi\,\partial\eta}\right)^2 - \left(\frac{\partial^2 w}{\partial\xi^2}\right)\left(\frac{\partial^2 w}{\partial\eta^2}\right) \right] \right\} d\eta
$$
$$
+ 4\Gamma_0 L \int_0^1 d\xi \int_{\frac{L}{C}\xi\tan\beta}^{1+\frac{L}{C}\xi\tan\beta} (\eta^2 - \eta)[\delta(\xi - 0)]w(\eta,\xi)\,d\eta
$$

In considering the integration of the last expression above note from the properties of delta functions that for $a < d < b$:

$$
\int_a^b g(x)[\delta(x - d)]\,dx = g(d) \qquad (6.121)
$$

Hence we have for the last expression

$$
4\Gamma_0 L \int_0^1 \left[\int_{\frac{L}{C}\xi\tan\beta}^{1+\frac{L}{C}\xi\tan\beta} (\eta^2 - \eta)[\delta(\xi - 0)]w(\eta,\xi)\,d\eta \right] d\xi
$$
$$
= 4L\Gamma_0 \int_0^1 [\delta(\xi - 0)]\left[\int_{\frac{L}{C}\xi\tan\beta}^{1+\frac{L}{C}\xi\tan\beta} (\eta^2 - \eta)w(\eta,\xi)\,d\eta \right] d\xi
$$
$$
= 4L\Gamma_0 \left(\int_0^1 (\eta^2 - \eta)w(\eta,0)\,d\eta \right)
$$

Now substituting for w in the total potential energy expression using Eq. (6.118) we get, on using the above result:

$$
\pi = D\left\{ B^2\left[\frac{18}{5}\frac{\tan^2\beta}{LC} + \frac{9\tan\beta}{2L^2} + \frac{2C}{L^3} + \frac{24(1-v)}{5LC} \right] \right.
$$
$$
\left. + \frac{6A^2 C}{L^3} + AB\left[\frac{9\tan\beta}{L^2} + \frac{6C}{L^3} \right] \right\} - \frac{2}{3}L\Gamma_0(2A + B)
$$

By extremizing with respect to A and B we obtain a pair of algebraic equations which when solved yield

$$
A = \frac{2L^4\Gamma_0\lambda}{3CD}
$$
$$
B = \frac{4L^4\Gamma_0}{9D(3L\tan\beta + 2C)}(1 - 6\lambda)
$$
$$
\lambda = \frac{72L^2\tan^2\beta + 45CL\tan\beta + 10C^2 + 96L^2(1-v)}{27L^2\tan^2\beta + 60C^2 + 576L^2(1-v)}
$$

To obtain the chordwise rotation of the tip chord we proceed as follows:

$$
(\text{Rotation})_{\text{Tip}} = \frac{\partial w(0, y)}{\partial y} = \frac{1}{C}\frac{\partial w(0, \eta)}{\partial\eta} = 2\frac{B}{C}
$$

For the case where $\beta = 45°$ we get

$$
\text{Rotation} = -\frac{40L^5\Gamma_0}{CD(201L^2 + 20C^2 - 192L^2 v)} \qquad (6.122)
$$

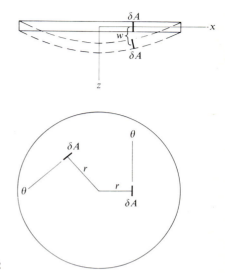

FIGURE 6.22

6.12 IMPROVED THEORY—AXISYMMETRIC CIRCULAR PLATES

In the classical theory of plates presented thus far in this chapter we have neglected transverse shear, i.e., we have set it equal to zero despite the fact that we knew from equilibrium considerations that it was not zero. We then justified this step in Sec. 6.6 by showing that such stresses are small for thin plates ($L/h \gg 10$). At this time we shall present an improved theory which takes into account transverse shear. This addition in our treatment of plates is analogous to the Timoshenko beam theory presented in Chap. 4.

We shall consider here the case of the axisymmetric circular plate.[1] The theory is due to Mindlin. We proceed by introducing a function $\psi(r)$ to be determined later such that

$$u_r = -z\psi(r)$$
$$u_\theta = 0$$
$$u_z = w(r) \qquad (6.123)$$

Examining Fig. 6.22 in conjunction with the above equations we see that the assumed displacement field permits transverse area elements normal to r, such as areas δA, to displace vertically a distance w and at the same time, since $\partial u_r/\partial z = -\psi(r)$, to rotate

---

[1] We shall consider the rectangular plate in this regard in Chap. 7 when we consider the vibration of plates.

about an axis in the transverse direction, θ, an amount dependent only on r (through the function ψ). This is still a simple displacement field. The strains for this displacement field (see Problem 1.17) can then be given as follows:

$$\epsilon_{rr} = \frac{\partial u_r}{\partial r} = -z\frac{d\psi}{dr} = -z\psi'$$

$$\epsilon_{\theta\theta} = \frac{1}{r}\frac{\partial u_\theta}{\partial \theta} + \frac{u_r}{r} = -\frac{z}{r}\psi$$

$$\epsilon_{zz} = \frac{\partial u_z}{\partial z} = 0$$

$$\epsilon_{r\theta} = \frac{1}{2}\left(\frac{1}{r}\frac{\partial u_r}{\partial \theta} + \frac{\partial u_\theta}{\partial r} - \frac{u_\theta}{r}\right) = 0$$

$$\epsilon_{rz} = \frac{1}{2}\left(\frac{\partial u_z}{\partial r} + \frac{\partial u_r}{\partial z}\right) = \tfrac{1}{2}(w' - \psi)$$

$$\epsilon_{\theta z} = \frac{1}{2}\left(\frac{\partial u_\theta}{\partial z} + \frac{1}{r}\frac{\partial u_z}{\partial \theta}\right) = 0 \qquad (6.124)$$

We now formulate the strain energy for the plate as follows for linear elastic behavior

$$U = \tfrac{1}{2}\int_0^{2\pi}\int_{r_o}^{r_1}\int_{-h/2}^{h/2} (\tau_{\theta\theta}\epsilon_{\theta\theta} + \tau_{rr}\epsilon_{rr} + \tau_{zz}\epsilon_{zz} + 2\tau_{r\theta}\epsilon_{r\theta}$$
$$+ 2\tau_{rz}\epsilon_{rz} + 2\tau_{\theta z}\epsilon_{\theta z})\, d\theta r\, dr\, dz$$

where r_o and r_1 are the inner and outer radii of the plate. (We assume for now there may be a hole at the center of the plate of radius r_o.) We thus have, after retaining only nonzero terms and integrating with respect to θ:

$$U = \pi\int_{r_o}^{r_1}\int_{-h/2}^{h/2} (\tau_{rr}\epsilon_{rr} + \tau_{\theta\theta}\epsilon_{\theta\theta} + 2\tau_{rz}\epsilon_{rz})r\, dr\, dz \qquad (6.125)$$

We are thus retaining transverse shear effects. Now employ Hooke's law for plane stress given as

$$\tau_{rr} = \frac{E}{1 - v^2}(\epsilon_{rr} + v\epsilon_{\theta\theta}) \qquad (a)$$

$$\tau_{\theta\theta} = \frac{E}{1 - v^2}(\epsilon_{\theta\theta} + v\epsilon_{rr}) \qquad (b)$$

$$\tau_{rz} = 2G\epsilon_{rz} \qquad (c) \quad (6.126)$$

Note that the transverse shear stress, as a result of our assumptions, is uniform with respect to thickness. We shall later take into account the fact that the transverse shear stress and strain do vary over the thickness. Substituting into Eq. (6.125) we have:

$$U = \pi \int_{r_o}^{r_1} \int_{-h/2}^{h/2} \left\{ \frac{E}{1-v^2}[(\epsilon_{rr}^2 + v\epsilon_{\theta\theta}\epsilon_{rr}) + (\epsilon_{\theta\theta}^2 + v\epsilon_{rr}\epsilon_{\theta\theta})] + 4G\epsilon_{rz}^2 \right\} r \, dr \, dz$$

Taking the first variation we get:

$$\delta^{(1)}U = \pi \int_{r_o}^{r_1} \int_{-h/2}^{h/2} \left\{ \frac{E}{1-v^2}[(\epsilon_{rr} + v\epsilon_{\theta\theta})\, \delta\epsilon_{rr} \right.$$
$$\left. + (\epsilon_{\theta\theta} + v\epsilon_{rr})\, \delta\epsilon_{\theta\theta}] + 8G\epsilon_{rz}\, \delta\epsilon_{rz} \right\} r \, dr \, dz$$

Noting Eq. (6.126) we can replace the coefficients of the strain variations with stresses to form:

$$\delta^{(1)}U = 2\pi \int_{r_o}^{r_1} \int_{-h/2}^{h/2} (\tau_{rr}\, \delta\epsilon_{rr} + \tau_{\theta\theta}\, \delta\epsilon_{\theta\theta} + 2\tau_{rz}\, \delta\epsilon_{rz}) r \, dr \, dz$$

Now replacing the strains using Eq. (6.124) we get:

$$\delta^{(1)}U = 2\pi \int_{r_o}^{r_1} \int_{-h/2}^{h/2} \left[-z\tau_{rr}\, \delta\psi' - \frac{z}{r}\tau_{\theta\theta}\, \delta\psi + \tau_{rz}(\delta w' - \delta\psi) \right] r \, dr \, dz \qquad (6.127)$$

We now introduce the following nomenclature which is familiar from previous work:

$$M_r = \int_{-h/2}^{h/2} z\tau_{rr}\, dz$$

$$M_\theta = \int_{-h/2}^{h/2} z\tau_{\theta\theta}\, dz$$

$$Q = \int_{-h/2}^{h/2} \tau_{rz}\, dz$$

Integrating with respect to z in Eq. (6.127) while noting that ψ and w are not functions of z and using the above results we get:

$$\delta^{(1)}U = 2\pi \int_{r_o}^{r_1} \left[-M_r\, \delta\psi' - \frac{M_\theta}{r}\, \delta\psi + Q\, \delta(w' - \psi) \right] r \, dr$$

Note next that:

$$\delta^{(1)}V = -2\pi \int_{r_o}^{r_1} q\, \delta w\, r \, dr$$

For the variation of total potential energy $(U + V)$ we then have:

$$\delta^{(1)}(U + V) = 2\pi \int_{r_o}^{r_1} \left[-M_r \, \delta\psi' - \frac{M_\theta}{r} \delta\psi + Q \, \delta(w' - \psi) - q \, \delta w \right] r \, dr = 0$$

This may be written as:

$$\int_{r_o}^{r_1} \left[-rM_r \frac{d}{dr}(\delta\psi) - M_\theta \, \delta\psi + rQ \frac{d}{dr}(\delta w) - rQ \, \delta\psi - rq \, \delta w \right] dr = 0$$

Integrating the first and third terms by parts we get on collecting terms:

$$\int_{r_o}^{r_1} \left[(rM_r)' - M_\theta - rQ \right] \delta\psi \, dr + \int_{r_o}^{r_1} \left[-(rQ)' - rq \right] \delta w \, dr + (-rM_r) \, \delta\psi \Big|_{r_o}^{r_1}$$

$$+ (rQ) \, \delta w \Big|_{r_o}^{r_1} = 0$$

The following then are the Euler–Lagrange equations:

$$rM_r' + M_r - M_\theta - rQ = 0$$

'or

$$M_r' + \frac{M_r - M_\theta}{r} = Q \qquad (6.128)$$

And:

$$[-(rQ)' - rq] = 0$$

or

$$\frac{1}{r} \frac{d}{dr}(rQ) = -q \qquad (6.129)$$

The boundary conditions are at $r = r_o, r_1$:

$$rM_r = 0 \quad \text{OR} \quad \psi \text{ PRESCRIBED}$$

$$rQ = 0 \quad \text{OR} \quad w \text{ PRESCRIBED} \qquad (6.130)$$

In order to express the plate equations in terms of displacements note from Eq. (6.126(a)) and (6.124) that:

$$M_r = \int_{-h/2}^{h/2} z\tau_{rr} \, dz = \int_{-h/2}^{h/2} \frac{E}{1 - v^2}(\epsilon_{rr} + v\epsilon_{\theta\theta})z \, dz$$

$$= \frac{E}{1 - v^2}\left(-\psi' - v\frac{\psi}{r} \right) \int_{-h/2}^{h/2} z^2 \, dz$$

$$= -D\left(\psi' + v\frac{\psi}{r} \right)$$

Similarly:

$$M_\theta = -D\left(\frac{\psi}{r} + v\psi'\right)$$

As for Q we shall now introduce the correction mentioned earlier to account for variation of transverse shear over the thickness of the plate. Thus, as we did in Timoshenko's improved theory, we proceed as follows:

$$Q = \int_{-h/2}^{h/2} \tau_{rz}\, dz = k(\tau_{rz})_{z=0} h$$

We are accordingly assuming here that a constant factor k properly chosen can account for the nonuniformity of shear across the thickness everywhere in the plate. Now using Hooke's law and Eq. (6.124) we have for the above equation:

$$Q = kG(w' - \psi)h \qquad (6.131)$$

We can now express Eqs. (6.128) and (6.129) as follows using preceding results for M_v, M_θ, and Q:

$$-D\frac{d}{dr}\left(\psi' + v\frac{\psi}{r}\right) + \frac{D}{r}\left(-\psi' - v\frac{\psi}{r} + v\psi' + \frac{\psi}{r}\right) = hkG(w' - \psi)$$

$$\frac{1}{r}\frac{d}{dr}[hkGr(w' - \psi)] = -q$$

The first of the above equations may be simplified in form to

$$-D\left[\psi'' + \frac{\psi'}{r} - \frac{\psi}{r^2}\right] = kG(w' - \psi)h \qquad (6.132)$$

As for the second equation, we have on integrating:

$$hkGr(w' - \psi) = -\int_{r_o}^{r} q(\xi)\xi\, d\xi + C_1$$

To evaluate C_1, we restrict the discussion at this time to a whole plate (no hole)— hence $r_0 = 0$. Accordingly setting $r_o = 0$ in the above equation and excluding a point load at the center we see that $C_1 = 0$. We can solve for ψ from the above equation as follows

$$\psi = w' + \frac{1}{hkGr}\int_{0}^{r} q(\xi)\xi\, d\xi \qquad (6.133)$$

Hence once we know w we can readily determine ψ from the above equation. We shall accordingly concentrate on Eq. (6.132) with the view towards eliminating ψ

in an expeditious manner. For this purpose we return to Eq. (6.131) and solve for ψ as follows

$$\psi = w' - \frac{Q}{kGh} \qquad (6.134)$$

Substituting into Eq. (6.132) we get the following result

$$-D\left[w''' - \frac{Q''}{hkG} + \frac{w''}{r} - \frac{Q'}{rkGh} - \frac{w'}{r^2} + \frac{Q}{hkGr^2} \right] = Q$$

Rearranging the terms we get

$$-D\left[w''' + \frac{w''}{r} - \frac{w'}{r^2} \right] = Q - \frac{D}{hkG}\left[Q'' + \frac{Q'}{r} - \frac{Q}{r^2} \right] \qquad (6.135)$$

Now multiply the above equation by r and then apply the operator $-\dfrac{1}{r}\dfrac{d}{dr}$ as follows:

$$D\frac{1}{r}\frac{d}{dr}\left[rw''' + w'' - \frac{w'}{r} \right] = D\left[\frac{1}{r}\frac{d}{dr} + \frac{d^2}{dr^2} \right]\left[\frac{1}{r}\frac{d}{dr} + \frac{d^2}{dr^2} \right]w$$

$$= D\nabla^2\nabla^2 w = D\nabla^4 w$$

$$= -\frac{1}{r}\frac{d}{dr}(rQ) + \frac{D}{hkG}\frac{1}{r}\frac{d}{dr}\left(rQ'' + Q' - \frac{Q}{r} \right) \qquad (6.136)$$

But from Eq. (6.129), $-\dfrac{1}{r}\dfrac{d}{dr}(rQ) = q$. Also note from Eq. (6.129) that

$$\nabla^2 q = \frac{1}{r}\frac{d}{dr}\left[r\frac{dq}{dr} \right] = \frac{1}{r}\frac{d}{dr}\left[r\frac{d}{dr}\left(-\frac{1}{r}\frac{d}{dr}(rQ) \right) \right]$$

$$= -\frac{1}{r}\frac{d}{dr}\left[r\frac{d}{dr}\left(Q' + \frac{Q}{r} \right) \right] = -\frac{1}{r}\frac{d}{dr}\left[rQ'' + Q' - \frac{Q}{r} \right]$$

We can then write Eq. (6.136) as follows using the above results:

$$D\nabla^4 w = q - \frac{D}{hkG}\nabla^2 q$$

therefore

$$\boxed{ D\nabla^4 w + \frac{D}{hkG}\nabla^2 q - q = 0 } \qquad (6.137)$$

We thus have an equation in terms of w only, including now the simplification due to the k factor.

To illustrate the use of the theory presented we consider the case of uniform loading q_0 on the *clamped* circular plate. On consulting Eq. (6.137) it is immediately clear that the differential equation in this case is exactly the same for the improved theory as that for classical theory. However, the *boundary conditions are different*. For comparison purposes we set forth the boundary conditions for the classical and improved theories.[1] Thus at $r = a$ we require in each case

$$\text{classical theory:} \quad w = \frac{dw}{dv} = 0$$

$$\text{improved theory:} \quad w = \psi = 0 \qquad (6.138)$$

where ψ, you will recall from Sec. 6.12, is the rotation about a circumferential line of area elements having normals in the radial direction. Furthermore ψ is given in terms of w and q by Eq. (6.133).

We may consider the solution for the improved theory in two parts owing to the linearity of the equations. First there is the bending deflection (no shear effects) alone which has been solved in Sec. 6.10

$$w_B = \frac{q_0 a^4}{64D}\left[1 - \left(\frac{r}{a}\right)^2\right]^2$$

For the deflection w_S due to transverse shear only we will first heuristically propose such a solution and then we will demonstrate that the total deflection function $w_B + w_S$ solves the boundary value problem for the improved theory. Consider accordingly a portion of the plate having a radius r. From equilibrium considerations we know that

$$Q = \frac{q_0 r}{2} \qquad (6.139)$$

But recall that

$$Q = \int_{-h/2}^{h/2} \tau_{rz}\, dz = k(\tau_{rz})_{z=0}h$$

Hence, employing Eq. (6.139):

$$(\tau_{rz})_{z=0} = \frac{Q}{kh} = \frac{q_0 r}{2hk} \qquad (6.140)$$

Let us consider shear deformation at the centerplane of the plate. An element is shown in Fig. 6.23(*a*) at the centerplane in the undeformed geometry. In the deformed geometry we assume, as in our study of beams, that the vertical sides of the element remain vertical while the horizontal sides stay parallel to the centerplane in the deformed geometry as has been shown in Fig. 6.23(*b*). From this diagram we conclude that:

$$\gamma_{rz} = -\alpha = -\frac{dw_S}{dr}$$

[1] For the case of the *simply-supported* circular plate we have at the edge:

$$\text{classical theory} \quad w = w'' + v\frac{w'}{r} = 0$$

$$\text{improved theory} \quad w = \psi' + v\frac{\psi}{r} = 0$$

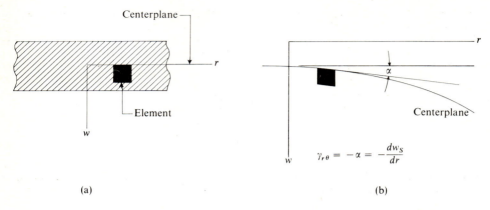

$$\gamma_{r\theta} = -\alpha = -\frac{dw_S}{dr}$$

(a)   (b)

FIGURE 6.23

Hence, using Hooke's law:

$$(\tau_{rz})_{z=0} = -G\frac{dw_S}{dr} \qquad (6.141)$$

Substituting the above result into Eq. (6.140) we get:

$$\frac{dw_S}{dr} = -\frac{q_0 r}{2kGh}$$

Integrating we get:

$$w_S = \frac{q_0}{4kGh}(-r^2 + C_1') = \frac{q_0 a^2}{4kGh}\left(C_1 - \frac{r^2}{a^2}\right) \qquad (6.142)$$

where C_1 is the constant of integration. It is now a simple matter to show that w_S satisfies the homogeneous part of the differential equation which, for the axisymmetric case involving cylindrical coordinates, is given as follows:

$$\nabla^4 w = \left(\frac{\partial^2}{\partial r^2} + \frac{1}{r}\frac{\partial}{\partial r}\right)\left(\frac{\partial^2}{\partial r^2} + \frac{1}{r}\frac{\partial}{\partial r}\right)w = 0$$

The proposed solution

$$w = w_B + w_S = \frac{q_0 a^4}{64D}\left[1 - \left(\frac{r}{a}\right)^2\right]^2 + \frac{a^2 q_0}{4kGh}\left(C_1 - \frac{r^2}{a^2}\right)$$

then satisfies the entire differential equation

$$\nabla^4 w = q_0$$

for the problem at hand.[1]

We now proceed to the boundary conditions. Clearly to have $w_S = 0$ when $r = a$ it is necessary that $C_1 = 1$. As for the second condition we require that (see Eqs. (6.138) and (6.133)):

$$\psi(a) = w'(a) + \frac{1}{hkG}\frac{1}{a}\int_0^a q_0\xi\,d\xi = 0$$

---

[1] Note that the equation is actually an ordinary differential equation for the problem at hand.

Since $w'_B(a) = 0$ (the classical condition for bending) we need only include $w_S(a)$ in the above formulation. We may directly see on substitution and on carrying out the integration that the above condition is also satisfied. Thus the proposed solution leads to an exact solution of the boundary value problem here. It will be useful to introduce $D \, (= Eh^3/12(1 - v^2))$ and to replace G by $E/2(1 + v)$ in w_S as follows:

$$w_S = \frac{q_0 a^2}{\{4kE/[(2)(1 + v)]\}(D)[12(1 - v^2)/Eh^3]h} \left[1 - \left(\frac{r}{a}\right)^2 \right] = \frac{q_0 a^2 h^2}{24k(1 - v)D} \left[1 - \left(\frac{r}{a}\right)^2 \right]$$

The solution then is

$$w = \frac{qa^4}{64D} \left[1 - \left(\frac{r}{a}\right)^2 \right]^2 + \frac{qa^2 h^2}{(24)k(1 - v)D} \left[1 - \left(\frac{r}{a}\right)^2 \right]$$

As for k, we know from the exact solution of plates that τ_{rz} varies parabolically across the thickness of the plate[1] and so we take $k = \frac{4}{3}$ to correspond to this result. One can show that this result then corresponds to the exact solution from the theory of elasticity.[2]

6.13   CLOSURE

In this chapter we have used the variational approach to formulate a useful boundary value problem for plates. We then used variational methods for developing approximate solutions to these boundary value problems. Much of the work paralleled what was done in Chap. 4 on beams and Chap. 5 on torsion. You will recall in particular that we introduced in the discussion of torsion a method for finding the lower limit of the quadratic functional I for the so-called von Neumann problem. This was the Trefftz method. From this we can get a sequence of functions v_n that converges to the solution under certain circumstances. However, we used the method to find an upper bound to the torsional rigidity thereby affording us, with the aid of the Ritz method, a means to bracket the correct value of this quantity. We thus entered into the difficult area of error analysis. Now for the case of clamped plates and simply-supported plates there is an extension to the Trefftz method discussed earlier namely the method of Rafalson. Here we can similarly find a lower bound of the quadratic functional to go with an upper bound associated with the Ritz method. In this approach a sequence w_n can be found that converges to the correct solution under certain appropriate conditions. We shall not get into this development in this text other than to point out its existence since this takes us beyond the level we have prescribed for the book.

We have restricted ourselves entirely to static problems of beams and plates up to this time. In the following chapter we shall investigate simple features of the vibration of beams and plates and thereby we shall introduce the eigenvalue problem. You will see that variational methods are again extremely useful for such problems.

[1] See Timoshenko and Goodier: "Theory of Elasticity," McGraw-Hill Book Co., p. 351.
[2] See Love: "Mathematical Theory of Elasticity," Dover Publications, p. 435.

READING

LANGHAAR, H. L.: "Energy Methods in Applied Mechanics," John Wiley and Sons, 1962.

MANSFIELD, E. H.: "The Bending and Stretching of Plates," Macmillan, N.Y., 1964.

MORLEY, L.: "Skew Plates and Structures," Macmillan, N.Y., 1963.

RIVELLO, R. M.: "Theory and Analysis of Flight Structures," McGraw-Hill Book Co., 1969.

TIMOSHENKO, S. AND WOINOWSKY-KREIGER, S.: "Theory of Plates," McGraw-Hill Book Co., 1959.

WAY, S.: "Plates," in Flügge, W., "Handbook of Engineering Mechanics," McGraw-Hill Book Co., 1962.

PROBLEMS

6.1 Derive Eq. (6.32) using Green's theorem and starting from Eq. (6.31).

6.2 Derive the following expression for the complementary strain energy of a rectangular plate $a \times b \times h$.

$$U^* = \tfrac{1}{2} \int_0^a \int_0^b [M_x{}^2 - 2vM_xM_y + M_y{}^2 + 2(1 + v)M_{xy}{}^2] \, dx \, dy$$

6.3 An *orthotropic continuum* has three planes of symmetry with respect to elastic properties. For plane stress, generalized Hooke's law for such a material is given as:

$$\tau_{xx} = C_{11}\epsilon_{xx} + C_{12}\epsilon_{yy}$$

$$\tau_{yy} = C_{12}\epsilon_{xx} + C_{22}\epsilon_{yy}$$

$$\tau_{xy} = 2G_{12}\epsilon_{xy}$$

(Note there are four elastic moduli.) Formulate the equations of equilibrium as well as the boundary conditions for an *orthotropic plate* in terms of w using rectangular coordinates. Demonstrate that your equation degenerates to the case of the isotropic plate. Use whatever results of Sec. 6.3 that are valid for this case.

6.4 Develop a reciprocal theorem between the bending deflection and the transverse load on a plate. That is, show that:

$$\iint [q^{(1)}(x,y) \, w^{(2)}(x,y) - q^{(2)}(x,y) \, w^{(1)}(x,y)] \, dx \, dy = 0$$

where superscripts (1) and (2) refer to different loadings and corresponding deflection conditions on the same plate.

Hint: Start with the reciprocal theorem (Problem 1.22)

$$\iiint \tau_{ij}{}^{(1)}\epsilon_{ij}{}^{(2)} \, dv = \iiint \tau_{ij}{}^{(2)}\epsilon_{ij}{}^{(1)} \, dv \qquad (a)$$

Employ for plane stress in Eq. (*a*) the strain displacement relations, the stress intensity resultant function and finally Eq. (6.50).

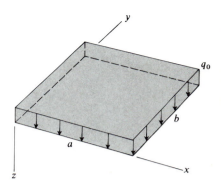

FIGURE 6.24 z

6.5 Consider an orthotropic plate (see Problem 6.3) where the planes of symmetry are the coordinate planes of a cylindrical coordinate system. Express Eqs. (6.4) for the case of deformation of an axisymmetric plate. Using the strain displacement relations for polar coordinates (see Problem 1.17) show that the strain energy for an orthotropic, axisymmetric, circular plate is given, using data of the previous problem, as:

$$U = \pi \int_{r_0}^{r_1} \left\{ D_{11} \left(\frac{d^2 w}{dr^2} \right)^2 + 2D_{12} \frac{1}{r} \frac{dw}{dr} \frac{d^2 w}{dr^2} + D_{22} \left(\frac{1}{r} \frac{dw}{dr} \right)^2 \right\} r \, dr$$

where

$$D_{ij} = \tfrac{1}{12} C_{ij} h^3$$

Show that the equation of equilibrium for such a plate is:

$$D_{11} \left(r \frac{d^2 w}{dr^2} \right)'' - D_{12} \left(\frac{1}{r} \frac{dw}{dr} \right)' = r \, q(r)$$

6.6 Derive an exact power series solution to a simply-supported rectangular plate (see Fig. 6.24) loaded uniformly with loading q_0. Show that the solution can be given as follows:

$$w(x,y) = \frac{16 q_0}{\pi^6} \frac{12(1 - v^2)}{E h^3} \sum_{m(\text{odd})}^{\infty} \sum_{n(\text{odd})}^{\infty} \frac{\sin \dfrac{m \pi x}{a} \sin \dfrac{n \pi y}{b}}{mn^5 \left[1 + \left(\dfrac{m/a}{n/b} \right)^2 \right]^2}$$

Show that for a plate where $b/a \to 0$, the above solution degenerates to that of a beam uniformly loaded by loading p_0 and given as:

$$w(y) = \frac{4 p_0 b^4}{\pi^5 EI} \sum_{n(\text{odd})}^{\infty} \frac{\sin (n \pi y / b)}{n^5}$$

where I is the moment of inertia per unit width. Also $(q_0)(1) = p_0$ and $v = 0$ in the development. Finally take $a/m = \infty$.

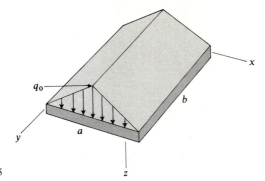

FIGURE 6.25

Hint:

$$\sum_{m(\text{odd})}^{\infty} \frac{\sin m\pi x}{m} = \frac{\pi}{4}$$

Thus we have derived the deflection of a beam by considering that of a rectangular plate by letting one dimension become very large compared to the other dimension.

6.7   Consider the rectangular plate shown in Fig. 6.25 with a triangular load. This plate is simply-supported at $x = 0$ and $x = a$ and is fixed at $y = 0$ and $y = b$. Using the Lévy method show that

$$q_m = \frac{8q_0}{Dm^2\pi^2} \sin \frac{m\pi}{2}.$$

Next determine the constants $(C_1)_m, (C_2)_m, (C_3)_m$, and $(C_4)_m$ and show that for the following data

$$h = 1 \text{ in.}$$
$$a = 2 \text{ ft.}$$
$$b = 2 \text{ ft.}$$
$$E = 30 \times 10^6 \text{ psi}$$
$$v = 0.3$$
$$q_0 = 100 \text{ psi}$$

we get the following results:

	$m = 1$	$m = 3$	$m = 5$	$m = 7$
$(C_1)_m$	-1.1681×10^{-3}	9.6494×10^{-9}	0	0
$(C_2)_m$	-7.2063×10^{-3}	1.14779×10^{-5}	-5.3596×10^{-7}	7.118158×10^{-8}
$(C_3)_m$	4.2838×10^{-4}	-4.36159×10^{-9}	0	0
$(C_4)_m$	-9.9130×10^{-3}	5.40473×10^{-5}	-4.20943×10^{-6}	7.826824×10^{-7}

Finally determine the deflection at the center of the plate to be as follows:

	$m = 1$	$m = 3$	$m = 5$	$m = 7$
w	0.015492	0.015479	0.01548	0.01548

Use a computer for these calculations.

6.8 Using the Ritz method find an approximate solution for the central deflection of a rectangular plate simply supported at $x = 0$ and $x = a$ and clamped at $y = 0$ and $y = b$. This plate is loaded by a triangular loading $q = q_0(x/a)$ (see Fig. 6.17) and it thus poses the identical problem as solved in Example 6.1 by the method of Lévy. For a square plate the exact answer is:

$$w\left(\frac{a}{2},\frac{b}{2}\right) = 0.0009586\frac{q_0 a^4}{D}$$

and for a rectangular plate $b/a = 2$ we have

$$w\left(\frac{a}{2},\frac{b}{2}\right) = 0.004222\frac{q_0 a^4}{D}$$

Compare your results with the above.

6.9 Find an approximate solution for the center-point deflection of a simply-supported square plate under the triangular loading

$$q(x,y) = q_0\frac{x}{a}$$

Compare with the exact answer:

$$w\left(\frac{a}{2},\frac{a}{2}\right) = 0.00203\frac{q_0 a^4}{D}$$

6.10 Find an approximate Ritz solution for the deflection at the center of a simply-supported, square plate loaded by moment intensities (moment per unit length) M_0 along the edges $y = 0$ and $y = b$ so as to cause downward deflection of the plate. Use a function that satisfies the kinematic boundary conditions of the problem. Compare with the exact solution for this problem:

$$w\left(\frac{a}{2},\frac{a}{2}\right) = 0.0368\frac{M_0 a^2}{D}$$

6.11 Consider a rectangular plate simply supported at the edges $x = 0$ and $y = 0$ while freely supported at the edges $x = a$ and $y = b$ (see Fig. 6.26). The load is a force at $x = a$, $y = b$. Find an approximate solution that satisfies the boundary conditions. What corner force does this solution generate at $x = a$, $y = b$? Comment on result.

6.12 Consider a clamped rectangular plate ($0 \leq x \leq a$, $0 \leq y \leq b$) with a concentrated load at the center. Find an approximate solution for the deflection and compare it with the

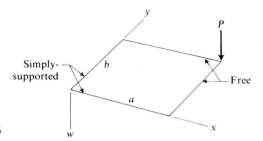

FIGURE 6.26

following accurate results:

$$\text{Square plate } (a = b) \qquad w_{max} = 0.00560 \frac{Pa^2}{D}$$

$$\text{Rectangular plate } \left(a = \frac{b}{2} \right) \qquad w_{max} = 0.00722 \frac{Pa^2}{D}$$

6.13 Using Green's theorem, the boundary condition (6.104) and the vector identity:

$$\nabla \cdot (\varphi \mathbf{F}) = \varphi \nabla \cdot \mathbf{F} + \mathbf{F} \cdot \nabla \varphi$$

show that Eq. (6.105) can be rewritten as Eq. (6.106). Show that Eq. (6.107) is an exact solution to the clamped elliptic plate problem.

6.14 Show that the Gaussian curvature for an axisymmetric, isotropic simply-supported circular plate is

$$\left[\frac{1}{r} w'(r) w''(r) \right] \qquad (a)$$

Now show that the total potential energy is given for such a plate (radius "*a*") as follows:

$$(U + V) = 2\pi \int_0^a \left\{ \frac{D}{2} \left[\left(w''(r) + \frac{1}{r} w'(r) \right)^2 - \frac{2(1 - v)}{r} w'(r) w''(r) \right] - wq \right\} r \, dr \qquad (b)$$

The boundary conditions for this problem may be given as:

$$\text{at } r = a \qquad w = 0 \qquad M_r = 0$$

Noting that when $\theta = 0$ we have

$$M_r = M_x$$

show that (see Eq. (6.12))

$$M_r = -D \left[\frac{\partial^2 w}{\partial r^2} + \frac{v}{r} \frac{\partial w}{\partial r} \right]$$

for the axisymmetric case. Hence the boundary conditions at $r = a$ are:

$$w = w'' + \frac{v}{r} w' = 0$$

6.15 Consider a simply-supported circular plate loaded by a moment intensity M_0 uniformly around the edge. Referring to Problem 6.14 and starting with an approximate function $w = A + Br^2$, satisfy the kinematic boundary conditions of the problem and then find a Ritz approximate solution. Finally show that the obtained solution actually satisfies the natural boundary conditions of the problem.

6.16 The solutions for the deflection of the clamped, circular plate of radius "a" under a uniform loading q_0 has been shown to be (Eq. (6.109))

$$w_q = \frac{q_0 a^4}{64D}\left[1 - \left(\frac{r}{a}\right)^2\right]^2$$

For the pure bending of a simply-supported circular plate we have from the previous problem the result:

$$w_{M_0} = \frac{M_0 a^2}{2(1 + v)D}\left[1 - \left(\frac{r}{a}\right)^2\right]$$

Obtain from these the solution for a simply-supported plate under a uniform loading q_0. (See Problem 6.14 for information as to boundary conditions for a circular plate.)

6.17 Find an approximate solution for the clamped circular orthotropic plate of radius "a" (see Problem 6.3) for a centrally located concentrated load. What is the ratio of maximum deflections between the orthotropic plate and the isotropic plate for the same coordinate function you have used?

6.18 Using the results of Problem 6.14 obtain an approximate solution for the deflection of a simply-supported circular plate of radius "a" loaded by a ring loading of total value P at radius b where $0 \le b \le a$. The exact solution for $r = 0$ is:

$$w(0) = \frac{P}{8\pi D}\left[\frac{3 + v}{2(1 + v)}(a^2 - b^2) - b^2 \ln \frac{a}{b}\right]$$

For $b/a = \frac{1}{2}$ and $v = 0.3$ compare solutions at $r = 0$. Use a polynomial for the coordinate function that satisfies the boundary conditions of the problem.

6.19 Do the swept wing problem of Sec. 6.11 for $\beta = 45°$ with a triangular load along the chord given as

$$w = \Gamma_0\left(\frac{y}{C} - \frac{L/2 \tan \beta}{C}\right)$$

at position $x = L/2$. Find the rotation of the tip chord.

6.20 Using Eqs. (6.53), (6.54), and (6.58) we assume the following stress distribution for axisymmetric deformation:

$$\tau_{rr} = \frac{M_r z}{h^3/12}$$

$$\tau_{\theta\theta} = \frac{M_\theta z}{h^3/12}$$

$$\tau_{rz} = \frac{3}{2}\frac{Q_r}{h}\left[1 - \left(\frac{z}{h/2}\right)^2\right]$$

$$\tau_{zz} = \frac{3}{4}q(r)\left[\frac{2}{3} - \frac{z}{h/2} + \frac{1}{3}\left(\frac{z}{h/2}\right)^3\right]$$

In addition we assume a displacement field as given by Eq. (6.123) in the Mindlin improved theory of plates. Using the Reissner functional given as:

$$R = \iiint (\tau_{ij}\epsilon_{ij} - U_0^*)\, dv - \int_{r_0}^{r_1} \int_0^{2\pi} . q(r,\theta)\, w(r,\theta) r\, dr\, d\theta$$

where U_0^* is the complementary strain energy density function, derive a plate theory including both shear *and* normal stress effects. *Hint:* Employing strain displacement relations for the cylindrical coordinates (see Problem 1.17) and neglecting a term τ_{zz}^2, show first that:

$$R = 2\pi \int_{r_0}^{r_1} \left\{ -M_r \psi'(r) + Q_r(w' - \psi) - \frac{1}{r}M_\theta\psi - \frac{M_r^2}{2D_1} - \frac{M_\theta^2}{2D_1} + v\frac{M_r M_\theta}{D_1} \right.$$

$$\left. -\frac{vq}{D_1}\frac{h^2}{10}[M_r + M_\theta] - \frac{3}{5}\frac{Q_r^2}{Gh} - qw \right\} r\, dr$$

where $D_1 = Eh^3/12$. Next consider ψ, w, M_r, M_θ and Q_r as the extremization variables to reach the following results:

$$\text{Equilibrium}\begin{cases} rQ_r = -M_\theta + \dfrac{d}{dr}(rM_r) \\[2mm] \dfrac{d}{dr}(rQ_r) = -qr \end{cases}$$

$$\begin{matrix}\text{Stress}\\ \text{Displacement}\\ \text{Relations}\end{matrix}\begin{cases} M_r - vM_\theta = -D_1\psi'(r) - \dfrac{vqh^2}{10} \\[2mm] M_\theta - vM_r = -D_1\dfrac{\psi}{r} - \dfrac{vqh^2}{10} \\[2mm] Q_r = \frac{5}{6}Gh(w' - \psi) \end{cases}$$

The boundary conditions are:

$$rM_r = 0 \quad \text{or} \quad \psi \text{ specified}$$

$$rQ_r = 0 \quad \text{or} \quad w \text{ specified}$$

6.21 Show that the contribution to the bending energy of plates of the Gaussian curvature vanishes (a) for smooth-edge plates that are clamped and (b) for polygonal plates with only zero displacement required at the edges.

*6.22 Use the Reissner principle to derive the Mindlin-type improved plate theory for axisymmetric deformation of a circular plate loaded by surface loading $q(r)$, edge moments per unit length M_0, and edge loading per unit length Q_0, downward. Obtain all appropriate differential equations, stress-displacement relations, and all boundary conditions.

*6.23 Solve the pure bending problem for a circular plate using the Mindlin-type theory, i.e., a simply-supported circular plate loaded by uniform edge moment per unit length M_0 at the edge $r = a$.

6.24 Obtain the exact solution and an approximate solution for the axisymmetric deflection of a clamped circular plate under the loading

$$q(r) = q_0 \left[1 - \left(\frac{r}{a} \right)^2 \right]^2 .$$

7

DYNAMICS OF BEAMS AND PLATES

7.1 INTRODUCTION

Up to this time we have considered only the case of structural bodies in static equilibrium. We now examine variational aspects of the dynamics of beams and plates. Our procedure will be first to present Hamilton's principle since this principle underlies much of what we do in this chapter. In part A we shall center our attention on beams, first deriving the equations of motion from Hamilton's principle, and then considering both exact and approximate solutions for free vibrations. For the latter calculations we shall feature the Rayleigh and Rayleigh–Ritz methods. The same pattern is then followed in Part B for the study of plates. In Part C of this chapter we examine the Rayleigh quotient, used in earlier parts of the chapter, in a more general manner and develop strong supporting arguments for some physically inspired assertions made earlier concerning the Rayleigh and the Rayleigh–Ritz methods. The very powerful maximum theorem and "mini-max" theorem from the calculus of variations are presented. Thus this section is a more theoretical exposition. It is supportive of earlier work as well as basic to new work on stability in Chap. 9.

7.2 HAMILTON'S PRINCIPLE

Those who have studied dynamics of particles and rigid bodies at the intermediate level should recall Hamilton's principle as employed for discrete systems. The functional of interest is

$$\int_{t_1}^{t_2} (T - V)\, dt$$

where T is the total-kinetic energy of the bodies and V is the potential energy of the forces. The variable t is of course the time. Using $q_1, q_2, \ldots, q_n$ as the generalized coordinates and assuming they are independent, the Euler–Lagrange equations then yield the well-known *Lagrange's* equations of motion given as follows in terms of the Lagrangian $L = T - V$:

$$\frac{d}{dt}\left(\frac{\partial L}{\partial \dot{q}_i}\right) - \frac{\partial L}{\partial q_i} = 0 \qquad i = 1, 2, \ldots, n$$

These equations are then the starting point for Lagrangian mechanics.

To develop Hamilton's principle from the method of virtual work we employ the D'Alembert principle which, you will recall, sets forth Newton's law in the following form:

$$\mathbf{F} - m\mathbf{a} = \mathbf{0}$$

The vector $-m\mathbf{a}$ is considered a force with the result that the above equation has a form corresponding to the equations of statics. For a continuous deformable body the inertia force for the D'Alembert principle is given as $-\rho\, d^2\mathbf{u}/dt^2$ and is a body force distribution playing the same role as $\mathbf{B}$ in the discussion of Sec. 3.2. Accordingly Eq. (3.3), which is valid for static conditions, may be generalized to include dynamic conditions if the integral

$$-\iiint_V \rho \frac{d^2 u_i}{dt^2}\, \delta u_i\, dv$$

is added to the left side of the equation. We thus have at time t for a virtual displacement field δu_i consistent with the constraints:

$$-\iiint_V \rho \frac{d^2 u_i}{dt^2} \delta u_i \, dv + \iiint_V B_i \, \delta u_i \, dv + \iint_{S_1} T_i^{(v)} \, \delta u_i \, ds = \iiint_V \tau_{ij} \, \delta \epsilon_{ij} \, dv \qquad (7.1)$$

We may introduce V such that:

$$V = -\iiint_V B_i u_i \, dv - \iint_{S_1} T_i^{(v)} u_i \, ds$$

The second and third terms on the right side of Eq. (7.1) can then be given as:

$$\iiint_V B_i \, \delta u_i \, dv + \iint_{S_1} T_i^{(v)} \, \delta u_i \, ds = -\delta^{(1)} V$$

Similarly we assume the existence of a strain energy density function U_0 for the body such that $\tau_{ij} = \partial U_0 / \partial \epsilon_{ij}$. The right side of Eq. (7.1) becomes $\delta^{(1)} U$ and we may give this equation in the following form:

$$-\iiint_V \rho \frac{d^2 u_i}{dt^2} \delta u_i \, dv - \delta^{(1)} V - \delta^{(1)} U = 0$$

The variables u_i, V, and U in the above equation may be functions of time and so the statement applies for time t. We now integrate with respect to time from the limit t_1, (denoting the onset of the motion) to time t_2. We thus have:

$$-\int_{t_1}^{t_2} \iiint_V \rho \frac{d^2 u_i}{dt^2} \delta u_i \, dv \, dt - \int_{t_1}^{t_2} \delta^{(1)} V \, dt - \int_{t_1}^{t_2} \delta^{(1)} U \, dt = 0$$

We now interchange integration operations in the first expression and the variation and integration operations in the remaining terms as follows:

$$-\iiint_V \left(\rho \int_{t_1}^{t_2} \frac{d^2 u_i}{dt^2} \delta u_i \, dt \right) dv - \delta^{(1)} \int_{t_1}^{t_2} V \, dt - \delta^{(1)} \int_{t_1}^{t_2} U \, dt = 0 \qquad (7.2)$$

Next we integrate by parts the bracketed integral in the first term. Thus:

$$\int_{t_1}^{t_2} \frac{d^2 u_i}{dt^2} \delta u_i \, dt = \frac{du_i}{dt} \delta u_i \bigg|_{t_1}^{t_2} - \int_{t_1}^{t_2} \frac{du_i}{dt} \frac{d}{dt} (\delta u_i) \, dt$$

We adopt the rule here that $\delta u_i = 0$ at $t = t_1$ and $t = t_2$. That is, we take *the virtual displacement fields to be zero at the time limits t_1 and t_2.* The above statement may then be written as follows:

$$\int_{t_1}^{t_2} \frac{d^2 u_i}{dt^2} \delta u_i \, dt = -\int_{t_1}^{t_2} \dot{u}_i \, \delta \dot{u}_i \, dt = -\int_{t_1}^{t_2} \delta^{(1)} \left(\frac{\dot{u}_i^{\,2}}{2} \right) dt$$

Substituting back into Eq. 7.2 we have:

$$\iiint_V \frac{\rho}{2} \int_{t_1}^{t_2} \delta^{(1)} \dot{u}_i{}^2 \, dt \, dv - \delta^{(1)} \int_{t_1}^{t_2} V \, dt - \delta^{(1)} \int_{t_1}^{t_2} U \, dt = 0$$

Again considering the first expression we have:

$$\iiint_V \frac{\rho}{2} \int_{t_1}^{t_2} \delta^{(1)} \dot{u}_i{}^2 \, dt \, dv = \int_{t_1}^{t_2} \iiint_V \frac{\rho}{2} \delta^{(1)} (\dot{u}_i{}^2) \, dv \, dt$$

$$= \int_{t_1}^{t_2} \delta^{(1)} \left(\iiint_V \frac{\rho \dot{u}_i{}^2}{2} \, dv \right) dt$$

$$= \int_{t_1}^{t_2} \delta^{(1)} T \, dt = \delta^{(1)} \left(\int_{t_1}^{t_2} T \, dt \right)$$

where T is the kinetic energy of the body at time t. We now may give Hamilton's principle in the following manner:

$$\delta^{(1)} \left\{ \int_{t_1}^{t_2} (T - V - U) \, dt \right\} = 0$$

Noting that $(V + U)$ is the *total potential energy* π, we then get the following familiar form for the Hamilton principle

$$\boxed{\delta^{(1)} \int_{t_1}^{t_2} (T - \pi) \, dt = \delta^{(1)} \int_{t_1}^{t_2} L \, dt = 0} \qquad (7.3)$$

where L, still called the Lagrangian, is now $T - \pi$. *Hamilton's principle* states that *of all the paths of admissible configurations that the body can take as it goes from configuration 1 at time t_1 to configuration 2 at time t_2, the path that satisfies Newton's law at each instant during the interval (and is thus the actual locus of configurations) is the path that extremizes the time integral of the Lagrangian during the interval.*

We shall directly illustrate the use of Hamilton's principle in the following section wherein we consider the vibrating beam.

Part A

BEAMS

7.3 EQUATIONS OF MOTION FOR VIBRATING BEAMS

We shall return to Chap. 4 to generalize the assumed kinematics of a simple beam to include time variation. Omitting axially applied external forces we thus have from Eq. (4.4):

$$u_1 = -z \frac{\partial w(x,t)}{\partial x}$$

$$u_2 = 0$$

$$u_3 = w(x,t) \qquad (7.4)$$

We will need the Lagrangian for use in Hamilton's principle and so we proceed by considering the kinetic energy of the beam. Thus we have using the preceding results:

$$
\begin{aligned}
T = \frac{1}{2}\int_0^L \iint_A \rho \dot{u}_i \dot{u}_i \, dy \, dz \, dx &= \frac{1}{2}\int_0^L \iint_A \rho \left(\frac{\partial w}{\partial t}\right)^2 dx \, dy \, dz \\
&\quad + \frac{1}{2}\int_0^L \iint_A \rho \left(\frac{\partial^2 w}{\partial t \, \partial x}\right)^2 z^2 \, dx \, dy \, dz \\
&= \int_0^L \frac{\rho A}{2}\left(\frac{\partial w}{\partial t}\right)^2 dx + \int_0^L \frac{\rho I}{2}\left(\frac{\partial^2 w}{\partial t \, \partial x}\right)^2 dx
\end{aligned}
\tag{7.5}
$$

The kinetic energy is thus seen to be composed of two parts. The first represents the kinetic energy due to translatory motion in the vertical direction z. The second expression involves half the product of an angular velocity $\partial(\partial w/\partial x)/\partial t$, squared, of a beam element dx about the y axis times the mass moment of inertia about the y axis for this element, $(\rho I \, dx)$. Thus the second expression represents the kinetic energy due to rotation of the beam elements. For most problems, *particularly those involving thin beams*, this contribution to T may be neglected since it is usually very small for long slender beams compared to the other term. In those cases we say that we are neglecting the *rotatory inertia* for the problem.[1] We shall here neglect this contribution and so we shall use for T the following result:

$$
T = \int_0^L \frac{\rho A}{2}\left(\frac{\partial w}{\partial t}\right)^2 dx
\tag{7.6}
$$

As for the strain energy of the beam we include at this time only that energy which is due to bending. You will recall from Chap. 4 (see development of Eq. (4.8)) that

$$
U = \int_0^L \frac{EI}{2}\left(\frac{\partial^2 w}{\partial x^2}\right)^2 dx
\tag{7.7}
$$

The Lagrangian that we shall employ is then:

$$
\begin{aligned}
L &= \int_0^L \frac{\rho A}{2}\left(\frac{\partial w}{\partial t}\right)^2 dx - \int_0^L \frac{EI}{2}\left(\frac{\partial^2 w}{\partial x^2}\right)^2 dx + \int_0^L q(x,t)\, w(x,t)\, dx \\
&= \int_0^L \left[\frac{\rho A}{2}(\dot{w})^2 - \frac{EI}{2}(w_{xx})^2 + qw\right] dx
\end{aligned}
$$

---

[1] This should be easily seen by noting that ρA is proportional to the depth of the beam while ρI is proportional to the depth cubed. However, it should be noted that the rotatory inertia may not be negligible even for thin beams when complex mode shapes (to be described soon) at high frequencies are involved.

where $\dot{w} = \partial w / \partial t$ and $w_{xx} = \partial^2 w / \partial x^2$. Hamilton's principle then requires that:

$$\delta^{(1)} \int_{t_1}^{t_2} \int_0^L \left[\frac{\rho A}{2} (\dot{w})^2 - \frac{EI}{2} (w_{xx})^2 + qw \right] dx\, dt = 0 \qquad (7.8)$$

Denoting the integrand as F and choosing a one parameter family of admissible functions $\tilde{w}(x,t) = w(x,t) + \epsilon\eta(x,t)$ wherein $\eta(x,t_1) = \eta(x,t_2) = 0$, as required by Hamilton's principle, we have for Eq. 7.8:[1]

$$\left\{ \frac{d}{d\epsilon} \left(\int_{t_1}^{t_2} \int_0^L \tilde{F}\, dx\, dt \right) \right\}_{\epsilon=0} = 0$$

therefore

$$\int_{t_1}^{t_2} \int_0^L \left(\frac{\partial F}{\partial \dot{w}} \dot{\eta} + \frac{\partial F}{\partial w_{xx}} \eta_{xx} + \frac{\partial F}{\partial w} \eta \right) dx\, dt = 0 \qquad (7.9)$$

Integrate the first expression by parts with respect to time. Thus:

$$\int_{t_1}^{t_2} \int_0^L \frac{\partial F}{\partial \dot{w}} \frac{\partial \eta}{\partial t}\, dx\, dt = \int_0^L \left[\eta \frac{\partial F}{\partial \dot{w}} \right]_{t_1}^{t_2} dx - \int_{t_1}^{t_2} \int_0^L \eta \frac{\partial}{\partial t} \left(\frac{\partial F}{\partial \dot{w}} \right) dx\, dt$$

Since $\eta = 0$ at t_1 and t_2 we may delete the first expression on the right side of the above equation. Now integrate the second expression in Eq. 7.9 twice by parts with respect to x as follows

$$\int_{t_1}^{t_2} \int_0^L \frac{\partial F}{\partial w_{xx}} \eta_{xx}\, dx\, dt = \int_{t_1}^{t_2} \left[\frac{\partial F}{\partial w_{xx}} \eta_x \right]_0^L dt - \int_{t_1}^{t_2} \int_0^L \frac{\partial}{\partial x} \frac{\partial F}{\partial w_{xx}} \eta_x\, dx\, dt$$

$$= \int_{t_1}^{t_2} \left[\frac{\partial F}{\partial w_{xx}} \eta_x \right]_0^L dt - \int_{t_1}^{t_2} \left[\frac{\partial}{\partial x} \left(\frac{\partial F}{\partial w_{xx}} \right) \eta \right]_0^L dt$$

$$+ \int_{t_1}^{t_2} \int_0^L \frac{\partial^2}{\partial x^2} \left(\frac{\partial F}{\partial w_{xx}} \right) \eta\, dx\, dt$$

Substituting these results back into Eq. (7.9) we get:

$$\int_{t_1}^{t_1} \int_0^L \left\{ -\frac{\partial}{\partial t} \left(\frac{\partial F}{\partial \dot{w}} \right) + \frac{\partial^2}{\partial x^2} \left(\frac{\partial F}{\partial w_{xx}} \right) + \frac{\partial F}{\partial w} \right\} \eta\, dx\, dt + \int_{t_1}^{t_2} \left[\frac{\partial F}{\partial w_{xx}} \eta_x \right]_0^L dt$$

$$- \int_{t_1}^{t_2} \left[\frac{\partial}{\partial x} \left(\frac{\partial F}{\partial w_{xx}} \right) \eta \right]_0^L dt = 0$$

---

[1] Since we derived Hamilton's principle using the delta operator we shall work here with a single parameter family representation and will ask you to develop the same results using the operator approach.

Since η and η_x could possibly be zero at $x = 0$, and $x = L$, we conclude from the above formulation that the following equation

$$\frac{\partial}{\partial t}\left(\frac{\partial F}{\partial \dot{w}}\right) - \frac{\partial^2}{\partial x^2}\left(\frac{\partial F}{\partial w_{xx}}\right) - \frac{\partial F}{\partial w} = 0 \qquad (7.10)$$

must necessarily be satisfied along the beam. Inserting the expression $[(\rho A/2)\dot{w}^2 - (EI/2)(w_{xx})^2 + qw]$ for F, we then get the basic equation of motion for the beam

$$\boxed{\rho A\ddot{w} + \frac{\partial^2}{\partial x^2}\left[EI\left(\frac{\partial^2 w}{\partial x^2}\right)\right] = q} \qquad (7.11)$$

The boundary conditions at $x = 0$ and $x = L$ clearly are

$$\text{EITHER} \quad \frac{\partial F}{\partial w_{xx}} = 0 \qquad \text{OR} \quad w_x \text{ IS PRESCRIBED}$$

$$\text{EITHER} \quad \frac{\partial}{\partial x}\left(\frac{\partial F}{\partial w_{xx}}\right) = 0 \quad \text{OR} \quad w \text{ IS PRESCRIBED} \qquad (7.12)$$

Inserting the function F and going back to the usual notation for derivatives we get for these end conditions

$$\text{EITHER} \quad \frac{\partial^2 w}{\partial x^2} = 0 \qquad \text{OR} \quad \frac{\partial w}{\partial x} \text{ IS PRESCRIBED}$$

$$\text{EITHER} \quad \frac{\partial}{\partial x}\left(EI\frac{\partial^2 w}{\partial x^2}\right) = 0 \quad \text{OR} \quad w \text{ IS PRESCRIBED} \qquad (7.13)$$

It is to be noted that the boundary conditions can be time-dependent.

We shall now investigate the free vibration of beams using the simplified equations presented in this section. Later we shall present an improved theory accounting for transverse shear and rotatory inertia when we consider the vibrations of the so-called "Timoshenko beam."

7.4 FREE VIBRATIONS OF A SIMPLY-SUPPORTED BEAM

With no external loads the differential equation of motion for the beam is given as:

$$\frac{\partial^2}{\partial x^2}\left[EI\frac{\partial^2 w}{\partial x^2}\right] + \rho A\frac{\partial^2 w}{\partial t^2} = 0 \qquad (7.14)$$

We employ a *separation of variables* approach by expressing w as the product of a function W of x and a function T of t. Thus:

$$w = W(x)T(t) \qquad (7.15)$$

Substitution into Eq. (7.14) and dividing by WT gives us the following result:

$$-\frac{1}{\rho A}\left\{\frac{\partial^2}{\partial x^2}\left(EI\frac{\partial^2 w/\partial x^2}{W}\right)\right\} = \left[\frac{\partial^2 T/\partial t^2}{T}\right] \qquad (7.16)$$

Since each side is separately a function of a different variable we set each side equal to a constant $-\omega^2$, in the familiar manner for this technique. We get two ordinary differential equations as a result:

$$\frac{d^2 T}{dt^2} + \omega^2 T = 0 \qquad (a)$$

$$\frac{d^2}{dx^2}\left[EI\frac{d^2 W}{dx^2}\right] - (\rho A \omega^2)W = 0 \qquad (b) \quad (7.17)$$

The general solution to the equations is given as follows for the case where EI and ρA are constant

$$W = G \cosh kx + B \sinh kx + C \cos kx + D \sin kx \qquad (a)$$

$$T = E \sin \omega t + F \cos \omega t \qquad (b) \quad (7.18)$$

where

$$k = \left(\frac{\rho A \omega^2}{EI}\right)^{1/4} \qquad (7.19)$$

and where G, H, C, D, E, and F are integration constants.

We may satisfy the end conditions for this problem by using at each end a natural boundary condition $\partial^2 w/\partial x^2 = 0$ to give a zero moment there (see Eq. (4.16)), and a kinematic condition $w = 0$. We thus have the following equations as a result of imposing these conditions on the solution WT as given by Eqs. (7.18):

$$G + H = 0$$

$$k^2(G - H) = 0$$

$$G \cosh kL + B \sinh kL + H \cos kL + D \sin kL = 0$$

$$k^2(G \cosh kL + B \sinh kL - H \cos kL - D \sin kL) = 0$$

The first two equations immediately indicate on inspection that $G = H = 0$. We thus have for the remaining equations:

$$B \sinh kL + D \sin kL = 0$$

$$B \sinh kL - D \sin kL = 0 \qquad (7.20)$$

For a nontrivial solution the determinant formed by the constants B and D in the above equations must be zero. This results in the following equation called the *frequency or characteristic equation*:

$$(\sin kL)(\sinh kL) = 0 \qquad (7.21)$$

We shall rule out the possibility that $k = 0$ to satisfy this equation. Our reason for this is that $k = 0$ implies $\omega = 0$ from Eq. (7.19), and a zero value of ω requires from Eq. (7.18(b)) that T is a constant so that $w = WT$ is then independent of time. This can only mean here that the bar is at rest[1] and we have a trivial result. The only other possibility then is that

$$kL = n\pi \qquad n = 1, 2, \ldots \qquad (7.22)$$

so as to get the sine function equal to zero. We thus have an infinite discrete set of possible values for ω^2 which from Eqs. (7.22) and (7.19) are given as follows:

$$\omega_n = \left(\frac{n\pi}{L}\right)^2 \left(\frac{EI}{\rho A}\right)^{1/2} \qquad n = 1, 2, \ldots \qquad (7.23)$$

These are the *natural frequencies* or *eigenvalues* of the beam. Substituting the allowed values of kL, namely $n\pi$, into Eqs. (7.20) we see that the constant B must now be zero to satisfy these equations. We then have as possible solutions for W:

$$W_n = D_n\left[\sin\frac{n\pi x}{L}\right] \qquad n = 1, 2, \ldots \qquad (7.24)$$

The function $\sin n\pi x/L$ is termed the *n*th *mode shape* or the *n*th *eigenfunction* corresponding *respectively* to the *n*th *natural frequency* or the *n*th *eigenvalue* of the beam. A possible solution w_n for the problem is accordingly

$$w_n(x,t) = \left(\sin\frac{n\pi x}{L}\right)[A_n \cos \omega_n t + B_n \sin \omega_n t] \qquad (7.25)$$

wherein D_n has been incorporated with the arbitrary constants in Eq. (7.18(b)) to form A_n and B_n. Note with the aid of a phasor diagram we may replace the second bracketed expression with a single harmonic function as follows:

$$w_n(x,t) = \left(\sin\frac{n\pi x}{L}\right)[C_n \cos(\omega_n t + \alpha_n)] \qquad (7.26)$$

where C_n and α_n are arbitrary constants replacing A_n and B_n.[2] From this we can see that the centerline deformation of the beam has the shape of the sinusoid $\sin n\pi x/L$

---

[1] Since $w = WT = \text{Const.}$ is also a possible solution, the condition $k = 0$ can then represent a rigid-body movement. This movement can be of interest in the beam with free end conditions—the so-called free–free beam. However, in this case we have simple pin supports at the ends and rigid-body movement must be ruled out.

[2] Note that $C_n = \sqrt{A_n{}^2 + B_n{}^2}$ and $\alpha_n = \tan^{-1} A_n/B_n$.

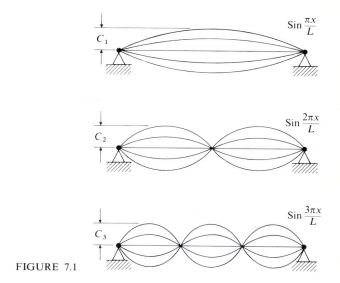

FIGURE 7.1

with an amplitude q_n which is a harmonic function of time given as:

$$q_n = C_n \cos(\omega_n t + \alpha_n)$$

This motion is shown for $n = 1,2,3$ in Fig. 7.1. The motion depicted by Eqs. (7.25) and (7.26) is that of the beam moving from one extreme configuration for the nth mode to the other (as denoted in the diagram for the first three modes) with a frequency ω_n. The general vibration of the beam is a superposition of all such modes of vibration each having constants C_n and α_n (or A_n and B_n). That is:

$$w(x,t) = \sum_{n=1}^{\infty} \left(\sin \frac{n\pi x}{L}\right)[C_n \cos(\omega_n t + \alpha_n)] \qquad (a)$$

$$w(x,t) = \sum_{n=1}^{\infty} \left(\sin \frac{n\pi x}{L}\right)[A_n \cos \omega_n t + B_n \sin \omega_n t] \qquad (b) \quad (7.27)$$

The infinite set of constants C_n and α_n or A_n and B_n must be determined so that the initial conditions of the problem are satisfied, namely

$$w(x,0) = \phi(x)$$

$$\frac{\partial w}{\partial t}(x,0) = \psi(x) \qquad (7.28)$$

where ϕ and ψ are given functions. Thus using Eq. (7.27(b)) we have:

$$\phi = \sum_{n=1}^{\infty} \sin \frac{n\pi x}{L}[A_n] \qquad (a)$$

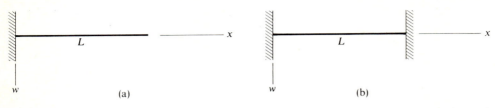

(a)

(b)

FIGURE 7.2

$$\psi = \sum_{n=1}^{\infty} \sin \frac{n\pi x}{L} [\omega_n] B_n \qquad (b)$$

The technique, you will recall, is to multiply in each case by $\sin (n\pi x/L)$ and integrate from 0 to L. Noting that[1]

$$\int_0^L \sin \frac{n\pi x}{L} \sin \frac{m\pi x}{L} dx = (L/2) \, \delta_{mn}$$

where δ_{mn} is the Kronecker delta, we may readily solve for A_m and B_m as follows:

$$A_m = \frac{2}{L} \int_0^L \phi(x) \sin \frac{m\pi x}{L} dx \qquad (a)$$

$$B_m = \frac{2}{L\omega_n} \int_0^L \psi(x) \sin \frac{m\pi x}{L} dx \qquad (b) \quad (7.29)$$

The free vibrations of the simply-supported beam are now fully determined.

We may readily solve for free vibrations of beams under different end conditions. As exercises you will be asked to verify the following results:

(a)   Cantilevered Beam (Fig. 7.2(a))

$$\omega_n = k_n^2 \left(\frac{EI}{\rho A} \right)^{1/2} \qquad (a)$$

where k_n is determined by the following frequency equation:

$$\cos k_n L \cosh k_n L = -1 \qquad (b)$$

The eigenfunctions are:

$$W_n(x) = \left\{ \cosh k_n x - \cos k_n x - \frac{\cos k_n L + \cosh k_n L}{\sin k_n L + \sinh k_n L} [\sinh k_n x - \sin k_n x] \right\} \qquad (c) \quad (7.30)$$

---

[1] We will discuss orthogonality of eigenfunctions later in the chapter in greater generality.

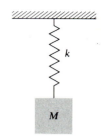

FIGURE 7.3

(b)   Beam Clamped at Both Ends (Fig. 7.2(b)) [1]

$$\omega_n{}^2 = k_n{}^2\left(\frac{EI}{\rho A}\right) \qquad (a)$$

where k_n is determined by the following frequency equation:

$$\cos kL \cosh kL = 1 \qquad (b) \qquad (7.31)$$

The eigenfunctions are:

$$W_n(x) = \left\{\cosh k_n x - \cos k_n x - \frac{\cos k_n L - \cosh k_n L}{\sin k_n L - \sinh k_n L}[\sinh k_n x - \sin k_n x]\right\}$$

Since the knowledge of the lowest natural frequency of a beam is of importance in engineering applications, we shall now present a powerful technique for approximating this frequency of beams. This technique will be particularly useful when the beams do not have uniform geometry and mechanical properties (that is, I and/or E are not constant but are functions of position) so that analytical approaches of the type presented here are not straightforward and simple.

7.5   RAYLEIGH'S METHOD FOR BEAMS

We shall now present an approximate method for determining the natural frequency ω_1 for beams. Recall from the free vibration of an elementary spring-mass system (see Fig. 7.3) that the potential energy of the spring has its greatest value when the kinetic energy of the mass is zero and vice versa. Because of conservation of mechanical energy one can then say: [2]

$$T_{\max} = U_{\max}$$

---

[1] For a free–free beam, the frequency equation is the same as that for a beam clamped at both ends (i.e., the clamped–clamped beam). See exercises.

[2] The same results are obtainable by equating mean kinetic and potential energies.

therefore

$$\tfrac{1}{2}M\dot{x}_{max}^2 = \tfrac{1}{2}kx_{max}^2 \qquad (7.32)$$

Since the system is vibrating sinusoidally we can say furthermore:

$$x = A \sin \omega t$$

therefore

$$\dot{x} = A\omega \cos \omega t$$

Accordingly we conclude:

$$x_{max} = A; \qquad \dot{x}_{max} = A\omega$$

Now going back to Eq. (7.32) we have:

$$\tfrac{1}{2}M(A\omega)^2 = \tfrac{1}{2}k(A^2)$$

therefore

$$\omega = \sqrt{\frac{k}{M}}$$

We are thus able to determine the natural frequency ω.

For a beam vibrating in a natural mode $W_n(x)$ with frequency ω_n we have similarly a kinetic energy

$$T = \int_0^L \frac{\rho A}{2}(\dot{w}_n^2)\, dx$$

and a strain energy:

$$U = \int_0^L \frac{EI}{2}(w_n'')^2\, dx$$

Since $w_n = [W_n]C_n \sin(\omega_n t + \alpha_n)$ we have,

$$T = \tfrac{1}{2}\int_0^L (\rho A)[C_n^2\omega_n^2 \cos^2(\omega_n t + \alpha_n)W_n^2]\, dx$$

$$U = \tfrac{1}{2}\int_0^L EI\, C_n^2(W_n'')^2 \sin^2(\omega_n t + \alpha_n)\, dx$$

For reasons of conservation of mechanical energy we equate the maximum values of T and U, as in the earlier case, to form:

$$T_{max} = \tfrac{1}{2}\omega_n^2 C_n^2 \int_0^L \rho A W_n^2\, dx = U_{max} = \tfrac{1}{2}C_n^2 \int_0^L EI(W_n'')^2\, dx$$

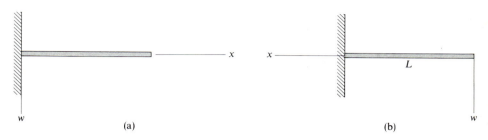

FIGURE 7.4

From this we may solve for $\omega_n{}^2$ as follows:

$$\omega_n{}^2 = \frac{\int_0^L (EI)(W_n'')^2 \, dx}{\int_0^L (\rho A)(W_n)^2 \, dx} \qquad (7.33)$$

The expression on the right side is the so-called *Rayleigh quotient* applied to beams. With the correct eigenfunction W_n in the quotient, clearly we get the proper eigenvalue $\omega_n{}^2$. We now demonstrate that we may get good approximate results for eigenvalue $\omega_1{}^2$ if in place of W_1 for the first mode, we use a function, $\tilde{W}_1$, that on the one hand satisfies the boundary conditions of the problem and on the other hand reasonably resembles the first mode function W_1. Since the mode shape W_1 corresponding to the lowest eigenvalue is usually the only one that can readily be approximated as described above (one can use the *statical deflection curve* for such purposes), we have accordingly restricted the approach to this case. Thus by using $\tilde{W}_1$ in the Rayleigh quotient we obtain an approximate eigenvalue $\tilde{\omega}_1{}^2$. We will show (Part C) that Rayleigh's quotient is a functional which has an *extreme* with respect to admissible functions $\tilde{W}_1$ satisfying the boundary conditions, the extremal function being the eigenfunction W_1. This means that by using a function $\tilde{W}_1$ in the Rayleigh quotient such that $(\tilde{W}_1 - W_1)$ has a small average value over the beam we will result in even a *smaller* value for $(\tilde{\omega}_1{}^2 - \omega_1{}^2)$. This is the reason for the success of the Rayleigh method.[1] Furthermore we shall now demonstrate and later prove (Part C) that the approximation of ω_1 is always *from above*, i.e., $(\tilde{\omega}_1{}^2 - \omega_1{}^2) \geq 0$.

EXAMPLE 7.1   We now illustrate the use of the Rayleigh quotient by estimating the fundamental frequency of a cantilever beam (see Fig. 7.4(a)). We choose the following function to approximate the fundamental mode:

$$W_{\text{app}} = \frac{\rho A}{24EI}[x^4 - 4Lx^3 + 6L^2x^2] \qquad (7.34)$$

---

[1] This is analogous to an extremal condition for a function $y(x)$ at $x = a$. A small departure from the extremal position "a" (the departure being analogous to $(\tilde{W}_1 - W_1)_{av}$) results in an even smaller change in $y(x)$ (the change being analogous to $(\tilde{\omega}_1{}^2 - \omega_1{}^2)$).

You may readily show that this is the static deflection equation for the cantilever beam and accordingly satisfies the boundary conditions of the problem. Substituting this function into the Rayleigh quotient gives us

$$\omega = \frac{3.53}{L^2} \sqrt{\frac{EI}{\rho A}} \qquad (7.35)$$

Now going back to the exact solution of the cantilever problem we may show by a process of trial and error, with the aid of a plot, that the smallest root of the frequency equation (Eq. 7.30(b)) is:

$$k_1 L = 1.875$$

Hence from Eq. (7.30(a)) we have

$$\omega_1 = \frac{(1.875)^2}{L^2} \sqrt{\frac{EI}{\rho A}} = \frac{3.52}{L^2} \sqrt{\frac{EI}{\rho A}} \qquad (7.36)$$

The closeness of the result speaks for itself.

As a second approximation we assume:

$$W_{\text{app}} = a\left(1 - \cos\frac{\pi x}{2L}\right) \qquad (7.37)$$

where "a" is an undetermined constant. This function satisfies both the geometric boundary condition at the clamped end at $x = 0$ (i.e., $W(0, t) = \partial W(0, t)/\partial x = 0$) and also satisfies one of the two natural boundary conditions at the free end ($\partial^2 W(L, t)/\partial x^2 = 0$). Now substituting this approximation into the Rayleigh quotient we get the following result:[1]

$$\omega = \frac{(1.915)^2}{L^2} \cdot \sqrt{\frac{EI}{\rho A}} = \frac{3.667}{L^2} \sqrt{\frac{EI}{\rho A}}$$

While still a reasonably good approximation it is not as good as the one found using the static deflection curve. Note that in both cases we have approximations from above as noted (without proof) earlier. ////

As a final step in this section we now investigate the significance of the rotatory inertia term neglected in the development of the equations of motion of the beam in Sec. 7.3. We consider for this purpose the simply-supported beam whose natural frequencies we have already determined. The Rayleigh quotient with the rotatory inertia included becomes (as you may demonstrate):

$$\omega^2 = \frac{EI \int_0^L [W''(x)]^2 \, dx}{\rho A \int_0^L [W(x)]^2 \, dx + \rho I \int_0^L [W'(x)]^2 \, dx} \qquad (7.38)$$

---

[1] This example demonstrates that we can still get acceptable results if not all the natural boundary conditions are satisfied at an endpoint of the beam.

We take as the approximate function $\sin \pi x/L$ which is the exact fundamental mode shape of the solution with no rotatory inertia.[1] By a simple calculation we get:

$$\omega^2 = \frac{EI\pi^4/\rho AL^4}{[1 + (\pi^2/12)(h/L)^2]}$$

The numerator represents the fundamental frequency from the elementary theory. We then conclude that for long thin beams ($(h/L) \ll 1$) the effect of rotatory inertia is only a slight reduction of the fundamental frequency from the elementary result. For such beams and for low-order mode shapes we are then justified in neglecting rotatory inertia. (In a later section we shall consider rotatory inertia and transverse shear in the Timoshenko beam.)

7.6 RAYLEIGH–RITZ METHOD FOR BEAMS

In the previous section we used the Rayleigh quotient to give an upper bound to ω_1 the fundamental frequency of vibrations of a beam. Although we presented the Rayleigh quotient in terms of beams the principles set forth apply to any free vibration of a linear elastic body. Indeed we shall use the Rayleigh quotient later in this chapter for estimating the fundamental natural frequency of certain plate problems and in part C of the chapter we shall present a rather general discussion of the Rayleigh quotient with supporting arguments for the conclusions of the previous section, substantiated up to now only by examples.

We shall now employ the Ritz approach of Chap. 3 in conjunction with the Rayleigh quotient to form the very valuable Rayleigh–Ritz method. Instead of employing a single function ϕ for the Rayleigh quotient we shall now select a set of n linearly independent functions[2] ϕ_i, each satisfying the boundary conditions of the problem, to represent W_{app} as a parameter-laden sum as follows

$$W_{\text{app}} = A_1\phi_1 + \cdots + A_n\phi_n \qquad (7.39)$$

where A_i are undetermined constants. Now substitute this approximate function into the Rayleigh quotient and denoting it as Λ^2 we get:

$$\Lambda^2 = \frac{\displaystyle\int_0^L EI\left[\sum_{i=1}^n A_i\phi_i''\right]^2 dx}{\displaystyle\int_0^L \rho A\left[\sum_{i=1}^n A_i\phi_i\right]^2 dx} \qquad (7.40)$$

[1] Actually the exact solution for the fundamental mode with rotatory inertia is the same as the exact solution for the simpler theory.

[2] A linearly independent set of functions ϕ_i can sum up to zero identically ($A_i\phi_i = 0$) only if all the coefficients A_i are zero.

Introducing the following notation:

$$a_{ij} = \int_0^L EI\phi_i''\phi_j'' \, dx$$

$$D_{ij} = \int_0^L \rho A \phi_i \phi_j \, dx \qquad (7.41)$$

we get for Λ^2:

$$\Lambda^2 = \frac{\displaystyle\sum_{i=1}^{n}\sum_{j=1}^{n} a_{ij} A_i A_j}{\displaystyle\sum_{i=1}^{n}\sum_{j=1}^{n} D_{ij} A_i A_j} \qquad (7.42)$$

Now we extremize Λ^2 with respect to the coefficients A_i as follows:

$$\frac{\partial(\Lambda^2)}{\partial A_i} = 0$$

therefore

$$\frac{2\displaystyle\sum_{j=1}^{n} a_{ij} A_j}{\displaystyle\sum_{i=1}^{n}\sum_{j=1}^{n} D_{ij} A_i A_j} - \frac{2\left(\displaystyle\sum_{i=1}^{n}\sum_{j=1}^{n} a_{ij} A_i A_j\right)\left(\displaystyle\sum_{j=1}^{n} D_{ij} A_j\right)}{\left(\displaystyle\sum_{j=1}^{n}\sum_{i=1}^{n} D_{ij} A_i A_j\right)^2} = 0 \qquad i = 1, 2, \ldots, n$$

Canceling $2/\sum_{i=1}^{n}\sum_{j=1}^{n} D_{ij} A_i A_j$ and employing Eq. (7.42) in the second term we get:

$$\sum_{j=1}^{n} a_{ij} A_j - \Lambda^2 \sum_{j=1}^{n} D_{ij} A_j = 0$$

therefore

$$\sum_{j=1}^{n} [a_{ij} - \Lambda^2 D_{ij}] A_j = 0 \qquad i = 1, 2, \ldots, n \qquad (7.43)$$

We thus have a homogeneous system of n equations for the n constants A_j. For a nontrivial solution the determinant of the coefficients must be zero. We thus get the frequency equation familiar from vibrations of discrete systems:

$$|a_{ij} - \Lambda^2 D_{ij}| = 0 \qquad (7.44)$$

The satisfaction of the above determinant will yield an nth-order equation in Λ^2 having n roots $\Lambda_1^2, \Lambda_2^2, \ldots, \Lambda_n^2$ which we arrange in ascending order of magnitude. For each root Λ_j^2 we can determine from Eq. (7.43) the ratios of a set of constants

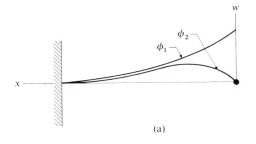
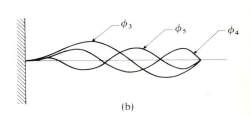

FIGURE 7.5

$A_i^{(j)}$ so that we form a function $W_{app}^{(j)}$ as follows:

$$W_{app}^{(j)} = \sum_{i=1}^{n} A_i^{(j)} \phi_i \qquad j = 1, 2, \ldots, n \qquad (7.45)$$

We will demonstrate now by example and later prove that the roots Λ_j^2 *are approximations from above of the first n natural frequencies of the system.* And Eq. (7.45) represents *approximations of the first n mode shapes of the system.* (The latter will not be nearly as good approximations as the former.)

What we are doing here is representing a bar with infinite number of degrees of freedom by one with n degrees of freedom. Naturally the more functions ϕ_i that are employed, the better should the approximation Λ_j^2 be for the natural frequencies. The method is most successful when we have a general notion of the shape of the modes so that the functions ϕ_i can be chosen to reasonably resemble them.

EXAMPLE 7.2   We go back to the cantilever beam of Example 7.1 (see Fig. 7.4(*b*)) and we shall employ for the Rayleigh–Ritz method two functions:[1]

$$\phi_1 = \left(1 - \frac{x}{L}\right)^2$$

$$\phi_2 = \frac{x}{L}\left(1 - \frac{x}{L}\right)^2$$

These functions both satisfy the kinematic boundary conditions of the problem at the support and have a resemblance to what can be expected for the first and second mode shapes. (See Fig. 7.5(*a*).) We form the approximate function W_{app} as follows:

$$W_{app} = A_1\left(1 - \frac{x}{L}\right)^2 + A_2\frac{x}{L}\left(1 - \frac{x}{L}\right)^2$$

[1] As in the Rayleigh method, this example shows that good results can often be found using functions ϕ_i that do not satisfy natural boundary conditions at one endpoint of the beam.

The constants a_{ij} and D_{ij} are then evaluated as follows:

$$a_{11} = EI \int_0^L (\phi_1'')^2 \, dx = EI \int_0^L \left(\frac{2}{L^2}\right)^2 dx = \frac{4EI}{L^3}$$

$$a_{21} = a_{12} = EI \int_0^L (\phi_1'')(\phi_2'') \, dx = EI \int_0^L \left(\frac{2}{L^2}\right)\left(\frac{2}{L^2}\right)\left(3\frac{x}{L} - 2\right) dx$$

$$= -\frac{2EI}{L^3}$$

$$a_{22} = EI \int_0^L (\phi_2'')^2 \, dx = EI \int_0^L \left(\frac{2}{L}\right)^2 \left(3\frac{x}{L} - 2\right)^2 dx = \frac{4EI}{L^3}$$

$$D_{11} = \rho A \int_0^1 \left(1 - \frac{x}{L}\right)^4 dx = \frac{\rho A}{5}\left(1 - \frac{x}{L}\right)^5 L \bigg|_0^L = \frac{\rho LA}{5}$$

$$D_{21} = D_{12} = \rho A \int_0^L \left(1 - \frac{x}{L}\right)^2 \left(\frac{x}{L}\right)\left(1 - \frac{x}{L}\right)^2 dx = \frac{\rho AL}{30}$$

$$D_{22} = \rho A \int_0^L \left(1 - \frac{x}{L}\right)^4 \left(\frac{x}{L}\right)^2 dx = \frac{\rho AL}{105}$$

The frequency determinant then becomes:

$$\begin{vmatrix} \left[\frac{4EI}{L^3} - \Lambda^2\left(\frac{\rho LA}{5}\right)\right] & \left[\frac{-2EI}{L^3} - \Lambda^2\left(\frac{\rho AL}{30}\right)\right] \\ \left[\frac{-2EI}{L^3} - \Lambda^2\left(\frac{\rho AL}{30}\right)\right] & \left[\frac{4EI}{L^3} - \Lambda^2\left(\frac{\rho AL}{105}\right)\right] \end{vmatrix} = 0$$

We then get in carrying out the determinant:

$$\left[\frac{4EI}{L^3} - \Lambda^2\left(\frac{\rho LA}{5}\right)\right]\left[\frac{4EI}{L^3} - \Lambda^2\left(\frac{\rho AL}{105}\right)\right] - \left[\frac{2EI}{L^3} + \Lambda^2\left(\frac{\rho AL}{30}\right)\right]^2 = 0$$

Letting $4EI/L^3 = \alpha$ and $\rho LA = \beta$ we have:

$$\left(\alpha - \frac{\beta}{5}\Lambda^2\right)\left(\alpha - \frac{\beta}{105}\Lambda^2\right) - \left(\frac{\alpha}{2} + \Lambda^2\frac{\beta}{30}\right)^2 = 0$$

We get the following equation in Λ from above:

$$\beta^2(0.000794)\Lambda^4 - \alpha\beta(0.243)\Lambda^2 + 0.75\alpha^2 = 0$$

The roots are:

$$\Lambda_1^2 = \frac{\alpha}{\beta}(3.15) = \frac{4EI}{\rho L^4 A}(3.15) = 12.60\frac{EI}{\rho L^4 A}$$

$$\Lambda_2^2 = \frac{\alpha}{\beta}(303) = \frac{4EI}{\rho L^4 A}(303) = 1212\frac{EI}{\rho L^4 A}$$

The results for the first two modes as developed exactly are:

$$\omega_1{}^2 = k_1{}^4\left(\frac{EI}{\rho AL^4}\right) = (1.875)^4\frac{EI}{\rho L^4 A} = 12.30\frac{EI}{\rho L^4 A}$$

$$\omega_2{}^2 = k_2{}^4\left(\frac{EI}{\rho AL^4}\right) = (4.694)^4\frac{EI}{\rho L^4 A} = 483\frac{EI}{\rho L^4 A}$$

While there is a very close approximation from above for the first mode we get a very poor result for the natural frequency of the second mode.

We now shall attempt an improvement in the evaluation of ω_2 by employing six functions having the following form:

$$\phi_1 = \left(1 - \frac{x}{L}\right)^2$$

$$\phi_2 = \frac{x}{L}\left(1 - \frac{x}{L}\right)^2$$

$$\phi_3 = \left(\frac{x}{L} - 0.5\right)\left(\frac{x}{L}\right)\left(1 - \frac{x}{L}\right)^2$$

$$\phi_4 = \left(\frac{x}{L} - 0.75\right)\left(\frac{x}{L} - 0.25\right)\left(\frac{x}{L}\right)\left(1 - \frac{x}{L}\right)^2$$

$$\phi_5 = \left(\frac{x}{L} - 0.2\right)\left(\frac{x}{L} - 0.5\right)\left(\frac{x}{L} - 0.8\right)\left(\frac{x}{L}\right)\left(1 - \frac{x}{L}\right)^2$$

$$\phi_6 = \left(\frac{x}{L} - 0.18\right)\left(\frac{x}{L} - 0.34\right)\left(\frac{x}{L} - 0.6\right)\left(\frac{x}{L} - 0.84\right)\frac{x}{L}\left(1 - \frac{x}{L}\right)^2$$

Figure 7.5(b) illustrates some of the new functions chosen. Notice they have increasing numbers of nodal points as is to be expected for ascending modes. The results are given (with the aid of a computer) in Table 7.1 where the first column indicates the number of functions used from the above list and where $\gamma = EI/\rho AL^4$. Notice the convergence from above of the results particularly for the first four modes. In Table 7.2 we show a comparison of the results next to the exact results. We can get a reasonably good result up to the fourth eigenvalue. It is clear from the computation here that it requires many functions and much work to get higher eigenvalues.

Table 7.1

n	$\dfrac{A_1{}^2}{\gamma}$	$\dfrac{A_2{}^2}{\gamma}$	$\dfrac{A_3{}^2}{\gamma}$	$\dfrac{A_4{}^2}{\gamma}$	$\dfrac{A_5{}^2}{\gamma}$	$\dfrac{A_6{}^2}{\gamma}$
2	12.480	1211.5				
3	12.400	494.3	13,958.1			
4	12.362	491.0	4012.8	79,296.4		
5	12.362	485.5	3999.3	16,517.2	316,640	
6	12.355	485.5	3780.7	16,507.3	51,790	1,030,864

Table 7.2

Approximate solutions		Exact solution	
$A_1^2/\gamma =$	12.36	$\omega_1^2/\gamma =$	12.36
$A_2^2/\gamma =$	486	$\omega_2^2/\gamma =$	485.48
$A_3^2/\gamma =$	3,781	$\omega_3^2/\gamma =$	3,807
$A_4^2/\gamma =$	16,507	$\omega_4^2/\gamma =$	14,617
$A_5^2/\gamma =$	51,791	$\omega_5^2/\gamma =$	39,944
$A_6^2/\gamma =$	1,030,684	$\omega_6^2/\gamma =$	173,881

////

7.7 THE TIMOSHENKO BEAM

As a final step in our study of the dynamics of beams we go to the Timoshenko beam where you will recall we include the effect of both transverse shear and rotatory inertia. From Eq. (4.30) for no axial force we assume the following displacement field for such beams

$$u_1(x,y,z,t) = -z\psi(x,t)$$

$$u_2(x,y,z,t) = 0$$

$$u_3(x,y,z,t) = w(x,t) \tag{7.46}$$

where the functions are now dependent on time. The non-zero strains are then:

$$\epsilon_{xx} = -z\frac{\partial\psi}{\partial x} \tag{a}$$

$$\epsilon_{xz} = \frac{1}{2}\left(\frac{\partial w}{\partial x} - \psi\right) \qquad (b) \tag{7.47}$$

Note that ϵ_{xz} is thus taken as constant over a section in the above approximation. Considering the cross-section of the beam to be rectangular $h \times b$ (see Fig. 7.6) as in the development of Chap. 4 and using the same approximation $\tau_{xx} = E\epsilon_{xx}$ as was employed earlier, we get for the bending moment at a section the following result with the aid of Eq. (7.47(a)):

$$M = \int_{-h/2}^{h/2} \tau_{xx}zb\,dz = -EI\frac{\partial\psi}{\partial x} \tag{7.48}$$

We can give the shear force V_s using Hooke's law, $\tau_{xz} = 2\epsilon_{xz}G$ and Eq. (7.47(b)) as follows:

$$V_s = \int_{-h/2}^{h/2} \tau_{xz}b\,dz = kGA\left(\frac{\partial w}{\partial x} - \psi\right) \tag{7.49}$$

where k is the shear constant that we insert as a correction factor to account for the fact that τ_{xz} is not (in reality) uniform over the height of the section.

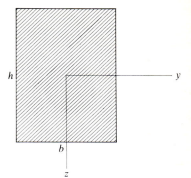

FIGURE 7.6

We may now express the energies that will be needed for Hamilton's principle. For kinetic energy we have:

$$T = \frac{1}{2} \int_{-h/2}^{h/2} \int_{-b/2}^{b/2} \int_0^L \rho \dot{u}_i \dot{u}_i \, dx \, dy \, dz$$

$$= \frac{1}{2} \int_{-h/2}^{h/2} \int_{-b/2}^{b/2} \int_0^L \rho \left[z^2 \left(\frac{\partial \psi}{\partial t} \right)^2 + \left(\frac{\partial w}{\partial t} \right)^2 \right] dx \, dy \, dz$$

$$= \frac{1}{2} \int_0^L \left[\rho I \left(\frac{\partial \psi}{\partial t} \right)^2 + \rho A \left(\frac{\partial w}{\partial t} \right)^2 \right] dx \qquad (7.50)$$

Note: we now retain the kinetic energy due to rotation of elements of the beam. As for the strain energy plus the potential energy of the loading q we employ Eq. (4.37) as follows:

$$\pi = U + V = \int_0^L \left[\frac{EI}{2} \left(\frac{\partial \psi}{\partial x} \right)^2 + \frac{kGA}{2} \left(\frac{\partial w}{\partial x} - \psi \right)^2 - qw \right] dx \qquad (7.51)$$

Accordingly, Hamilton's principle is given as

$$\delta^{(1)} \int_{t_1}^{t_2} L \, dt = \delta^{(1)} \int_{t_1}^{t_2} \int_0^L \left\{ \frac{1}{2} \left[\rho I \left(\frac{\partial \psi}{\partial t} \right)^2 + \rho A \left(\frac{\partial w}{\partial t} \right)^2 \right] - \frac{EI}{2} \left(\frac{\partial \psi}{\partial x} \right)^2 \right.$$

$$\left. - \frac{kGA}{2} \left(\frac{\partial w}{\partial x} - \psi \right)^2 \right] + qw \right\} dx \, dt = 0$$

The integrand F is composed of two functions to be varied, w and ψ. Carrying out the first variation we get:

$$\int_{t_1}^{t_2} \int_0^L \left\{ \rho I \frac{\partial \psi}{\partial t} \frac{\partial \, \delta \psi}{\partial t} + \rho A \frac{\partial w}{\partial t} \frac{\partial \, \delta w}{\partial t} - EI \frac{\partial \psi}{\partial x} \frac{\partial \, \delta \psi}{\partial x} \right.$$

$$\left. - kGA \left(\frac{\partial w}{\partial x} - \psi \right) \left(\frac{\partial \, \delta w}{\partial x} \right) + kGA \left(\frac{\partial w}{\partial x} - \psi \right) \delta \psi + q \, \delta w \right\} dx \, dt = 0 \qquad (7.52)$$

Integrating the first two terms by parts with respect to t and the third and fourth terms with respect to x we get:

$$\int_{t_1}^{t_2} \int_0^L \left\{ -\frac{\partial}{\partial t}(\rho I \dot{\psi})\, \delta\psi - \frac{\partial}{\partial t}(\rho A \dot{w})\, \delta w + \frac{\partial}{\partial x}\left(EI\frac{\partial\psi}{\partial x}\right)\delta\psi \right.$$
$$\left. + \frac{\partial}{\partial x}\left[kGA\left(\frac{\partial w}{\partial x} - \psi\right)\right]\delta w + kGA\left(\frac{\partial w}{\partial x} - \psi\right)\delta\psi + q\,\delta w \right\} dx\, dt$$
$$+ \int_0^L \rho I \frac{\partial\psi}{\partial t}\, \delta\psi \Big]_{t_1}^{t_2} dx + \int_0^L \rho A \frac{\partial w}{\partial t}\, \delta w \Big]_{t_1}^{t_2} dx$$
$$- \int_{t_1}^{t_2} EI\frac{\partial\psi}{\partial x}\, \delta\psi \Big]_{x_1}^{x_2} dt - \int_{t_1}^{t_2} kGA\left(\frac{\partial w}{\partial x} - \psi\right)\delta w \Big]_{x_1}^{x_2} dt = 0 \qquad (7.53)$$

Noticing that $\delta\psi = \delta w = 0$ at the times t_1 and t_2 we then get on grouping terms in the above equation:

$$\int_{t_1}^{t_2} \int_0^L \left[\left\{ -\frac{\partial}{\partial t}(\rho I \dot{\psi}) + \frac{\partial}{\partial x}\left(EI\frac{\partial\psi}{\partial x}\right) + kGA\left(\frac{\partial w}{\partial x} - \psi\right) \right\} \delta\psi \right.$$
$$\left. + \left\{ -\frac{\partial}{\partial t}(\rho A \dot{w}) + \frac{\partial}{\partial x}\left[kGA\left(\frac{\partial w}{\partial x} - \psi\right)\right] + q \right\} \delta w \right.$$
$$- \int_{t_1}^{t_2} EI\frac{\partial\psi}{\partial x}\, \delta\psi \Big]_{x_1}^{x_2} dt - \int_{t_1}^{t_2} kGA\left(\frac{\partial w}{\partial x} - \psi\right)\delta w \Big]_{x_1}^{x_2} dt = 0$$

The Euler–Lagrange equations are accordingly:

$$-\frac{\partial}{\partial t}(\rho A \dot{w}) + \frac{\partial}{\partial x}[(kGA)(w_x - \psi)] + q = 0$$

$$-\frac{\partial}{\partial t}(\rho I \dot{\psi}) + \frac{\partial}{\partial x}(EI\psi_x) + (kGA)(w_x - \psi) = 0 \qquad (7.54)$$

For constant values of E, I, G, A, etc., we then have:

$$\rho A \ddot{w} - kGA(w_{xx} - \psi_x) - q = 0 \qquad (a)$$
$$\rho I \ddot{\psi} - EI\psi_{xx} - kGA(w_x - \psi) = 0 \qquad (b) \quad (7.55)$$

The boundary conditions then become:

At the ends

$$\text{EITHER} \quad (w_x - \psi) = 0 \quad \text{OR} \quad w \text{ IS SPECIFIED}$$
$$\text{(i.e., } \eta_1 = 0)$$
$$\text{EITHER} \quad \psi_x = 0 \quad \text{OR} \quad \psi \text{ IS SPECIFIED}$$
$$\text{(i.e., } \eta_2 = 0) \qquad (7.56)$$

Noting Eqs. (7.48) and (7.49) we may express the above results in terms of V_s and M as follows

$$\frac{\partial V_s}{\partial x} = \rho A \ddot{w} - q$$

$$V_s - \frac{\partial M}{\partial x} = \rho I \ddot{\psi} \qquad (7.57)$$

where at the ends:

EITHER $V_s = 0$ OR w IS SPECIFIED

EITHER $M = 0$ OR ψ IS SPECIFIED (7.58)

We can uncouple Eqs. (7.55) in terms of the transverse displacement w:

$$EI\frac{\partial^4 w}{\partial x^4} + \rho A \frac{\partial^2 w}{\partial t^2} - \rho I \left(1 + \frac{E}{kG}\right)\frac{\partial^4 w}{\partial t^2\,\partial x^2} + \frac{\rho^2 I}{kG}\frac{\partial^4 w}{\partial t^4}$$

$$= q + \frac{\rho I}{kGA}\frac{\partial^2 q}{\partial t^2} - \frac{EI}{kGA}\frac{\partial^2 q}{\partial x^2} \qquad (7.59)$$

We shall now consider the free vibrations of a beam using Timoshenko beam theory. The appropriate form of Eq. (7.59) with $q = 0$ is restated as follows:

$$EI\frac{\partial^4 w}{\partial x^4} + \rho A \frac{\partial^2 w}{\partial t^2} - \rho I \left(1 + \frac{E}{kG}\right)\frac{\partial w^4}{\partial t^2 \partial x^2} + \frac{\rho^2 I}{kG}\frac{\partial^4 w}{\partial t^4} = 0 \qquad (7.60)$$

We assume a separation of variables for the above equation in the form

$$w(x,t) = W(x)\cos \omega t \qquad (7.61)$$

That is, we assume the bar is oscillating at a frequency ω with a mode shape $W(x)$. Substituting the above solution into Eq. (7.60) we get for W:

$$EIW^{\text{IV}} - \rho A \omega^2 W + \rho I \left(1 + \frac{E}{kG}\right)\omega^2 W'' + \frac{\rho^2 I}{kG}\omega^4 W = 0 \qquad (7.62)$$

The general solution to this ordinary differential equation has the form:

$$W = A \sin \lambda x + B \cos \lambda x + C \sinh \lambda x + D \cosh \lambda x$$

We take for the present discussion the case where the beam is simply-supported at both ends. We shall employ as boundary conditions for w the correct kinematical condition $w = 0$ at the ends and as an approximation will use here the classical natural boundary conditions $w'' = 0$ at the ends. These boundary conditions can be satisfied by taking $B = C = D = 0$ and having $\lambda = n\pi/L$ for $n = 1, 2, \ldots$. Solutions from these steps can then be given as follows:

$$W_n = A_n \sin \frac{n\pi x}{L} \qquad (7.63)$$

By substituting the above result back into Eq. 7.62 we may then determine the allowable values of ω_n to be associated with the eigenfunctions $\sin n\pi x/L$. Thus we get (on cancelling $A_n \sin n\pi x/L$):

$$EI\left(\frac{n\pi}{L}\right)^4 - \rho A \omega_n^2 - \rho I\left(1 + \frac{E}{kG}\right)\omega_n^2 \left(\frac{n\pi}{L}\right)^2 + \frac{\rho^2 I}{kG}\omega_n^4 = 0 \qquad (7.64)$$

Dividing through by $-EI/L^4$ and rearranging terms we get:

$$-\frac{\rho^2 L^4}{EkG}\omega_n^4 + \left[\frac{\rho A L^4}{EI} + \frac{\rho L^2}{E}\left(1 + \frac{E}{kG}\right)(n\pi)^2\right]\omega_n^2 - (n\pi)^4 = 0 \qquad (7.65)$$

Now introduce the following terms

$$\Omega_n^2 = \text{(dimensionless frequency)}^2 = \frac{\rho I}{EA}\omega_n^2 \quad (a)$$

$$r^2 = \text{(radius of gyration)}^2 = \frac{I}{A} \qquad (b) \quad (7.66)$$

into Eq. (7.65). We get:

$$-\frac{E}{kG}\left(\frac{L}{r}\right)^4 \Omega_n^4 + \left[\left(\frac{L}{r}\right)^4 + \left(\frac{L}{r}\right)^2\left(1 + \frac{E}{kG}\right)(n\pi)^2\right]\Omega_n^2 - (n\pi)^4 = 0 \qquad (7.67)$$

We would next like to show how our results can be altered so that we can omit shear and rotatory inertia and thus get back to earlier results for the case of the simply-supported beam according to classical beam theory. If we go back to Eqs. (7.50) and (7.51) we see that by letting $k = 0$ and letting $\rho I = 0$ we get the energy quantities used for the simple beam theory. We cannot use this procedure in Eq. (7.67) because of the various operations performed in arriving at the equation (such as dividing through by k). We can proceed *formally*, however, to get to the simple beam theory. If we set equal to zero in Eq. (7.67) the expression $-(E/kG)(L/r)^4$ as well as $(L/r)^2 \times (1 + E/kG)$ we arrive at the following result:

$$\left(\frac{L}{r}\right)^4 \Omega_n^2 - (n\pi)^4 = 0$$

therefore

$$\Omega_n^2 = \frac{n^4 \pi^4 r^4}{L^4} \qquad (7.68)$$

From Eq. (7.66(a)) we can solve for ω_n using the above result. Replacing r^2 by I/A we then have:

$$\omega_n^2 = \frac{(n\pi)^4}{L^4}\frac{EI}{\rho A} \qquad (7.69)$$

We recognize these frequencies as the natural frequency given by the simple beam theory found in Sec. 7.4 (see Eq. (7.23)). It should be clear that of the expressions deleted, the ones containing the term k relate to transverse shear and so if we wish formally to include *only rotatory inertia* we set equal to zero only the expressions

$$-\frac{E}{kG}\left(\frac{L}{r}\right)^4 \quad \text{and} \quad \left(\frac{L}{r}\right)^2\frac{E}{kG}.$$

We then have from Eq. (7.67) the result:

$$\left[\left(\frac{L}{r}\right)^4 + \left(\frac{L}{r}\right)^2(n\pi)^2\right]\Omega_n^2 - (n\pi)^4 = 0$$

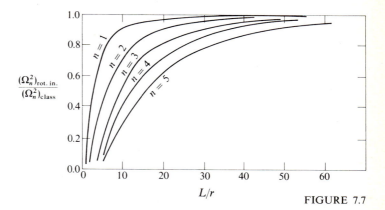

$$\frac{(\Omega_n^2)_{\text{rot. in.}}}{(\Omega_n^2)_{\text{class}}}$$

$$L/r$$

FIGURE 7.7

Hence the dimensionless frequency involving only rotatory inertia becomes

$$[\Omega_n^2]_{\substack{\text{rot} \\ \text{inertia}}} = \frac{(n\pi)^4}{(L/r)^4[1 + (n\pi)^2(r/L)^2]} = \frac{(\Omega_n^2)_{\text{class}}}{[1 + (n\pi)^2(r/L)^2]} \tag{7.70}$$

where $(\Omega_n^2)_{\text{class}}$ represents the dimensionless frequency for the classical beam in accordance with Eq. (7.68). To indicate the effects of rotatory inertia we have plotted $(\Omega_n^2)_{\text{rot inertia}}/(\Omega_n^2)_{\text{class}}$ as a function of (L/r) for various values of n as shown in Fig. 7.7. For small values of n and for large slenderness ratio L/r there is little difference between the classical frequency and that which includes rotatory inertia. However, for short stubby beams or for higher modes, the rotatory inertia can be significant.

Now if we wish only to include *shear effects* we again go back to Eq. (7.67) and this time we formally delete the term $(L/r)^2(n\pi)^2$ as well as the term $-(E/kG)(L/r)^4\Omega_n^4$ since the latter contains *both* shear effects and rotatory inertia as can be seen on tracing steps back to the original energy expressions for T and $U + V$. The frequency equation now becomes:

$$\left[\left(\frac{L}{r}\right)^4 + \left(\frac{E}{kG}\right)\left(\frac{L}{r}\right)^2(n\pi)^2\right](\Omega_n^2)_{\text{shear}} - (n\pi)^4 = 0$$

therefore

$$(\Omega_n^2)_{\text{shear}} = \frac{(n\pi)^4}{(L/r)^4[1 + (E/kG)(r/L)^2(n\pi)^2]}$$

$$= \frac{(\Omega_n^2)_{\text{class}}}{[1 + (E/kG)(r/L)^2(n\pi)^2]} \tag{7.71}$$

A plot of $(\Omega_n^2)_{\text{shear}}/(\Omega_n^2)_{\text{class}}$ versus L/r for $E/kG = 3.06$ and for various values of n has been included in Fig. 7.8. Note that for large values of n and for small values of L/r we get pronounced shear effects that must be taken into account.

We now turn to the full equation for Ω^4 including both shear and rotatory inertia effects. The frequency equation is too complicated to give useful information as it stands so we will proceed by giving Ω^2 as follows:

$$\Omega^2 = \Omega_{\text{class}}^2 + \delta^2 \tag{7.72}$$

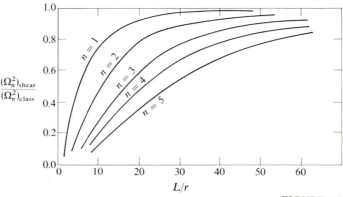

FIGURE 7.8

wherein we will in subsequent calculations neglect higher order terms in δ^2. This approach will be valid when there is not a large effect from rotatory inertia and shear. Accordingly substituting Eq. (7.72) into the frequency equation (7.67) we get:

$$-\left(\frac{E}{kG}\right)\left(\frac{L}{r}\right)^4(\Omega_{\text{class}}{}^4 + 2\delta^2\Omega_{\text{class}}{}^2) + \left[\left(\frac{L}{r}\right)^4 + \left(\frac{L}{r}\right)^2\left(1 + \frac{E}{kG}\right)(n\pi)^2\right][\Omega_{\text{class}}{}^2 + \delta^2] - (n\pi)^4 = 0$$

Solving for δ^2 from the above equation we have

$$\delta^2 = \frac{\left(\dfrac{E}{kG}\right)\left(\dfrac{L}{r}\right)^4\Omega_{\text{class}}{}^4 - \left[\left(\dfrac{L}{r}\right)^4 + \left(\dfrac{L}{r}\right)^2\left(1 + \dfrac{E}{kG}\right)(n\pi)^2\right]\Omega_{\text{class}}{}^2 + (n\pi)^4}{-2\dfrac{E}{kG}\left(\dfrac{L}{r}\right)^4\Omega_{\text{class}}{}^2 + \left[\left(\dfrac{L}{r}\right)^4 + \left(\dfrac{L}{r}\right)^2\left(1 + \dfrac{E}{kG}\right)(n\pi)^2\right]} \qquad (7.73)$$

Hence going back to Eq. (7.72) and using Eq. (7.73) we have:

$$\Omega^2 = \Omega_{\text{class}}{}^2 + \delta^2 = \Omega_{\text{class}}{}^2\left[1 + \frac{\delta^2}{\Omega_{\text{class}}{}^2}\right]$$

$$= \Omega_{\text{class}}{}^2\left[1 + \frac{\left(\dfrac{E}{kG}\right)\left(\dfrac{L}{r}\right)^4\Omega_{\text{class}}{}^2 - \left[\left(\dfrac{L}{r}\right)^4 + \left(\dfrac{L}{r}\right)^2\left(1 + \dfrac{E}{kG}\right)(n\pi)^2\right] + \dfrac{(n\pi)^4}{\Omega_{\text{class}}{}^2}}{-2\dfrac{E}{kG}\left(\dfrac{L}{r}\right)^4\Omega_{\text{class}}{}^2 + \left[\left(\dfrac{L}{r}\right)^4 + \left(\dfrac{L}{r}\right)^2\left(1 + \dfrac{E}{kG}\right)(n\pi)^2\right]}\right]$$

$$= \Omega_{\text{class}}{}^2\left[\frac{-\dfrac{E}{kG}\left(\dfrac{L}{r}\right)^4\Omega_{\text{class}}{}^2 + (n\pi)^4/\Omega_{\text{class}}{}^2}{-2\dfrac{E}{kG}\left(\dfrac{L}{r}\right)^4\Omega_{\text{class}}{}^2 + \left[\left(\dfrac{L}{r}\right)^4 + \left(\dfrac{L}{r}\right)^2\left(1 + \dfrac{E}{kG}\right)(n\pi)^2\right]}\right]$$

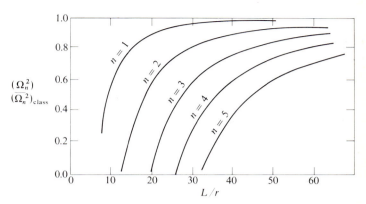

FIGURE 7.9

Using Eq. (7.68) for Ω_n^2 in the bracketed expression we get:

$$\Omega^2 = \Omega_{\text{class}}^2 \left[\frac{-\dfrac{E}{kG}\left(\dfrac{L}{r}\right)^4 (n\pi)^4 \left(\dfrac{r}{L}\right)^4 + (n\pi)^4 \left(\dfrac{L}{r}\right)^4 (1/(n\pi)^4)}{-2\dfrac{E}{kG}\left(\dfrac{L}{r}\right)^4 (n\pi)^4 \left(\dfrac{r}{L}\right)^4 + \left[\left(\dfrac{L}{r}\right)^4 + \left(\dfrac{L}{r}\right)^2 \left(1 + \dfrac{E}{kG}\right)(n\pi)^2\right]} \right]$$

Canceling terms and multiplying numerator and denominator by $(r/L)^4$ we get:

$$\Omega^2 = \Omega_{\text{class}}^2 \left[\frac{1 - \dfrac{E}{kG}(n\pi)^4 \left(\dfrac{r}{L}\right)^4}{1 + \left(1 + \dfrac{E}{kG}\right)(n\pi)^2 \left(\dfrac{r}{L}\right)^2 - 2\dfrac{E}{kG}(n\pi)^4 \left(\dfrac{r}{L}\right)^4} \right] \tag{7.74}$$

For the case where $r/L < 1$ we may give the following simplification by dropping terms with $(r/L)^4$:

$$\Omega^2 \approx \Omega_{\text{class}}^2 \left[\frac{1}{1 + (1 + E/kG)(n\pi)^2 (r/L)^2} \right]$$

$$\approx \Omega_{\text{class}}^2 \left[1 - \left(1 + \frac{E}{kG}\right)\left(\frac{r}{L}\right)^2 (n\pi)^2 \right] \tag{7.75}$$

We have plotted in Fig. 7.9 $\Omega^2/\Omega_{\text{class}}^2$ versus (L/r) for various values of n for a value E/kG given as 3.06.

Part B

PLATES

7.8 EQUATIONS OF MOTION FOR PLATES

We now proceed with a discussion for plates that parallels the discussion on beams. Accordingly we shall first develop the equations of motion and boundary conditions for classical plate theory using Hamilton's principle.

The kinetic energy of the plate is needed. For the classical case this results from vertical movement of elements of the plate only (rotatory inertia will be discussed later). Thus we have

$$T = \tfrac{1}{2} \iint_R h\rho(\dot{w})^2 \, dx \, dy \qquad (7.76)$$

where R denotes the transverse area of the plate. The total potential energy has been given by Eq. (6.30) as:

$$(U + V) = \frac{D}{2} \iint_R \left\{ (\nabla^2 w)^2 + 2(1 - v)\left[\left(\frac{\partial^2 w}{\partial x \, \partial y}\right)^2 - \frac{\partial^2 w}{\partial x^2}\frac{\partial^2 w}{\partial y^2} \right] \right\} dx \, dy$$

$$- \iint_R qw \, dx \, dy \qquad (7.77)$$

Hamilton's principle then requires that

$$\delta^{(1)}\left[\int_{t_1}^{t_2} \iint_R \left\{ \frac{h\rho}{2}(\dot{w})^2 - \frac{D}{2}\left[(\nabla^2 w)^2 + 2(1 - v)\left[\left(\frac{\partial^2 w}{\partial x \, \partial y}\right)^2 - \frac{\partial^2 w}{\partial x^2}\frac{\partial^2 w}{\partial y^2} \right]\right] \right. \right.$$

$$\left. \left. - qw \right\} dx \, dy \, dt \right] = 0 \qquad (7.78)$$

Carrying out the first variation using the operator approach we get:

$$\int_{t_1}^{t_2} \iint_R \left[\rho h \frac{\partial w}{\partial t}\frac{\partial \, \delta w}{\partial t} - D\left\{ \nabla^2 w \nabla^2 \, \delta w + 2(1 - v)\left[\frac{\partial^2 w}{\partial x \, \partial y}\frac{\partial^2 \, \delta w}{\partial x \, \partial y} \right. \right. \right.$$

$$\left. \left. \left. - \frac{1}{2}\frac{\partial^2 w}{\partial x^2}\frac{\partial^2 \, \delta w}{\partial y^2} - \frac{1}{2}\frac{\partial^2 w}{\partial y^2}\frac{\partial^2 \, \delta w}{\partial x^2} \right] \right\} - q \, \delta w \right] dx \, dy \, dt = 0 \qquad (7.79)$$

Noting that $\nabla^2 \, \delta w = \partial^2 \, \delta w/\partial x^2 + \partial^2 \, \delta w/\partial y^2$ we next integrate by parts with respect to time for the first term in the integrand and then successively with respect to the spatial coordinates for all but the last term. We get:

$$\int_{t_1}^{t_2} \iint_R \left[-h\rho \frac{\partial^2 w}{\partial t^2} \, \delta w - D\left\{ \frac{\partial^2}{\partial x^2}\nabla^2 w + \frac{\partial^2}{\partial y^2}\nabla^2 w + 2(1 - v)\left[\frac{\partial^4 w}{\partial x^2 \, \partial y^2} \right. \right. \right.$$

$$\left. \left. \left. - \frac{1}{2}\frac{\partial^4 w}{\partial x^2 \, \partial y^2} - \frac{1}{2}\frac{\partial^4 w}{\partial x^2 \, \partial y^2} \right] \right\} - q \, \delta w \right] dx \, dy \, dt$$

$$+ \int_{t_1}^{t_2}\left[\oint_\Gamma [\cdots] \, dx + \oint_\Gamma [\cdots] \, dy \right] dt = 0 \qquad (7.80)$$

The line integrals themselves are identical to those developed in the statical development of plates in Chap. 6 since the dynamic term from the kinetic energy makes no contribution to the line integrals. We shall not accordingly have to dwell on the line integrals here but will be able to use the results of Sec. 6.4. Collecting terms in the surface integrals we may write the above equation as follows:

$$\int_{t_1}^{t_2}\left[\iint_R\left\{-h\rho\ddot{w}+D\left[-\frac{\partial^2}{\partial x^2}(\nabla^2 w)-\frac{\partial^2}{\partial y^2}(\nabla^2 w)+q\right]\right\}\delta w\,dx\,dy\right.$$
$$\left.+\oint_\Gamma[\cdots]\,dx+\oint_\Gamma[\cdots]\,dy\right]dt=0 \tag{7.81}$$

This becomes:

$$\int_{t_1}^{t_2}\left[\iint_R\left\{-h\rho\ddot{w}-D\nabla^4 w+q\right\}\delta w\,dx\,dy+\oint_\Gamma[\cdots]\,dx\right.$$
$$\left.+\oint_\Gamma[\cdots]\,dy\right]dt=0 \tag{7.82}$$

We can conclude from the above that the differential equation of motion is

$$D\nabla^4 w+\rho h\ddot{w}=q \tag{7.83}$$

The boundary conditions stemming from the line integrals are given by Eq. (6.44) and are rewritten below. On the boundary

$$\text{EITHER } M_\nu=0 \qquad \text{OR}\quad \frac{\partial w}{\partial\nu}\text{ IS PRESCRIBED} \qquad (a)$$

$$\text{EITHER } Q_\nu+\frac{\partial M_{\nu s}}{\partial s}=0 \quad \text{OR}\quad w\text{ IS PRESCRIBED} \qquad (b)\quad (7.84)$$

where ν is the outward normal direction from the boundary. The natural boundary conditions can be readily given in terms of w by employing Eq. (6.15) and (6.12) in (a), and with Eqs. (6.15), (6.16), (6.17), (6.18) and (6.12) in (b).

7.9 FREE VIBRATIONS OF A SIMPLY-SUPPORTED PLATE

As a way of introducing certain concepts concerning the vibration of plates and for providing "exact" data to be used later for comparing results from approximate methods, we now consider the free vibrations of a thin plate measuring $a\times b\times h$ and simply-supported on all edges (see Fig. 7.10).

The differential equation for w for this case is Eq. (7.83) with $q=0$. We shall express this equation in the following simple form

$$\beta^2\nabla^4 w+\ddot{w}=0 \tag{7.85}$$

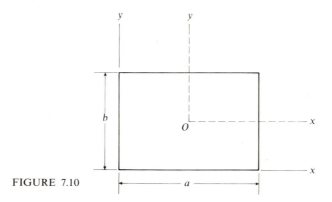

FIGURE 7.10

where:

$$\beta^2 = \left(\frac{D}{\rho h}\right) \qquad (7.86)$$

As is the usual procedure we attempt a separation of variables as follows:

$$w = W(x,y)T(t) \qquad (7.87)$$

This results in the usual way in an ordinary differential equation in time and a partial differential equation in spatial coordinates (x,y). Thus

$$\frac{\ddot{T}}{T} = -\omega^2 \qquad \text{therefore } \ddot{T} + \omega^2 T = 0 \qquad (a)$$

$$\frac{\beta^2 \nabla^4 W}{W} = \omega^2 \qquad \text{therefore } \nabla^4 W - \left(\frac{\omega^2}{\beta^2}\right) W = 0 \qquad (b) \quad (7.88)$$

where ω^2 is the separation parameter. The solution to Eq. (7.88(a)) is immediately seen to be:

$$T = A \cos \omega t + B \sin \omega t \qquad (7.89)$$

As for Eq. (7.88(b)) we again proceed by a separation of variable process. However, here because of the simplicity of the problem we shall be able a priori to choose functions $X(x)$ and $Y(y)$ for a product solution $W(x,y)$. We shall consider the product $X(x)Y(y)$ as follows:

$$W(x,y) = X(x)Y(y) = \sin\frac{n\pi x}{a} \sin\frac{m\pi y}{b} \qquad (7.90)$$

Recall from Sec. 6.4 (static bending of simply-supported rectangular plates) that this product satisfies the complete boundary conditions for a simply-supported plate, i.e.,

$$W = M_x = 0 \quad \text{at} \quad x = 0, a \qquad (7.91)$$

$$W = M_y = 0 \quad \text{at} \quad y = 0, b \qquad (7.92)$$

Now substitute this product into Eq. (7.88(b)) to find

$$\left[\left(\frac{n\pi}{a}\right)^4 + 2\left(\frac{n\pi}{a}\right)^2\left(\frac{m\pi}{b}\right)^2 + \left(\frac{m\pi}{b}\right)^4 - \frac{\omega_{nm}^2}{\beta^2} \right] \sin\frac{n\pi x}{a} \sin\frac{m\pi y}{b} = 0 \qquad (7.93)$$

where the subscripts n, m have been used for the separation variable ω^2 to associate it with the particular product having integers n and m. We then conclude that:

$$\omega_{nm}^2 = \beta^2\left[\left(\frac{n\pi}{a}\right)^4 + 2\left(\frac{n\pi}{a}\right)^2\left(\frac{m\pi}{b}\right)^2 + \left(\frac{m\pi}{b}\right)^4 \right]$$

therefore

$$\omega_{nm}^2 = \beta^2\pi^4\left[\left(\frac{n}{a}\right)^2 + \left(\frac{m}{b}\right)^2 \right]^2 \qquad (7.94)$$

From the above steps we see that ω_{nm} is a natural frequency (or eigenvalue) for a mode shape (or eigenfunction) W_{nm} given as $\sin(n\pi x/a)\sin(m\pi y/b)$. Thus we can say for the (nm)th mode:

$$\omega_{nm} = \beta\pi^2\left[\left(\frac{n}{a}\right)^2 + \left(\frac{m}{b}\right)^2 \right] \qquad (a)$$

$$W_{nm} = \sin\frac{n\pi x}{a} \sin\frac{m\pi y}{b} \qquad (b) \quad (7.95)$$

The free vibration of the plate is a superposition of all the modes with proper amplitudes and properly phased together to satisfy the *initial conditions* of the problem. Thus we can say:

$$w = \sum_{m=1}^{\infty} \sum_{n=1}^{\infty} \left(\sin\frac{n\pi x}{a} \sin\frac{m\pi y}{b} \right)\left(A_{nm}\sin\omega_{nm}t + B_{nm}\cos\omega_{nm}t \right) \qquad (7.96)$$

where the double infinity of constants A_{nm} and B_{nm} are determined to satisfy the conditions

$$w(x,y,0) = \phi(x,y)$$

$$\frac{\partial w(x,y,0)}{\partial t} = \psi(x,y) \qquad (7.97)$$

with ϕ and ψ as known functions. Accordingly, submitting Eq. (7.96) to the above conditions we get:

$$\sum_{n=1}^{\infty} \sum_{m=1}^{\infty} \sin \frac{n\pi x}{a} \sin \frac{m\pi y}{b} B_{nm} = \phi(x,y) \qquad (a)$$

$$\sum_{n=1}^{\infty} \sum_{m=1}^{\infty} \left(\sin \frac{n\pi x}{a} \right)\left(\sin \frac{m\pi y}{b} \right) \omega_{nm} A_{nm} = \psi(x,y) \qquad (b) \quad (7.98)$$

Now making use of the fact that

$$\int_0^a \sin \frac{r\pi x}{a} \sin \frac{s\pi x}{a} \, dx = \frac{a}{2}\delta_{rs}$$

$$\int_0^b \sin \frac{p\pi y}{b} \sin \frac{q\pi y}{b} \, dy = \frac{b}{2}\delta_{pq} \qquad (7.99)$$

we note that

$$\int\int_R (W_{rs})(W_{pq}) \, dx \, dy = \frac{ab}{4}[\delta_{rs}\, \delta_{pq}] \qquad (7.100)$$

This is the orthogonality property of the eigenfunction for this problem. Now multiplying Eq. (7.98(a)) by $\sin(p\pi x/a) \sin(q\pi y/b)$ and integrating over R

$$\int\int_R \left[\sum_{m=1}^{\infty} \sum_{n=1}^{\infty} \sin \frac{n\pi x}{a} \sin \frac{m\pi y}{b} \sin \frac{p\pi x}{a} \sin \frac{q\pi y}{b} B_{nm} \right] dx \, dy$$

$$= \int\int_R \phi(x,y) \sin \frac{p\pi x}{a} \sin \frac{q\pi y}{b} \, dx \, dy$$

Using the orthogonality property given by Eq. (7.100) we may solve for B_{pq} from the above formulation in a direct manner. Thus:

$$B_{pq} = \frac{4}{ab} \int\int_R \phi(x,y) \sin \frac{p\pi x}{a} \sin \frac{q\pi y}{b} \, dx \, dy \qquad (a)$$

Similarly we have for A_{pq}:

$$A_{pq} = \frac{4}{ab\omega_{pq}} \int\int \psi(x,y) \sin \frac{p\pi x}{a} \sin \frac{q\pi y}{b} \, dx \, dy \qquad (b) \quad (7.101)$$

We have thus solved the problem of the rectangular simply-supported freely vibrating plate. It is to be pointed out that there are no such simple solutions available for the clamped rectangular plate and the free rectangular plate. This is partly due to the fact that the governing equation (Eq. 7.88) for such plates is not separable in rectangular coordinates. We shall be able to investigate, in the problems, solutions of the circular plates since for cylindrical coordinates the basic plate equation is separable.

7.10 RAYLEIGH'S METHOD FOR PLATES

We now present Rayleigh's method for determining an approximation from above for the fundamental frequency of a plate. The basis for the method is as given in Sec. 7.5 where we discussed beams. We assume the plate is vibrating freely with frequency ω_1 in the *fundamental mode*. We can then represent the motion as follows:

$$w = W_1(x,y) \cos \omega_1 t \qquad (7.102)$$

Since we do not have dissipation we equate the maximum kinetic energy of the system to the maximum strain energy of the system. That is:

$$T_{max} = U_{max} \qquad (7.103)$$

To find T_{max} note that we can give T as follows, using Eq. (7.102):

$$T = \tfrac{1}{2} \iint_R h\rho(\dot{w})^2 \, dA = \frac{\omega_1{}^2 \sin^2 \omega_1 t}{2} \iint h\rho(W_1)^2 \, dA$$

Clearly for T_{max} we take $\sin^2 \omega_1 t = 1$ to get:

$$T_{max} = \tfrac{1}{2}h\omega_1{}^2\rho \iint_R W_1{}^2 \, dA$$

As for U we use Eq. (7.102) in Eq. (7.77) with q set equal to zero. Thus:

$$U = \frac{D}{2} \cos^2 \omega_1 t \iint_R \left\{ (\nabla^2 W_1)^2 + 2(1 - v)\left[\left(\frac{\partial^2 W_1}{\partial x \, \partial y} \right)^2 - \frac{\partial^2 W_1}{\partial x^2} \frac{\partial^2 W_1}{\partial y^2} \right] \right\} dA$$

For U_{max} we set $\cos^2 \omega_1 t = 1$. Thus:

$$U_{max} = \frac{D}{2} \iint_R \left\{ (\nabla^2 W_1)^2 + 2(1 - v)\left[\left(\frac{\partial^2 W_1}{\partial x \, \partial y} \right)^2 - \frac{\partial^2 W_1}{\partial x^2} \frac{\partial^2 W_1}{\partial y^2} \right] \right\} dA$$

Substituting into Eq. (7.103) we can now form Rayleigh's quotient as follows by solving for $\omega_1{}^2$:

$$\omega_1{}^2 = \frac{D \iint_R \left\{ (\nabla^2 W_1)^2 + 2(1 - v)\left[\left(\frac{\partial^2 W_1}{\partial x \, \partial y} \right)^2 - \frac{\partial^2 W_1}{\partial x^2\cdot} \frac{\partial^2 W_1}{\partial y^2} \right] \right\} dA}{h\rho \iint_R W_1{}^2 \, dA} \qquad (7.104)$$

It will be shown that Rayleigh's quotient given above is a functional which, like the corresponding functional for beams, is a minimum for the lowest eigenfunction with respect to functions $\tilde{W}_1$ that satisfy the boundary conditions of the problem. Hence a function $\tilde{W}_1$ satisfying boundary conditions and a good approximation to the fundamental mode shape will yield on substitution into Eq. (7.104) a value of $\tilde{\omega}_1{}^2$

which will be an even better approximation of $\omega_1{}^2$. Furthermore this approximation will be *from above*. We have here a powerful method, then, of approximating the very important fundamental frequency of a plate—a tool which is particularly valuable considering the paucity of solutions available for the plate.

We now illustrate this approach for the case of the simply-supported rectangular plate, which was solved in the previous section, in order to illustrate the method and to compare approximate results with the exact results for this case.

EXAMPLE 7.3 We again examine the rectangular simply-supported flat plate having dimensions $a \times b \times h$. It will be simplest here to place the reference at the center of the plate as has been shown in Fig. 7.10 (see dashed reference).

To approximate the first mode we use the following even function:

$$w_{\text{app}} = \left[1 - \left(\frac{x}{a/2}\right)^2\right]\left[1 - \left(\frac{y}{b/2}\right)^2\right] = \left(1 - \frac{4x^2}{a^2}\right)\left(1 - \frac{4y^2}{b^2}\right) \qquad (a)$$

This function satisfies the kinematic boundary conditions $w = 0$ but not the natural boundary conditions $\partial^2 w/\partial v^2 = 0$. We now substitute the above function into Eq. (7.104) to find ω_1. After straightforward computations we get:

$$\omega_1{}^2 = \frac{40D}{h\rho}\left(\frac{3}{a^4} + \frac{200}{a^2b^2} + \frac{3}{b^4}\right) \qquad (b)$$

We compute $\omega_1{}^2$ for various values of a/b and compare with the exact solution (Eq. (7.94)). The results are given below in Table 7.3 where we use the frequency parameter $(h\rho a^4/D)\omega^2 = k^2$. We see that for $a/b = 1$ the ratio of frequencies $\omega_1/\omega_{\text{exact}}$ is:

$$\frac{\omega_1}{\omega_{\text{exact}}} = \sqrt{\frac{440}{389}} = 1.06 \qquad (c)$$

In an effort to improve the accuracy of the computation we now employ a function that satisfies both the kinematic $w = 0$ and the natural boundary conditions

Table 7.3

a/b	1	2	3
$k_1{}^2$	440	2,890	11,640
k_{exact}^2	389	2,435	9,740

$\partial^2 w / \partial v^2$. Thus consider the following even function:

$$W_1 = \left(x^4 - \frac{3a^2}{2}x^2 + \frac{5a^4}{16}\right)\left(y^4 - \frac{3b^2}{2}y^2 + \frac{5}{16}b^4\right) \qquad (d)$$

You may readily demonstrate that this function satisfies the aforestated boundary conditions. Substituting into Eq. (7.104) we then get on solving for $\omega_1{}^2$:

$$\omega_1{}^2 = 98.7 \frac{D}{h\rho}\left(\frac{1}{a^2} + \frac{1}{b^2}\right)^2 \qquad (e)$$

This result now coincides, within the computational accuracy employed, with the exact solution presented earlier. ////

7.11 RAYLEIGH–RITZ METHOD FOR PLATES

We now set forth the Rayleigh–Ritz method as applied to plates. We will be able thereby to improve the estimate of the fundamental frequency and if we can estimate higher mode shapes we will also be able to get approximations from above of higher natural frequencies. As was indicated in Sec. 7.6 on beams, we proceed by giving W_{app} in terms of a set of linearly independent functions ϕ_i which satisfy the boundary conditions of the problem. That is, let

$$W_{\text{app}}(x,y) = A_1\phi_1(x,y) + A_2\phi_2(x,y) + \cdots + A_n\phi_n(x,y) \qquad (7.105)$$

where A_i are the undetermined coefficients. Now substitute this result into the Rayleigh quotient for plates as follows:

$$\Lambda^2 = \frac{D\iint_R \left\{\left(\sum_{i=1}^{n} A_i\nabla^2\phi_i\right)^2 + 2(1-v)\left[\left(\sum_{i=1}^{n} A_i\frac{\partial^2\phi_i}{\partial x\,\partial y}\right)^2 - \left(\sum_{i=1}^{n} A_i\frac{\partial^2\phi_i}{\partial x^2}\right)\left(\sum_{j=1}^{n} A_j\frac{\partial^2\phi_j}{\partial y^2}\right)\right]\right\}\,dA}{h\rho\iint_R \left(\sum_{i=1}^{n} A_i\phi_i\right)^2\,dA} \qquad (7.106)$$

We next introduce the following notation, replacing x and y by x_1 and x_2 respectively:

$$c_{ij} = h\rho\iint_R \phi_i\phi_j\,dA$$

$$a_{ijpq} = D\iint_R \left(\frac{\partial^2\phi_i}{\partial x_p{}^2}\right)\left(\frac{\partial^2\phi_j}{\partial x_q{}^2}\right)\,dA \qquad \begin{cases} i,j = 1,2,\ldots,n \\ p,q = 1,2 \end{cases}$$

$$b_{ij} = D\iint_R \left(\frac{\partial^2\phi_i}{\partial x_1\,\partial x_2}\right)\left(\frac{\partial^2\phi_j}{\partial x_1\,\partial x_2}\right)\,dA \qquad (7.107)$$

Then we can give the Rayleigh quotient as:[1]

$$
\Lambda^2 = \frac{\displaystyle\sum_{j=1}^{n}\sum_{i=1}^{n}\left\{A_i A_j\left[\sum_{q=1}^{2}\sum_{p=1}^{2}a_{ijpq} + 2(1-v)(b_{ij}-a_{ij12})\right]\right\}}{\displaystyle\sum_{i=1}^{n}\sum_{j=1}^{n}A_i A_j c_{ij}} \tag{7.108}
$$

Now extremizing with respect to A_i we have n equations of the form:

$$
\frac{\partial \Lambda^2}{\partial A_i} = 0 = \frac{2\displaystyle\sum_{j=1}^{n}A_j\left[\sum_{q=1}^{2}\sum_{p=1}^{2}a_{ijpq} + 2(1-v)(b_{ij}-a_{ij12})\right]}{\displaystyle\sum_{i=1}^{n}\sum_{j=1}^{n}A_i A_j c_{ij}}
$$

$$
- \frac{\left\{\displaystyle\sum_{k=1}^{n}\sum_{j=1}^{n}A_k A_j\left[\sum_{q=1}^{2}\sum_{p=1}^{2}a_{kjpq} + 2(1-v)(b_{kj}-a_{kj12})\right]\right\}\left(2\displaystyle\sum_{j=1}^{n}A_j c_{ij}\right)}{\left(\displaystyle\sum_{k=1}^{n}\sum_{j=1}^{n}A_k A_j c_{kj}\right)^2}
$$

$$
i = 1, 2, \ldots, n
$$

Note that i has become a free index in the first fraction above and in the last bracketed quantity in the numerator of the second fraction. In order to avoid confusion we have, as a result, switched from i to k in the remaining expressions of the second fraction where i has played and would continue to play the role of a dummy index. Using Eq. (7.108) and canceling

$$
2\bigg/\left[\sum_{k=1}^{n}\sum_{j=1}^{n}A_k A_j c_{kj}\right]
$$

we get

$$
\sum_{j=1}^{n}A_j\left[\sum_{q=1}^{2}\sum_{p=1}^{2}a_{ijpq} + 2(1-v)(b_{ij}-a_{ij12})\right] - \Lambda^2\left(\sum_{j=1}^{n}A_j c_{ij}\right) = 0
$$

$$
i = 1, 2, \ldots, n
$$

Hence

$$
\sum_{j=1}^{n}\left[\sum_{q=1}^{2}\sum_{p=1}^{2}a_{ijpq} + 2(1-v)(b_{ij}-a_{ij12}) - \Lambda^2 c_{ij}\right]A_j = 0
$$

$$
i = 1, 2, \ldots, n
$$

[1] If any expressions in this equation are not clear you may justify them in your mind by using function $A_1\phi_1 + A_2\phi_2$ to test the formulations.

Using the following notation:

$$\alpha_{ij} = \sum_{q=1}^{2} \sum_{p=1}^{2} a_{ijpq} + 2(1 - v)(b_{ij} - a_{ij12}) \qquad (7.109)$$

We get

$$\sum_{i=1}^{n} [\alpha_{ij} - \Lambda^2 c_{ij}]A_j = 0 \qquad i = 1, 2, \ldots, n \qquad (7.110)$$

We thus get a system of equations of the form given by Eq. (7.43) for beams. We know that a necessary condition for a nontrivial result is that:

$$|\alpha_{ij} - \Lambda^2 c_{ij}| = 0 \qquad (7.111)$$

This gives us an algebraic equation of degree n in Λ^2. The n roots will then be approximations from above of the first n natural frequencies of the problem. Also the ratios of the A_i for a given Λ^2 can be determined from Eq. (7.110) thus establishing an approximate eigenfunction.

We illustrate the Rayleigh–Ritz method in the following example. In part C of the chapter we shall justify the above statement.

EXAMPLE 7.4   We now consider the calculation of higher-order eigenvalues for the vibration of a simply-supported rectangular plate shown in Fig. 7.10 with the origin at the center.

As a first step, we shall choose a system of n coordinate functions that satisfy both the kinematic boundary condition $w = 0$ and the natural boundary condition $\partial^2 w/\partial v^2 = 0$ of the problem and which reasonably resemble what may be expected to be the first n even eigenfunctions. We shall formulate the coordinate functions ϕ_i as products of two functions—one of which is a function of x alone and the other a function of y alone. That is:

$$\phi_i(x,y) = g(x)h(y) \qquad (a)$$

We will require accordingly for the boundary conditions that:

$$\phi_i\left(\frac{a}{2}, y\right) = \left(\frac{\partial^2 \phi_i}{\partial x^2}\right)_{(a/2,y)} = 0 \qquad \text{therefore} \quad g\left(\frac{a}{2}\right) = g''\left(\frac{a}{2}\right) = 0$$

$$\phi_i\left(-\frac{a}{2}, y\right) = \left(\frac{\partial^2 \phi_i}{\partial x^2}\right)_{(-a/2,y)} = 0 \quad \text{therefore} \quad g\left(-\frac{a}{2}\right) = g''\left(-\frac{a}{2}\right) = 0$$

$$\phi_i\left(x, \frac{b}{2}\right) = \left(\frac{\partial^2 \phi_i}{\partial y^2}\right)_{(x,b/2)} = 0 \quad \text{therefore} \quad h\left(\frac{b}{2}\right) = h''\left(\frac{b}{2}\right) = 0$$

$$\phi_i\left(x, -\frac{b}{2}\right) = \left(\frac{\partial^2 \phi_i}{\partial y^2}\right)_{(x,-b/2)} = 0 \quad \text{therefore} \quad h\left(-\frac{b}{2}\right) = h''\left(-\frac{b}{2}\right) = 0 \qquad (b)$$

FIGURE 7.11

Since we know the shape of the mode shapes of the rectangular simply-supported plate, we shall set forth approximations for the first five even mode shapes to show how we can approximate the eigenvalues for these modes.

For instance consider the fundamental mode shape of the plate. If we start with the even function

$$g_1 = C_1 + C_2 x^2 + C_3 x^4$$

we may choose one constant arbitrarily and then choose the other two constants to satisfy the boundary conditions on g at $x = -a/2$ given by Eq. (b). Because the function is even we then also satisfy the boundary conditions at $x = +a/2$. Since we can here exclude nodal points (see Fig. 7.11) over the interval $-a/2 < x < a/2$ we can thus set forth a function in the x direction approximating the fundamental mode in that direction. We have for $g_1(x)$ for these requirements:

$$g_1(x) = -5 + \frac{24x^2}{a^2} - \frac{16x^4}{a^4} \qquad (c)$$

Similarly for approximating the fundamental mode shape in the y direction we express $h_1(y)$ as:

$$h_1(y) = -5 + \frac{24y^2}{b^2} - \frac{16y^4}{b^4} \qquad (d)$$

Hence we can give ϕ_1 as:

$$\phi_1 = [g_1(x)][h_1(y)]$$

We know that the next even mode shape will have two nodal points in either the x or y direction (see Fig. 7.12) and so we consider the even function:

$$C_1 + C_2 x^2 + C_3 x^4 + C_4 x^6$$

Choose one constant arbitrarily and then compute the other constants to satisfy the boundary conditions (Eq. (b)) at $x = -a/2$ plus the requirement of having a nodal point at $x = a/6$ (and hence also at $-a/6$). The following function denoted as $g_3(x)$ satisfies the above requirements:

$$g_3(x) = 19 - \frac{804x^3}{a^4} + \frac{4496x^4}{a^4} - \frac{6336x^6}{a^4} \qquad (e)$$

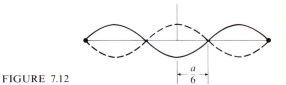

FIGURE 7.12

Similarly for the y direction we can form $h_3(y)$ as follows:

$$h_3(y) = 19 - \frac{804y^2}{b^2} + \frac{4496y^4}{b^4} - \frac{6336y^6}{b^6} \qquad (f)$$

As for Fig. 7.13 we proceed by considering a function of the form:

$$C_1 + C_2x^2 + C_3x^4 + C_4x^6 + C_5x^8$$

Choosing one constant arbitrarily, then satisfying the boundary conditions at $x = a/2$, and finally setting the function equal to zero at $x = a/10$ and $x = 3a/10$ (the nodal points) we may establish a set of C's. Thus denoting the function as $g_5(x)$ we have:

$$g_5(x) = 10.4 - \frac{1228.7x^2}{a^2} + \frac{19735.8x^4}{a^4} - \frac{94200x^6}{a^6} + \frac{137000x^8}{a^8} \qquad (k)$$

Similarly we may form a function $h_5(y)$ as follows:

$$h_5(y) = 10.4 - \frac{1228.7y^2}{b^2} + \frac{19735.8y^4}{b^4} - \frac{94200y^6}{b^6} + \frac{137000y^8}{b^8} \qquad (i)$$

We may now approximate the first five even mode shapes in the following manner:

$$\phi_1 = g_1h_1$$
$$\phi_2 = g_1h_3$$
$$\phi_3 = g_3h_3$$
$$\phi_4 = g_1h_5$$
$$\phi_5 = g_3h_5$$

Now compute α_{ij} and C_{ij} using the above functions. We may solve Eq. (7.111) successively using ϕ_1 (i.e., take $n = 1$); then using ϕ_1 and ϕ_2 ($n = 2$), and so forth until all five functions are employed ($n = 5$). We then get successively approximations from above of the eigenvalue corresponding to the first symmetric mode (for $n = 1$); then the eigenvalues of the first two symmetric modes (for $n = 2$); etc. These results

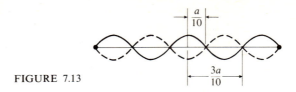

FIGURE 7.13

have been shown in Table 7.4 for $a/b = 1$ and for $a/b = 2$. A Poisson ratio of 0.5 has been used for simplicity.

We thus see that when *we can approximate the eigenfunctions reasonably closely* we may reach very good approximations from above of the corresponding eigenvalues by the Rayleigh–Ritz method. In an exercise you will be asked to make calculations such as those above for the odd modes of the simply-supported rectangular plate.

////

7.12 TRANSVERSE SHEAR AND ROTATORY INERTIA— MINDLIN PLATE THEORY

In Sec. 6.12 we examined an improved theory for the axisymmetric circular plate wherein the effect of transverse shear was taken into account. We now consider the rectangular plate and take into account the effects of transverse shear and rotatory

Table 7.4

$a/b = 1$

n	$\Lambda_1{}^2$	$\Lambda_2{}^2$	$\Lambda_3{}^2$	$\Lambda_4{}^2$	$\Lambda_5{}^2$
1	389.969				
2	389.803	10,213		multiply each term by $(D/\rho ha^4)$	
3	389.802	10,213	32,787		
4	389.802	9,747	32,787	80.713	
5	389.802	9,747	32,136	80,711	131,154
exact	389.635	9,741	31,560	65,848	112,604

$a/b = 2$

n	$\Lambda_1{}^2$	$\Lambda_2{}^2$	$\Lambda_3{}^2$	$\Lambda_4{}^2$	$\Lambda_5{}^2$
1	2437.811				
2	2435.475	140,616		multiply each term by $(D/\rho ha^4)$	
3	2435.475	140,615	205,723		
4	2435.474	133,444	205,723	1,227,430	
5	2435.474	133,444	197,865	1,227,427	1,403,654
exact	2435.219	133,352	197,257	993,667	1,157,313

inertia while formulating the free vibration problem for this case. The theory is due to Mindlin.

We start in the usual way by proposing a displacement field. Deleting stretching effects we have:

$$u_1 = -z\psi_x(x,y,t) = -z\psi_x$$
$$u_2 = -z\psi_y(x,y,t) = -z\psi_y$$
$$u_3 = w(x,y,t) = w \tag{7.112}$$

where ψ_x and ψ_y are components of the vector field $\boldsymbol{\psi}$. The deformation differs from the classical case (Eqs. 6.4) in that functions ψ_i, yet to be determined, replace $w_{,i}$ in u_1 and u_2. This means we are maintaining the assumption that line elements originally normal to the midplane remain straight on deformation (no warping), but we are abandoning the assumption that such line elements remain normal to the midplane, after deformation. The assumption of no warping is not correct and we shall later introduce means of correcting for it. The strain field for the assumed displacement field follows immediately as:

$$\epsilon_{xx} = -z\frac{\partial \psi_x}{\partial x}$$

$$\epsilon_{yy} = -z\frac{\partial \psi_y}{\partial y}$$

$$\epsilon_{xy} = -\frac{z}{2}\left(\frac{\partial \psi_x}{\partial y} + \frac{\partial \psi_y}{\partial x}\right)$$

$$\epsilon_{xz} = \frac{1}{2}\left(\frac{\partial w}{\partial x} - \psi_x\right)$$

$$\epsilon_{yz} = \frac{1}{2}\left(\frac{\partial w}{\partial y} - \psi_y\right) \tag{7.113}$$

We shall consider the terms ψ_x and ψ_y as variables in the ensuing discussion. Next we shall ascertain the resultant intensity functions. Thus for M_x we have:

$$M_x = \int_{-h/2}^{h/2} \tau_{xx}z\,dz = \int_{-h/2}^{h/2} \frac{E}{1-v^2}(\epsilon_{xx} + v\epsilon_{yy})z\,dz$$

wherein we have assumed a plane stress distribution in employing Hooke's law. Now using Eq. (7.113) in the above formulation we get:

$$M_x = \frac{E}{1-v^2}\int_{-h/2}^{h/2} z^2\left(-\frac{\partial \psi_x}{\partial x} - v\frac{\partial \psi_y}{\partial y}\right)dz$$

$$= -\frac{Eh^3}{12(1-v^2)}\left[\frac{\partial \psi_x}{\partial x} + v\frac{\partial \psi_y}{\partial y}\right] = -D\left[\frac{\partial \psi_x}{\partial x} + v\frac{\partial \psi_y}{\partial y}\right] \tag{7.114}$$

Similarly we have for the other moment resultant functions:

$$M_y = -D\left[\frac{\partial \psi_y}{\partial y} + v\frac{\partial \psi_x}{\partial x}\right] \qquad (a)$$

$$M_{xy} = -(1 - v)\left(\frac{D}{2}\right)\left(\frac{\partial \psi_x}{\partial y} + \frac{\partial \psi_y}{\partial x}\right) \qquad (b) \quad (7.115)$$

We now consider the resultant shear force intensity function. For Q_x we have:

$$Q_x = \int_{-h/2}^{h/2} \tau_{xz} \, dz = k\tau_{xz}h \qquad (7.116)$$

where we have introduced a shear factor k to correct for the error stemming from Eq. (7.113) that τ_{xz} is a constant over the thickness of the plate (i.e., is not a function of z). Here is where warping effects, mentioned earlier, are taken into account. Now using Hooke's law and Eq. (7.113) we have:

$$Q_x = khG\left(\frac{\partial w}{\partial x} - \psi_x\right) \qquad (7.117)$$

Similarly we get for Q_y:

$$Q_y = khG\left(\frac{\partial w}{\partial y} - \psi_y\right) \qquad (7.118)$$

Since we shall employ Hamilton's principle we next turn towards evaluation of the various energies involved in order to be able to express the Lagrangian function in terms of our plate variables. First we consider the kinetic energy T. Thus:

$$T = \frac{1}{2}\iint_R \int_{-h/2}^{h/2} \rho \dot{u}_i \dot{u}_i \, dz \, dA$$

$$= \frac{1}{2}\iint_R \int_{-h/2}^{h/2} \rho \left\{ z^2\left(\frac{\partial \psi_x}{\partial t}\right)^2 + z^2\left(\frac{\partial \psi_y}{\partial t}\right)^2 + \left(\frac{\partial w}{\partial t}\right)^2 \right\} dz \, dA$$

Note we are now retaining rotatory inertia effects. Integrating with respect to z we get:

$$T = \frac{1}{2}\iint_R \left\{ \frac{\rho h^3}{12}\left(\frac{\partial \psi_x}{\partial t}\right)^2 + \frac{\rho h^3}{12}\left(\frac{\partial \psi_y}{\partial t}\right)^2 + \rho h\left(\frac{\partial w}{\partial t}\right)^2 \right\} dA \qquad (7.119)$$

As for the strain energy U we have:

$$U = \frac{1}{2}\iint_R \int_{-h/2}^{h/2} [\tau_{xx}\epsilon_{xx} + \tau_{yy}\epsilon_{yy} + \tau_{zz}\epsilon_{zz} + 2\tau_{xy}\epsilon_{xy} + 2\tau_{xz}\epsilon_{xz}$$

$$+ 2\tau_{yz}\epsilon_{yz}] \, dz \, dA$$

We now assume a plane stress state for τ_{xx} and τ_{yy} (i.e., we drop τ_{zz}) and employ Hooke's law in the above equation[1] for τ_{xx}, τ_{yy}, and τ_{xy}:

$$U = \frac{1}{2} \int\int_R \int_{-h/2}^{h/2} \left\{ \frac{E}{1-v^2}(\epsilon_{xx} + v\epsilon_{yy})\epsilon_{xx} + \frac{E}{1-v^2}(\epsilon_{yy} + v\epsilon_{xx})\epsilon_{yy} \right.$$

$$\left. + 4G\epsilon_{xy}{}^2 + 2\tau_{xz}\epsilon_{xz} + 2\tau_{yz}\epsilon_{yz} \right\} dA\,dz$$

Next use Eqs. 7.113 to replace the strains in the above equation. We get:

$$U = \frac{1}{2} \int\int_R \int_{-h/2}^{h/2} \left\{ \frac{E}{1-v^2}z^2\left[\left(\frac{\partial\psi_x}{\partial x}\right)^2 + \left(\frac{\partial\psi_y}{\partial y}\right)^2 + 2v\frac{\partial\psi_x}{\partial x}\frac{\partial\psi_y}{\partial y}\right] \right.$$

$$\left. + Gz^2\left(\frac{\partial\psi_x}{\partial y} + \frac{\partial\psi_y}{\partial x}\right)^2 + \tau_{xz}\left(\frac{\partial w}{\partial x} - \psi_x\right) + \tau_{yz}\left(\frac{\partial w}{\partial y} - \psi_y\right) \right\} dA\,dz$$

Integrate with respect to z over the thickness of the plate

$$U = \frac{1}{2} \int\int_R \left\{ \frac{Eh^3}{12(1-v)}\left[\left(\frac{\partial\psi_x}{\partial x}\right)^2 + \left(\frac{\partial\psi_y}{\partial y}\right)^2 + 2v\left(\frac{\partial\psi_x}{\partial x}\right)\left(\frac{\partial\psi_y}{\partial y}\right)\right] \right.$$

$$\left. + \frac{Gh^3}{12}\left(\frac{\partial\psi_x}{\partial y} + \frac{\partial\psi_y}{\partial x}\right)^2 + kh\tau_{xz}\left(\frac{\partial w}{\partial x} - \psi_x\right) + kh\tau_{yz}\left(\frac{\partial w}{\partial y} - \psi_y\right) \right\} dA$$

where we have used Eq. (7.116) and the corresponding equation for τ_{yz} keeping in mind that $(\partial w/\partial y - \psi_y)$ and $(\partial w/\partial x - \psi_x)$ are not functions of z. Introducing D and replacing τ_{xz} and τ_{yz} using Hooke's law and Eq. (7.113) we get:

$$U = \frac{1}{2} \int\int_R \left\{ D\left[\left(\frac{\partial\psi_x}{\partial x}\right)^2 + \left(\frac{\partial\psi_y}{\partial y}\right)^2 + 2v\left(\frac{\partial\psi_x}{\partial x}\right)\left(\frac{\partial\psi_y}{\partial y}\right)\right] \right.$$

$$\left. + \frac{Gh^3}{12}\left(\frac{\partial\psi_x}{\partial y} + \frac{\partial\psi_y}{\partial x}\right)^2 + khG\left[\left(\frac{\partial w}{\partial x} - \psi_x\right)^2 + \left(\frac{\partial w}{\partial y} - \psi_y\right)^2\right] \right\} dA \qquad (7.120)$$

Finally noting that $V = -\int\int_R wq\,dA$ we may state Hamilton's principle as follows:

$$\delta^{(1)}\int_{t_1}^{t_2} L\,dt = 0 = \delta^{(1)}\int_{t_1}^{t_2} \left(\frac{1}{2}\right)\int\int_R \left\{ \frac{\rho h^3}{12}\left(\frac{\partial\psi_x}{\partial t}\right)^2 + \frac{\rho h^3}{12}\left(\frac{\partial\psi_y}{\partial t}\right)^2 + \rho h\left(\frac{\partial w}{\partial t}\right)^2 \right.$$

$$- D\left[\left(\frac{\partial\psi_x}{\partial x}\right)^2 + \left(\frac{\partial\psi_y}{\partial y}\right)^2 + 2v\left(\frac{\partial\psi_x}{\partial x}\right)\left(\frac{\partial\psi_y}{\partial y}\right)\right]$$

$$- \frac{Gh^3}{12}\left(\frac{\partial\psi_x}{\partial y} + \frac{\partial\psi_y}{\partial x}\right)^2 - kGh\left[\left(\frac{\partial w}{\partial x} - \psi_x\right)^2 \right.$$

$$\left. \left. + \left(\frac{\partial w}{\partial y} - \psi_y\right)^2\right] + 2wq \right\} dA\,dt = 0$$

---

[1] We are thus using a simplified constitutive law at this juncture. We have alluded to this kind of simplification procedure on a number of earlier occasions.

Carrying out the variation operation using the δ operator we then get:

$$\int_{t_1}^{t_2}\int\int_R \left\{ \left(\frac{\rho h^3}{12}\right)\left(\frac{\partial \psi_x}{\partial t}\right)\left(\frac{\partial\, \delta\psi_x}{\partial t}\right) + \left(\frac{\rho h^3}{12}\right)\left(\frac{\partial \psi_y}{\partial t}\right)\left(\frac{\partial(\delta\psi_y)}{\partial t}\right) + \rho h\left(\frac{\partial w}{\partial t}\right)\left(\frac{\partial\, \delta w}{\partial t}\right) \right.$$

$$- D\left[\left(\frac{\partial \psi_x}{\partial x}\right)\left(\frac{\partial\, \delta\psi_x}{\partial x}\right) + \left(\frac{\partial \psi_y}{\partial y}\right)\left(\frac{\partial\, \delta\psi_y}{\partial y}\right) + v\left(\frac{\partial \psi_x}{\partial x}\right)\left(\frac{\partial\, \delta\psi_y}{\partial y}\right) \right.$$

$$\left. + v\left(\frac{\partial \psi_y}{\partial y}\right)\left(\frac{\partial\, \delta\psi_x}{\partial x}\right) \right] - \frac{Gh^3}{12}\left(\frac{\partial \psi_x}{\partial y} + \frac{\partial \psi_y}{\partial x}\right)\left(\frac{\partial\, \delta\psi_x}{\partial y} + \frac{\partial\, \delta\psi_y}{\partial x}\right)$$

$$- kGh\left[\left(\frac{\partial w}{\partial x} - \psi_x\right)\left(\frac{\partial\, \delta w}{\partial x} - \delta\psi_x\right) + \left(\frac{\partial w}{\partial y} - \psi_y\right)\left(\frac{\partial\, \delta w}{2\partial y} - \delta\psi_y\right) \right]$$

$$\left. + q\, \delta w \right\} dA\, dt = 0 \tag{7.121}$$

We shall next integrate by parts with respect to time for the first three terms keeping in mind that all variations at the limits t_1 and t_2 are zero. And for all but the last of the other terms we employ Green's theorem having the following form (see Eq. (I.52) Appendix I) with $a_{vx}\, ds = dy$ and $a_{vy}\, ds = -dx$ in accordance with Eq. (6.39) and Fig. 6.9:

$$\int\int_R G\frac{\partial H}{\partial x}\, dx\, dy = -\int\int_R H\frac{\partial G}{\partial x}\, dx\, dy + \oint_\Gamma GH\, dy$$

$$\int\int_R G\frac{\partial H}{\partial y}\, dx\, dy = -\int\int_R H\frac{\partial G}{\partial y}\, dx\, dy - \oint_\Gamma GH\, dx \tag{7.122}$$

We then get the following result:

$$\int_{t_1}^{t_2}\int\int_R \left\{ -\frac{\rho h^3}{12}\frac{\partial^2 \psi_x}{\partial t^2}\delta\psi_x - \frac{\rho h^3}{12}\frac{\partial^2 \psi_y}{\partial t^2}\delta\psi_y - \rho h\frac{\partial^2 \omega}{\partial t^2}\delta w \right.$$

$$+ D\left(\frac{\partial^2 \psi_x}{\partial x^2}\delta\psi_x + \frac{\partial^2 \psi_y}{\partial y^2}\delta\psi_y + v\frac{\partial^2 \psi_x}{\partial x\, \partial y}\delta\psi_y + v\frac{\partial^2 \psi_y}{\partial x\, \partial y}\delta\psi_x\right)$$

$$+ \frac{Gh^3}{12}\left(\frac{\partial^2 \psi_x}{\partial y^2}\delta\psi_x + \frac{\partial^2 \psi_x}{\partial y\, \partial x}\delta\psi_y + \frac{\partial^2 \psi_y}{\partial x\, \partial y}\delta\psi_x + \frac{\partial^2 \psi_y}{\partial x^2}\delta\psi_y\right)$$

$$+ kGh\left(\frac{\partial^2 w}{\partial x^2}\delta w + \frac{\partial w}{\partial x}\delta\psi_x - \frac{\partial \psi_x}{\partial x}\delta w - \psi_x\delta\psi_x + \frac{\partial^2 w}{\partial y^2}\delta w + \frac{\partial w}{\partial y}\delta\psi_y\right.$$

$$\left. - \frac{\partial \psi_y}{\partial y}\delta w - \psi_y\delta\psi_y\right) + q\, \delta w \right\} dA\, dt + \int_{t_1}^{t_2}\oint_\Gamma \left\{ -D\left[\frac{\partial \psi_x}{\partial x}\delta\psi_x\, dy\right.\right.$$

$$\left.\left. - \frac{\partial \psi_y}{\partial y}\delta\psi_y\, dx - v\frac{\partial \psi_x}{\partial x}\delta\psi_y\, dx + v\frac{\partial \psi_y}{\partial y}\delta\psi_x\, dy\right] \right.$$

$$-\frac{Gh^3}{12}\left(-\frac{\partial\psi_x}{\partial y}\delta\psi_x\,dx + \frac{\partial\psi_x}{\partial y}\delta\psi_y\,dy - \frac{\partial\psi_y}{\partial x}\delta\psi_x\,dx + \frac{\partial\psi_y}{\partial x}\delta\psi_y\,dy\right)$$

$$-kGh\left(\frac{\partial w}{\partial x}\delta w\,dy - \psi_x\,\delta w\,dy - \frac{\partial w}{\partial y}\delta w\,dx + \psi_y\,\delta w\,dx\right)\Bigg\}\,dt = 0$$

We now collect terms in the above formulation with respect to the variation terms:

$$\int_{t_1}^{t_2}\int\int_R\left[\left\{-\frac{\rho h^3}{12}\frac{\partial^2\psi_x}{\partial t^2} + D\left(\frac{\partial^2\psi_x}{\partial x^2} + v\frac{\partial^2\psi_y}{\partial x\,\partial y}\right) + \frac{Gh^3}{12}\left(\frac{\partial^2\psi_x}{\partial y^2} + \frac{\partial^2\psi_y}{\partial x\,\partial y}\right)\right.\right.$$

$$\left.+\,kGh\left(\frac{\partial w}{\partial x} - \psi_x\right)\right\}\delta\psi_x + \left\{-\frac{\rho h^3}{12}\left(\frac{\partial^2\psi_y}{\partial t^2}\right) + D\left(\frac{\partial^2\psi_y}{\partial y^2} + v\frac{\partial^2\psi_x}{\partial x\,\partial y}\right)\right.$$

$$\left.+\,\frac{Gh^3}{12}\left(\frac{\partial^2\psi_y}{\partial x^2} + \frac{\partial^2\psi_x}{\partial y\,\partial x}\right) + kGh\left(\frac{\partial w}{\partial y} - \psi_y\right)\right\}\delta\psi_y$$

$$+\left\{-\rho h\frac{\partial^2 w}{\partial t^2} + kGh\left(\frac{\partial^2 w}{\partial x^2} - \frac{\partial\psi_x}{\partial x} + \frac{\partial^2 w}{\partial y^2} - \frac{\partial\psi_y}{\partial y} + q\right)\right\}\delta w\Bigg]\,dA\,dt$$

$$+\int_{t_1}^{t_2}\oint_\Gamma\left[\left\{-D\left(\frac{\partial\psi_x}{\partial x}dy + v\frac{\partial\psi_y}{\partial y}dy\right) + \frac{Gh^3}{12}\left(\frac{\partial\psi_x}{\partial y}dx + \frac{\partial\psi_y}{\partial x}dx\right)\right\}\delta\psi_x\right.$$

$$+\left\{D\left(\frac{\partial\psi_y}{\partial y}dx + v\frac{\partial\psi_x}{\partial x}dx\right) - \frac{Gh^3}{12}\left(\frac{\partial\psi_x}{\partial y}dy + \frac{\partial\psi_y}{\partial x}dy\right)\right\}\delta\psi_y$$

$$-\left.\left\{(kGh)\left(\frac{\partial w}{\partial x}dy - \psi_x\,dy - \frac{\partial w}{\partial y}dx + \psi_y\,dx\right)\right\}\delta w\right]\,dt = 0 \qquad (7.123)$$

In accordance with previous remarks it is clear that each of the coefficients in the integral $\int_{t_1}^{t_2}\int\int_R$ for the variations must be zero. We consider the coefficient for $\delta\psi_x$ now. We have then:

$$-\frac{\rho h^3}{12}\frac{\partial^2\psi_x}{\partial t^2} + D\frac{\partial}{\partial x}\left(\frac{\partial\psi_x}{\partial x} + v\frac{\partial\psi_y}{\partial y}\right) + \frac{Gh^3}{12}\frac{\partial}{\partial y}\left(\frac{\partial\psi_x}{\partial y} + \frac{\partial\psi_y}{\partial x}\right)$$

$$+\,kGh\left(\frac{\partial w}{\partial x} - \psi_x\right) = 0$$

Noting that

$$D = \frac{Eh^3}{12(1 - v^2)}$$

and that

$$G = \frac{E}{2(1 + v)}$$

we may replace the term $Gh^3/12$ by $(D/2)(1 - v)$ in the third expression above. Now observing Eqs. (7.114), (7.115(b)), and (7.117) we get the following result:

$$-\frac{\rho h^3}{12}\frac{\partial^2 \psi_x}{\partial t^2} - \frac{\partial M_x}{\partial x} - \frac{\partial M_{xy}}{\partial y} + Q_x = 0$$

We may carry out similar operations for the coefficients of $\delta\psi_y$ and δw in the integral $\int_{t_1}^{t_2}\iint_R$ of Eq. (7.123). The resulting equations then may be stated as follows:

$$Q_x = \frac{\partial M_x}{\partial x} + \frac{\partial M_{xy}}{\partial y} + \frac{\rho h^3}{12}\frac{\partial^2 \psi_x}{\partial t^2}$$

$$Q_y = \frac{\partial M_{xy}}{\partial x} + \frac{\partial M_y}{\partial y} + \frac{\rho h^3}{12}\frac{\partial^2 \psi_y}{\partial t^2}$$

$$\frac{\partial Q_x}{\partial x} + \frac{\partial Q_y}{\partial y} + q = \rho h\frac{\partial^2 w}{\partial t^2} \qquad (7.124)$$

If we delete the time derivatives we get a set having a familiar form from our work in static classical plate theory (Sec. 6.3). Keep in mind, however, that the shear effects are built into the system by the way we have formulated the resultant intensity functions Q_x and Q_y. It will be more useful at this time to use equations of motion in terms of displacement functions and so we go to Eq. (7.123) and set the coefficients of the variations to zero in the first integral and replace $Gh^3/12$ by $(D/2)(1 - v)$ to obtain:

$$D\left[\frac{\partial^2 \psi_x}{\partial x^2} + \frac{1-v}{2}\frac{\partial^2 \psi_x}{\partial y^2} + \frac{1+v}{2}\frac{\partial^2 \psi_y}{\partial x\,\partial y}\right] + kGh\left(\frac{\partial w}{\partial x} - \psi_x\right) - \frac{\rho h^3}{12}\frac{\partial^2 \psi_x}{\partial t^2} = 0$$

$$D\left[\frac{1-v}{2}\frac{\partial^2 \psi_y}{\partial x^2} + \frac{\partial^2 \psi_y}{\partial y^2} + \frac{1+v}{2}\frac{\partial^2 \psi_x}{\partial x\,\partial y}\right] + kGh\left(\frac{\partial w}{\partial y} - \psi_y\right) - \frac{\rho h^3}{12}\frac{\partial^2 \psi_y}{\partial t^2} = 0$$

$$-kGh\left(\frac{\partial^2 w}{\partial x^2} + \frac{\partial^2 w}{\partial y^2} - \frac{\partial \psi_x}{\partial x} - \frac{\partial \psi_y}{\partial y}\right) + \rho h\frac{\partial^2 w}{\partial t^2} = q \qquad (7.125)$$

We may eliminate the functions ψ_i from the above equations by tedious but straightforward means to arrive at the following equation for w:

$$\left(\nabla^2 - \frac{\rho}{kG}\frac{\partial^2}{\partial t^2}\right)\left(D\nabla^2 - \frac{\rho h^3}{12}\frac{\partial^2}{\partial t^2}\right)w + \rho h\frac{\partial^2 w}{\partial t^2}$$

$$= \left(1 - \frac{D}{kGh}\nabla^2 + \frac{\rho h^2}{12kG}\frac{\partial^2}{\partial t^2}\right)q \qquad (7.126)$$

To delete rotatory inertia in the above equation we simply set the expressions

$$\frac{\rho h^3}{12}\frac{\partial^2 w}{\partial t^2} \quad \text{and} \quad \frac{\rho h^2}{12kG}\frac{\partial^2 q}{\partial t^2}$$

equal to zero. We then have:

$$D\left(\nabla^2 - \frac{\rho}{kG}\frac{\partial^2}{\partial t^2}\right)\nabla^2 w + \rho h\frac{\partial^2 w}{\partial t^2} = \left(1 - \frac{D}{kGh}\nabla^2\right)q \qquad (7.127)$$

And if we wish to delete only the effects of shear deformation we let $1/kGh$ go to zero. We then get:

$$\left(D\nabla^2 - \frac{\rho h^3}{12}\frac{\partial^2}{\partial t^2}\right)\nabla^2 w + \rho h\frac{\partial^2 w}{\partial t^2} = q \qquad (7.128)$$

If both modifications are made simultaneously we get the classical plate equation:

$$D\nabla^4 w + \rho h\frac{\partial^2 w}{\partial t^2} = q \qquad (7.129)$$

Now we get back to the line integral of Eq. (7.123). We collect terms as follows in setting this line integral equal to zero:

$$\int_{t_1}^{t_2}\oint_{\Gamma}\left\{\delta\psi_x\,dy\left[-D\left(\frac{\partial\psi_x}{\partial x} + v\frac{\partial\psi_y}{\partial y}\right)\right] - \delta\psi_x\,dx\left[-\frac{Gh^3}{12}\left(\frac{\partial\psi_x}{\partial y} + \frac{\partial\psi_y}{\partial x}\right)\right]\right.$$

$$+ \delta\psi_y\,dy\left[-\frac{Gh^3}{12}\left(\frac{\partial\psi_x}{\partial y} + \frac{\partial\psi_y}{\partial x}\right)\right] - \delta\psi_y\,dx\left[-D\left(\frac{\partial\psi_y}{\partial y} + v\frac{\partial\psi_x}{\partial x}\right)\right]$$

$$\left. + \delta w\,dy\left[-kGh\left(\frac{\partial w}{\partial x} - \psi_x\right)\right] - \delta w\,dx\left[-kGh\left(\frac{\partial w}{\partial y} - \psi_y\right)\right]\right\}dt = 0 \qquad (7.130)$$

Using Eqs. (7.114), (7.115), (7.117), and (7.118) this becomes:

$$\int_{t_1}^{t_2}\oint_{\Gamma}[M_x\,\delta\psi_x\,dy - M_y\,\delta\psi_y\,dx + M_{xy}\,\delta\psi_y\,dy - M_{xy}\,\delta\psi_x\,dx$$

$$- Q_x\,\delta w\,dy + Q_y\,\delta w\,dx]\,dt = 0$$

Again in accordance with Eq. (6.39) (see also Fig. 6.9) we replace dx by $-a_{vy}\,ds$ and dy by $a_{vx}\,ds$ to get:

$$\int_{t_1}^{t_2}\oint_{\Gamma}[M_x a_{vx}\,\delta\psi_x + M_y a_{vy}\,\delta\psi_y + M_{xy}a_{vx}\,\delta\psi_y + M_{xy}a_{vy}\,\delta\psi_x$$

$$- Q_x a_{vx}\,\delta w - Q_y a_{vy}\,\delta w]\,ds\,dt = 0 \qquad (7.131)$$

Using v and s as coordinates we can say:

$$\psi_x = a_{vx}\psi_v - a_{vy}\psi_s$$
$$\psi_y = a_{vy}\psi_v + a_{vx}\psi_s \qquad (7.132)$$

where ψ_v and ψ_s are components of $\boldsymbol{\psi}$ in the v and s directions respectively. Solving for ψ_v and ψ_s we get:

$$\psi_v = a_{vx}\psi_x + a_{vy}\psi_y$$
$$\psi_s = -a_{vy}\psi_x + a_{vx}\psi_y \qquad (7.133)$$

Now going back to Eq. (7.131) we have:

$$\int_{t_1}^{t_2} \oint_\Gamma [M_x a_{vx}(a_{vx}\,\delta\psi_v - a_{vy}\,\delta\psi_s) + M_y a_{vy}(a_{vy}\,\delta\psi_v + a_{vx}\,\delta\psi_s)$$
$$+ M_{xy}[a_{vx}(a_{vy}\,\delta\psi_v + a_{vx}\,\delta\psi_s) + a_{vy}(a_{vx}\,\delta\psi_v - a_{vy}\,\delta\psi_s)]$$
$$- (Q_x a_{vx} + Q_y a_{vy})\,\delta w]\,ds\,dt = 0$$

Collecting terms:

$$\int_{t_1}^{t_2} \oint_\Gamma \{[M_x a_{vx}{}^2 + M_y a_{vy}{}^2 + 2a_{vx}a_{vy}M_{xy}]\,\delta\psi_v + [-M_x a_{vx}a_{vy} + M_y a_{vx}a_{vy}$$
$$+ M_{xy}(a_{vx}{}^2 - a_{vy}{}^2)]\,\delta\psi_s - (Q_x a_{vx} + Q_y a_{vy})\,\delta w\}\,ds\,dt = 0$$

Going back to Eqs. (6.15) and (6.16) we see that the above equation can be given as follows:

$$\int_{t_1}^{t_2} \oint [M_v\,\delta\psi_v + M_{vs}\,\delta\psi_s - Q_v\,\delta w]\,ds\,dt = 0 \qquad (7.134)$$

We conclude from the above statement that along the boundary of the plate:

$$\text{EITHER} \quad M_v = 0 \quad \text{OR} \quad \psi_v \text{ IS SPECIFIED} \qquad (a)$$
$$\text{EITHER} \quad M_{vs} = 0 \quad \text{OR} \quad \psi_s \text{ IS SPECIFIED} \qquad (b)$$
$$\text{EITHER} \quad Q_v = 0 \quad \text{OR} \quad w \text{ IS SPECIFIED} \qquad (c) \quad (7.135)$$

We get here three conditions on the edge of the plate as opposed to the two conditions in classical plate theory that required considerable discussion.

We have thus posed the entire boundary value problem. We can solve Eq. (7.126) for w and then going back to Eqs. (7.125) we can get ψ_x and ψ_y. The natural boundary conditions have to do with the resultant force, moment, and torque intensities imposed at the edges. These can then be written as conditions on w, ψ_x, and ψ_y on the boundaries by employing Eqs. (6.15)–(6.16) as well as Eqs. (7.114), (7.115),

(7.117), and (7.118). As for the factor k, we have as yet not had to specify it in this theory. Actually for isotropic materials we find that the factor 5/6 gives good results in many situations.

We shall not attempt full solutions of the boundary value problem that has been developed. Instead we shall consider the case of a simply-supported rectangular plate and we shall examine the natural frequencies of vibration for this case when rotatory inertia and shear are included. Accordingly for a plate $a \times b \times h$ (see Fig. 7.10) we assume that there is a free vibration in a mode. We assume w to have the following form for this case:

$$w = A_{mn} \sin \frac{m\pi x}{a} \sin \frac{n\pi y}{b} \cos \omega_{mn} t \qquad (7.136)$$

Note that the function specifies that w is zero on the boundaries for integer values of n and m and so we satisfy the kinematic conditions for w given by Eq. (7.135(c)). Since we shall only seek ω_{mn} we shall not need to solve the entire boundary value problem and hence we shall not get involved with ψ_x, ψ_y and the boundary conditions involving these functions. Accordingly we substitute Eq. (7.136) into Eq. (7.126) with q set equal to zero to get the following result on canceling $(A_{mn} \sin (m\pi x/a) \sin (n\pi y/b) \cos \omega_{mn} t)$:

$$\left\{ -\left(\frac{m\pi}{a}\right)^2 - \left(\frac{n\pi}{b}\right)^2 + \frac{\rho\omega_{mn}^2}{kG} \right\} \left\{ -D \left[\left(\frac{m\pi}{a}\right)^2 + \left(\frac{n\pi}{a}\right)^2 \right] + \frac{\rho h^3 \omega_{mn}^2}{12} \right\} - \rho h \omega_{mn}^2 = 0 \qquad (7.137)$$

To simplify to the classical theory we can show that $\rho h^3/12$ and $1/kGh$ must be set equal to zero. Thus, when this is done above we get

$$\frac{\rho h \omega_{mn}^2}{D} = \left[\left(\frac{m\pi}{a}\right)^2 + \left(\frac{n\pi}{b}\right)^2 \right]^2 \qquad (7.138)$$

which is precisely the result reached earlier in Sec. 7.9 for the classical theory. We now introduce the following dimensionless parameters:

$$(\Omega_{mn})^2 = \frac{\rho h^3(1 - v^2)}{12E} \omega_{mn}^2 \qquad (a)$$

$$S^2 = \frac{1}{12} \left(\frac{h}{a}\right)^2 \qquad (b)$$

$$\lambda_{mn} = \frac{a/m}{b/n} \qquad \text{(aspect ratio)} \qquad (c) \quad (7.139)$$

Substituting for ω_{mn} in Eq. (7.138) using Eq. (7.139(a)) and replacing D by $Eh^2/[12(1 - v^2)]$ we get, on using Eqs. (7.139(b) and (c)), the classical value of Ω_{mn}^2:

$$[\Omega_{mn}^2]_{\text{class}} = (m\pi)^4 S^4 [1 + \lambda_{mn}^2]^2 \qquad (7.140)$$

And in going back to the improved theory (Eq. (7.137)) the aforementioned substitutions lead to the following result:

$$\Omega_{mn}^2 - \{\Omega_{mn}^2 - \Gamma_{mn} S^2 (m\pi)^2\} \left\{ \frac{E'}{kG} \Omega_{mn}^2 - (m\pi)^2 S^2 \Gamma_{mn} \right\} = 0 \qquad (a)$$

where:

$$\Gamma_{mn} = 1 + \lambda_{mn}^2 \quad \text{and} \quad E' = \frac{E}{1 - v^2} \qquad (b) \quad (7.141)$$

To obtain a simplified solution to Eq. (7.141) for Ω_{mn}^2 we let:

$$\Omega_{mn}^2 = (\Omega_{mn}^2)_{\text{class}} + \delta^2 \qquad (7.142)$$

The success of this procedure rests on the condition that δ^4 is much smaller than $(\Omega_{mn}^2)_{\text{class}}$ or δ^2. Substituting Eq. (7.142) into Eq. (7.141) we obtain the result:

$$(\Omega_{mn}^2)_{\text{class}} + \delta^2 - \{(\Omega_{mn}^2)_{\text{class}} + \delta^2 - \Gamma_{mn}S^2(m\pi)^2\}\left\{\frac{E'}{kG}[(\Omega_{mn}^2)_{\text{class}} + \delta^2] - (m\pi)^2(S^2)\Gamma_{mn}\right\} = 0$$

We get on carrying out the multiplication and dropping terms involving δ^4:

$$(\Omega_{mn}^2)_{\text{class}} + \delta^2 - (\Omega_{mn}^4)_{\text{class}}\frac{E'}{kG} - (\Omega_{mn}^2)_{\text{class}}\frac{E'}{kG}\delta^2 - \delta^2(\Omega_{mn}^2)_{\text{class}}\frac{E'}{kG}$$

$$+ (m\pi)^2 S^2\Gamma_{mn}(\Omega_{mn}^2)_{\text{class}} + (m\pi)^2(S^2)(\Gamma_{mn})\delta^2$$

$$+ \Gamma_{mn}S^2(m\pi)^2\frac{E'}{kG}(\Omega_{mn}^2)_{\text{class}} + \Gamma_{mn}S^2(m\pi)^2\frac{E'}{kG}\delta^2$$

$$- (m\pi)^4 S^4 \Gamma_{mn}^2 = 0$$

Solving for δ^2 we have:

$$\delta^2 = \frac{-(\Omega_{mn}^2)_{\text{class}} + \dfrac{E'}{kG}(\Omega_{mn}^4)_{\text{class}} - (m\pi)^2 S^2\Gamma_{mn}\left(1 + \dfrac{E'}{kG}\right)(\Omega_{mn}^2)_{\text{class}} + (m\pi)^4 S^4\Gamma_{mn}}{1 + (m\pi)^2 S^2\Gamma_{mn}\left(1 + \dfrac{E'}{kG}\right) - 2(\Omega_{mn}^2)_{\text{class}}\dfrac{E'}{kG}}$$

We may now give (Ω_{mn}^2) as follows:

$$(\Omega_{mn}^2) = (\Omega_{mn}^2)_{\text{class}} + \delta^2 = (\Omega_{mn}^2)_{\text{class}}\left[\frac{-(\Omega_{mn}^2)_{\text{class}}\dfrac{E'}{kG} + (m\pi)^4 S^4\pi^2/(\Omega_{mn}^2)_{\text{class}}}{1 + (m\pi)^2 S^2\Gamma_{mn}\left(1 + \dfrac{E'}{kG}\right) - 2(\Omega_{mn}^2)_{\text{class}}\dfrac{E'}{kG}}\right]$$

Replace $(\Omega_{mn}^2)_{\text{class}}$ in the bracketed expression above by $(m\pi)^4 S^4\Gamma_{mn}^2$ from Eqs. (7.140) and (7.141(b)):

$$(\Omega_{mn}^2) = (\Omega_{mn}^2)_{\text{class}}\frac{1 - (m\pi)^4 S^4\Gamma_{mn}^2\dfrac{E'}{kG}}{1 + (m\pi)^2 S^2\Gamma_{mn}\left(1 + \dfrac{E'}{kG}\right) - 2(m\pi)^4 S^4\Gamma_{mn}^2\dfrac{E'}{kG}}$$

For thin plates $S^2 \ll 1$. We can then drop the last terms in the numerator and denominator and, expanding what is left as a power series, we can then say as an approximation:

$$\Omega_{mn}^2 = (\Omega_{mn}^2)_{\text{class}}\left[1 - \left(1 + \frac{E'}{kG}\right)S^2(m\pi)^2\Gamma_{mn}\right] \qquad (7.143)$$

We have now the means of ascertaining natural frequencies of the simply-supported plate including rotatory inertia and shear.

Part C
GENERAL CONSIDERATIONS

7.13 THE EIGENFUNCTION–EIGENVALUE PROBLEM RESTATED

In the consideration of the vibrations of beams and plates undertaken in Parts A and B of this chapter we proceeded by the familiar separation of variable technique to arrive at an equation of the form

$$L(W) = \omega^2 M(W) \qquad (7.144)$$

where L and M are differential operators acting on the dependent variable $W(x,y)$. Also there were linear homogeneous boundary conditions of the form

$$N(W) = 0 \qquad (7.145)$$

where N is a linear differential operator. Thus in the case of the beam we see from Eq. (7.17(b)) that:

$$L = \frac{d^2}{dx^2}(EI)\frac{d^2}{dx^2} \qquad (a)$$

$$M = \rho A \qquad (b) \ (7.146)$$

And in the case of classical plates we see from Eq. (7.85) that:

$$L = \nabla^4 \qquad (a)$$

$$M = \frac{1}{\beta^2} \qquad (b) \ (7.147)$$

These are examples of eigenvalue-eigenfunction problems. The eigenvalue-eigenfunction problem then asks the question: *What function W, satisfying the given boundary conditions of the problem, when acted on by the operator L gives a result which is equal to some number ω^2 times the result of the M operator acting on this function W.* The function W that satisfies the aforestated condition is called an *eigenfunction* while the accompanying value ω^2 is called the *eigenvalue*. In the problems examined in Parts A and B of the chapter we found that the ω^2, for the problems considered, formed an infinite discrete set of numbers giving the natural frequencies of the vibration problem. The eigenfunctions for the problem were the so-called mode shapes for free vibration. (It is possible that several functions W may correspond to a single eigenvalue ω^2 and we then have a *degeneracy* present. We shall not get involved here in such considerations.)

In the two problems undertaken (the simply-supported beam and the simply-supported plate) we found that integrals of the product of two eigenfunctions over the domain of the problem were zero when these eigenfunctions were different from each other. This was a useful property in establishing the constants of integration to satisfy the initial conditions of the problem. Actually you will soon see that this condition is a consequence of a property of the function called *orthogonality*. We now define orthogonal eigenfunctions as those satisfying the equation $L(W) = \omega^2 M(W)$ for which the following conditions hold:

$$\iiint W_i L(W_j)\, dv = 0 \qquad \text{if } i \neq j \quad (a)$$

$$\iiint W_i M(W_j)\, dv = 0 \qquad \text{if } i \neq j \quad (b) \quad (7.148)$$

We will now show that the eigenfunctions W_i are orthogonal in the above sense if the operators L and M are *symmetric*.[1] Using the more universal notation λ_i for ω_i^2 we consider eigenfunctions W_i and W_j satisfying the differential equation as follows:

$$L(W_i) = \lambda_i M(W_i) \qquad (a)$$

$$L(W_j) = \lambda_j M(W_j) \qquad (b) \quad (7.149)$$

Hence

$$W_j L(W_i) = \lambda_i W_j M(W_i) \qquad (a)$$

$$W_i L(W_j) = \lambda_j W_i M(W_j) \qquad (b) \quad (7.150)$$

where we have multiplied $(7.149(a))$ by W_j and $(7.149(b))$ by W_i. Integrating over the domain of the problem and subtracting we then get the result:

$$\iiint_D [W_j L(W_i) - W_i L(W_j)]\, dv = \iiint_D [\lambda_i W_j M(W_i) - \lambda_j W_i M(W_j)]\, dv$$

Now employing the self-adjoint property of the operators we get:

$$0 = (\lambda_i - \lambda_j) \iiint_D W_i M(W_j)\, dv$$

Clearly if $\lambda_i \neq \lambda_j$, i.e., if the eigenvalues are different, then we conclude that:

$$\iiint_D W_i M(W_j)\, dv = 0 \qquad (7.151)$$

Now going back to Eq. 7.150(a) we integrate over the domain and note the above conclusion. We then must conclude that

$$\iiint_D W_j L(W_i)\, dv = 0 \qquad (7.152)$$

---

[1] A symmetric or self-adjoint operator S, you will recall from Chap. 3, is one for which $(u, Sv) = (v, Su)$.

thus completing the proof. Now we have already shown in Chap. 3 that the operators L and M are self-adjoint for beams and plates under certain boundary conditions (see Sec. 3.11). We leave it for you to show accordingly that the simply-supported beam and the simply-supported plate examined earlier have boundary conditions that satisfy the conditions presented in Chap. 3 and thereby have orthogonal eigenfunctions.

7.14   THE RAYLEIGH QUOTIENT IN TERMS OF OPERATORS

We will now develop Rayleigh's quotient directly from the differential equation for W in the case of a beam. Thus we have for the beam from Eq. (7.17(b)):

$$\frac{d^2}{dx^2}\left[EI\frac{d^2W}{dx^2}\right] = \rho A\omega^2 W$$

Now multiply both sides of the equation by W. Integrating over the domain of the beam we get

$$\int_0^L W\frac{d^2}{dx^2}\left(EI\frac{d^2W}{dx^2}\right)dx = \omega^2\int_0^L \rho A W^2\,dx \qquad (7.153)$$

Next integrate the left side of the equation by parts twice as follows:

$$\int_0^L W\frac{d^2}{dx^2}\left(EI\frac{d^2W}{dx^2}\right)dx = \int_0^L EI\left(\frac{d^2W}{dx^2}\right)^2 dx$$

$$+ W\frac{d}{dx}\left(EI\frac{d^2W}{dx^2}\right)\Bigg|_0^L - EI\frac{dW}{dx}\frac{d^2W}{dx^2}\Bigg|_0^L$$

For simply-supported or built-in end conditions it is clear that the last two expressions are zero on the right side of the above equation and so we use this result in Eq. (7.153) to solve for ω^2 as follows:

$$\omega^2 = \frac{\int_0^L EI(d^2W/dx^2)^2\,dx}{\int_0^L \rho A W^2\,dx} \qquad (7.154)$$

But this is *Rayleigh's quotient* for the problem (see Eq. (7.33)). Let us express the above equation using the differential operators. Thus going back to Eq. (7.153) we can conclude that:

$$\omega^2 = \frac{\int_0^L WL(W)\,dx}{\int_0^L WM(W)\,dx}$$

Indeed if we go back to Eq. (7.144), multiply by W, and integrate over the domain we get the above result directly.

In the general case, we can give the Rayleigh quotient as follows, using the notation of Chap. 3:

$$\omega^2 = \frac{(W,LW)}{(W,MW)} \qquad (7.155)$$

Thus using L and M for a plate in Eq. (7.155) and employing Green's theorem plus the boundary conditions which we have formulated for plates, we can establish Eq. (7.104) which was developed earlier using conservation of energy considerations. We shall see in the following sections that the Rayleigh quotient as given by Eq. (7.155) (we shall denote it at times at $R(W)$) has very useful properties of a mathematical nature which permit the approximation of the eigenvalues.

7.15 STATIONARY VALUES OF THE RAYLEIGH QUOTIENT

We now prove a very important theorem concerning stationary properties of the Rayleigh quotient. This theorem will form the basis for supporting the statement made earlier concerning the direction of approximation for the Rayleigh–Ritz method and will lend credence to the process.

Let $(\phi_1, \phi_2, \ldots, \phi_{k-1})$ be the first nonzero $(k-1)$ eigenfunctions for the eigenvalue problem

$$L(W) = \omega^2 M(W) \qquad (7.156)$$

where the boundary conditions are such that L and M are self adjoint. We shall extremize the Rayleigh quotient for the above case with respect to functions ψ which satisfy the boundary conditions of the problem and which are *orthogonal* to the aforestated $(k-1)$ eigenfunctions. We will prove that the *minimum of the Rayleigh quotient with respect to ψ is the kth eigenvalue and the extremizing function ψ is simply ϕ_k, the kth eigenfunction.*

As a first step we express ψ as an infinite series expansion in the nonzero eigenfunctions:

$$\psi = \sum_{n=1}^{\infty} C_n \phi_n \qquad (7.157)$$

where C_n are computed in the familiar manner making use of the orthogonality property of ϕ_n. The $(k-1)$ orthogonality conditions then may be expressed as follows:

$$\iiint_D \phi_i L(\psi) \, dv = \iiint_D \phi_i L\left(\sum_{n=1}^{\infty} C_n \phi_n\right) dv = \sum_{n=1}^{\infty} C_n \iiint_D \phi_i L(\phi_n) \, dv = 0$$

$$\iiint_D \phi_i M(\psi) \, dv = \iiint_D \phi_i M\left(\sum_{n=1}^{\infty} C_n \phi_n\right) dv = \sum_{n=1}^{\infty} C_n \iiint_D \phi_i M(\phi_n) \, dv = 0$$

$$i = 1, 2, \ldots, (k-1)$$

Now consider $i = 1$. We have from above making use of the orthogonal property of the eigenfunctions:

$$C_1 \iiint_D \phi_1 M(\phi_1) \, dv = 0$$

But $\iiint \phi_1 M(\phi_1) \, dv$ represents a kinetic energy in this discussion and so it must be positive-definite. Assuming as we did that ϕ_1 is not everywhere zero then clearly $C_1 = 0$. In a similar manner we can conclude that $C_2 = C_3 = \cdots C_{k-1} = 0$. Accordingly we have from the orthogonality requirements on ψ the result that:

$$C_1 = C_2 = \cdots C_{k-1} = 0$$

We now examine the Rayleigh quotient. For this purpose we note on using Eq. (7.157) for ψ

$$\iiint_D \psi L(\psi) \, dv = \iiint_D \sum_{n=k}^{\infty} C_n \phi_n \sum_{j=k}^{\infty} C_j L(\phi_j) \, dv$$

$$= \sum_{n=k}^{\infty} \sum_{j=k}^{\infty} C_n C_j \iiint_D \phi_n L(\phi_j) \, dv$$

$$= \sum_{n=k}^{\infty} C_n^{\ 2} \iiint_D \phi_n L(\phi_n) \, dv$$

$$= \sum_{n=k}^{\infty} C_n^{\ 2} \omega_n^{\ 2} \iiint_D \phi_n M(\phi_n) \, dv \qquad (a) \ (7.158)$$

where we have used the orthogonality property of the eigenfunctions in the next to the last expression and Eq. (7.156) in the last expression. Also using Eq. (7.157) for ψ and again using the orthogonality property of the eigenfunctions we get:

$$\iiint_D \psi M(\psi) \, dv = \iiint_D \sum_{n=k}^{\infty} C_n \phi_n \sum_{j=k}^{\infty} C_j M(\phi_j) \, dv$$

$$= \sum_{n=k}^{\infty} \sum_{j=k}^{\infty} C_n C_j \iiint_D \phi_n M(\phi_j) \, dv$$

$$= \sum_{n=k}^{\infty} C_n^{\ 2} \iiint_D \phi_n M(\phi_n) \, dv \qquad (b) \ (7.158)$$

We now introduce the notation

$$\kappa_n = \iiint_D \phi_n M(\phi_n)\,dv \qquad (7.159)$$

so that the Rayleigh quotient R can be given as follows, making use of Eqs. (7.158(a)) and (7.158(b)):

$$R = \frac{\sum_{n=k}^{\infty} C_n{}^2 \kappa_n \omega_n{}^2}{\sum_{n=k}^{\infty} C_n{}^2 \kappa_n} \qquad (7.160)$$

We can rewrite the above equation as follows:[1]

$$R = \omega_k{}^2 + \frac{\kappa_{k+1}C_{k+1}{}^2(\omega_{k+1}{}^2 - \omega_k{}^2) + \kappa_{k+2}C_{k+2}{}^2(\omega_{k+2}{}^2 - \omega_k{}^2) + \cdots}{\sum_{n=k}^{\infty} C_n{}^2 \kappa_n} \qquad (7.161)$$

Since $\omega_{k+1} > \omega_k$, etc., and since κ must be positive-definite (kinetic energy) we may conclude that the minimum value for R is $\omega_k{}^2$—the kth eigenvalue. This occurs when $C_{k+1}, C_{k+2}, \ldots$ are all zero. Thus the only nonzero coefficient for this condition is C_k and so going back to Eq. (7.157) we see that the function ψ is simply ϕ_k up to a proportionality constant. *Thus the minimizing function ψ for the Rayleigh quotient is the kth eigenfunction.*

Of course if $k = 1$, then the function ψ does not require orthogonality restrictions vis a vis eigenfunctions; it merely must satisfy the boundary conditions of the problem. Thus we see here that *any* such function when substituted into the Rayleigh quotient will yield a result λ_1 which will be greater than or equal to $\omega_1{}^2$ the lowest eigenvalue. And by using a parameter-laden set of functions to express ψ we can adjust the constants to give the smallest Rayleigh quotient for that family of functions and thereby come closer to the lowest eigenvalue $\omega_1{}^2$ *from above*. Finally, since the eigenfunction ϕ_1 is an extremal function for the Rayleigh quotient, clearly a function ψ *reasonably close* to ϕ_1 should give a λ *very close* to $\omega_1{}^2$. These remarks then give the bases for Rayleigh's method of finding the lowest eigenvalue.

The question now arises as to the basis of the Rayleigh–Ritz method for finding eigenvalues *above* the lowest eigenvalue. We now re-examine the Rayleigh–Ritz method itself in a more formal manner in order to prepare the ground work for assessing its validity.

---

[1] You may easily justify the formulation by recombining the terms under a common denominator and carrying out necessary cancellation of terms in the numerator.

7.16 RAYLEIGH–RITZ METHOD RE-EXAMINED

We shall now develop the Rayleigh–Ritz method for finding eigenvalues above the lowest eigenvalue by a method that is closely tied to the theorem presented in the previous section.

In the usual manner we consider the linear combination of n functions ψ_i that satisfy the boundary conditions of the problem to form W as follows:

$$W = C_1\psi_1 + \cdots + C_n\psi_n \qquad (7.162)$$

By limiting the possible deflections of the body to what is describable by the expression given above, we have tacitly built in hidden constraints such that the body instead of having an infinite number of degrees of freedom will soon be seen to have only n degrees of freedom. We are in effect promulgating a new system to represent the original system. We can find the lowest eigenvalue of this hypothetical system by extremizing the Rayleigh quotient. Clearly $\Lambda_1{}^2$ will be some sort of approximation of the eigenvalue $\omega_1{}^2$ of the actual problem. We can determine the C's[1] and thus establish the corresponding eigenfunction for our hypothetical system. Thus:

$$W^{(1)} = C_1^{(1)}\psi_1 + \cdots + C_n^{(1)}\psi_n \qquad (7.163)$$

The superscript indicates that we are dealing with the first mode. Presumably $W^{(1)}$ must then be some approximation of the first eigenfunction of the actual problem. To get the second eigenvalue of the hypothetical system we employ the minimum theorem of the previous section by minimizing the Rayleigh quotient using admissible functions W given by Eq. (7.162) but now subject to the requirement that they be orthogonal to $W^{(1)}$. This generates $\Lambda_2{}^2$ and a second eigenfunction for the hypothetical system given in the form:

$$W^{(2)} = C_1^{(2)}\psi_1 + \cdots + C_n^{(2)}\psi_n$$

The procedure is continued in this manner. Thus for the kth eigenvalue we adjust the C's to extremize the Rayleigh quotient and also so as to make W orthogonal to the $k - 1$ lower eigenfunctions of the hypothetical system. Let us now do this more formally. Accordingly for the process to approximate the kth eigenvalue from above we have

$$\Lambda_k{}^2 = \frac{\iiint_D WL(W)\,dv}{\iiint_D WM(W)\,dv} = \frac{\iiint_D \left(\sum_{i=1}^{n} C_i\psi_i\right) L\left(\sum_{j=1}^{n} C_j\psi_j\right) dv}{\iiint_D \left(\sum_{i=1}^{n} C_i\psi_i\right) M\left(\sum_{j=1}^{n} C_j\psi_j\right) dv}$$

---

[1] This can be done up to a multiplicative constant.

$$= \frac{\sum\limits_{i=1}^{n}\sum\limits_{j=1}^{n} C_i C_j G_{ij}}{\sum\limits_{i=1}^{n}\sum\limits_{j=1}^{n} C_i C_j E_{ij}} \qquad (7.164)$$

where

$$G_{ij} = \iiint_D \psi_i L(\psi_j)\, dv$$

$$E_{ij} = \iiint_D \psi_i M(\psi_j)\, dv \qquad (7.165)$$

We have the following orthogonality requirement:

$$\iiint_D W^{(S)} M\left(\sum_{i=1}^{n} C_i \psi_i\right) dv = 0 \qquad S = 1, 2, \ldots, (k-1)$$

Replacing $W^{(S)}$ by $\sum_{j=1}^{n} C_j^{(S)} \psi_j$, we may rewrite the above equations as follows:

$$\iiint_D W^{(S)} M\left(\sum_{i=1}^{n} C_i \psi_i\right) dv = \iiint_D \left(\sum_{j=1}^{n} C_j^{(S)} \psi_j\right)\left(\sum_{i=1}^{n} C_i M(\psi_i)\right) dv = 0$$

therefore

$$\sum_{j=1}^{n}\sum_{i=1}^{n} C_j^{(S)} C_i E_{ij} = 0 \qquad S = 1, 2, \ldots, k-1 \qquad (7.166)$$

The above $k-1$ equations are *constraining equations* for the extremization process and so we extremize the function Λ_k^2* given as

$$\Lambda_k^2* = \frac{\sum\limits_{i=1}^{n}\sum\limits_{j=1}^{n} C_i C_j G_{ij}}{\sum\limits_{i=1}^{n}\sum\limits_{j=1}^{n} C_i C_j E_{ij}} - \sum_{S=1}^{k-1} \lambda^{(S)} \sum_{j=1}^{n}\sum_{i=1}^{n} C_j^{(S)} C_i E_{ij}$$

where $\lambda^{(S)}$ are Lagrange multipliers. To extremize Λ_k^2* we proceed as follows:

$$\frac{\partial \Lambda_k^2*}{\partial C_i} = 0 \qquad i = 1, 2, \ldots, n$$

therefore

$$\frac{2 \sum\limits_{j=1}^{n} C_j G_{ij}}{\sum\limits_{i=1}^{n}\sum\limits_{j=1}^{n} C_i C_j E_{ij}} - \frac{\left(\sum\limits_{i=1}^{n}\sum\limits_{j=1}^{n} C_i C_j G_{ij}\right)\left(2 \sum\limits_{j=1}^{n} C_j E_{ij}\right)}{\left(\sum\limits_{j=1}^{n}\sum\limits_{i=1}^{n} C_i C_j E_{ij}\right)^2} - \sum_{S=1}^{k-1} \lambda^{(S)} \sum_{j=1}^{n} C_j^{(S)} E_{ij} = 0$$

$$i = 1, 2, \ldots, n$$

Next multiply through by $\frac{1}{2}\sum_{i=1}^{n}\sum_{j=1}^{n}C_iC_jE_{ij}$ to get:

$$\sum_{j=1}^{n}C_jG_{ij} - \left(\frac{\sum_{i=1}^{n}\sum_{j=1}^{n}C_iC_jG_{ij}}{\sum_{i=1}^{n}\sum_{j=1}^{n}C_iC_jE_{ij}}\right)\left(\sum_{j=1}^{n}C_jE_{ij}\right)$$

$$-\left[\sum_{S=1}^{k-1}\lambda^{(S)}\sum_{j=1}^{n}C_j^{(S)}E_{ij}\right]\left[\frac{1}{2}\sum_{v=1}^{n}\sum_{p=1}^{n}C_vC_pE_{vp}\right] = 0$$

$$i = 1, 2, \ldots, n$$

Note we have changed dummy indices in the last term for convenience. Using Eq. (7.164) to bring in Λ_k^2 into the second expression of the above equation we get:

$$\sum_{j=1}^{n}C_jG_{ij} - \Lambda_k^2\sum_{j=1}^{n}C_jE_{ij}$$

$$-\frac{1}{2}\left[\sum_{S=1}^{k-1}\lambda^{(S)}\sum_{j=1}^{n}C_j^{(S)}E_{ij}\right]\left[\sum_{v=1}^{n}\sum_{p=1}^{n}C_vC_pE_{vp}\right] = 0$$

$$i = 1, \ldots, n \qquad (7.167)$$

Note first that *if the $(k-1)$ Lagrange multipliers are zero in value* then we have

$$\sum_{j=1}^{n}(G_{ij} - \Lambda_k^2E_{ij})C_j = 0 \qquad i = 1, 2, \ldots, n \qquad (7.168)$$

We will prove in Appendix IV that this is indeed the situation (the proof while straightforward is somewhat tedious and offers no new insight). Considering the C's as unknowns we then have n simultaneous homogeneous equations for the unknowns. For a nontrivial solution it is necessary that the determinant of the C's be zero. That is, dropping the subscript k for reasons soon to be apparent we have:

$$|G_{ij} - \Lambda^2E_{ij}| = 0 \qquad (7.169)$$

We will get n roots for Λ^2 which we order in magnitude as $\Lambda^{2(1)}, \Lambda^{2(2)}, \ldots, \Lambda^{2(n)}$. Note in particular that no matter what k is for the problem (assuming it is no greater than n) we arrive at the *same* equation (7.169) to be solved. That is, this equation applies for *all* the eigenvalues of the hypothetical system from 1 to n that we might be seeking to establish by this method. And so we interpret the n roots Λ^2 as the eigenvalues of the hypothetical system and hence as the sought-for approximations from above of the n eigenvalues of the actual problem. By using the proper operators L and M for the beam and the plate we can also see that the formulation of the Λ's above is *exactly the same formulation* that was used in Parts A and B respectively for the Rayleigh–Ritz method, thus justifying our use here of the same notation Λ as was used earlier. Note also that we can get a set of constants $C_i^{(k)}$ for each root Λ_k^2 by substituting Λ_k^2 into Eq. (7.168) and solving for the constants $C_i^{(k)}$. However, because the equations are homogeneous, we can solve uniquely only for the *ratios* of these constants. This gives the *shape* of the eigenfunctions of the hypothetical system and hence approximations of the actual eigenfunctions but not their amplitudes.

Now that we have presented the Rayleigh–Ritz method in a framework directly from the vital minimum theorem of Sec. 7.15 we will be able soon to show that the n values Λ^2 in ascending order of size as computed by the Rayleigh–Ritz method are approximations from above of the first n eigenvalues of the problem. For this step we need the use of the maximum–minimum principle which we now present.

7.17 MAXIMUM–MINIMUM PRINCIPLE

We now consider $k - 1$ functions $g_1, g_2, \ldots, g_{k-1}$ which are perfectly arbitrary except for simple integrability requirements in the domain of interest. Now as admissible functions ψ to extremize the Rayleigh quotient for a given eigenvalue-eigenfunction problem we shall consider those which satisfy the boundary conditions of the problem and satisfy the following $(k - 1)$ orthogonality conditions:

$$\iiint_D g_i M(\psi)\, dv = 0 \qquad i = 1, 2, \ldots, k - 1 \qquad (7.170)$$

We will now show that the *minimum value of the Rayleigh quotient with respect to ψ for the given set of functions g_i will be less than the kth eigenvalue of the problem and will equal the kth eigenvalue of the problem if the g's are taken as the first $k - 1$ eigen-functions ϕ_i of the problem*—thus getting back to the case presented in Sec. 7.15.

To prove this we first show that always there exists a function ψ_0 which is a linear combination of the first k eigenfunctions

$$\psi_0 = C_1 \phi_1 + \cdots C_k \phi_k \qquad (7.171)$$

and which satisfies Eq. (7.170) for any set of functions g_i. To show this, substitute Eq. (7.171) into Eq. (7.170) as follows:

$$\iiint_D g_i M\left(\sum_{j=1}^{k} C_j \phi_j \right) dv = \sum_{j=1}^{k} \iiint_D g_i C_j M(\phi_j)\, dv = 0$$

$$i = 1, 2, \ldots, (k - 1) \qquad (7.172)$$

Since i goes from 1 to $(k - 1)$ there are hence $k - 1$ equations. Considering the C's as k unknowns it is clear that we can always find, although not uniquely, a set of values for these constants which will satisfy the equation (7.172). That is, because we have more unknowns than equations here we can always determine a set of values for the unknowns that satisfy the equations.

Now employ the function ψ_0 in the Rayleigh quotient R. We may in this regard use the results stemming from Eqs. (7.158(a)) and (7.158(b)) (by summing now from 1 to k) as well as the definition given by Eq. (7.159). This leads to Eq. (7.160) which is made valid for our case here by using $n = 1$ to k giving the result:

$$R = \frac{C_1{}^2 \kappa_1 \omega_1{}^2 + C_2{}^2 \kappa_2 \omega_2{}^2 + \cdots + C_k{}^2 \kappa_k \omega_k{}^2}{C_1{}^2 \kappa_1 + C_2{}^2 \kappa_2 + \cdots + C_k{}^2 \kappa_k} \qquad (7.173)$$

Remembering that $\omega_1{}^2 \leq \omega_2{}^2 \ldots \leq \omega_k{}^2$ and noting that the κ's are positive-definite we can conclude on replacing $\omega_1{}^2, \ldots, \omega_{k-1}{}^2$ by $\omega_k{}^2$ in the above equation that:

$$R \leq \frac{C_1{}^2 \kappa_1 \omega_k{}^2 + \cdots + C_k{}^2 \kappa_k \omega_k{}^2}{C_1{}^2 \kappa_1 + \cdots + C_k{}^2 \kappa_k} = \omega_k{}^2 \qquad (7.174)$$

Thus we see that the Rayleigh quotient for ψ_0, which is a member of the family of admissible functions ψ, *is less than or equal* to the kth eigenvalue of the problem. Certainly the *minimum* of R with respect to the *whole* admissible class of functions ψ must then be less than or equal to $\omega_k{}^2$. The equality is achieved according to the minimum theorem of Sec. 7.15 when the functions g_i used for orthogonalization are the first $(k - 1)$ eigenfunctions of the problem. The theorem is now proved.

What is the relation between the minimum principle presented earlier in Sec. 7.15 (and used to justify Rayleigh's method for the first mode) and the present maximum–minimum principle? We shall use for this explanation some plots whose detailed meaning is not of concern here. Thus in Fig. 7.14 we have shown a region I which is meant to give all the values of the Rayleigh quotient when the admissible functions are made orthogonal to the first $(k - 1)$ *eigenfunctions* of the problem. We identify the family of functions involved in the orthogonality conditions as $\Theta(\ )$. Thus for region I we have indicated $\Theta(\phi_i)$ in the diagram. Notice that the *minimum* value for this family of Rayleigh quotients is $\omega_k{}^2$ in accordance with the *minimum principle*. Next consider a system of Rayleigh quotients for which the admissible functions have been made orthogonal to a set of $(k - 1)$ functions g_i (not equal to ϕ_i). This set of functions is denoted as $\Theta(g_i)$ and the system of Rayleigh quotients is shown in the diagram as region II. Notice now that the minimum value of R is less than $\omega_k{}^2$ in accordance with the *maximum–minimum* principle of this section. By changing the functions g_i we may form other regions in our plot such as regions III and IV and thereby increase the minimum value of the Rayleigh quotient as has been shown in the diagram. One could then conceivably continually increase the minimum values of R for each family Θ by continuously changing the family of functions g_i (see dashed line on the diagram). The maximum value of such minimums is attained when Θ consists of the eigenfunctions ϕ_i.

We shall next employ the maximum–minimum principle to show that $\Lambda_k{}^2 \geq \omega_k{}^2$, thus justifying the earlier assertions of estimating from above by the Rayleigh–Ritz method.

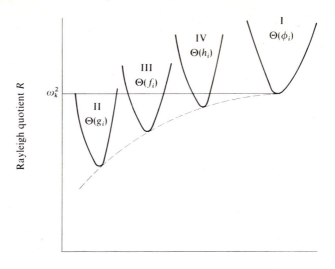

FIGURE 7.14

7.18 JUSTIFICATION OF THE ESTIMATION FROM ABOVE ASSERTION OF THE RAYLEIGH–RITZ METHOD

We would like to show here that $\Lambda_k^2 \geq \omega_k^2$ where Λ_k^2 is calculated from the Rayleigh–Ritz procedure as used in parts A and B and as presented more formally in part C of this chapter.

Consider two systems A and B identical in every way except that system B has additional hidden constraints making it a "stiffer" system. Such constraints we have pointed out, occur as a result of simplifications made in the manner in which system B deforms. Such simplifications occur when we replace a body having infinite number of degrees of freedom by a substitute having a finite number of degrees of freedom as we did in the Rayleigh–Ritz method described in Sec. 7.16. The admissible functions $\bar{\psi}_i$ for minimizing the Rayleigh quotient for B will be a *subset* of the set ψ_i for system A. Taking ϕ_i to be the *eigenfunctions* of system A, we can say, on minimizing the Rayleigh quotient with respect to the complete set ψ_i while maintaining orthogonality with respect to the first $k - 1$ eigenfunctions ϕ_i of system A:

$$\omega_k^2 = [R_{\min}(\psi_i)]_{\Theta(\phi_i)} \qquad (7.175)$$

The above result is directly from the minimum theorem. If now we use *subset* $\bar{\psi}_i$ to extremize the Rayleigh quotient above we clearly get a result which is *greater than or equal to* the value of the minimum Rayleigh quotient using the complete set ψ_i.

Thus we can say:

$$\omega_k{}^2 \leq [R_{\min}(\overline{\psi}_i)]_{\Theta(\phi_i)} \qquad (7.176)$$

Now in the Rayleigh–Ritz method as pointed out we effectively replace the system with infinite degrees of freedom by one having as many degrees of freedom as the number of coordinate functions used. We shall consider B to be such a system formed from system A. The eigenfunctions of system B are the so-called *approximate eigenfunctions* of system A coming out of the Rayleigh–Ritz method. We denote them as $(\phi_i)_{\mathrm{app}}$. Thus using the minimum theorem again for system B and denoting $\Lambda_k{}^2$ as the kth eigenvalue for system B:

$$\Lambda_k{}^2 = [R_{\min}(\overline{\psi}_i)]_{\Theta(\phi_i)_{\mathrm{app}}} \qquad (7.177)$$

Now we compare the right sides of (7.176) and (7.177). We can use the maximum–minimum principle, noting that $(\phi_i)_{\mathrm{app}}$ and *not* ϕ_i are the *eigenfunctions* for system B,[1] to conclude that:

$$[R_{\min}(\overline{\psi}_i)]_{\Theta(\phi_i)} \leq [R_{\min}(\overline{\psi}_i)]_{\Theta(\phi_i)_{\mathrm{app}}} \qquad (7.178)$$

Accordingly, using Eq. (7.176) to replace the left side of the above inequality and Eq. 7.177 to replace the right side of the above inequality we have:

$$\omega_k{}^2 \leq \Lambda_k{}^2$$

We have thus accomplished our goal since $\omega_k{}^2$ may be considered the eigenvalues of the problem and $\Lambda_k{}^2$ clearly are the approximate eigenvalues of the problem found by the Rayleigh–Ritz method.

7.19 CLOSURE

We began this chapter by presenting Hamilton's principle which permits us to set forth differential equations of motion. In particular we formulated the equations of motion for the simple beam and the Timoshenko beam in Part A and then, in Part B, we did the same for simple plates and the Mindlin plate. Once these equations were developed, we concentrated on finding approximate solutions for natural frequencies of these bodies via Rayleigh's method and the Rayleigh–Ritz method. At first we employed conservation of energy as the underlying basis for the methods. We found in our examples that the approximations for the natural frequencies via these methods were always from above. In Part C of the chapter we generalized the arguments of the preceding sections to encompass the eigenvalue–eigenfunction problem thereby presenting the Rayleigh quotient and its properties. With the aid of the minimum

---

[1] ϕ_i corresponds to the functions g_i of the maximum–minimum principle.

theorem and the minimum–maximum theorem we were able to justify that the approximations for eigenfunctions via the Rayleigh and Rayleigh–Ritz methods had to be from above. The generalizations of Part C will permit us to employ the Rayleigh and Rayleigh–Ritz methods for the stability problems of columns and plates in Chap. 9.

You will recall in the chapter on torsion that we got approximations from above for the torsional rigidity via the Ritz method, and in addition got approximations from below via the Trefftz method—thus bracketing the correct torsional rigidity. We wish to point out here that it is possible also to find approximations from below of eigenvalues, thus permitting a bracketing of the correct eigenvalues when results from the Rayleigh and Rayleigh–Ritz methods are used. The method is due to A. Weinstein and involves the use of weakened or relaxed boundary conditions. Detailed discussion of the method however, is beyond the scope of this text.[1]

Up to this time we have considered only the case of small deformation, i.e., we have maintained geometric linearity. On several occasions, however, we considered nonlinear constitutive laws. In the following chapter we examine certain aspects of nonlinear geometric behavior. In doing so we shall continue to use a linear constitutive law, namely Hooke's law.

READING

DEN HARTOG, J. P.: "Mechanical Vibrations," McGraw-Hill Book Co., 1947.
GOULD, S. H.: "Variational Methods for Eigenvalue Problems," University of Toronto Press.
MEIROVITCH, L.: "Analytical Methods in Vibrations," Macmillan Co., N.Y., 1970.
MIKHLIN, S. G.: "Variational Methods of Mathematical Physics," Macmillan Co., N.Y., 1964.
NOWACKI, W.: "Dynamics of Elastic Systems," Chapman and Hall, London, 1963.
RAYLEIGH, J. W. S.: "The Theory of Sound," Dover.
TEMPLE, G., AND BICKLEY, W. G.: "Rayleigh's Principle," Dover Publications.
TIMOSHENKO, S.: "Vibration Problems in Engineering," Van Nostrand Co., 1955.
TONG, K. N.: "Theory of Mechanical Vibration," John Wiley and Sons., N.Y., 1960.
VOLTERRA, E., AND ZACHMANOGLOU, E. C.: "Dynamics of Vibrations," Merrill Books, Inc., 1965.

PROBLEMS

7.1 Derive Hamilton's principle for a system using the δ-operator approach rather than using the single-parameter family approach as we have done in Sec. 7.3.

7.2 Verify Eqs. (7.30(a)) and (7.30(b)).

7.3 Derive Eq. (7.38).

[1] For a detailed exposition of Weinstein's method see "Variational Methods for Eigenvalue Problems," by S. H. Gould, University of Toronto Press, 1966.

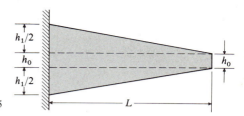

FIGURE 7.15

7.4   Consider the non-uniform cantilever beam shown in Fig. 7.15. Using a simple parabolic shape for the fundamental mode show that Rayleigh's quotient gives as the fundamental natural frequency:

$$\omega = \sqrt{\frac{30}{12}\left[\frac{(h_0 + h_1)^4 - h_0{}^4}{h_1{}^2 + 6h_0 h_1}\right]\frac{E}{\rho L^4}}$$

Compare your result (after suitable adjustments) with the corresponding result for a uniform cantilever beam

$$\omega = 1.015\frac{h_0}{L^2}\sqrt{\frac{E}{\rho}}$$

and that of a pure triangle,

$$\omega = 1.535\frac{h_1}{L^2}\sqrt{\frac{E}{\rho}}$$

(This result is due to Kirchhoff.)

7.5   Consider a cantilever beam with a mass M attached firmly at a position x^* (see Fig. 7.16) from the base ($x = 0$). What is the Rayleigh quotient for transverse vibration? Find an approximation to the fundamental natural frequency. Is it possible that the added mass could not affect natural frequencies of higher modes? Explain. Where should the bar be placed to decrease as much as possible the fundamental frequency?

7.6   Two identical cantilever beams A and B are pinned together at what would be their free ends (see Fig. 7.17). Using functions W_A and W_B for each cantilever, what are the proper boundary conditions these functions must satisfy? Find two functions that satisfy all boundary conditions except one condition at the pin. Now with the aid of the Rayleigh quotient find the fundamental natural frequency approximately.

7.7   Derive the differential equation and boundary conditions for the torsional oscillation of a bar in the absence of warping in terms of the rotation angle $\theta(z,t)$ and using the polar moment of inertia of the cross-vectors $I_p = \iint (x^2 + y^2)\, dA$. Also discuss under what conditions the effect of the geometric cross section of the bar will not be a factor in the solution.

7.8   Do the first part of Problem 7.7 this time including the effects of two rigid cylinders, respectively of uniform mass per unit length ρ_1 and ρ_2 and polar moments of inertia of the cross section I_1 and I_2, placed at the ends $z = 0$ and $z = L$ of the bar.

7.9   Derive the differential equations and boundary conditions for longitudinal vibrations of a uniform bar. Also obtain the appropriate Rayleigh quotient for such a problem.

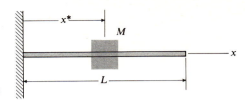

FIGURE 7.16

7.10 In the preceding problem devise approximate eigenfunctions that satisfy the end conditions of

(a) Both ends free, i.e., $\tau_{xx} = E(\partial u/\partial x) = 0$.

(b) Fixed at both ends, i.e., $u = 0$.

(c) Fixed at $x = 0$ and free at $x = L$.

Using the Rayleigh quotient find the lowest eigenvalues for these three cases.

7.11 Find the Rayleigh quotient for torsional vibration of Problem 7.8. What is the effect of the added masses on the free vibration frequency?

7.12 Using a cosine function of z so as to approximate a mode shape with a single nodal point along the shaft, approximate from the Rayleigh quotient the fundamental frequency of the system presented in Problems 7.8 and 7.11. Show that if $I_1, I_2, < I_p$ (recall I_1 and I_2 are polar moments of inertia of the added rigid cylinders and I_p is the polar moment of inertia of the connecting shaft) that

$$\omega \approx \frac{\pi}{L}\sqrt{\frac{G}{\rho}}$$

Note that this corresponds to weak (or no) coupling with the added masses at the boundary conditions, which are ordinarily inhomogeneous.

7.13 We have derived in an earlier problem (Chap. 3) the strain energy of a stretched membrane to be

$$U = \frac{S}{2}\iint_A \left[\left(\frac{\partial w}{\partial x}\right)^2 + \left(\frac{\partial w}{\partial y}\right)^2\right] dx\,dy = \frac{S}{2}\iint_A (\nabla w)^2\,dA \qquad (a)$$

where S is the force per unit length along the membrane. Show from Hamilton's principle that the governing equation for the motion of the membrane is:

$$\nabla^2 w = (\gamma/S)\frac{\partial^2 w}{\partial t^2}$$

where $\gamma = \rho t$, t being the membrane thickness. Determine the boundary conditions for the problem. Show that the Rayleigh quotient for the dynamic behavior of the stretched membrane is:

$$\omega^2 = \frac{S\iint [(\partial W/\partial x)^2 + (\partial W/\partial y)^2]\,dx\,dy}{\gamma \iint W^2\,dx\,dy}$$

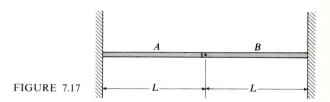

FIGURE 7.17

7.14   Using Rayleigh's quotient find the fundamental frequency of a stretched rectangular membrane supported at $x = 0, a$ and $y = 0, b$. The exact solution is given as

$$\omega_{11}^2 = \frac{S}{\gamma}\pi^2\left(\frac{1}{a^2} + \frac{1}{b^2}\right).$$

7.15   With the aid of Eq. (a) of Problem 7.13 derive the Rayleigh quotient for a circular membrane for axisymmetric deformation. Use this result to obtain an approximation of the fundamental natural frequency whose exact value is 2.404 $(\sqrt{S/\gamma})/a$ where a is the radius, S is the tensile force per unit length and $\gamma = \rho t$.

7.16   With the aid of Eq. (b) of Problem 6.14 formulate the Rayleigh quotient for axisymmetric circular plates. Now obtain an approximation of the lowest frequency of vibration of a clamped circular plate. The exact result is given as

$$\omega = 2.955\sqrt{\frac{Eh^2}{\rho a^4(1 - v^2)}}$$

7.17   Obtain an approximate result for the lowest natural frequency of axisymmetric vibration of an annulus clamped at the inner edge $(r = b)$ and free at the outer edge $(r = a)$. Southwell has given the following exact results for different ratios of b/a:

b/a	0.276	0.642	0.840
ω^*	1.81	7.23	23.4

where

$$\omega^* = \omega\sqrt{\frac{(1 - v^2)\rho}{E}}\left(\frac{a^2}{h}\right)$$

The numbers in the table assume $v = 0.30$.

7.18   Consider a free-free beam. Simplify the Rayleigh quotient employing $\zeta = x/L$ and $k = \omega L^2\sqrt{\rho A/EI}$ such that:

$$k^2 = \frac{\int_0^1 [W''(\zeta)]^2 \, d\zeta}{\int_0^1 [W(\zeta)]^2 \, d\zeta}$$

Find the first two natural frequencies and corresponding eigenfunctions for the free-free beam. Do this by employing a function W reflecting possible rigid body motion, plus possible symmetric deformation, plus possible antisymmetric deformation. The following function is suggested:

$$W(\zeta) = A + B\zeta + C \sin \pi\zeta + D \sin 2\pi\zeta$$

Show that:

$$\begin{cases} k_1 = 22.6 \\ \phi_1 = -\dfrac{2}{\pi} + \sin \pi\zeta \end{cases}$$

$$\begin{cases} k_2 = 64 \\ \phi_2 = -\dfrac{3}{\pi} + \dfrac{6}{\pi}\zeta + \sin 2\pi\zeta \end{cases}$$

Furthermore show that the approximate eigenfunctions are orthogonal. Exact solutions for this problem are $k_1 = 22.4$ and $k_2 = 61.7$.

7.19 Using the operator approach give the Rayleigh quotient for torsional oscillation of a uniform shaft if the equation of torsional motion is:

$$\frac{\partial^2 u}{\partial t^2} = \frac{G}{\rho} \frac{\partial^2 \theta}{\partial z^2}$$

7.20 Using the operator approach what is the Rayleigh quotient for longitudinal vibration of a uniform bar if the equation of longitudinal motion is given as:

$$\frac{\partial^2 u}{\partial t^2} = \frac{E}{\rho} \frac{\partial^2 u}{\partial x^2}$$

8

NONLINEAR ELASTICITY

$$\epsilon_{ij} = \tfrac{1}{2}(u_{i,j} + u_{j,i}) \qquad (8.1)$$

a simplification of the Green strain tensor given in Chap. 1 as:

$$\epsilon_{ij} = \tfrac{1}{2}(u_{i,j} + u_{j,i} + u_{k,i}u_{k,j}) \qquad (8.2)$$

8.1 INTRODUCTION

Up to this time we have restricted the discussion to so-called small deformation. This permitted the employment of the strain tensor of the form:

$$\epsilon_{ij} = \tfrac{1}{2}(u_{i,j} + u_{j,i}) \qquad (8.1)$$

a simplification of the Green strain tensor given in Chap. 1 as:

$$\epsilon_{ij} = \tfrac{1}{2}(u_{i,j} + u_{j,i} + u_{k,i}u_{k,j}) \qquad (8.2)$$

As a result of the small deformation restriction we employed the *original undeformed geometry* in conjunction with the actual stresses (the latter of course exist on the deformed geometry) for the equations of equilibrium. This procedure permits the solution with engineering precision of many important classes of problems. There are, however, problems for which a more exact approach must be taken. One important class of such problems, to be undertaken in the following chapter, is that of elastic stability. Accordingly we shall consider the case of large deformation wherein the kinematics of deformation as well as the formulation of equations of equilibrium must be re-examined. As can be seen immediately from Eq. (8.2) this introduces nonlinearities into the theory from a geometric source—hence the title of the chapter. The constitutive law that we use is yet another way of introducing nonlinearities. We shall use, as we have done in most of the text, the linear, isotropic, Hooke's law. Since we shall use Eq. (8.2) for strain terms ϵ_{ij} we shall introduce next the following notation

$$e_{ij} = \tfrac{1}{2}(u_{i,j} + u_{j,i}) \qquad (8.3)$$

where e_{ij} is still a second-order tensor but with no longer the same physical interpretation as for small deformation. Its terms are now referred to as the *strain parameters*. Thus we have for ϵ_{ij}:

$$\epsilon_{ij} = e_{ij} + \tfrac{1}{2}u_{k,i}u_{k,j} \qquad (8.4)$$

As in Chap. 1, the *rotation parameters* ω_{ij} are given as follows:

$$\omega_{ij} = \tfrac{1}{2}(u_{i,j} - u_{j,i}) \qquad (8.5)$$

8.2   KINEMATICS OF POINTS AND LINE SEGMENTS

Let us reconsider the deformation of a body with no restriction as to the amount of deformation. The position of a material point P is denoted by coordinates x_i in the undeformed state for a reference X_i (see Fig. 8.1), while for the material point in the deformed geometry P' we employ ξ_i measured again relative to X_i. These coordinates (they may also be considered as components of position vectors) are related as follows, using the displacement field u_i:

$$\xi_i = x_i + u_i \qquad (8.6)$$

Now the displacement field may be expressed in terms of the x_i coordinates or the ξ_i coordinates—i.e., in terms of the initial or the final geometry. These are Lagrange and Euler viewpoints discussed in Chap. 1. We shall continue to use the *Lagrange viewpoint* in this chapter. Hence we can express the differential form of the above equation as follows:

$$d\xi_i = dx_i + u_{i,j}\,dx_j = (\delta_{ij} + u_{i,j})\,dx_j \qquad (8.7)$$

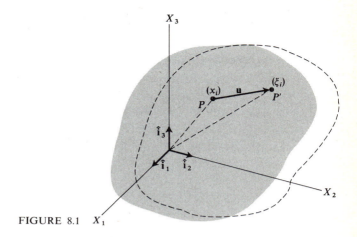

FIGURE 8.1 X_1

Using e_{ij} and ω_{ij} we can give the above equation in the following manner:

$$d\xi_i = (\delta_{ij} + \omega_{ij} + e_{ij})\,dx_j \qquad (8.8)$$

An interpretation which one can ascribe to the differential $d\xi_i$ is that it gives the *component* along the X_i axis (see Fig. 8.2) of the segment $\overline{P'Q'}$ in the deformed geometry corresponding to segment $\overline{PQ}$ having components dx_i in the undeformed geometry. Hence if $\overline{PQ}$ is dx_1 and accordingly lies along the X_1 axis, i.e., $dx_2 = dx_3 = 0$, we get the following components for $(\overline{P'Q'})^{(1)}$ (the superscript $^{(1)}$ denotes that the segment was originally along the X_1 axis) from the above equation (see Fig. 8.3)

$$d\xi_1{}^{(1)} = (1 + e_{11})\,dx_1$$
$$d\xi_2{}^{(1)} = (\omega_{21} + e_{21})\,dx_1$$
$$d\xi_3{}^{(1)} = (\omega_{31} + e_{31})\,dx_1 \qquad (8.9)$$

Similarly we can show for increments $d\xi_i{}^{(2)}$ and $d\xi_i{}^{(3)}$ the following results:

$$d\xi_1{}^{(2)} = (\omega_{12} + e_{12})\,dx_2 \qquad d\xi_1{}^{(3)} = (\omega_{13} + e_{13})\,dx_3$$
$$d\xi_2{}^{(2)} = (1 + e_{22})\,dx_2 \qquad d\xi_2{}^{(3)} = (\omega_{23} + e_{23})\,dx_3$$
$$d\xi_3{}^{(2)} = (\omega_{32} + e_{32})\,dx_2 \qquad d\xi_3{}^{(3)} = (1 + e_{33})\,dx_3 \qquad (8.10)$$

Let us now express the length $(\overline{P'Q'})^{(1)}$. With the aid of Eq. 8.9 we get directly:

$$(\overline{P'Q'})^{(1)} = (d\xi_i{}^{(1)}\,d\xi_i{}^{(1)})^{1/2}$$
$$= [(1 + e_{11})^2 + (\omega_{21} + e_{21})^2 + (\omega_{31} + e_{31})^2]^{1/2}\,dx_1 \qquad (8.11)$$

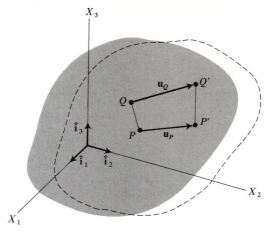

FIGURE 8.2

We now define the *elongation rate E_1* of the line segment dx_1 as follows:

$$E_1 = \frac{(\overline{P'Q'})^{(1)} - dx_1}{dx_1} = [(1 + e_{11})^2 + (e_{21} + \omega_{21})^2 + (e_{31} + \omega_{31})^2]^{1/2} - 1 \quad (a)$$

therefore

$$1 + E_1 = \sqrt{(1 + e_{11})^2 + (e_{21} + \omega_{21})^2 + (e_{31} + \omega_{31})^2} \quad (b) \quad (8.12)$$

You may readily demonstrate that we can give the above equation as follows:

$$1 + E_1 = [(\delta_{j1} + e_{j1} + \omega_{j1})(\delta_{j1} + e_{j1} + \omega_{j1})]^{1/2}$$

For a free index α, i.e., for any original coordinate direction x_α, we then have for the above equation (using now the convention that we do not sum over repeated *Greek* indices):

$$1 + E_\alpha = [(\delta_{j\alpha} + e_{j\alpha} + \omega_{j\alpha})(\delta_{j\alpha} + e_{j\alpha} + \omega_{j\alpha})]^{1/2}$$
$$= [1 + 2e_{\alpha\alpha} + e_{j\alpha}e_{j\alpha} + 2e_{j\alpha}\omega_{j\alpha} + \omega_{j\alpha}\omega_{j\alpha} + 2\omega_{\alpha\alpha}]^{1/2} \quad (8.13)$$

We thus have means of computing the deformed lengths of the *orthogonal triad* of *elements dx_1, dx_2,* and *dx_3* in terms of the strain and rotation parameters. Thus, from the first part of Eq. (8.12(a)) we have:

$$(\overline{P'Q'})^{(\alpha)} = (1 + E_\alpha) dx_\alpha \quad (8.14)$$

The next consideration is to determine the *directions* of these segments in the deformed geometry.

For this purpose we denote as $\hat{\mathbf{r}}^{(\alpha)}$ the *unit vector* in the *deformed geometry* giving the direction of the segment corresponding to the coordinate increment $\overline{PQ}^{(\alpha)}$

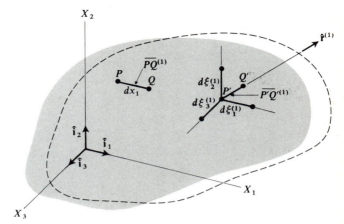

FIGURE 8.3

along X_α in the undeformed geometry. Thus we have for $\hat{\mathbf{r}}^{(\alpha)}$ the result

$$\mathbf{r}^{(\alpha)} = \frac{d\xi_1^{(\alpha)}}{(P'Q')^{(\alpha)}}\hat{\mathbf{i}}_1 + \frac{d\xi_2^{(\alpha)}}{(P'Q')^{(\alpha)}}\hat{\mathbf{i}}_2 + \frac{d\xi_3^{(\alpha)}}{(P'Q')^{(\alpha)}}\hat{\mathbf{i}}_3 \qquad (8.15)$$

where you will recall the *superscript* (α) on $\overline{P'Q'}$ simply identifies the original axis of the element. In Fig. 8.3 we have shown the details for $\hat{\mathbf{r}}^{(1)}$ corresponding to segment $\overline{PQ}^{(1)}$. Now employing Eqs. (8.9), (8.10), and (8.14) in Eq. (8.15) we get:

$$
\begin{bmatrix} \hat{\mathbf{r}}^{(1)} \\[2mm] \hat{\mathbf{r}}^{(2)} \\[2mm] \hat{\mathbf{r}}^{(3)} \end{bmatrix}
=
\begin{bmatrix}
\dfrac{1+e_{11}}{1+E_1} & \dfrac{\omega_{21}+e_{21}}{1+E_1} & \dfrac{\omega_{31}+e_{31}}{1+E_1} \\[3mm]
\dfrac{\omega_{12}+e_{12}}{1+E_2} & \dfrac{1+e_{22}}{1+E_2} & \dfrac{\omega_{32}+e_{32}}{1+E_2} \\[3mm]
\dfrac{\omega_{13}+e_{13}}{1+E_3} & \dfrac{\omega_{23}+e_{23}}{1+E_3} & \dfrac{1+e_{33}}{1+E_3}
\end{bmatrix}
\begin{bmatrix} \hat{\mathbf{i}}_1 \\[2mm] \hat{\mathbf{i}}_2 \\[2mm] \hat{\mathbf{i}}_3 \end{bmatrix}
\qquad (8.16(a))
$$

We have here a transformation matrix that transforms a set of orthogonal unit vectors $\hat{\mathbf{i}}_i$ into a set of *non-orthogonal unit vectors* $\hat{\mathbf{r}}^{(i)}$. The loss of orthogonality above is of course a result of the deformation. This has been shown in Fig. 8.4. Yet another way of expressing Eq. (8.16(a)) is as follows, as you may yourself verify:

$$\hat{\mathbf{r}}^{(\alpha)} = \left[\frac{\delta_{\alpha k} + \dfrac{\partial u_k}{\partial x_\alpha}}{1+E_\alpha}\right]\hat{\mathbf{i}}_k \qquad (8.16(b))$$

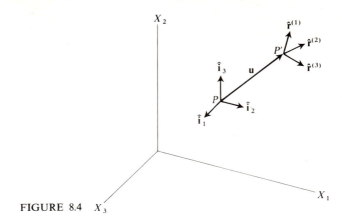

FIGURE 8.4 X_3

Knowing how the line segments dx_1, dx_2, and dx_3 deform at a point, i.e., knowing their extension and rotation, we can now consider local deformation at a point and thus make some interpretations concerning the meaning of strain ϵ_{ij} and the meaning of rotation ω_{ij} for finite deformation. Before proceeding to this step consider a line $\overline{PQ}$ having components dx_1, dx_2, and dx_3, i.e., a line segment having *any* orientation in the undeformed geometry. Then from Eqs. (8.9) and (8.10) we have for $d\xi_i$, the components of $\overline{P'Q'}$, the following result:

$$d\xi_1 = d\xi_1^{(1)} + d\xi_1^{(2)} + d\xi_1^{(3)}$$
$$= (1 + e_{11})\,dx_1 + (\omega_{12} + e_{12})\,dx_2 + (\omega_{13} + e_{13})\,dx_3$$
$$d\xi_2 = d\xi_2^{(1)} + d\xi_2^{(2)} + d\xi_2^{(3)}$$
$$= (\omega_{21} + e_{21})\,dx_1 + (1 + e_{22})\,dx_2 + (\omega_{23} + e_{23})\,dx_3$$
$$d\xi_3 = d\xi_3^{(1)} + d\xi_3^{(2)} + d\xi_3^{(3)}$$
$$= (\omega_{31} + e_{31})\,dx_1 + (\omega_{32} + e_{32})\,dx_2 + (1 + e_{33})\,dx_3 \qquad (8.17)$$

We can solve for dx_i in terms of $d\xi_i$ by straightforward algebra using Cramer's rule to get

$$dx_i = \frac{1}{D}\alpha_{ij}\,d\xi_j \qquad (8.18)$$

where:

$$\alpha_{11} = (1 + e_{22})(1 + e_{33}) - e_{23}^2 + \omega_{32}^2$$
$$\alpha_{12} = (e_{13} + \omega_{13})(e_{23} + \omega_{32}) - (1 + e_{33})(e_{12} - \omega_{21})$$
$$\alpha_{13} = (e_{12} - \omega_{21})(e_{23} - \omega_{32}) - (1 + e_{22})(e_{13} + \omega_{13})$$
$$\alpha_{21} = (e_{23} - \omega_{32})(e_{13} - \omega_{13}) - (1 + e_{33})(e_{12} + \omega_{21})$$

$$\alpha_{22} = (1 + e_{11})(1 + e_{33}) - e_{13}{}^2 + \omega_{13}{}^2$$

$$\alpha_{23} = (e_{13} + \omega_{13})(e_{12} + \omega_{21}) - (1 + e_{11})(e_{23} - \omega_{32})$$

$$\alpha_{31} = (e_{12} + \omega_{21})(e_{23} - \omega_{23}) - (1 + e_{22})(e_{13} - \omega_{13})$$

$$\alpha_{32} = (e_{12} - \omega_{21})(e_{13} - \omega_{13}) - (1 + e_{11})(e_{23} + \omega_{32})$$

$$\alpha_{33} = (1 + e_{11})(1 + e_{22}) - e_{12}{}^2 + \omega_{21}{}^2 \qquad (8.19)$$

and where:

$$D = \begin{vmatrix} 1 + e_{11} & \omega_{12} + e_{12} & \omega_{13} + e_{13} \\ \omega_{21} + e_{21} & 1 + e_{22} & \omega_{23} + e_{23} \\ \omega_{31} + e_{31} & \omega_{32} + e_{32} & 1 + e_{33} \end{vmatrix} \qquad (8.20)$$

We shall make use of the above formulae later in the discussion.

8.3 INTERPRETATION OF STRAIN AND ROTATION TERMS

Now that we can give the deformation effects on a set of orthogonal vanishingly small line elements at a point, we are ready to discuss physical connotations of the general Green strain tensor presented in Chap. 1.

First we consider the *diagonal terms* of ϵ_{ij}, the so-called *normal strains*. For this purpose examine the rate of change of length of any vanishingly small material line segment $\overline{PQ}$ which is of length ds in the undeformed geometry and of length ds^* in the deformed geometry. Denoting this elongation rate as $E_{\overline{PQ}}$ we have:

$$E_{\overline{PQ}} = \frac{ds^* - ds}{ds} \qquad (a)$$

Hence:

$$\frac{ds^*}{ds} = 1 + E_{\overline{PQ}} \qquad (b) \quad (8.21)$$

Now go back to Eq. 1.41(a), which we now express as follows

$$(ds^{*2} - ds^2) = (ds^* - ds)(ds^* + ds) = 2\epsilon_{ij}\, dx_i\, dx_j \qquad (8.22)$$

where you will recall dx_i are the components of ds in the undeformed geometry. Now divide both sides by ds^2 to reach the following result:

$$\left(\frac{ds^*}{ds} - 1\right)\left(\frac{ds^*}{ds} + 1\right) = 2\epsilon_{ij}\left(\frac{dx_i}{ds}\right)\left(\frac{dx_i}{ds}\right)$$

Hence using Eqs. (8.21) we can say

$$E_{\overline{PQ}}(E_{\overline{PQ}} + 2) = 2\epsilon_{ij}\lambda_i\lambda_j \qquad (8.23)$$

where $\lambda_i = dx_i/ds$ and $\lambda_j = dx_j/ds$. Now consider the case where $\overline{PQ}$ is along the X_1 axis. Then $\lambda_1 = 1$ and $\lambda_2 = \lambda_3 = 0$. Also, $E_{\overline{PQ}}$ corresponds to E_1 of the previous section. Then we get from the above:

$$E_1(1 + \tfrac{1}{2}E_1) = \epsilon_{11} \qquad (a)$$

therefore

$$E_1 = \sqrt{1 + 2\epsilon_{11}} - 1 \qquad (b) \quad (8.24)$$

We see that the normal strain ϵ_{11} is *simply-related to the stretching rate of a vanishingly small line element originally in the X_1 direction.* We can similarly relate the other two normal strains.

Now we consider the *off-diagonal terms* of the strain tensor. For this purpose we examine the angle between unit vectors $\hat{\mathbf{r}}^{(1)}$ and $\hat{\mathbf{r}}^{(2)}$ of the previous section. Thus:

$$\cos{(\hat{\mathbf{r}}^{(1)}, \hat{\mathbf{r}}^{(2)})} = \hat{\mathbf{r}}^{(1)} \cdot \hat{\mathbf{r}}^{(2)}$$

$$= \frac{(1 + e_{11})(\omega_{12} + e_{12}) + (\omega_{21} + e_{21})(1 + e_{22}) + (\omega_{31} + e_{31})(\omega_{32} + e_{32})}{(1 + E_1)(1 + E_2)}$$

where we have used Eq. (8.16(a)) in carrying out the dot product. By carrying out the products in the numerator and by then replacing the terms using their definitions (Eqs. (8.3) and (8.5)) we can then conclude from Eq. (8.2) that the numerator is simply $2\epsilon_{12}$. Thus we have:

$$\cos{(\hat{\mathbf{r}}^{(1)}, \hat{\mathbf{r}}^{(2)})} = \frac{2\epsilon_{12}}{(1 + E_1)(1 + E_2)}$$

We may also express the above equation as:

$$\cos{(\hat{\mathbf{r}}^{(1)}, \hat{\mathbf{r}}^{(2)})} = \cos{\left(\frac{\pi}{2} - \phi_{12}\right)} = \sin{\phi_{12}} = \frac{2\epsilon_{12}}{(1 + E_1)(1 + E_2)} \qquad (8.25)$$

The angle ϕ_{12} is the change from a right angle of the line segments dx_1, dx_2 as a result of deformation and is called the *shear angle*. We thus conclude that the *sine of the shear angle is directly proportional to the shear strain for a given set of elements orthogonal in the undeformed geometry.*

Note that each grouping of terms in the expression for ϵ_{ij} (see Eq. (8.2)) is a second-order tensor. Thus, we have shown directly in Chap. 1 that $u_{i,j}$ and $u_{j,i}$ are second-order tensors. The expression $u_{k,i}u_{k,j}$ is the inner product of two second-order tensors and is hence also a second-order tensor. We can conclude that ϵ_{ij} is a second-order tensor. Clearly the tensor is symmetric. We can conclude from the preceding remarks, accordingly, that there are a set of principal axes at each point in the body and that the off-diagonal terms are zero for such axes. Since these terms have been shown to be proportional to the sines of the change in right angles of the undeformed elements dx_i, then clearly the axes corresponding to principal axes *remain orthogonal* to each other during deformation.

Let us next turn to the *rotation parameter* ω_{ij}. You will recall from the earlier work for small deformation, that we defined rotation of a body element about an axis as the average rotation about this axis of all line segments in the body. We maintain this definition for large deformation. Whereas the rotation angles about reference axes for small deformation were equal to the non-zero terms of ω_{ij}, for large deformation we can show that the angles of rotation are *proportional* to the corresponding ω_{ij} terms with the proportionality factor involving the strain parameter.[1]

As for the *strain parameter* e_{ij} we have already shown in Chap. 1 that a diagonal component such as e_{11} represents the *component* of elongation in the X_1 direction of a segment *originally* in the X_1 direction. Recall that the small deformation assumption permitted us in Chap. 1 to use the *elongation itself* rather than the aforementioned component. As for the off-diagonal terms of e_{ij} we cannot give, for finite deformation, any meaningful physical interpretation.

8.4 VOLUME CHANGE DURING DEFORMATION

We next examine volume change for large deformation. Consider a rectangular parallelopiped having sides dx_1, dx_2, and dx_3 in the undeformed geometry. Then, using Eq. (8.9) we see that the element dx_1 of the rectangular parallelopiped becomes a directed line segment $d\mathbf{R}^{(1)}$ in the deformed geometry whose components are $d\xi_1^{(1)}$, $d\xi_2^{(1)}$, and $d\xi_3^{(1)}$ such that:

$$d\mathbf{R}^{(1)} = d\xi_1^{(1)}\hat{\mathbf{i}}_1 + d\xi_2^{(1)}\hat{\mathbf{i}}_2 + d\xi_3^{(1)}\hat{\mathbf{i}}_3$$
$$= [(1 + e_{11})\,dx_1\hat{\mathbf{i}}_1 + (\omega_{21} + e_{21})\,dx_1\hat{\mathbf{i}}_2 + (\omega_{31} + e_{31})\,dx_1\hat{\mathbf{i}}_3]$$

For the other sides of the rectangular parallelopiped we have similarly from Eqs. (8.10):

$$d\mathbf{R}^{(2)} = [(\omega_{12} + e_{12})\,dx_2\hat{\mathbf{i}}_1 + (1 + e_{22})\,dx_2\hat{\mathbf{i}}_2 + (\omega_{32} + e_{32})\,dx_2\hat{\mathbf{i}}_3]$$
$$d\mathbf{R}^{(3)} = [(\omega_{13} + e_{13})\,dx_3\hat{\mathbf{i}}_1 + (\omega_{23} + e_{23})\,dx_3\hat{\mathbf{i}}_2 + (1 + e_{33})\,dx_3\hat{\mathbf{i}}_3]$$

---

[1] See Novozhilov, V. V.: "Foundations of the Non-Linear Theory of Elasticity," Graylock Press, Rochester, 1953.

The volume of the body element in the deformed geometry V^* then becomes:

$$V^* = d\mathbf{R}^{(1)} \cdot d\mathbf{R}^{(2)} \times d\mathbf{R}^{(3)} = \begin{vmatrix} 1 + e_{11} & \omega_{21} + e_{21} & \omega_{31} + e_{31} \\ \omega_{12} + e_{12} & 1 + e_{22} & \omega_{32} + e_{32} \\ \omega_{13} + e_{13} & \omega_{23} + e_{23} & 1 + e_{33} \end{vmatrix} dx_1 dx_2 dx_3$$

Using D to denote V^*/V where $V = dx_1\, dx_2\, dx_3$ we get:

$$D = \begin{vmatrix} 1 + e_{11} & \omega_{21} + e_{21} & \omega_{31} + e_{31} \\ \omega_{12} + e_{12} & 1 + e_{22} & \omega_{32} + e_{32} \\ \omega_{13} + e_{13} & \omega_{23} + e_{23} & 1 + e_{33} \end{vmatrix} \tag{8.26}$$

Now using the definitions of e_{ij} and ω_{ij} in terms of u_i we may express Eq. (8.26) as follows:

$$D = \begin{vmatrix} 1 + \dfrac{\partial u_1}{\partial x_1} & \dfrac{\partial u_2}{\partial x_1} & \dfrac{\partial u_3}{\partial x_1} \\[2mm] \dfrac{\partial u_1}{\partial x_2} & 1 + \dfrac{\partial u_2}{\partial x_2} & \dfrac{\partial u_3}{\partial x_2} \\[2mm] \dfrac{\partial u_1}{\partial x_3} & \dfrac{\partial u_2}{\partial x_3} & 1 + \dfrac{\partial u_3}{\partial x_3} \end{vmatrix} \tag{8.27}$$

One may directly show that by squaring the above equation we get, on introducing the strain terms, the following result:

$$D^2 = \begin{vmatrix} 1 + 2\epsilon_{11} & \epsilon_{21} & \epsilon_{31} \\ \epsilon_{21} & 1 + 2\epsilon_{22} & \epsilon_{32} \\ \epsilon_{31} & \epsilon_{32} & 1 + 2\epsilon_{33} \end{vmatrix} \tag{8.28}$$

Carrying out the determinant and considering now that we are using *principal axes* we reach the result:

$$D^2 = (1 + 2\epsilon_1)(1 + 2\epsilon_2)(1 + 2\epsilon_3) \tag{8.29}$$

If we now define *relative change* in volume Δ as follows

$$\Delta = \frac{V^* - V}{V} = D - 1 \tag{8.30}$$

we have from Eq. (8.29):

$$\Delta = \sqrt{(1 + 2\epsilon_1)(1 + 2\epsilon_2)(1 + 2\epsilon_3)} - 1 \tag{8.31}$$

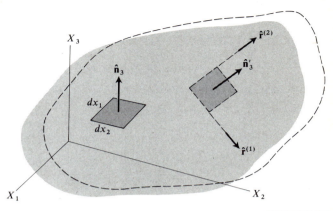

FIGURE 8.5

Finally using Eq. (8.24(*b*)) we get

$$\Delta = (1 + E_1)(1 + E_2)(1 + E_3) - 1 \qquad (8.32)$$

where the E's are elongation rates along the principal axes.

8.5 CHANGES OF AREA ELEMENTS DURING LARGE DEFORMATION

We now consider the orientation in the deformed geometry of an area at point P which in the undeformed geometry has sides $dx_1 \, dx_2$ having normal $\hat{\mathbf{n}}_3$. We denote the normal to the area element in the *deformed geometry* as $\hat{\mathbf{n}}'_3$ (see Fig. 8.5). Using the unit vectors $\hat{\mathbf{r}}^{(1)}$ and $\hat{\mathbf{r}}_2^{(2)}$ for the orientation of dx_1 and dx_2 in the deformed geometry we have:

$$\hat{\mathbf{n}}'_3 = \frac{\hat{\mathbf{r}}^{(1)} \times \hat{\mathbf{r}}^{(2)}}{\sin{(\hat{\mathbf{r}}^{(1)}, \hat{\mathbf{r}}^{(2)})}} \qquad (8.33)$$

Note that $\hat{\mathbf{n}}'_3$ is not collinear with $\hat{\mathbf{r}}^{(3)}$ and recall that the latter vector need not be orthogonal to the vectors $\hat{\mathbf{r}}^{(1)}$ and $\hat{\mathbf{r}}^{(2)}$. To get the *components* of $\hat{\mathbf{n}}'_3$ (these are the cosines $\cos{(X_1, \hat{\mathbf{n}}'_3)}$, $\cos{(X_2, \hat{\mathbf{n}}'_3)}$ and $\cos{(X_3, \hat{\mathbf{n}}'_3)}$) we simply take successively the dot product of $\hat{\mathbf{n}}'_3$ with $\hat{\mathbf{i}}_1$, $\hat{\mathbf{i}}_2$, and $\hat{\mathbf{i}}_3$. Thus for $\cos{(X_1, \hat{\mathbf{n}}'_3)}$ we get:

$$\cos{(X_1, \hat{\mathbf{n}}'_3)} = \hat{\mathbf{i}}_1 \cdot \hat{\mathbf{n}}'_3 = \frac{1}{\sin{(\hat{\mathbf{r}}^{(1)}, \hat{\mathbf{r}}^{(2)})}} \hat{\mathbf{i}}_1 \cdot \hat{\mathbf{r}}^{(1)} \times \hat{\mathbf{r}}^{(2)}$$

Using components for $\hat{\mathbf{r}}^{(1)}$ and $\hat{\mathbf{r}}^{(2)}$ from Eq. (8.16(a)) we get, making use of the determinantal representation of the scalar triple product:

$$\cos(X_1, \hat{\mathbf{n}}_3') = \frac{1}{\sin(\hat{\mathbf{r}}^{(1)}, \hat{\mathbf{r}}^{(2)})} \begin{vmatrix} 1 & 0 & 0 \\ \dfrac{1 + e_{11}}{1 + E_1} & \dfrac{\omega_{21} + e_{21}}{1 + E_1} & \dfrac{\omega_{31} + e_{31}}{1 + E_1} \\ \dfrac{\omega_{12} + e_{12}}{1 + E_2} & \dfrac{1 + e_{22}}{1 + E_2} & \dfrac{\omega_{32} + e_{32}}{1 + E_2} \end{vmatrix}$$

$$= \frac{1}{\sin(\hat{\mathbf{r}}^{(1)}, \hat{\mathbf{r}}^{(2)})}$$
$$\left[\frac{(\omega_{21} + e_{21})(\omega_{32} + e_{32}) - (\omega_{31} + e_{31})(1 + e_{22})}{(1 + E_1)(1 + E_2)} \right] \tag{8.34}$$

As for the term $\sin(\hat{\mathbf{r}}_1^{(1)}, \hat{\mathbf{r}}_2^{(2)})$ we note from Eq. (8.25) using a trigonometric identity that

$$\sin(\hat{\mathbf{r}}^{(1)}, \hat{\mathbf{r}}^{(2)}) = \cos\phi_{12} = \sqrt{1 - \frac{4\epsilon_{12}^2}{(1 + E_1)^2(1 + E_2)^2}}$$

$$= \frac{\sqrt{(1 + 2\epsilon_{11})(1 + 2\epsilon_{22}) - 4\epsilon_{12}^2}}{(1 + E_1)(1 + E_2)} \tag{8.35}$$

where we have used Eq. (8.24) in the last step of the above formulations. Substituting for $\sin(\hat{\mathbf{r}}^{(1)}, \hat{\mathbf{r}}^{(2)})$ in Eq. (8.34) we get:

$$\cos(X_1, \hat{\mathbf{n}}_3') = \frac{(\omega_{21} + e_{21})(\omega_{32} + e_{32}) - (\omega_{31} + e_{31})(1 + e_{22})}{\sqrt{(1 + 2\epsilon_{11})(1 + 2\epsilon_{22}) - 4\epsilon_{12}^2}}$$

Now returning to Eq. (8.19) we see that the numerator in the above equation is simply α_{31}. Thus we get

$$\cos(X_1, \hat{\mathbf{n}}_3') = \frac{\alpha_{31}}{\sqrt{(1 + 2\epsilon_{11})(1 + 2\epsilon_{22}) - 4\epsilon_{12}^2}}$$

We may similarly find $\cos(X_2, \hat{\mathbf{n}}_3')$ and $\cos(X_3, \hat{\mathbf{n}}_3')$ so that we may state:

$$\begin{bmatrix} \cos(X_1, \hat{\mathbf{n}}_3') \\ \cos(X_2, \hat{\mathbf{n}}_3') \\ \cos(X_3, \hat{\mathbf{n}}_3') \end{bmatrix} = \frac{1}{\sqrt{(1 + 2\epsilon_{11})(1 + 2\epsilon_{22}) - 4\epsilon_{12}^2}} \begin{bmatrix} \alpha_{31} \\ \alpha_{32} \\ \alpha_{33} \end{bmatrix} \tag{8.36}$$

We may also proceed in a similar manner to get the direction cosines for $\hat{\mathbf{n}}_1'$ and $\hat{\mathbf{n}}_2'$.

The results are as given below:

$$
\begin{bmatrix}
\cos(X_1, \hat{\mathbf{n}}_1') \\
\cos(X_2, \hat{\mathbf{n}}_1') \\
\cos(X_3, \hat{\mathbf{n}}_1')
\end{bmatrix}
= \frac{1}{\sqrt{(1 + 2\epsilon_{33})(1 + 2\epsilon_{22}) - 4\epsilon_{23}}}
\begin{bmatrix}
\alpha_{11} \\
\alpha_{12} \\
\alpha_{13}
\end{bmatrix}
\tag{8.37}
$$

$$
\begin{bmatrix}
\cos(X_1, \hat{\mathbf{n}}_2') \\
\cos(X_2, \hat{\mathbf{n}}_2') \\
\cos(X_2, \hat{\mathbf{n}}_2')
\end{bmatrix}
= \frac{1}{\sqrt{(1 + 2\epsilon_{11})(1 + 2\epsilon_{33}) - 4\epsilon_{13}^2}}
\begin{bmatrix}
\alpha_{21} \\
\alpha_{22} \\
\alpha_{23}
\end{bmatrix}
\tag{8.38}
$$

Now that we have the *direction* of the faces of a rectangular parallelopiped when it is deformed, we next consider the *areas of these faces in the deformed geometry.* Since the side dx_1 has a new length $(1 + E_1)\,dx_1$ and side dx_2 has a new length $(1 + E_2)\,dx_2$ in accordance with Eq. (8.14), we have for the area A_3^* in the deformed geometry of the area element $dx_1\,dx_2$:

$$
A_3^* = (1 + E_1)\,dx_1(1 + E_2)\,dx_2 \sin(\hat{\mathbf{r}}^{(1)}, \hat{\mathbf{r}}^{(2)})
$$

Denoting the area $dx_1\,dx_2$ as A_3 we then have:

$$
\left(\frac{A_3^*}{A_3}\right) = (1 + E_1)(1 + E_2) \sin(\hat{\mathbf{r}}^{(1)}, \hat{\mathbf{r}}^{(2)})
\tag{8.39}
$$

Using Eq. (8.35) for $\sin(\hat{\mathbf{r}}^{(1)}, \hat{\mathbf{r}}^{(2)})$ we get:

$$
\left(\frac{A_3^*}{A_3}\right) = \sqrt{(1 + 2\epsilon_{11})(1 + 2\epsilon_{22}) - 4\epsilon_{12}^2}
\tag{8.40}
$$

Similarly we can get:

$$
\left(\frac{A_2^*}{A_2}\right) = \sqrt{(1 + 2\epsilon_{11})(1 + 2\epsilon_{33}) - 4\epsilon_{13}^2} \tag{a}
$$

$$
\left(\frac{A_1^*}{A_1}\right) = \sqrt{(1 + 2\epsilon_{22})(1 + 2\epsilon_{33}) - 4\epsilon_{23}^2} \tag{b} \quad (8.41)
$$

In the next section we will simplify the strain tensor for use in Chap. 9.

8.6 SIMPLIFICATION OF STRAIN

In our work in stability we shall be concerned with problems wherein the strain components ϵ_{ij} will be small compared to unity. Also the rotation of body elements will be small compared to unity but *large* compared to the strain terms. That is:

$$
\epsilon_{ij} \ll \omega_{ij} \ll 1 \tag{8.42}
$$

We wish to simplify the theory for such a situation.

Referring to Eq. (8.16(a)) we examine the angle between $\hat{\mathbf{r}}^{(1)}$ and $\hat{\mathbf{i}}_1$, i.e., the angle between the deformed orientation of segment dx_1 and dx_1 itself. We have:

$$\hat{\mathbf{r}}^{(1)} \cdot \hat{\mathbf{i}}_1 = \frac{1 + e_{11}}{1 + E_1} = \cos \phi_1 \qquad (8.43)$$

The order of magnitude of ϕ_1 will be comparable to the order of magnitude of the rotation parameters for the situation depicted by Eq. (8.42). Hence ϕ_1 *will be much smaller than unity.* Expanding $\cos \phi_1$ as a power series and retaining terms up to $\phi_1{}^2$ we get:

$$\frac{1 + e_{11}}{1 + E_1} \approx 1 - \frac{\phi_1{}^2}{2}$$

Now replace the denominator of the left side in the above equation using Eq. (8.24(b)) for this purpose. We then get:

$$\frac{1 + e_{11}}{\sqrt{1 + 2\epsilon_{11}}} \approx 1 - \frac{\phi_1{}^2}{2}$$

Because the strain components are assumed small, we may expand the expression $1/\sqrt{1 + 2\epsilon_{11}}$ as a power series retaining only terms ϵ_{11} up to the first power. Thus we get:

$$(1 + e_{11})(1 - \epsilon_{11}) \approx 1 - \frac{\phi_1{}^2}{2}$$

therefore

$$e_{11} - \epsilon_{11} - e_{11}\epsilon_{11} \approx -\frac{\phi_1{}^2}{2}$$

We may drop $e_{11}\epsilon_{11}$ when comparing magnitudes with e_{11} and so the above approximation becomes:

$$\epsilon_{11} - e_{11} \approx \frac{\phi_1{}^2}{2} \qquad (8.44)$$

Similarly we can show:

$$\epsilon_{22} - e_{22} \approx \frac{\phi_2{}^2}{2}$$

$$\epsilon_{33} - e_{33} \approx \frac{\phi_3{}^2}{2} \qquad (8.45)$$

Thus for the assumption contained in inequality (8.42) we conclude that the normal strain parameters differ from the normal strains in the order of magnitude of the square of the rotation angles. We may express the above results as follows:

$$\epsilon_{\alpha\beta} - e_{\alpha\beta} \approx \frac{\phi_\alpha \phi_\beta}{2} \qquad \text{for } \alpha = \beta \qquad (8.46)$$

Actually the above relation can be shown to apply when $\alpha \neq \beta$. We shall here merely demonstrate this for a simple case where we have:

$$u_1 = u_1(x_2, x_3) \qquad u_2 = 0 \qquad u_3 = 0$$

Then from Eq. (8.2) we get for ϵ_{23}:

$$\epsilon_{23} = e_{23} + \frac{1}{2}\frac{\partial u_1}{\partial x_2}\frac{\partial u_1}{\partial x_3}$$

But for small rotations it is not difficult to see that $\partial u_1/\partial x_2 \approx \phi_3$ and $\partial u_1/\partial x_3 \approx \phi_2$ giving us the result:

$$\epsilon_{23} - e_{23} \approx \frac{\phi_2 \phi_3}{2}$$

With these conclusions we now examine the component ϵ_{11}. Thus from Eq. (8.2) we have:

$$\epsilon_{11} = \frac{\partial u_1}{\partial x_1} + \frac{1}{2}\left(\frac{\partial u_1}{\partial x_1}\frac{\partial u_1}{\partial x_1} + \frac{\partial u_2}{\partial x_1}\frac{\partial u_2}{\partial x_1} + \frac{\partial u_3}{\partial x_1}\frac{\partial u_3}{\partial x_1}\right)$$

You may yourself demonstrate that the above expression can be expressed as follows in terms of e_{ij} and ω_{ij}:

$$\epsilon_{11} = e_{11} + \tfrac{1}{2}(e_{11}{}^2 + e_{12}{}^2 + 2e_{12}\omega_{21} + \omega_{21}{}^2 + e_{13}{}^2 - 2e_{13}\omega_{31} + \omega_{31}{}^2) \qquad (8.47)$$

Now consider the term $e_{11}{}^2$. We have, on considering Eq. (8.46) for $\alpha = \beta = 1$, solving for $e_{11}{}^2$ and then squaring:

$$e_{11}{}^2 \approx \epsilon_{11}{}^2 - \epsilon_{11}\phi_1{}^2 + \frac{\phi_1{}^4}{4}$$

In accordance with the order of magnitude hierarchy of Eq. (8.42) note that $\epsilon_{11}{}^2 \ll \omega_{31}{}^2$ or $\omega_{21}{}^2$. Since, furthermore, ω_{31} and ω_{21} are of the order of magnitude of ϕ_1, we can thus conclude from the above equation that $e_{11}{}^2$ may be neglected in Eq. (8.47) when compared to $\omega_{31}{}^2$ and $\omega_{21}{}^2$. This is similarly true for $e_{12}{}^2$ and $e_{13}{}^2$. Now consider the expression $e_{12}\omega_{21}$. We can say using Eq. (8.46) for $\alpha = 1, \beta = 2$:

$$e_{12}\omega_{21} \approx \left(\epsilon_{12} - \frac{\phi_1\phi_2}{2}\right)\omega_{21} = \epsilon_{12}\omega_{21} - \frac{\phi_1\phi_2}{2}\omega_{21}$$

Again we see that this expression may be dropped in Eq. (8.47) since from inequality (8.42) $\epsilon_{12}\omega_{21} \ll \omega_{21}{}^2$ and $(\phi_1\phi_2\omega_{21})/2 \ll \omega_{21}{}^2$ (since the ϕ's are of the order of magnitude of ω's). This is also true for the term $e_{13}\omega_{31}$. We are thus left in Eq. (8.47) with:

$$\epsilon_{11} = e_{11} + \tfrac{1}{2}(\omega_{21}{}^2 + \omega_{31}{}^2)$$

We may form similar relations for the other strains leading us to the result:

$$\epsilon_{\alpha\beta} = e_{\alpha\beta} + \tfrac{1}{2}\omega_{k\alpha}\omega_{k\beta} \tag{8.48}$$

We will make use later of this formulation for strain. Note, that although it is considerably simpler than the general expression (8.2) it is still nevertheless nonlinear. If we can neglect products of rotation angles, i.e., when $\epsilon_{ij} \approx e_{ij}$, then we get back to the case of linear classical elasticity studied up to this time.

8.7 STRESS AND THE EQUATIONS OF EQUILIBRIUM

Consider a rectangular parallelopiped in the *deformed geometry* having sides $d\xi_i$. Then clearly from our work in Chap. 1 we have for equilibrium

$$\frac{\partial \tau_{11}}{\partial \xi_1} + \frac{\partial \tau_{12}}{\partial \xi_2} + \frac{\partial \tau_{13}}{\partial \xi_3} + B_1^* = 0$$

$$\frac{\partial \tau_{21}}{\partial \xi_1} + \frac{\partial \tau_{22}}{\partial \xi_2} + \frac{\partial \tau_{23}}{\partial \xi_3} + B_2^* = 0$$

$$\frac{\partial \tau_{31}}{\partial \xi_1} + \frac{\partial \tau_{32}}{\partial \xi_2} + \frac{\partial \tau_{33}}{\partial \xi_3} + B_3^* = 0 \tag{8.49}$$

where

$$\tau_{12} = \tau_{21}, \quad \text{etc.} \tag{8.50}$$

Note that B_i^* is the body force vector per unit volume in the deformed geometry. Using stress vectors $\overset{*}{\mathbf{T}}{}^{(\xi_1)}$, $\overset{*}{\mathbf{T}}{}^{(\xi_2)}$, and $\overset{*}{\mathbf{T}}{}^{(\xi_3)}$ (see Fig. 8.6) we can give Eq. (8.49) as follows:

$$\frac{\partial \overset{*}{\mathbf{T}}{}^{(\xi_1)}}{\partial \xi_1} + \frac{\partial \overset{*}{\mathbf{T}}{}^{(\xi_2)}}{\partial \xi_2} + \frac{\partial \overset{*}{\mathbf{T}}{}^{(\xi_3)}}{\partial \xi_3} + \mathbf{B}^* = 0$$

therefore

$$\frac{\partial}{\partial \xi_i}[\overset{*}{\mathbf{T}}{}^{(\xi_i)}] + \mathbf{B}^* = 0 \tag{8.51}$$

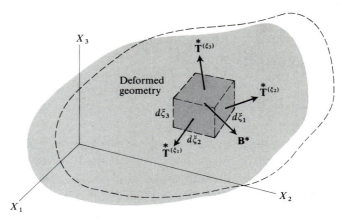

FIGURE 8.6

In-the linear theory because of the small deformation restriction, we used the *undeformed geometry* when relating stresses, rather than the deformed geometry as is properly done above. Because boundary conditions are often expressed in terms of the undeformed geometry we wish as a first step to transform the above equation to the Lagrangian coordinates x_i. Thus using the chain rule for differentiation we get for Eq. (8.51):

$$\frac{\partial \overset{*}{\mathbf{T}}{}^{(\xi_j)}}{\partial x_k}\frac{\partial x_k}{\partial \xi_j} + \mathbf{B}^* = \mathbf{0} \qquad (8.52)$$

From Eq. (8.18) we can determine $\partial x_k/\partial \xi_j$ as follows

$$\frac{\partial x_k}{\partial \xi_j} = \frac{1}{D}\alpha_{kj}$$

and so Eq. (8.52) becomes:

$$\frac{\partial \overset{*}{\mathbf{T}}{}^{(\xi_j)}}{\partial x_k}\alpha_{kj} + D\mathbf{B}^* = \mathbf{0}$$

We may rewrite the above equation as follows:

$$\frac{\partial}{\partial x_k}[\overset{*}{\mathbf{T}}{}^{(\xi_j)}\alpha_{kj}] - \overset{*}{\mathbf{T}}{}^{(\xi_j)}\frac{\partial}{\partial x_k}(\alpha_{kj}) + D\mathbf{B}^* = \mathbf{0}$$

By carrying out the expression $\partial(\alpha_{kj})/\partial x_k$ using the terms in Eq. (8.19) you can demonstrate that this expression is zero for all values of the index j. Accordingly we have for

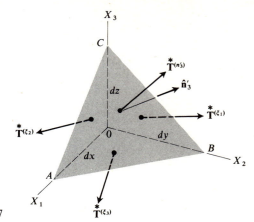

FIGURE 8.7

the above equation the result:

$$\frac{\partial}{\partial x_k}[\overset{*}{\mathbf{T}}{}^{(\xi_j)}\alpha_{kj}] + D\mathbf{B}^* = 0 \qquad (8.53)$$

We wish next to express the stress vectors given in the above equation *in terms of stress vectors on surfaces in the deformed geometry arising from the deformation of a rectangular parallelopiped in the original geometry.* For this purpose examine the bracketed quantity above for the case $k = 3$. We get, using Eq. (8.36) to replace the α's:

$$\overset{*}{\mathbf{T}}{}^{(\xi_j)}\alpha_{3j} = \overset{*}{\mathbf{T}}{}^{(\xi_1)}\alpha_{31} + \overset{*}{\mathbf{T}}{}^{(\xi_2)}\alpha_{32} + \overset{*}{\mathbf{T}}{}^{(\xi_3)}\alpha_{33}$$

$$= \sqrt{(1 + 2\epsilon_{11})(1 + 2\epsilon_{22}) - 4\epsilon_{12}{}^2}\,[\overset{*}{\mathbf{T}}{}^{(\xi_1)}\cos(X_1,\hat{\mathbf{n}}_3')$$

$$+ \overset{*}{\mathbf{T}}{}^{(\xi_2)}\cos(X_2,\hat{\mathbf{n}}_3') + \overset{*}{\mathbf{T}}{}^{(\xi_3)}\cos(X_3,\hat{\mathbf{n}}_3')] \qquad (8.54)$$

To interpret the bracketed expression, we have shown in Fig. 8.7 an interface $\overline{ABC}$ in the deformed geometry of the body having orientation $\hat{\mathbf{n}}_3'$.[1] A stress vector $\overset{*}{\mathbf{T}}{}^{(n'_3)}$ is shown on this interface and other interfaces are drawn to form a tetrahedron with the latter interfaces parallel to the coordinate planes. From equilibrium considerations we see that:

$$\overset{*}{\mathbf{T}}{}^{(n'_3)}\overline{ABC} = \overset{*}{\mathbf{T}}{}^{(\xi_1)}\overline{COB} + \overset{*}{\mathbf{T}}{}^{(\xi_2)}\overline{AOC} + \overset{*}{\mathbf{T}}{}^{(\xi_3)}\overline{AOB}$$

---

[1] This interface then has resulted from the deformation of an element originally in the x_1x_2 plane.

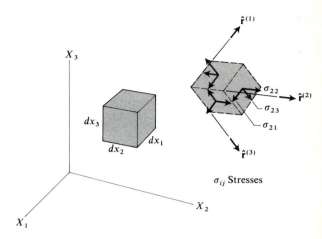

FIGURE 8.8

therefore

$$\overset{*}{\mathbf{T}}{}^{(n_3')} = \overset{*}{\mathbf{T}}{}^{(\xi_1)} \cos(X_1, \hat{\mathbf{n}}_3') + \overset{*}{\mathbf{T}}{}^{(\xi_2)} \cos(X_2, \hat{\mathbf{n}}_3') + \overset{*}{\mathbf{T}}{}^{(\xi_3)} \cos(X_3, \hat{\mathbf{n}}_3') \qquad (8.55)$$

Hence we can give Eq. (8.54) as follows:[1]

$$\overset{*}{\mathbf{T}}{}^{(\xi_j)} \alpha_{3j} = \sqrt{(1 + 2\epsilon_{11})(1 + 2\epsilon_{22}) - 4\epsilon_{12}{}^2} \, [\overset{*}{\mathbf{T}}{}^{(n_3')}]$$

Finally using Eq. (8.40) to replace the root in the above equation we have:

$$\overset{*}{\mathbf{T}}{}^{(\xi_j)} \alpha_{3j} = \left(\frac{A_3^*}{A_3} \right) \overset{*}{\mathbf{T}}{}^{(n_3')}$$

We can get similar expressions for $\overset{*}{\mathbf{T}}{}^{(\xi_j)} \alpha_{2j}$ and $\overset{*}{\mathbf{T}}{}^{(\xi_j)} \alpha_{1j}$ so that Eq. (8.53) can be given as follows

$$\sum_{\beta=1}^{3} \frac{\partial}{\partial x_\beta} \left[\left(\frac{A_\beta^*}{A_\beta} \right) \overset{*}{\mathbf{T}}{}^{(n_\beta')} \right] + \frac{V^*}{V} \mathbf{B}^* = 0 \qquad (8.56)$$

where we have replaced D by V^*/V in the last term on the left side of the equation. Now decompose $\overset{*}{\mathbf{T}}{}^{(n_\beta')}$ into three component stresses parallel to the $\hat{\mathbf{r}}^{(1)}$, $\hat{\mathbf{r}}^{(2)}$, and $\hat{\mathbf{r}}^{(3)}$

---

[1] We have thus related the traction vectors for the rectangular parallelopiped in the *deformed geometry*, namely $\overset{*}{\mathbf{T}}{}^{(\xi_j)}$, with a traction vector $\overset{*}{\mathbf{T}}{}^{(n_3')}$ on an interface which in the *undeformed geometry* may be considered a face of a rectangular parallelopiped, namely one with normal $\hat{\mathbf{n}}_3$.

directions. This generates nine stress terms which we denote as σ_{ij}. Thus for surface having normal $\hat{\mathbf{n}}'_1$ we have $\sigma_{11}, \sigma_{12}, \sigma_{13}$; for $\hat{\mathbf{n}}'_2$ we have $\sigma_{21}, \sigma_{22}, \sigma_{23}$; and for $\hat{\mathbf{n}}'_3$ we have $\sigma_{31}, \sigma_{32}, \sigma_{33}$.[1] This has been shown in Fig. 8.8. Note the first subscript corresponds to the subscript of $\hat{\mathbf{n}}'$, i.e., to the direction of the normal of the area element in the undeformed geometry.

We are thus now dealing with an element in the deformed geometry which is a vanishingly small rectangular parallelopiped in the undeformed geometry. We can then give Eq. (8.56) as follows:

$$\sum_{\beta=1}^{3} \frac{\partial}{\partial x_\beta}\left[\left(\frac{A_\beta^*}{A_\beta}\right)\sigma_{\beta j}\hat{\mathbf{r}}^{(j)}\right] + \frac{V^*}{V}\mathbf{B}^* = 0 \qquad (8.57)$$

We now examine the component of the above equation in the x_1 direction. That means we consider the following scalar equation:

$$\sum_{\beta=1}^{3} \frac{\partial}{\partial x_\beta}\left[\frac{A_\beta^*}{A_\beta}\sigma_{\beta j}\hat{r}_1^{(j)}\right] + \frac{V^*}{V}B_1^* = 0 \qquad (8.58)$$

But from Eq. (8.16(a)) we see that:

$$\begin{bmatrix} \hat{r}_1^{(1)} \\[2mm] \hat{r}_1^{(2)} \\[2mm] \hat{r}_1^{(3)} \end{bmatrix} = \begin{bmatrix} \dfrac{1 + e_{11}}{1 + E_1} \\[3mm] \dfrac{\omega_{12} + e_{12}}{1 + E_2} \\[3mm] \dfrac{\omega_{13} + e_{13}}{1 + E_3} \end{bmatrix} \qquad (8.59)$$

Substituting into Eq. (8.58) we get:

$$\sum_{\beta=1}^{3} \frac{\partial}{\partial x_\beta}\left[\frac{A_\beta^*}{A_\beta}\left(\sigma_{\beta 1}\frac{1 + e_{11}}{1 + E_1} + \sigma_{\beta 2}\frac{\omega_{12} + e_{12}}{1 + E_2} + \sigma_{\beta 3}\frac{\omega_{13} + e_{13}}{1 + E_3}\right)\right] + \frac{V^*}{V}B_1^* = 0$$

Introducing next the notation

$$\bar{\sigma}_{\beta\gamma} = \left[\frac{A_\beta^*}{A_\beta}\frac{\sigma_{\beta\gamma}}{1 + E_\gamma}\right] \qquad (8.60)$$

we can give the above equation as:

$$\sum_{\beta=1}^{3} \frac{\partial}{\partial x_\beta}\left[\bar{\sigma}_{\beta 1}(1 + e_{11}) + \bar{\sigma}_{\beta 2}(\omega_{12} + e_{12}) + \bar{\sigma}_{\beta 3}(\omega_{13} + e_{13})\right] + \frac{V^*}{V}B_1^* = 0$$

Now the expressions $(1 + e_{11})$, $(\omega_{12} + e_{12})$, and $(\omega_{13} + e_{13})$ are readily seen to be components of $(\delta_{1k} + \omega_{1k} + e_{1k})$ for k equal successivily to 1, 2, and 3. Hence we

---

[1] Note that $\hat{\mathbf{n}}'_1$, $\hat{\mathbf{n}}'_2$, and $\hat{\mathbf{n}}'_3$ in general will not be orthogonal to each other.

have:

$$\sum_{\gamma=1}^{3}\sum_{\beta=1}^{3}\frac{\partial}{\partial x_{\beta}}[\bar{\sigma}_{\beta\gamma}(\delta_{1\gamma}+e_{1\gamma}+\omega_{1\gamma})]+\frac{V^{*}}{V}B_{1}^{*}=0$$

To get other components from Eq. (8.57) for the x_2 and x_3 directions we leave it to you to show that we need only change the subscript (1) in the above equation to (2) and to (3) respectively. Hence we may give the Eq. (8.58) in terms of its components in the x_i direction as follows:

$$\sum_{\gamma=1}^{3}\sum_{\beta=1}^{3}\frac{\partial}{\partial x_{\beta}}[\bar{\sigma}_{\beta\gamma}(\delta_{i\gamma}+e_{i\gamma}+\omega_{i\gamma})]+\frac{V^{*}}{V}B_{i}^{*}=0 \qquad (8.61)$$

We can now go over to dummy indices in place of summing over Greek indices to give the above equation as follows:

$$\frac{\partial}{\partial x_{k}}[\bar{\sigma}_{kj}(\delta_{ij}+e_{ij}+\omega_{ij})]+\frac{V^{*}}{V}B_{i}^{*}=0 \qquad (8.62)$$

We have here the equations for equilibrium. The quantities $\bar{\sigma}_{kj}$ are *pseudo* stresses and are called the *Kirchhoff stress components*. We will later show that $\bar{\sigma}_{kj}$ is symmetric; hence from Eq. (8.60) we can conclude that the actual stresses σ_{kj} are not symmetric. This is not a violation of the equilibrium condition as to moments since: (1) these stresses are on the deformed geometry of the original rectangular parallelopiped (see Fig. 8.8) which generally is no longer rectangular and: (2) the component stresses are not orthogonal to each other.

8.8 SIMPLIFICATION OF EQUATIONS

We shall simplify the equations of equilibrium given in Sec. 8.7 to forms that we find useful in certain classes of problems. First we will assume that the *elongations and shears are small compared to unity*. Then from Eqs. (8.40) and (8.41) we can say that:

$$\frac{A_{1}^{*}}{A_{1}}\approx1 \qquad \frac{A_{2}^{*}}{A_{2}}\approx1 \qquad \frac{A_{3}^{*}}{A_{3}}\approx1$$

Also, from Eq. (8.29) we can then conclude for this case:

$$D=\frac{V^{*}}{V}\approx1 \qquad (8.63)$$

Now we can conclude further that (see Eq. (8.60))

$$\bar{\sigma}_{\beta\gamma}=\frac{A_{\beta}^{*}}{A_{\beta}}\frac{\sigma_{\beta\gamma}}{1+E_{\gamma}}\approx\sigma_{\beta\gamma} \qquad (8.64)$$

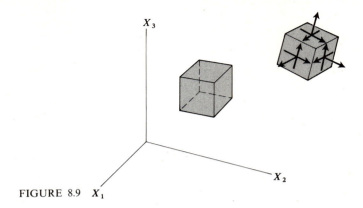

FIGURE 8.9 X_1

The equations of equilibrium thus become,

$$\frac{\partial}{\partial x_k}[\sigma_{kj}(\delta_{ij} + e_{ij} + \omega_{ij})] + B_i = 0 \qquad (8.65)$$

Here we are *neglecting effects of changing shape of an element in the undeformed geometry but are accounting for rotations of the element*. This has been shown in Fig. 8.9. Note that the element is taken with the same shape and size in the deformed geometry with the result that the *actual stresses are now orthogonal and given in terms of the original area values*. The stresses of course must now be symmetric from considerations of moments. That is:

$$\sigma_{\alpha\beta} = \sigma_{\beta\alpha}$$

We next assume that the *rotations* are also much smaller than unity but are generally considerably larger than the elongations and shears. That is,

$$\epsilon_{ij} \ll \omega_{ij} \ll 1 \qquad (8.66)$$

We have shown earlier that for such a case:

$$e_{ij} = \epsilon_{ij} + \frac{\phi_i\phi_j}{2} \qquad (8.67)$$

Then going to Eq. (8.65) we see that e_{ij} may be considered small compared to ω_{ij} in the bracketed expression since both ϵ_{ik} and $\phi_i\phi_k/2$ above are much smaller than ω_{ik}. Since we can expect that e_{ij} and ω_{ij}, being related gradient components of the displacement field, will vary with position x_i at approximately the same rate, we can then delete e_{ij} from Eq. (8.65). We thus get the following result

$$\frac{\partial}{\partial x_:}[\tau_{jk}(\delta_{ik} + \omega_{ik})] + B_i = 0 \qquad (8.68)$$

where we have now returned to the usual notation for stress as a result of the fact that there is now *no essential difference between the stresses measured for the original geometry and those for the deformed geometry*. Note that even though rotations are small their effect is to a degree accounted for. This is in contrast to the linear case where we neglect the rotation terms completely to get from above the familiar equations of equilibrium of the form:

$$\frac{\partial}{\partial x_j} \tau_{ij} + B_i = 0 \qquad (8.69)$$

We now have presented a general picture of strain and stress and have introduced some simplifications that constitute a departure from linear behaviour. We next present the principle of virtual work for the nonlinear case.

8.9 PRINCIPLE OF VIRTUAL WORK

We will derive in this section a version of the principle of virtual work that will include large deformations. We shall start as we did in Chap. 3 by considering the virtual work performed by the applied forces expressed in terms of the *deformed geometry*. Thus:

$$W_{\text{virt}} = \int \int \int_{V^*} \mathbf{B}^* \cdot \delta \mathbf{u} \, dv^* + \oint\oint_{S^*} \mathbf{T}^* \cdot \delta \mathbf{u} \, dS^* \qquad (8.70)$$

where * indicates deformed geometry. As usual $\delta \mathbf{u}$ is consistent with the constraints. In the first integral we first replace dV^* by $D \, dV$ in accordance with Sec. 8.4. Now the limits of integration become the undeformed geometry of the body. Also using Cartesian components in this integral relative to $X_1 X_2 X_3$ we get:

$$W_{\text{virt}} = \int \int \int_{V} DB_i^* \, \delta u_i \, dv + \oint\oint_{S^*} \mathbf{T}^* \cdot \delta \mathbf{u} \, dS^* \qquad (8.71)$$

We now consider the surface integral. For this purpose we have shown a tetrahedron in the undeformed geometry which deforms so that its surface dS becomes dS^* in the deformed geometry (see Fig. 8.10). Note that the areas comprising the *original* tetrahedron parallel to the coordinate faces are denoted as dS_1, dS_2, and dS_3 and they become in the deformed geometry the elements dS_1^* with direction $\hat{\mathbf{n}}_1'$, dS_2^* with direction $\hat{\mathbf{n}}_2'$, etc. The traction on the surfaces of the tetrahedron in the deformed geometry are shown as $\overset{*}{\mathbf{T}}^{(n')}$, $\overset{*}{\mathbf{T}}^{(n_1')}$, $\overset{*}{\mathbf{T}}^{(n_2')}$, and $\overset{*}{\mathbf{T}}^{(n_3')}$. It is clear that:[1]

$$\mathbf{T}^* \, dS^* = \overset{*}{\mathbf{T}}^{(n')} \, dS^* = \sum_{\alpha=1}^{3} \overset{*}{\mathbf{T}}^{(n_\alpha')} \, dS_\alpha^* \qquad (8.72)$$

---

[1] Note again that n is being used to give orientations in the undeformed geometry and n' is being used to give orientations in the deformed geometry with the subscripts of the latter tying back to normal directions of the element when it *was* in the undeformed geometry.

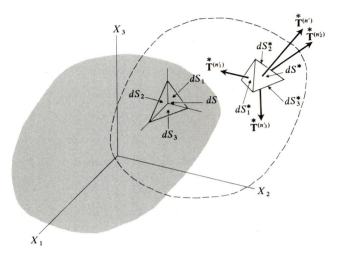

FIGURE 8.10

Also we can say for dS_α^* that:

$$dS_\alpha^* = \frac{A_\alpha^*}{A_\alpha} dS_\alpha \qquad (8.73)$$

But dS_α can be replaced as follows

$$dS_\alpha = n_\alpha \, dS$$

where n_α are the direction cosines of the normal to dS in the *undeformed* geometry. We can thus express Eq. (8.72) in the following way upon utilizing the preceding equations:

$$\mathbf{T}^* \, dS^* = \sum_{\alpha=1}^{3} \overset{*}{\mathbf{T}}{}^{(n_\alpha')} \frac{A_\alpha^*}{A_\alpha} n_\alpha \, dS \qquad (8.74)$$

Finally we express the stress vector $\overset{*}{\mathbf{T}}{}^{(n_\alpha')}$ in terms of components parallel to the directions $\hat{\mathbf{r}}^{(1)}$, $\hat{\mathbf{r}}^{(2)}$, and $\hat{\mathbf{r}}^{(3)}$ as has been done in earlier sections. That is:

$$\overset{*}{\mathbf{T}}{}^{(n_\alpha')} = \sum_{\beta=1}^{3} \sigma_{\alpha\beta} \hat{\mathbf{r}}^{(\beta)} = \sum_{\beta=1}^{3} \sigma_{\alpha\beta} \left(\frac{\delta_{k\beta} + \partial u_k / \partial x_\beta}{1 + E_\beta} \right) \hat{\mathbf{i}}_k \qquad (8.75)$$

where we have in the last step decomposed $\hat{\mathbf{r}}^{(\beta)}$ into components along the $X_1 X_2 X_3$ reference using Eq. (8.16(*b*)). Eq. (8.74) then may be given as:

$$\mathbf{T}^* \, dS^* = \sum_{\alpha=1}^{3} \sum_{\beta=1}^{3} \sigma_{\alpha\beta} \frac{A_\alpha^*}{A_\alpha} n_\alpha \, dS \frac{\delta_{k\beta} + \partial u_k / \partial x_\beta}{1 + E_\beta} \hat{\mathbf{i}}_k$$

$$= \sum_{\alpha=1}^{3} \sum_{\beta=1}^{3} \bar{\sigma}_{\alpha\beta} \left(\delta_{k\beta} + \frac{\partial u_k}{\partial x_\beta} \right) n_\alpha \hat{\mathbf{i}}_k \, dS$$

Now going back to the expression for virtual work (Eq. (8.71)) we can say on interchanging α and β indices and *assuming* that $\bar{\sigma}_{\alpha\beta}$ is symmetric:

$$W_{\text{virt}} = \int\int\int_V DB_k^* \, \delta u_k \, dv + \oiint_S \sum_{\alpha=1}^{3} \sum_{\beta=1}^{3} \bar{\sigma}_{\alpha\beta} \left(\delta_{k\alpha} + \frac{\partial u_k}{\partial x_\alpha} \right) n_\beta \, \delta u_k \, dS$$

Applying now the divergence theorems to the surface integral we get:

$$W_{\text{virt}} = \int\int\int_V \left\{ DB_k^* \, \delta u_k + \sum_{\alpha=1}^{3} \sum_{\beta=1}^{3} \frac{\partial}{\partial x_\beta} \left[\bar{\sigma}_{\alpha\beta} \left(\delta_{k\alpha} + \frac{\partial u_k}{\partial x_\alpha} \right) \delta u_k \right] \right\} dv$$

$$= \int\int\int_V \left\{ DB_k^* + \sum_{\alpha=1}^{3} \sum_{\beta=1}^{3} \frac{\partial}{\partial x_\beta} \left[\bar{\sigma}_{\alpha\beta} \left(\delta_{k\alpha} + \frac{\partial u_k}{\partial x_\alpha} \right) \right] \right\} \delta u_k \, dv$$

$$+ \int\int\int_V \sum_{\alpha=1}^{3} \sum_{\beta=1}^{3} \bar{\sigma}_{\alpha\beta} \left(\delta_{k\alpha} + \frac{\partial u_k}{\partial x_\alpha} \right) \frac{\partial \, \delta u_k}{\partial x_\beta} \, dv \qquad (8.76)$$

Now examining the integrand of the first expression on the right side of the above equation we see that it must be zero in order to satisfy the equations of equilibrium[1] (see Eq. 8.61) and so we have only the second expression to deal with. We then have for W_{virt} the following result, which we can now give with the dummy indices i and j:

$$W_{\text{virt}} = \int\int\int_V \left[\bar{\sigma}_{ij} \left(\delta_{ki} + \frac{\partial u_k}{\partial x_i} \right) \frac{\partial \, \delta u_k}{\partial x_j} \right] dv \qquad (8.77)$$

Let us for the moment consider the variation of the Green strain tensor. We have

$$\delta \epsilon_{ij} = \frac{1}{2} \left(\frac{\partial \, \delta u_i}{\partial x_j} + \frac{\partial \, \delta u_j}{\partial x_i} + \frac{\partial u_k}{\partial x_i} \frac{\partial \, \delta u_k}{\partial x_j} + \frac{\partial u_k}{\partial x_j} \frac{\partial \, \delta u_k}{\partial x_i} \right)$$

Taking $\bar{\sigma}_{ij}$ as symmetric then we may conclude:

$$\bar{\sigma}_{ij} \, \delta \epsilon_{ij} = \bar{\sigma}_{ij} \left(\frac{\partial \, \delta u_i}{\partial x_j} + \frac{\partial u_k}{\partial x_i} \frac{\partial \, \delta u_k}{\partial x_j} \right)$$

$$= \bar{\sigma}_{ij} \left(\delta_{ik} + \frac{\partial u_k}{\partial x_i} \right) \left[\frac{\partial \, \delta u_k}{\partial x_j} \right]$$

Now compare the right side of the above equation with the integrand of Eq. (8.77). They are identical and so we have:

$$W_{\text{virt}} = \int\int\int_V \bar{\sigma}_{ij} \, \delta \epsilon_{ij} \, dv \qquad (8.78)$$

---

[1] Note in this regard that $u_{k,\alpha} = e_{k\alpha} + \omega_{k\alpha}$.

Hence we have for the principle of virtual work:

$$\int \int \int_{V^*} \mathbf{B}^* \cdot \delta\mathbf{u} \, dv^* + \int \int_{S^*} \overset{*}{\mathbf{T}}{}^{(n')} \cdot \delta\mathbf{u} \, dS^* = \int \int \int_V \bar{\sigma}_{ij} \, \delta\epsilon_{ij} \, dv \qquad (8.79)$$

The similarity between Eq. (8.79) and the classical result needs no further comment.

8.10 TOTAL POTENTIAL ENERGY

In the previous section we set forth a relation between any kinematically admissible deformation field and any stress field $\bar{\sigma}_{ij}$ that satisfied equations of equilibrium. The stress and strain fields were valid for large deformations. No constitutive law was employed in arriving at this relation. We now *link* the fields $\bar{\sigma}_{ij}$ and ϵ_{ij} by a constitutive law. To do this we postulate a function Φ, which we shall denote as a *strain energy function*, such that:

$$\bar{\sigma}_{ij} = \frac{\partial \Phi}{\partial \epsilon_{ij}} \qquad (8.80)$$

Under such circumstances one sees that since ϵ_{ij} is *symmetric* then $\bar{\sigma}_{ij}$ must be symmetric. That is:

$$\bar{\sigma}_{ij} = \frac{\partial \Phi}{\partial \epsilon_{ij}} = \frac{\partial \Phi}{\partial \epsilon_{ji}} = \bar{\sigma}_{ji}$$

Substituting for $\bar{\sigma}_{ij}$ using Eq. (8.80) in Eq. (8.79) we get:

$$\int \int \int_{V^*} \mathbf{B}^* \cdot \delta\mathbf{u} \, dv^* + \oint \int_{S^*} \mathbf{T}^{(n')} \cdot \delta\mathbf{u} \, dS^* = \int \int \int_V \frac{\partial \Phi}{\partial \epsilon_{ji}} \, \delta\epsilon_{ij} \, dv = \int \int \int_V \delta\Phi \, dv \qquad (8.81)$$

We may express the above equation in the following way:

$$\delta^{(1)} \left\{ \int \int \int_V \Phi \, dv - \int \int \int_{V^*} \mathbf{B} \cdot \mathbf{u} \, dv^* - \oint \int_{S^*} \overset{*}{\mathbf{T}}{}^{(n')} \cdot \mathbf{u} \, dS^* \right\} = 0 \qquad (8.82)$$

The bracketed expression is the total potential energy for large deformations.

We now carry out the extremization process to show that we get the same equations of equilibrium as were developed in the earlier section. We shall also get proper boundary conditions which as yet we have not developed for large deformation. Thus using Eq. (8.2) to compute $\delta(\epsilon_{ij})$ and replacing dv^* by $D \, dv$ we get:

$$\int \int \int_V \left\{ \frac{\partial \Phi}{\partial \epsilon_{ij}} \left(\frac{1}{2} \right) (\delta u_{i,j} + \delta u_{j,i} + u_{k,i} \, \delta u_{k,j} + \delta u_{k,i} u_{k,j}) - D B_k^* \, \delta u_k \right\} dv$$

$$- \oint \int_{S^*} \overset{*}{T}{}^{(n')}_i \, \delta u_i \, dS^* = 0$$

We may shorten the above integral by making use of the symmetry of ϵ_{ij} and collecting terms to get:

$$\iiint_V \left[\frac{\partial \Phi}{\partial \epsilon_{ij}} (\delta u_{i,j} + u_{k,i}\, \delta u_{k,j}) - DB^* \delta u_k \right] dv - \oiint_{S^*} \overset{*}{T}_i^{(n')} \delta u_i\, dS^* = 0$$

Using Green's theorem for the first integral we get:

$$\iiint_V \left\{ -\frac{\partial}{\partial x_j} \left[\frac{\partial \Phi}{\partial \epsilon_{ij}} (\delta u_i + u_{k,i}\, \delta u_k) \right] - DB_k^*\, \delta u_k \right\} dv$$

$$+ \oiint_S \frac{\partial \Phi}{\partial \epsilon_{ij}} (\delta u_i + u_{k,i}\, \delta u_k) n_j\, dS - \oiint_{S^*} \overset{*}{T}_i^{(n')} \delta u_i\, dS^* = 0$$

Making use of Kronecker delta to extract δu_k in the first and second integrals we get:

$$\iiint_V \left\{ -\frac{\partial}{\partial x_j} \left[\frac{\partial \Phi}{\partial \epsilon_{ij}} (\delta_{ki} + u_{k,i}) \right] - DB_k^* \right\} \delta u_k\, dv$$

$$+ \oiint_S \frac{\partial \Phi}{\partial \epsilon_{ij}} (\delta_{ki} + u_{k,i}) n_j\, \delta u_k\, dS - \oiint_{S^*} \overset{*}{T}_i^{(n')} \delta u_i\, dS^* = 0 \qquad (8.83)$$

We see immediately from above that on replacing $u_{k,i}$ by $e_{ki} + \omega_{ki}$ in the integrand of the volume integral and by using Eq. (8.80) we have as the Euler–Lagrange equation:

$$\frac{\partial}{\partial x_j} [\bar{\sigma}_{ij}(\delta_{ki} + e_{ki} + \omega_{ki})] + DB_k^* = 0 \qquad (8.84)$$

We thus rederive the equations of equilibrium of Sec. 8.7 and in so doing show that the principle of total potential energy is sufficient for satisfying the equations of equilibrium for large deformations.

We may now simplify this equation as described in the previous section. It is to be noted, however, that we could not get the result developed in Sec. 8.8 namely:

$$\frac{\partial}{\partial x_j} [\tau_{ij}(\delta_{ki} + \omega_{ki})] + B_k = 0 \qquad (8.85)$$

by incorporating the approximation presented earlier

$$\epsilon_{ij} = e_{ij} + \tfrac{1}{2}\omega_{ki}\omega_{kj} \qquad (8.86)$$

into the variational approach.[1] That is, the pair of equations (8.85) and (8.86) are not variationally consistent. Thus we have to make for these equations certain

---

[1] Actually, when using Eq. (8.86) we get

$$\frac{1}{2} \frac{\partial}{\partial x_i} [\tau_{ij}(\delta_{kj} + \omega_{kj})] + \frac{1}{2} \frac{\partial}{\partial x_l} [\tau_{kl}(\delta_{lj} - \omega_{lj})] + B_k = 0$$

simplifications from *results* that may be developed from the variational process. At other times we can proceed as we have in the past by using the simplified kinematic conditions in the variational process to set forth variationally consistent results.

We now may go to the proper *boundary conditions* for the *large deformation* case. We have then from Eq. (8.83):

$$\oiint_S \frac{\partial \Phi}{\partial \epsilon_{ij}} (\delta_{ki} + u_{k,i}) n_j \, \delta u_k \, dS - \oiint_{S^*} \overset{*}{T}_k^{(n')} \, \delta u_k \, dS^* = 0 \qquad (8.87)$$

We may express the above equation furthermore as follows by replacing $u_{k,i}$:

$$\oiint_S \frac{\partial \Phi}{\partial \epsilon_{ij}} (\delta_{ki} + e_{ki} + \omega_{ki}) n_j \, \delta u_k \, dS - \oiint_{S^*} \overset{*}{T}_k^{(n')} \, \delta u_k \, dS^* = 0 \qquad (8.88)$$

For the so-called classical nonlinear case we simplify the above result in a manner consistent with steps taken in earlier work. Thus we *may neglect e_{ki} in the first integrand* and drop the stars and prime in the second integrand and *return to the original geometry for that integral.* We thus have:

$$\oiint_S \left[\frac{\partial \Phi}{\partial \epsilon_{ij}} (\delta_{ki} + \omega_{ki}) n_j - T_k^{(n)} \right] dS \, \delta u_k = 0 \qquad (8.89)$$

We can then conclude on the boundary that:[1]

$$\frac{\partial \Phi}{\partial \epsilon_{ij}} (\delta_{ki} + \omega_{ki}) n_j = T_k^{(n)} \qquad (8.90)$$

In the following section we shall follow the procedure used heretofore of making all kinematic simplifications first for use in the variational process to arrive at equations of equilibrium.

8.11 VON KÁRMÁN PLATE THEORY

We consider a thin plate (Fig. 8.11) loaded normal to its surface by the distribution $q(x,y)$ and loaded at the edge by a tangential force distribution $\bar{N}_{vs}$ per unit length of the edge and a normal force distribution $\bar{N}_v$ per unit length of the edge. The bounding curve of the plate is denoted as Γ while the thickness of the plate is denoted as h.

We now set forth assumptions as to the deformation of the body of the type discussed in previous sections. That is, we shall assume that strains and *rotation are both small compared to unity* so that we can ignore the effects of changes of geometry in the definition of stress components and in the limits of integration needed for

---

[1] If we neglect ω_{ki} note that we get back to the familiar Cauchy formula.

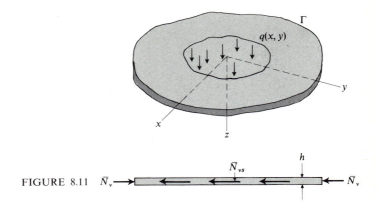

FIGURE 8.11 $\bar{N}_v$ [diagram] $\bar{N}_v$

work and energy considerations. Furthermore we will stipulate that *strains are much smaller generally than rotations* and we will employ Eq. (8.86) accordingly. Finally we use again the Kirchhoff assumption that *lines normal to the undeformed middle surface remain normal to this surface in the deformed geometry and are unextended after deformation.* This means that:

$$u_1(x_1,x_2,x_3) = u(x,y) - z\frac{\partial w(x,y)}{\partial x} \qquad (a)$$

$$u_2(x_1,x_2,x_3) = v(x,y) - z\frac{\partial w(x,y)}{\partial y} \qquad (b)$$

$$u_3(x_1,x_2,x_3) = w(x,y) \qquad (c) \quad (8.91)$$

where u, v, and w are the displacement components of the middle surface of the plate. We may now give the strain parameters and rotation parameters as follows:

$$e_{11} = \frac{\partial u}{\partial x} - z\frac{\partial^2 w}{\partial x^2} \qquad e_{12} = \frac{1}{2}\left(\frac{\partial u}{\partial y} + \frac{\partial v}{\partial x} - 2z\frac{\partial^2 w}{\partial x\,\partial y}\right)$$

$$e_{22} = \frac{\partial v}{\partial y} - z\frac{\partial^2 w}{\partial y^2} \qquad e_{13} = e_{23} = 0$$

$$e_{33} = \frac{\partial w}{\partial z} = 0 \qquad\qquad\qquad\qquad\qquad (8.92)$$

$$\omega_{12} = \frac{1}{2}\left(\frac{\partial u}{\partial y} - \frac{\partial v}{\partial x}\right)$$

$$\omega_{13} = -\frac{\partial w}{\partial x}$$

$$\omega_{23} = -\frac{\partial w}{\partial y} \qquad\qquad\qquad\qquad\qquad (8.93)$$

We now observe the following. The rotation parameter ω_{12} approximates a rotation component about the z axis as has been shown for the linear case in Chap. 1 while ω_{23} and ω_{13} approximate rotation components about axes parallel to the X and Y axes respectively in the midplane of the plate. For a *thin*, hence *flexible*, plate we can reasonably expect that:

$$\omega_{12} \ll \omega_{23}, \omega_{13} \qquad (8.94)$$

Neglecting ω_{12}, we now employ Eq. (8.86) with Eqs. (8.92) and (8.93) to find ϵ_{ij}.

$$\epsilon_{11} = \frac{\partial u}{\partial x} + \frac{1}{2}\left(\frac{\partial w}{\partial x}\right)^2 - z\frac{\partial^2 w}{\partial x^2}$$

$$\epsilon_{22} = \frac{\partial v}{\partial y} + \frac{1}{2}\left(\frac{\partial w}{\partial y}\right)^2 - z\frac{\partial^2 w}{\partial y^2}$$

$$\epsilon_{33} = \frac{1}{2}\left(\frac{\partial w}{\partial x}\right)^2 + \frac{1}{2}\left(\frac{\partial w}{\partial y}\right)^2$$

$$\epsilon_{12} = \frac{1}{2}\left(\frac{\partial u}{\partial y} + \frac{\partial v}{\partial x} - 2z\frac{\partial^2 w}{\partial x\,\partial y}\right) + \frac{1}{2}\frac{\partial w}{\partial x}\frac{\partial w}{\partial y}$$

$$\epsilon_{13} \cong \epsilon_{23} \cong 0 \qquad (8.95)$$

For a constitutive law we will employ Hooke's law between stress and strain for plane stress over the thickness of the plate. Thus we shall be concerned here only with $\epsilon_{11}, \epsilon_{22}$, and ϵ_{12}. Accordingly when we consider the total potential energy for the assumptions presented here, we can say:

$$\iiint_V \bar{\sigma}_{ij}\,\delta\epsilon_{ij}\,dv = \iiint_V \tau_{ij}\,\delta\epsilon_{ij}\,dv$$

$$= \iint_R \int_{-h/2}^{h/2} (\tau_{11}\,\delta\epsilon_{11} + 2\tau_{12}\,\delta\epsilon_{12} + \tau_{22}\,\delta\epsilon_{22})\,dz\,dA \qquad (8.96)$$

We employ Eq. (8.95) to replace the strain terms in the above expression which we now denote as $\delta^{(1)}U$:

$$\delta^{(1)}U = \iint_R \int_{-h/2}^{h/2} \left\{ \tau_{11}\left[\frac{\partial\delta u}{\partial x} + \left(\frac{\partial w}{\partial x}\right)\left(\frac{\partial\delta w}{\partial x}\right) - z\frac{\partial^2\delta w}{\partial x^2}\right] \right.$$

$$+ \tau_{12}\left[\frac{\partial\delta u}{\partial y} + \frac{\partial\delta v}{\partial x} - 2z\frac{\partial^2\delta w}{\partial x\,\partial y} + \frac{\partial w}{\partial x}\frac{\partial\delta w}{\partial y} + \frac{\partial w}{\partial y}\frac{\partial\delta w}{\partial x}\right]$$

$$\left. + \tau_{22}\left[\frac{\partial\delta v}{\partial y} + \frac{\partial w}{\partial y}\frac{\partial\delta w}{\partial y} - z\frac{\partial^2\delta w}{\partial y^2}\right]\right\} dA\,dz \qquad (8.97)$$

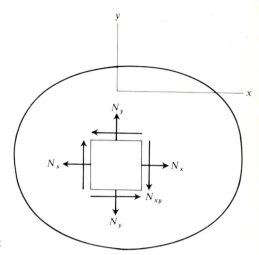

FIGURE 8.12

Next integrate with respect to z and use the resultant intensity functions presented in Chap. 6, namely M_x, M_y, and M_{xy} as well as the following new functions

$$N_x = \int_{-h/2}^{h/2} \tau_{xx}\, dz$$

$$N_y = \int_{-h/2}^{h/2} \tau_{yy}\, dz$$

$$N_{xy} = \int_{-h/2}^{h/2} \tau_{xy}\, dz$$

$$N_{yx} = \int_{-h/2}^{h/2} \tau_{yx}\, dz \qquad (8.98)$$

where $N_x(N_y)$ is a force in the x direction (y direction) measured per unit length in the y direction (x direction) and where $N_{xy}(N_{yx})$ is a force in the $x(y)$ direction per unit length in the $x(y)$ direction. We have shown such intensities in Fig. 8.12. These may be for practical purposes compared with normal and shear stresses and we may conclude that $N_{xy} = N_{yx}$. We then get for Eq. (8.97) the following result:

$$\delta^{(1)}U = \int\int_R \left\{ N_x\left[\frac{\partial \delta u}{\partial x} + \frac{\partial w}{\partial x}\left(\frac{\partial \delta w}{\partial x} \right) \right] - M_x\left[\frac{\partial^2 \delta w}{\partial x^2} \right] \right.$$

$$+ N_{xy}\left(\frac{\partial \delta u}{\partial y} + \frac{\partial \delta v}{\partial x} + \frac{\partial w}{\partial x}\frac{\partial \delta w}{\partial y} + \frac{\partial w}{\partial y}\frac{\partial \delta w}{\partial x} \right)$$

$$\left. - 2M_{xy}\frac{\partial^2 \delta w}{\partial x\, \partial y} + N_y\left(\frac{\partial \delta v}{\partial y} + \frac{\partial w}{\partial y}\frac{\partial \delta w}{\partial y} \right) - M_y\frac{\partial^2 \delta w}{\partial y^2} \right\} dx\, dy \qquad (8.99)$$

The first variation of the potential of the applied forces takes the form (noting that $\overline{N}_v$ is taken as positive in compression as shown in Fig. 8.11)

$$\delta^{(1)}V = -\iint_R q\,\delta w\,dx\,dy + \oint_\Gamma \overline{N}_v\,\delta u_v\,ds - \oint_\Gamma \overline{N}_{vs}\,\delta u_s\,ds \qquad (8.100)$$

where u_v and u_s are the in-plane displacements of the boundary of the plate in directions normal and tangential respectively to the boundary. We are using the undeformed geometry for the applied loads above rather than the deformed geometry—thereby restricting the result to reasonably small deformations. By using the above result for $\delta^{(1)}V$ and using Eq. (8.99) for $\delta^{(1)}U$ we may form $\delta^{(1)}\pi$. The total potential energy so formed then approximates the actual total potential energy for the kind of deformation restrictions as we embodied in Kirchhoff's assumptions. And, since we have used undeformed geometry for stresses and external loads, we are limited to small deformation as far as employing this functional. Finally because we employed Eq. 8.86 for strain we are assuming that strains are much smaller than rotations. Thus we have for the total potential energy principle:

$$\delta^{(1)}(\pi) = 0 = \iint_R \left\{ N_x \left[\frac{\partial \delta u}{\partial x} + \frac{\partial w}{\partial x}\frac{\partial \delta w}{\partial x} \right] - M_x \frac{\partial^2\,\delta w}{\partial x^2} \right.$$

$$+ N_{xy}\left(\frac{\partial \delta u}{\partial y} + \frac{\partial \delta v}{\partial x} + \frac{\partial w}{\partial x}\frac{\partial \delta w}{\partial y} + \frac{\partial w}{\partial y}\frac{\partial \delta w}{\partial x} \right)$$

$$\left. - 2M_{xy}\frac{\partial^2\,\delta w}{\partial x\,\partial y} + N_y\left(\frac{\partial \delta v}{\partial y} + \frac{\partial w}{\partial y}\frac{\partial \delta w}{\partial y} \right) - M_y \frac{\partial^2\,\delta w}{\partial y^2} \right\} dx\,dy$$

$$- \iint_R q\,\delta w\,dx\,dy + \oint_\Gamma \overline{N}_v\,\delta u_v\,ds - \oint_\Gamma \overline{N}_s\,\delta u_s\,ds \qquad (8.101)$$

We may proceed in a manner that should now be familiar to carry out the extremization process. We employ Green's theorem one or more times to get the δu's and δv's out from the partial derivatives. Then we proceed to simplify the expressions in the line integrals by noting from equilibrium (see Fig. 8.13) that

$$\overline{N}_v = N_x a_{vx}{}^2 + 2N_{xy}a_{vx}a_{vy} + N_y a_{vy}{}^2 \qquad (a)$$

$$\overline{N}_{vs} = (N_y - N_x)a_{vx}a_{vy} + N_{xy}(a_{vx}{}^2 + a_{vy}{}^2) \qquad (b) \;\; (8.102)$$

where a_{vx} and a_{vy} are the direction cosines of the outward normal of the boundary. Furthermore, simple vector projections permit us to say:

$$u_v = a_{vx}u + a_{vy}v$$

$$u_s = -a_{vy}u + a_{vx}v \qquad (8.103)$$

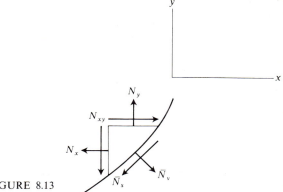

FIGURE 8.13

And from Eq. 6.37 we note additionally

$$\frac{\partial}{\partial x} = a_{vx}\frac{\partial}{\partial v} - a_{vy}\frac{\partial}{\partial s}$$

$$\frac{\partial}{\partial y} = a_{vy}\frac{\partial}{\partial v} + a_{vx}\frac{\partial}{\partial s} \qquad (8.104)$$

Finally Eqs. (6.16), (6.17), and (6.18) will be used and are now restated:

$$Q_v = Q_x a_{vx} + Q_y a_{vy} \qquad (a)$$

$$Q_x = \frac{\partial M_x}{\partial x} + \frac{\partial M_{xy}}{\partial y} \qquad (b)$$

$$Q_y = \frac{\partial M_y}{\partial y} + \frac{\partial M_{xy}}{\partial x} \qquad (c) \quad (8.105)$$

Now utilizing Eqs. (8.102) through (8.105) we may follow steps taken in Sec. 6.4 to rewrite Eq. (8.101) as follows:

$$\delta^{(1)}\pi = - \iint_R \left\{ \left(\frac{\partial N_x}{\partial x} + \frac{\partial N_{xy}}{\partial y} \right) \delta u + \left(\frac{\partial N_{xy}}{\partial x} + \frac{\partial N_y}{\partial y} \right) \delta v \right. $$

$$+ \left[\frac{\partial^2 M_x}{\partial x^2} + 2\frac{\partial^2 M_{xy}}{\partial x\, \partial y} + \frac{\partial^2 M_y}{\partial y^2} + \frac{\partial}{\partial x}\left(N_x \frac{\partial w}{\partial x} \right) + \frac{\partial}{\partial y}\left(N_{xy}\frac{\partial w}{\partial x} \right) \right.$$

$$\left. + \frac{\partial}{\partial x}\left(N_{xy}\frac{\partial w}{\partial y} \right) + \frac{\partial}{\partial y}\left(N_y \frac{\partial w}{\partial y} \right) + q \right] \delta w \Bigg\} dx\, dy$$

$$+ \oint_\Gamma (N_v + \bar{N}_v)\, \delta u_v\, ds + \oint_\Gamma (N_{vs} - \bar{N}_{vs})\, \delta u_s\, ds - \oint_\Gamma M_v \frac{\partial \delta w}{\partial v}\, ds$$

$$+ \oint_\Gamma \left(Q_v + \frac{\partial M_{vs}}{\partial s} + N_v \frac{\partial w}{\partial v} + N_{vs}\frac{\partial w}{\partial s} \right) \delta w\, ds - [M_{vs}\, \delta w]_\Gamma = 0 \qquad (8.106)$$

The last expression accounts for "corners" in the boundary as was discussed in Chap. 6. From the above equations we may now make a series of deductions. First in region R we can conclude that:

$$\frac{\partial N_x}{\partial x} + \frac{\partial N_{xy}}{\partial y} = 0 \qquad (a)$$

$$\frac{\partial N_{xy}}{\partial x} + \frac{\partial N_y}{\partial y} = 0 \qquad (b)$$

$$\frac{\partial^2 M_x}{\partial x^2} + 2\frac{\partial^2 M_{xy}}{\partial x\,\partial y} + \frac{\partial^2 M_y}{\partial y^2} + \frac{\partial}{\partial x}\left(N_x\frac{\partial w}{\partial x}\right) + \frac{\partial}{\partial y}\left(N_{xy}\frac{\partial w}{\partial x}\right)$$

$$+ \frac{\partial}{\partial x}\left(N_{xy}\frac{\partial w}{\partial y}\right) + \frac{\partial}{\partial y}\left(N_y\frac{\partial w}{\partial y}\right) + q = 0 \qquad (c) \quad (8.107)$$

The first two equations, above, clearly are identical to the equations of equilibrium for plane stress, as is to be expected. We shall use these equations now to simplify the third equation after we carry out the differentiation operator on the expression involving products. We are thus able to eliminate in this way expressions involving the first partial of w. We get:

$$\frac{\partial^2 M_x}{\partial x^2} + 2\frac{\partial^2 M_{xy}}{\partial x\,\partial y} + \frac{\partial^2 M_y}{\partial y^2} + N_x\frac{\partial^2 w}{\partial x^2} + 2N_{xy}\frac{\partial^2 w}{\partial x\,\partial y} + N_y\frac{\partial^2 w}{\partial y^2} + q = 0 \qquad (8.108)$$

Now, comparing Eq. (8.108) with the classical case (Eq. (6.20)), we see that we have here introduced nonlinear terms $[N_x(\partial^2 w/x^2) + 2N_{xy}(\partial^2 w/\partial x\,\partial y) + N_y(\partial^2 w/\partial y^2)]$ involving the in-plane force intensities as additional "transverse loading" terms.[1]

Considering next the remainder of Eq. (8.106) we can stipulate the following boundary conditions along Γ:

EITHER $N_v = -\bar{N}_v$ OR u_v IS SPECIFIED (a)

EITHER $N_{vs} = \bar{N}_{vs}$ OR u_s IS SPECIFIED (b)

EITHER $M_v = 0$ OR $\dfrac{\partial w}{\partial v}$ IS SPECIFIED (c)

EITHER $Q_v + \dfrac{\partial M_{vs}}{\partial s} + N_v\dfrac{\partial w}{\partial v} + N_{vs}\dfrac{\partial w}{\partial s} = 0$ OR w IS SPECIFIED (d)

At discontinuities $[M_{vs}\delta w]_\Gamma = 0$ (e) (8.109)

The last three conditions are familiar from earlier work on plates except that the effective shear force ($Q_v + \partial M_{vs}/\partial s$) is now augmented by projections of the inplane

---

[1] In the Russian literature these terms are often referred to as "reduced loads" of the in-plane forces or the "reduced forces."

plate forces at the plate edges. (It is to be pointed out that the negative sign for condition (8.109(a)) is a result of having chosen $\bar{N}_y$ positive in compression.)

The equations of equilibrium given by Eqs. (8.107(a)), (b) and Eq. (8.108) may be solved if a constitutive law is used. We will employ here the familiar Hooke's law for plane stress with the strain taken as per Eqs. (8.95). We will use the constitutive law to replace the resultant intensity functions by appropriate derivatives of the displacement field of the midplane of the plate as was done in the more simple undertaking in Chap. 6. Consider for example the quantity M_x. We have, using Hooke's law:

$$M_x = \int_{-h/2}^{h/2} z\tau_{xx}\, dz = \int_{-h/2}^{h/2} z\left(\frac{E}{1-v^2}\right)(\epsilon_{xx} + v\epsilon_{yy})\, dz$$

$$= \int_{-h/2}^{h/2} z\frac{E}{1-v^2}\left[\frac{\partial u}{\partial x} + \frac{1}{2}\left(\frac{\partial w}{\partial x}\right)^2 - z\frac{\partial^2 w}{\partial x^2} + v\frac{\partial v}{\partial y}\right.$$

$$\left. + \frac{v}{2}\left(\frac{\partial w}{\partial y}\right)^2 - vz\frac{\partial^2 w}{\partial y^2}\right]\, dz$$

Integrating and inserting limits we get

$$M_x = \left(\frac{E}{1-v^2}\right)\left(\frac{h^3}{12}\right)\left(-\frac{\partial^2 w}{\partial x^2} - v\frac{\partial^2 w}{\partial y^2}\right)$$

$$= -D\left[\frac{\partial^2 w}{\partial x^2} + v\frac{\partial^2 w}{\partial y^2}\right] \tag{8.110}$$

where D is the familiar bending rigidity given as $D = Eh^3/12(1-v^2)$. Similarly we have

$$M_y = -D\left[\frac{\partial^2 w}{\partial y^2} + v\frac{\partial^2 w}{\partial x^2}\right] \tag{a}$$

$$M_{xy} = -(1-v)D\frac{\partial^2 w}{\partial x\, \partial y} \tag{b}$$

$$N_x = C\left\{\left[\frac{\partial u}{\partial x} + \frac{1}{2}\left(\frac{\partial w}{\partial x}\right)^2\right] + v\left[\frac{\partial v}{\partial y} + \frac{1}{2}\left(\frac{\partial w}{\partial y}\right)^2\right]\right\} \tag{c}$$

$$N_y = C\left\{\left[\frac{\partial v}{\partial y} + \frac{1}{2}\left(\frac{\partial w}{\partial y}\right)^2\right] + v\left[\frac{\partial u}{\partial x} + \frac{1}{2}\left(\frac{\partial w}{\partial x}\right)^2\right]\right\} \tag{d}$$

$$N_{xy} = C\left(\frac{1-v}{2}\right)\left[\frac{\partial u}{\partial y} + \frac{\partial v}{\partial x} + \left(\frac{\partial w}{\partial x}\right)\left(\frac{\partial w}{\partial y}\right)\right] \tag{e} \quad (8.111)$$

where C is the *extensional rigidity* given as:

$$C = \frac{Eh}{1-v^2} \tag{8.112}$$

We could now substitute for the resultant intensity functions using the above relations and thus get the equations in terms of the displacement components of the midplane of the plate. However, we shall follow another route which leads to a somewhat less complicated system of equations.

Note, accordingly, that Eqs. (8.107(a)) and (b) will be individually satisfied if we use a function F defined like the Airy stress function of Chapter 1 as follows:

$$N_x = \frac{\partial^2 F}{\partial y^2} \qquad N_y = \frac{\partial^2 F}{\partial x^2} \qquad N_{xy} = -\frac{\partial^2 F}{\partial x\, \partial y} \qquad (8.113)$$

Then, replacing M_x, M_y, and M_{xy} in Eq. (8.108) using Eqs. (8.110) and (8.111(a)), (b) it is a simple matter to show that:

$$D\nabla^4 w = \frac{\partial^2 F}{\partial y^2}\frac{\partial^2 w}{\partial x^2} - 2\frac{\partial^2 F}{\partial x\, \partial y}\frac{\partial^2 w}{\partial x\, \partial y} + \frac{\partial^2 F}{\partial x^2}\frac{\partial^2 w}{\partial y^2} + q \qquad (8.114)$$

We now have a single partial differential equation with two dependent variables w and F. Since we are now studying in-plane effects using a *stress approach* we must assure the compatibility of the in-plane displacements. This will give us a second companion equation to go with the above equation. To do this, we shall seek to relate the strains ϵ_{xx}, ϵ_{yy}, and ϵ_{xy} at the midplane surface in such a way that when we employ Eq. (8.95) to replace the strains we end up with a result that does not contain the in-plane displacement components u and v. Thus you may readily demonstrate that substituting Eq. (8.95) into the expression

$$\left[\frac{\partial^2 \epsilon_{xx}}{\partial y^2} + \frac{\partial^2 \epsilon_{yy}}{\partial x^2} - 2\frac{\partial^2 \epsilon_{xy}}{\partial x\, \partial y}\right]_{z=0} \quad \text{gives} \quad \left(\frac{\partial^2 w}{\partial x\, \partial y}\right)^2 - \left(\frac{\partial^2 w}{\partial x^2}\right)\left(\frac{\partial^2 w}{\partial y^2}\right).$$

That is, we can say:

$$\left[\frac{\partial^2 \epsilon_{xx}}{\partial y^2} + \frac{\partial^2 \epsilon_{yy}}{\partial x^2} - 2\frac{\partial^2 \epsilon_{xy}}{\partial x\, \partial y}\right]_{z=0} = \left(\frac{\partial^2 w}{\partial x\, \partial y}\right)^2 - \left(\frac{\partial^2 w}{\partial x^2}\right)\left(\frac{\partial^2 w}{\partial y^2}\right)$$

Since the above equation ensures the proper relation of strains at the midplane surface to midplane displacement component w without explicitly involving in-plane displacement components u and v it serves as the desired compatibility equation for the strains at the midplane surface. We next express the compatibility equation in terms of the stress resultant intensity function as follows:

$$\frac{1}{Eh}\left[\frac{\partial^2 N_x}{\partial y^2} - v\frac{\partial^2 N_y}{\partial y^2} + \frac{\partial^2 N_y}{\partial x^2} - v\frac{\partial^2 N_x}{\partial x^2} + 2(1+v)\frac{\partial^2 N_{xy}}{\partial x\, \partial y}\right]$$

$$= \left(\frac{\partial^2 w}{\partial x\, \partial y}\right)^2 - \left(\frac{\partial^2 w}{\partial x^2}\right)\left(\frac{\partial^2 w}{\partial y^2}\right)$$

[To justify this equation simply substitute for the intensity function using Eqs. (8.111(c)), (d), (e) and Eq. (8.112) (we are thus using Hooke's law for plane stress here)]. Now replacing the intensity functions in terms of F using Eq. (8.113), we see that the left side of the equation is $(1/Eh)\nabla^4 F$. Hence the compatibility equation can be given as follows:

$$\nabla^4 F = Eh\left[\left(\frac{\partial^2 w}{\partial x\,\partial y}\right)^2 - \left(\frac{\partial^2 w}{\partial x^2}\right)\left(\frac{\partial^2 w}{\partial y^2}\right)\right] \qquad (8.115)$$

The above equation and Eq. (8.114) which we now rewrite

$$D\nabla^4 w = \frac{\partial^2 F}{\partial y^2}\frac{\partial^2 w}{\partial x^2} - 2\frac{\partial^2 F}{\partial x\,\partial y}\frac{\partial^2 w}{\partial x\,\partial y} + \frac{\partial^2 F}{\partial x^2}\frac{\partial^2 w}{\partial y^2} + q \qquad (8.116)$$

are the celebrated *von Kármán plate equations*. Note that they are still highly nonlinear. The equations furthermore have considerable mutual symmetry. This is brought out by defining the nonlinear operator L as follows:

$$L(p,q) = \frac{\partial^2 p}{\partial y^2}\frac{\partial^2 q}{\partial x^2} - 2\frac{\partial^2 p}{\partial x\,\partial y}\frac{\partial^2 q}{\partial x\,\partial y} + \frac{\partial^2 p}{\partial x^2}\frac{\partial^2 q}{\partial y^2} \qquad (8.117)$$

Then the von Kármán plate equations can be given as follows:

$$\nabla^4 F = -\frac{Eh}{2}L(w, w) \qquad (a)$$

$$D\nabla^4 w = L(F, w) + q \qquad (b) \quad (8.118)$$

In the following section we shall obtain an approximate solution to a "von Kármán plate" and compare it with the linear case.

8.12 AN EXAMPLE

As an example to illustrate the application of the theory presented in the previous section we consider now a clamped circular plate under the action of a uniform transverse load q_0. The radius of the plate we take as a while the thickness is t. As a first step we will want to transform the von Kármán plate equations to cylindrical coordinates. In this regard we can show that the operator L used above can be stated as follows for cylindrical coordinates:

$$L(w,F) = \frac{\partial^2 w}{\partial r^2}\left(\frac{1}{r}\frac{\partial F}{\partial r} + \frac{1}{r^2}\frac{\partial^2 F}{\partial \theta^2}\right) + \left(\frac{1}{r}\frac{\partial w}{\partial r} + \frac{1}{r^2}\frac{\partial^2 w}{\partial \theta^2}\right)\frac{\partial^2 F}{\partial r^2}$$

$$- 2\frac{\partial}{\partial r}\left(\frac{1}{r}\frac{\partial F}{\partial \theta}\right)\frac{\partial}{\partial r}\left(\frac{1}{r}\frac{\partial w}{\partial \theta}\right) \qquad (8.119)$$

Using the above formulation specialized for the case of axial symmetry and employing the dimensionless variable $\rho = r/a$ we may rewrite Eq. (8.118(b)) in the following form

$$D\nabla^4 w - \frac{1}{\rho}(F'w')' = q_0 a^4 \qquad (8.120)$$

where the primes and operators represent differentiation with respect to ρ. Similarly for Eq. (8.118(a)) we can say

$$\nabla^4 F = -\frac{Eh}{2}\frac{1}{\rho}[(w')^2]' \qquad (8.121)$$

Now using the fact that the Laplacian operator in cylindrical coordinates can be given as follows for axial symmetry

$$\nabla^2 = \frac{\partial^2}{\partial\rho^2} + \frac{1}{\rho}\frac{\partial}{\partial\rho}$$

we can say for Eq. (8.121)

$$\left(\frac{\partial^2}{\partial\rho^2} + \frac{1}{\rho}\frac{\partial}{\partial\rho}\right)(\nabla^2 F) = -\frac{Eh}{2}\frac{1}{\rho}[(w')^2]'$$

This may be written as follows

$$\left(\frac{\partial}{\partial\rho} + \frac{1}{\rho}\right)[(\nabla^2 F)'] = -\frac{Eh}{2}\frac{1}{\rho}[(w')^2]' \qquad (8.122)$$

You may demonstrate by direct substitution that

$$(\nabla^2 F)' = -\frac{Eh}{2}\frac{1}{\rho}(w')^2 \qquad (8.123)$$

is an integral of Eq. (8.122).

We will proceed by assuming that the transverse displacement w can be written as follows

$$w = w_0(1 - \rho^2)^2 \qquad (8.124)$$

where w_0, the maximum deflection occuring at the origin, is an undetermined constant. Now substitute into Eq. (8.123). We get

$$(\nabla^2 F)' = -\frac{Eh}{2}\frac{1}{\rho}[2w_0(1 - \rho^2)(-2\rho)]^2$$

$$= -8Ehw_0^2(\rho - 2\rho^3 + \rho^5)$$

The above equation can be written as follows

$$\nabla^2 F' - \frac{1}{\rho^2}F' = -8Ehw_0^2(\rho - 2\rho^3 + \rho^5) \qquad (8.125)$$

Now again by direct substitution you may directly demonstrate that the following is a particular solution for F' which satisfies the above equation

$$(F')_p = -\frac{Ehw_0^2}{6}(\rho^7 - 4\rho^5 + 6\rho^3) \qquad (8.126)$$

The complementary solution that may be demonstrated to satisfy Eq. (8.125) is given as follows

$$(F')_c = C_1\rho + \frac{C_2}{\rho} \qquad (8.127)$$

where C_1 and C_2 are arbitrary constants. To avoid $(F')_c \to \infty$ at the center $\rho = 0$ we set $C_2 = 0$. Combining Eqs. (8.126) and (8.127) we have

$$F' = -\frac{E h w_0^2}{6}(\rho^7 - 4\rho^5 + 6\rho^3) + C_1\rho \qquad (8.128)$$

To determine the value of C_1 we note that for the problem at hand we take $\bar{N}_v = 0$, i.e., the in-plane radial applied force is taken as zero at $\rho = 1$. To get N_v note next that for cylindrical coordinates:

$$N_x = \frac{\partial^2 F}{\partial y^2} = \frac{\partial^2 F}{\partial r^2}\sin^2\theta + 2\frac{\partial^2 F}{\partial\theta\,\partial r}\frac{\sin\theta\cos\theta}{r} + \frac{\partial F}{\partial r}\frac{\cos^2\theta}{r}$$

$$-2\frac{\partial F}{\partial\theta}\frac{\sin\theta\cos\theta}{r^2} + \frac{\partial^2 F}{\partial\theta^2}\frac{\cos^2\theta}{r^2} \qquad (8.129)$$

We can say that $N_v = (N_x)_{\theta=0}$. Accordingly, noting that $\partial F/\partial\theta = 0$ for axial symmetry, we can say from above:

$$N_v = (N_x)_{\theta=0} = \frac{1}{r}F'(r)$$

Now setting $N_v = 0$ at $\rho = 1$ we can then conclude that:

$$\left[\frac{1}{\rho}F'\right]_{\rho=1} = 0$$

Applying this condition to Eq. (8.128) we get for C_1 the following result.

$$C_1 = \frac{E h w_0^2}{2}$$

Hence the function $F'(\rho)$ given by Eq. (8.128) becomes

$$F' = -\frac{E h w_0^2}{6}(\rho^7 - 4\rho^5 + 6\rho^3 - 3\rho) \qquad (8.130)$$

We next employ the above result as well as the function w given by Eq. (8.124) as a Galerkin's integral for Eq. (8.120), in an effort to ascertain w_0. Thus

$$\int_0^1 \left\{ D\nabla^4[(w_0)(1-\rho^2)^2] + \frac{1}{\rho}\left[\frac{4}{3}E h w_0^3(5\rho^9 - 20\rho^7 + 30\rho^{15} - 18\rho^3 + 3\rho)\right] \right.$$
$$\left. - q_0 a^4 \right\}(1-\rho^2)^2\,d\rho = 0$$

We carry out the integration to reach the following cubic equation for w_0

$$\frac{(48)(0.496)}{64}\frac{(1-v^2)E h w_0^3}{E h^3} + w_0 = \frac{q_0 a^4}{64D}$$

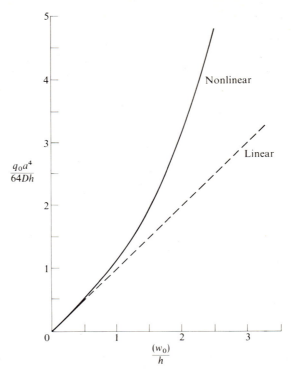

$$\frac{q_0 a^4}{64Dh}$$

$$\frac{(w_0)}{h}$$

FIGURE 8.14

where we have replaced D, in the first expression, by $Eh^3/12(1 - v^2)$. For the case where $v = 0.30$ the above equation can be given as follows

$$(0.346)\left(\frac{w_0}{h}\right)^3 + \left(\frac{w_0}{h}\right) = \bar{q}$$

where

$$\bar{q} = \frac{q_0 a^4}{64hD}$$

For comparison purposes the exact solution for (w_0/h) for the classical case is given as

$$\frac{w_0}{h} = \bar{q}$$

A plot of w_0/h vs. $\bar{q}$ is shown in Fig. 8.14 for the nonlinear and the classical cases. Although the graph has been extended beyond $w/h = 0(1)$ the nonlinear theory is generally not valid in this extended region. The linear theory is normally considered acceptable for w/h up to about 0.2.

The approach taken here clearly was one that generated an approximate solution. We guessed at a function w including an undetermined amplitude w_0 and found F' which satisfied the compatibility equations. We then found w_0 by satisfying in some average way the equation of equilibrium via the Galerkin integral.

READING

BIOT, M. A.: "Mechanics of Incremental Deformations," John Wiley and Sons, 1965.

GREEN, A. E., AND ADKINS, J. E.: "Large Elastic Deformations," Oxford University Press, 1970.

GREEN, A. E., AND ZERNA, W.: "Theoretical Elasticity," Oxford University Press, 1954.

MURNAGHAN, F. D.: "Finite Deformation of an Elastic Solid," Dover Publications, 1967.

NOVOZHILOV, V. V.: "Foundations of the Nonlinear Theory of Elasticity," Graylock Press, Rochester, 1953.

WASHIZU, K.: "Variational Methods in Elasticity and Plasticity," Pergamon Press, 1968.

9
ELASTIC STABILITY

9.1 INTRODUCTION

This chapter will be devoted to considerations of certain of the foundations underlying the theory of elastic stability. We will examine the meaning of stability especially within the context of variational methods and discuss various methods of obtaining stability bounds for various problems. As a recurring example problem we shall make use of the simple Euler column problem to indicate a variety of approaches. Using variational techniques we shall set forth approximate techniques for solving stability problems involving columns and plates.

In all of our work in this area we shall restrict ourselves to linear elastic materials while including the necessary geometrically nonlinear terms that are required to examine the stability of a state of equilibrium.

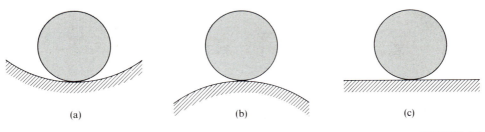

(a)   (b)   (c)

FIGURE 9.1

Part A
STABILITY OF RIGID BODY SYSTEMS

9.2 STABILITY

We say that a configuration in a state of equilibrium is *stable* if, after some slight disturbance causing a change of configuration, there then follows a return to the original configuration. Thus consider Fig. 9.1(*a*) showing a cylinder in a circular well. Clearly this equilibrium configuration is a stable one. In Fig. 9.1(*b*) we have shown an *unstable* equilibrium configuration. If perfectly balanced no change in configuration occurs. However an infinitesimal disturbance causes finite movement of the cylinder. Figure 9.1(*c*) represents what is called *neutral stability*. All positions of the cylinder are stable equilibrium positions. Furthermore there is no change in the total potential energy as the position is varied.

Note that the stable equilibrium illustrated above corresponds to a position of a local *minimum* for the total potential energy for the cylinder while the unstable equilibrium position corresponds to a position of a local *maximum* for the total potential energy of the cylinder. We may say for conservative systems that *a local minimum for the total potential energy corresponds to a stable equilibrium configuration.*

9.3 RIGID BODY PROBLEMS

Let us now consider the planar cylinder section shown in Fig. 9.2. Point C is the center of gravity of the cylinder section and R is the radius. In part (a) of the diagram we have shown an equilibrium position for the cylinder section ($\theta = 0$).

We note first that the total potential energy may be given as a function of θ as follows

$$\pi = W[(R - h) - (R - h)\cos\theta] \qquad (9.1)$$

where W is the weight of the cylinder and h is the distance from the center of gravity to the ground in the equilibrium configuration. If the angle θ is now given a variation

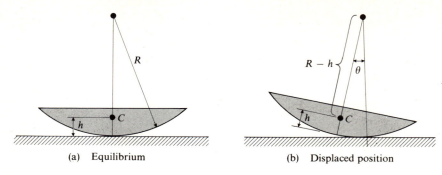

(a) Equilibrium   (b) Displaced position

FIGURE 9.2

$\delta\theta$, then the new potential energy can be given as follows:

$$\pi(\theta + \delta\theta) = W[(R - h) - (R - h)\cos(\theta + \delta\theta)]$$

We can rewrite both sides of the above equation as follows[1]

$$\pi(\theta) + \delta^{(T)}\pi(\theta) = W[(R - h) - (R - h)(\cos\theta\cos\delta\theta - \sin\theta\sin\delta\theta)] \qquad (9.2)$$

where $\delta^{(T)}\pi$, you will recall from Chap. 2, is the *total variation* of π. We shall now expand $\cos\delta\theta$ and $\sin\delta\theta$ as power series as follows:

$$\sin\delta\theta = \delta\theta - \frac{1}{3!}(\delta\theta)^3 + \cdots$$

$$\cos\delta\theta = 1 - \frac{1}{2!}(\delta\theta)^2 + \cdots$$

Equation (9.2) can then be written as follows:

$$\pi + \delta^{(T)}\pi = W[(R - h) - (R - h)\cos\theta] + W(R - h)(\sin\theta)\,\delta\theta$$

$$+ \frac{1}{2!}W(R - h)(\cos\theta)\delta\theta)^2$$

$$- \frac{1}{3!}W(R - h)(\sin\theta)(\delta\theta)^3 + \cdots$$

Noting Eq. (9.1) we can now give $\delta^{(T)}\pi$ as follows:

$$\delta^{(T)}\pi = W(R - h)\sin\theta\,\delta\theta + \frac{1}{2!}W(R - h)\cos\theta(\delta\theta)^2$$

$$- \frac{1}{3!}W(R - h)\sin\theta(\delta\theta)^3 + \cdots \qquad (9.3)$$

---

[1] Recall from Sec. 2.4 that:

$$F(\theta + \delta\theta) - F(\theta) = \delta^{(T)}F$$

We can also express $\delta^{(T)}\pi$ as a Taylor series expansion in the following way:

$$\delta^{(T)}\pi = \pi(\theta + \delta\theta) - \pi(\theta) = \frac{\partial \pi}{\partial \theta}\delta\theta + \frac{1}{2!}\frac{\partial^2 \pi}{\partial \theta^2}(\delta\theta)^2 + \frac{1}{3!}\frac{\partial^2 \pi}{\partial \theta^3}(\delta\theta)^3$$

$$+ \cdots \tag{9.4}$$

Note that the first expression on the extreme right side of Eq. (9.4) is the *first variation* of π, i.e., $\delta^{(1)}\pi$, and we state this formally:

$$\delta^{(1)}\pi = \frac{\partial \pi}{\partial \theta}\delta\theta$$

Also higher order variations of π such as the nth order variation will be given here as:

$$\delta^{(n)}\pi = \frac{\partial^n \pi}{\partial \theta^n}(\delta\theta)^n \tag{9.5}$$

Hence Eq. (9.4) can be given as follows:

$$\delta^{(T)}\pi = \sum_{n=1}^{\infty} \frac{1}{n!}\frac{\partial^n \pi}{\partial \theta^n}(\delta\theta)^n = \sum_{n=1}^{\infty} \frac{1}{n!}(\delta^{(n)}\pi)$$

Now comparing the above equation and Eq. (9.3) we note that:

$$\delta^{(1)}\pi = W(R - h)\sin\theta\,\delta\theta$$
$$\delta^{(2)}\pi = W(R - h)\cos\theta(\delta\theta)^2$$
$$\delta^{(3)}\pi = -W(R - h)\sin\theta(\delta\theta)^3 \quad \text{etc.} \tag{9.6}$$

To establish *equilibrium configurations* for the problem at hand *we set the first variation equal to zero* as we have done in the past. This requires that:

$$W(R - h)\sin\theta = 0$$

We see that $\theta = 0$ is the position of equilibrium as expected.

We now consider Eq. (9.3) for the case $\theta = 0$. We get:

$$\delta^{(T)}\pi = \frac{1}{2!}W(R - h)(\delta\theta)^2 + \cdots$$

For π to be a *minimum* at $\theta = 0$ for *stable equilibrium*, the expression $\delta^{(T)}\pi$ must be positive for $\pm\delta\theta$ as explained in Sec. 2.1. We accordingly require for stability that the dominant term in the above expansion

$$W(R - h)(\delta\theta)^2 = (\delta^{(2)}\pi)_{\text{eq.}} > 0$$

This condition is satisfied for $R > h$. Now decrease R until it reaches the value h. Clearly the second variation $(\delta^{(2)}\pi)_{\text{eq.}}$ becomes zero. Furthermore it is easily seen from Eq. (9.6) that *all variations are zero for this condition.* This means that there is zero change in the total potential energy for all configurations in the neighborhood of the equilibrium configuration. We have thus reached a case of *neutral stability.* We shall denote the value of the adjustable parameter that first renders the system *not stable* as the *critical value* of the parameter. Thus for this problem R is the adjustable parameter chosen and its critical value is h. We thus reach the conclusion that when the adjustable parameter R reaches its critical value h as it is decreased in value from $R > h$, we *cease* to have stable equilibrium. We shall next establish a criterion for finding the critical value. This will be useful in later computations involving adjustable loads.

Accordingly let us consider the second variation, $\delta^{(2)}(\pi)$.

$$\delta^{(2)}\pi = W(R - h)\cos\theta(\delta\theta)^2$$

Note that at the equilibrium position, $\theta = 0$, we pointed out that we had stable equilibrium for $R > h$. Also note above that the second variation of the total potential energy is then positive-definite[1] for a range of values of θ in the neighborhood of the equilibrium configuration. For decreasing R, note that the second variation *ceases* to be positive-definite exactly when $R = h$. We may make this observation. The body reaches a "critical" configuration and thus *ceases to have stability when the second variation of the total potential energy ceases to be positive-definite.* We may use this criterion as per the second variation of the total potential energy for finding the critical value of the adjustable parameter.

We now consider a second rigid-body problem so that we can introduce the concept of the *buckling load* and a second criterion for stability. Accordingly we have shown a rigid rod in a vertical position acted on by a force P (see Fig. 9.3(a)). Two identical linear springs having spring constant k connect the rod to pins in frictionless slots held by immovable walls. In Fig. 9.3(b) we have shown the rod in a deflected position measured by the angle θ. We shall first set forth the total potential energy of the system and then examine the first variation for possible equilibrium configurations. Thus we have for π:

$$\pi = 2(\tfrac{1}{2})(k)(L\sin\theta)^2 - PL(1 - \cos\theta) \tag{9.7}$$

The first variation then becomes:

$$\delta^{(1)}\pi = \frac{\partial\pi}{\partial\theta}\delta\theta = [2kL^2\sin\theta\cos\theta - PL\sin\theta]\delta\theta$$

---

[1] In this discussion a function $F(x_i)$ is said to be *positive definite* if the function is greater than zero in a neighborhood about $x_i = 0$ and can equal zero only at the position $x_i = 0$.

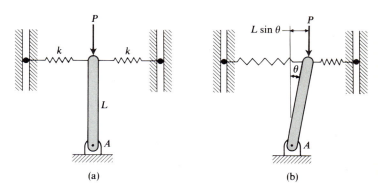

FIGURE 9.3

Replacing $2 \sin \theta \cos \theta$ by $\sin 2\theta$ in the first expression on the right side of the equation and setting the coefficient of $\delta\theta$ equal to zero for equilibrium we get:

$$[kL^2 \sin 2\theta - PL \sin \theta] = 0$$

For small values of θ we can express the above equation as follows:

$$(2kL^2 - PL)\theta = 0 \qquad (9.8)$$

Clearly $\theta = 0$ corresponds to a state of equilibrium for *any load P.*

Let us now discuss the stability of this configuration considering P to be the adjustable parameter just as R was the adjustable parameter in the last example. The second variation for the problem at hand becomes:

$$\delta^{(2)}\pi = \frac{\partial^2 \pi}{\partial \theta^2}(\delta\theta)^2 = (2kL^2 \cos 2\theta - PL \cos \theta)(\delta\theta)^2$$

For $\theta = 0$ we see that $\delta^{(2)}\pi$ is positive if the load P is less than $2kL$. Also there is a *range* of values for θ about $\theta = 0$ for a given load $P < 2kL$ for which $\delta^{(2)}\pi$ is positive. Thus if $P < 2kL$ the second variation is *positive–definite* at the equilibrium configuration $\theta = 0$. When $P = 2kL$ we have for the above equation

$$\delta^{(2)}\pi = PL(\cos 2\theta - \cos \theta)(\delta\theta)^2$$

and it becomes clear that the second variation is negative in the neighborhood of $\theta = 0$. The second variation has thus *ceased* to be positive-definite. This value of P clearly is the *critical* or, we say, the *buckling load,* for this system and is denoted as P_{cr}. Thus:

$$P_{\mathrm{cr}} = 2kL \qquad (9.9)$$

We can further conclude from our earlier remarks that the system ceases to be in stable equilibrium at $\theta = 0$ when P reaches the buckling load, P_{cr}.

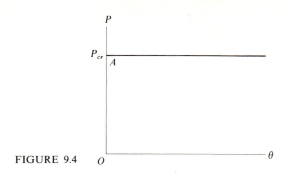

FIGURE 9.4

Note we have arrived at the buckling load by considering the positive-definiteness of the second variation. This is called an *energy approach* towards ascertaining the buckling load. We shall now present yet another way of arriving at the buckling load by consideration of the first variation of the total potential energy. Thus re-examining Eq. (9.8) we see that $\delta^{(1)}\pi = 0$ not only for $\theta = 0$ (which information we have used) but also when $P = P_{\mathrm{cr}}$. This means that under the buckling load there can be theoretically *any value of θ present for equilibrium*. A plot of θ versus P for equilibrium has been shown in Fig. 9.4. Note that the locus of such states includes the axis $\theta = 0$ and a horizontal line $P = P_{\mathrm{cr}}$. The point A where the path OA diverges into two paths is called a *bifurcation point*. Thus we see that the critical load occurs in this problem at the bifurcation point in such an equilibrium plot. (We will see in the next section that for elastic bodies there may be an infinity of bifurcation points corresponding to an infinite set of discrete values of load. Each such load is called a critical load; the *lowest critical* load is called additionally the *buckling load*.) We shall term a *trivial equilibrium configuration* as an equilibrium configuration corresponding to *zero external load*. The preceding discussion permits us to conclude that in the *presence of the critical loads we can have nontrivial equilibrium configurations*. Indeed the search for loads by which nontrivial equilibrium configurations are permitted is called the *equilibrium method* for finding critical loads.

In closing this section we may draw the following conclusions from the examples considered. First, the buckling load signals the upper limits of loading for stable equilibrium. Furthermore this load can be found by determining the value of loading at which the second variation of the total potential energy ceases to be positive-definite at an equilibrium configuration (the energy method). Or we can seek those values of load that permit nontrivial equilibrium configurations (equilibrium method).

In Part B we shall consider the elastic stability of columns wherein we can extend the conclusions reached above and indeed present, again by example, other notions pertinent to the understanding of elastic stability.

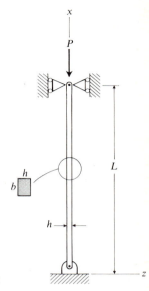

FIGURE 9.5

Part B
ELASTIC STABILITY OF COLUMNS

9.4   THE EULER LOAD; EQUILIBRIUM METHOD

We consider now the classic problem of the theory of elastic stability, dating back to the work of Euler, who in his studies on variational calculus, founded, as it were, the theory of elastic stability. This problem consists of an initially straight pin-ended, long, slender, column under the action of a compressive force P as is shown in Fig. 9.5.

We shall first compute the total potential energy as follows:

$$\pi = \frac{1}{2} \int_{-h/2}^{h/2} \int_0^L \tau_{xx}\epsilon_{xx}b \, dx \, dz + P \int_0^L \frac{du}{dx} dx \qquad (9.10)$$

Now employ the nonlinear strain term approximation presented in the previous chapter for which $\epsilon_{ij} \ll \omega_{ij} \ll 1$. Hence using Eq. (8.86) we have for ϵ_{xx}:

$$\epsilon_{xx} = e_{xx} + \frac{1}{2}\omega_{yx}\omega_{yx} + \frac{1}{2}\omega_{zx}\omega_{zx} \qquad (9.11)$$

Since $h < b$ (see Fig. 9.5) in this analysis, we will expect displacement of the column to occur primarily in the xz plane and so we expect that ω_{yx}, which is proportional to rigid body rotation of elements about the z axis, is negligible. Furthermore, we employ the assumption that lines normal to the centerline of the column remain normal to this line and are unextended after deformation. Hence using Eqs. (8.92) and (8.93) in Eq. (9.11) we can then give ϵ_{xx} as follows:

$$\epsilon_{xx} = \frac{du}{dx} + \frac{1}{2}\left(\frac{dw}{dx}\right)^2 - z\frac{d^2w}{dx^2} \qquad (9.12)$$

Now using Hooke's law to replace τ_{xx} in Eq. (9.10) by $E\epsilon_{xx}$ we get:

$$\pi = \frac{1}{2}\int_{-h/2}^{h/2}\int_0^L E\left[\frac{du}{dx} + \frac{1}{2}\left(\frac{dw}{dx}\right)^2 - z\left(\frac{d^2w}{dx^2}\right)\right]^2 b\,dx\,dz + P\int_0^L \frac{du}{dx}dx$$

In the first expression, integrate over z. Since only even powers of z yield non-zero integrations for the limits given we get:

$$\pi = \frac{EA}{2}\int_0^L\left[\frac{du}{dx} + \frac{1}{2}\left(\frac{dw}{dx}\right)^2\right]^2 dx + \frac{EI}{2}\int_0^L\left(\frac{d^2w}{dx^2}\right)^2 dx + P\int_0^L \frac{du}{dx}dx \qquad (9.13)$$

We now set the first variation equal to zero to establish the conditions for satisfying Newton's law—in this case to set up the conditions for equilibrium. Thus:

$$\delta^{(1)}\pi = 0 = EA\int_0^L\left[\frac{du}{dx} + \frac{1}{2}\left(\frac{dw}{dx}\right)^2\right]\left[\frac{d\delta u}{dx} + \left(\frac{dw}{dx}\right)\frac{d\delta w}{dx}\right]dx$$

$$+ EI\int_0^L \frac{d^2w}{dx^2}\frac{d^2\delta w}{dx^2}dx + P\int_0^L \frac{d\delta u}{dx}dx$$

We next integrate by parts to bring the δu and δw terms outside the various derivatives. Thus:

$$- EA\int_0^L\left\{\frac{d}{dx}\left[\frac{du}{dx} + \frac{1}{2}\left(\frac{dw}{dx}\right)^2\right]\right\}\delta u\,dx - EA\int_0^L \frac{d}{dx}\left\{\frac{dw}{dx}\left[\frac{du}{dx}\right.\right.$$

$$\left.\left. + \frac{1}{2}\left(\frac{dw}{dx}\right)^2\right]\right\}\delta w\,dx + EI\int_0^L \frac{d^4w}{dx^4}\delta w\,dx$$

$$+ \left\{P + EA\left[\frac{du}{dx} + \frac{1}{2}\left(\frac{dw}{dx}\right)^2\right]\right\}\delta u\,\Big|_0^L + \left\{EA\left(\frac{dw}{dx}\right)\left[\frac{du}{dx}\right.\right.$$

$$\left.\left. + \frac{1}{2}\left(\frac{dw}{dx}\right)^2\right] - EI\frac{d^3w}{dx^3}\right\}\delta w\,\Big|_0^L + EI\frac{d^2w}{dx^2}\delta\left(\frac{dw}{dx}\right)\Big|_0^L = 0 \qquad (9.14)$$

We conclude from above that:

$$\frac{d}{dx}\left[\frac{du}{dx} + \frac{1}{2}\left(\frac{dw}{dx}\right)^2\right] = 0$$

$$EA\frac{d}{dx}\left\{\frac{dw}{dx}\left[\frac{du}{dx} + \frac{1}{2}\left(\frac{dw}{dx}\right)^2\right]\right\} - EI\frac{d^4w}{dx^4} = 0 \qquad (9.15)$$

Note next that we may evaluate resultant intensity functions N and M as follows:

$$N = b \int_{-h/2}^{h/2} \tau_{xx}\, dz = b \int_{-h/2}^{h/2} E\left[\frac{du}{dx} + \frac{1}{2}\left(\frac{dw}{dx}\right)^2 - z\frac{d^2w}{dx^2}\right] dz$$

$$= EA\left[\frac{du}{dx} + \frac{1}{2}\left(\frac{dw}{dx}\right)^2\right] \qquad\qquad (a)$$

$$M = b \int_{-h/2}^{h/2} z\tau_{xx}\, dz = b \int_{-h/2}^{h/2} E\left[z\frac{du}{dx} + \frac{1}{2}z\left(\frac{dw}{dx}\right)^2 - z^2\frac{d^2w}{dx^2}\right] dz$$

$$= -EI\frac{d^2w}{dx^2} \qquad\qquad (b) \quad (9.16)$$

Substituting into Eq. (9.15) we get:

$$\frac{dN}{dx} = 0 \qquad (a)$$

$$\frac{d}{dx}\left[\frac{dw}{dx}N\right] + \frac{d^2M}{dx^2} = 0 \qquad (b) \quad (9.17)$$

We now consider the *boundary conditions.* From Eq. (9.14) we can say at the end points $x = 0, L$:

$$P + EA\left[\frac{du}{dx} + \frac{1}{2}\left(\frac{dw}{dx}\right)^2\right] = 0 \quad \text{OR} \quad u \text{ PRESCRIBED}$$

$$EA\left\{\frac{dw}{dx}\left[\frac{du}{dx} + \frac{1}{2}\left(\frac{dw}{dx}\right)^2\right]\right\} - EI\frac{d^3w}{dx^3} = 0 \quad \text{OR} \quad w \text{ PRESCRIBED}$$

$$EI\frac{d^2w}{dx^2} = 0 \quad \text{OR} \quad \frac{dw}{dx} \text{ PRESCRIBED} \qquad (9.18)$$

In terms of stress resultant intensity functions this gives us:

$$P = -N \quad \text{OR} \quad u \text{ PRESCRIBED} \qquad (a)$$

$$N\frac{dw}{dx} + \frac{dM}{dx} = 0 \quad \text{OR} \quad w \text{ PRESCRIBED} \qquad (b)$$

$$M = 0 \quad \text{OR} \quad \frac{dw}{dx} \text{ PRESCRIBED} \qquad (c) \quad (9.19)$$

Now go back to Eq. (9.17(a)). We see that N is a constant and from Eq. (9.19(a)) we note that the constant is $-P$. Accordingly Eq. (9.17(b)) then becomes:

$$-P\frac{d^2w}{dx^2} + \frac{d^2M}{dx^2} = 0$$

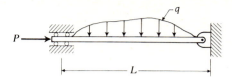

FIGURE 9.6

Next utilizing Eq. (9.16(b)) we may replace M by $-EI \cdot d^2w/dx^2$ in the above equation to get:

$$EI\frac{d^4w}{dx^4} + P\frac{d^2w}{dx^2} = 0 \qquad (9.20)$$

This is the *buckling equation* that we shall use. Notice that it is linear.

It is to be noted that we could have arrived directly at Eq. (9.20) by extremizing the following total potential energy functional:[1]

$$\pi = \frac{EI}{2}\int_0^L \left(\frac{d^2w}{dx^2}\right)^2 dx - \frac{P}{2}\int_0^L \left(\frac{dw}{dx}\right)^2 dx \qquad (9.21)$$

We can reach the above form directly from the total potential energy which was needed (see Eq. (9.13)) to generate Eqs. (9.17), by satisfying the following equation:

$$\frac{du}{dx} = -\frac{1}{2}\left(\frac{dw}{dx}\right)^2 \qquad (9.22)$$

Upon examining Eq. (9.12) we see that this step eliminates the stretching contribution from the study and renders the stability problem one of bending only. The compression that must accompany the application of load P, we thus conclude, has no effect on the stability problem.

Suppose next we have a *beam column* problem as has been shown in Fig. 9.6 where a loading $q(x)$ acts in addition to the axial load P. Thus the total potential energy given by Eq. (9.13) must be augmented by the term $-\int_0^L qw\,dx$. It is then clear that we may reach the following equation on extremizing this total potential energy functional and replacing N by $-P$:

$$EI\frac{d^4w}{dx^4} + P\frac{d^2w}{dx^2} = q(x) \qquad (9.23)$$

We thus have a generalization here of the earlier form (Eq. (9.20)).

---

[1] The first term represents the bending strain energy while the expression $\frac{1}{2}\int_0^L (dw/dx)^2\,dx$ represents an approximation of the contraction of the column along the x axis.

We now point out that despite the fact that Eq. (9.23) is linear, it nevertheless reflects the nonlinearity of its derivation since certain types of superpositions are not possible. For instance we could not superpose the effect of P and the effects of $q(x)$ separately. Thus we cannot superpose the solution of

$$EI\frac{d^4w}{dx^4} = q(x)$$

and

$$EI\frac{d^4w}{dx^4} + P\frac{d^2w}{dx^2} = 0$$

Clearly from simple physical reasoning, the effects of P on w will directly depend on the deflection that $q(x)$ induces.

Recall that the vanishing of the first variation for the rigid body problems produced *algebraic equations* from which we could deduce that for a given load, namely the critical load, equilibrium configurations other than the so-called trivial one were possible equilibrium states. In this case we have a *differential equation* resulting from the vanishing of the first variation. However, we may now ask for essentially the same information as we asked earlier; namely, for what loads P are there nontrivial configurations $w(x)$ which satisfy equilibrium conditions (i.e., our differential equation and boundary conditions). To pose such a question mathematically is to set forth an *eigenvalue problem* where the admissible values of P, namely the critical loads P_{cr}, are the *eigenvalues*, and the deflection shapes $w(x)$ corresponding to the loads are the *eigenfunctions*. Accordingly the various techniques discussed in Chap. 7 relative to the Rayleigh quotient are applicable here for determining, at least approximately, certain critical loads. We shall investigate this aspect of the problem later. Now we shall present an exact solution for the simple Euler column (Fig. 9.5).

You may readily verify that the solution to Eq. (9.20) can be given as follows:

$$w = A\sin kx + B\cos kx + Cx + D \qquad (a)$$

where

$$k = \sqrt{\frac{P}{EI}} \qquad (b) \quad (9.24)$$

Satisfaction of the boundary conditions, namely $w = d^2w/dx^2 = 0$ at $x = 0, L$ then gives us the following system of equations:

$$B + D = 0$$

$$A\sin kL + B\cos kL + CL + D = 0$$

$$B = 0$$

$$A\sin kL + B\cos kL = 0$$

Eigenfunctions

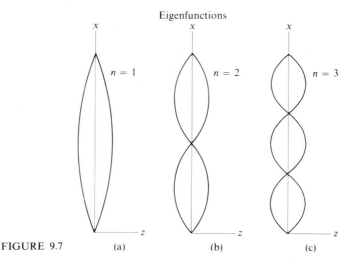

FIGURE 9.7   (a)   (b)   (c)

From these equations we obtain the following results:

$$B = C = D = 0 \qquad (a)$$

$$A \sin kL = 0 \qquad (b) \quad (9.25)$$

For a *nontrivial* solution we see that the following conditions must be met:

$$k = \frac{n\pi}{L} \qquad n = 1, 2, \ldots \qquad (9.26)$$

This generates for us a series of critical loads (the eigenvalues) given as follows:

$$P_{cr} = \frac{n^2 \pi^2 EI}{L^2} \qquad n = 1, 2, \ldots \qquad (9.27)$$

The smallest of these critical loads, i.e., the buckling load ($n = 1$) is called the *Euler load* and will be denoted as P_E. The mode shapes for the Euler load as well as higher critical loads are those of sinusoids and are shown in Fig. 9.7.

To understand the physical aspects of the Euler load and the other critical loads let us imagine that we are loading the column starting from $P = 0$ and plotting the ratio P/P_E, as we proceed, versus the amplitude of the deflection curve which we denote as w_0. Below the value $P/P_E = 1$, we know from our analysis that to satisfy all the boundary conditions (see Eq. (9.25(b))) $A = 0$ and we have a trivial solution. This means that $w_0 = 0$; the curve P/P_E vs. w_0 coincides with the P/P_E axis as has been shown in Fig. 9.8. At point (1), where $P/P_E = 1$, we may have from our analysis *any value* of A and hence any value of w_0 present. Above this point we are again

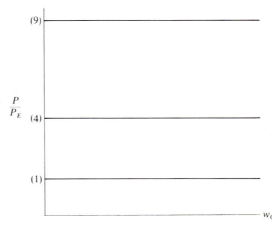

FIGURE 9.8

restricted to the trivial solution $A = 0$ and hence $w_0 = 0$. This point (1) is a bifurcation point as described earlier for the rigid-body problem. In this case we have an infinity of such points as P/P_E becomes successively 2^2, 3^2, etc. These clearly correspond to the series of critical loads.

In ascertaining the critical loads note that as in the rigid body examples we asked the question—are there loads for which there are nontrivial equilibrium configurations satisfying the boundary conditions of the problem? For the rigid body example undertaken earlier the question was answered by simple algebra; in the elastic problem we dealt with an eigenvalue problem. In later sections we will see that one can ask other queries of the system to generate information as to critical loads. However, the amount of information found is not always the same; it depends on the question asked and the system considered.

9.5 ENERGY METHODS

Up to this time we have employed the first variation of the total potential energy to set forth the eigenvalue problem which gave us the critical loads. Thus in the previous section we considered the Euler column and by seeking nontrivial solutions to the equilibrium equation we were able to ascertain the critical loads. We shall now demonstrate that we can generate the Euler load for this problem via other methods— the so-called energy methods, one of which was presented earlier for rigid bodies.

The second variation of the total potential energy can play a significant role in the study of elastic stability as was the case for rigid-body stability. You will recall from the latter considerations $\delta^{(2)}\pi$ ceased to be positive-definite at the buckling load. This ordinarily is the case for elastic bodies on which conservative forces act.

As a preliminary note in this regard let us recall that we can find all orders of variations of π via Eq. (9.5) when we have a single independent variable present. For s independent variables x_i we can find formulations for all orders of variations by first carrying out a Taylor series expansion for π, as we did in Eq. (9.4) for one variable. Thus:[1]

$$\pi(x_1 + \delta x_1, x_2 + \delta x_2, \ldots, x_s + \delta x_s) = \pi(x_1, x_2, \ldots, x_s)$$

$$+ \sum_{p=1}^{s} \frac{\partial \pi}{\partial x_p} \delta x_p + \frac{1}{2!} \sum_{p=1}^{s} \sum_{q=1}^{s} \frac{\partial^2 \pi}{\partial x_p \, \partial x_q} \delta x_p \, \delta x_q$$

$$+ \frac{1}{3!} \sum_{p=1}^{s} \sum_{q=1}^{s} \sum_{r=1}^{s} \frac{\partial^3 \pi}{\partial x_p \, \partial x_q \, \partial x_r} \delta x_p \delta x_q \delta x_r + \cdots$$

Hence:

$$\pi(x_1 + \delta x_1, x_2 + \delta x_2, \ldots, x_s + \delta x_s) - \pi(x_1, x_2, \ldots, x_s)$$

$$= \delta^{(T)} \pi = \sum_{p=1}^{s} \frac{\partial \pi}{\partial x_p} \delta x_p + \frac{1}{2!} \sum_{p=1}^{s} \sum_{q=1}^{s} \frac{\partial^2 \pi}{\partial x_p \, \partial x_q} \delta x_p \, \delta x_q$$

$$+ \frac{1}{3!} \sum_{p=1}^{s} \sum_{q=1}^{s} \sum_{r=1}^{s} \frac{\partial^3 \pi}{\partial x_p \, \partial x_q \, \partial x_r} \delta x_p \, \delta x_q \, \delta x_r + \cdots \qquad (9.28)$$

We can then say:

$$\delta^{(1)} \pi = \sum_{p=1}^{s} \frac{\partial \pi}{\partial x_p} \delta x_p$$

$$\delta^{(2)} \pi = \sum_{p=1}^{s} \sum_{q=1}^{s} \frac{\partial^2 \pi}{\partial x_p \, \partial x_q} \delta x_p \, \delta x_q \quad \text{etc.}$$

With this background, let us now go back to re-examine the pin-ended column. Thus we note that the general solution attained for the Euler column can be given as follows:

$$w = \sum_{n=1}^{\infty} A_n \sin \frac{n \pi x}{L}$$

The total potential energy, given by Eq. (9.21) for this case is then:

$$\pi = \frac{EIL}{4} \sum_{n=1}^{\infty} A_n^{\ 2} \left(\frac{n\pi}{L} \right)^4 - \frac{PL}{4} \sum_{n=1}^{\infty} A_n^{\ 2} \left(\frac{n\pi}{L} \right)^2$$

[1] Consult any standard advanced calculus book for Taylor series expansions in n variables.

We now find the total variation of π considering the A's to be independent variables. For this purpose we may employ Eq. (9.28) with s going to ∞. Since A_n appears to the power two, only the first two expressions on the right side of Eq. (9.28) give non-zero results. We then have by such a computation:

$$\delta^{(T)}\pi = \sum_{n=1}^{\infty} \left\{ \frac{EIL}{2}\left(\frac{n\pi}{L}\right)^4 - \frac{PL}{2}\left(\frac{n\pi}{L}\right)^2 \right\} A_n \delta A_n$$

$$+ \frac{1}{2!}\sum_{n=1}^{\infty} \left\{ \frac{EIL}{2}\left(\frac{n\pi}{L}\right)^4 - \frac{PL}{2}\left(\frac{n\pi}{L}\right)^2 \right\} (\delta A_n)^2$$

The second variation of π then may be seen to be:

$$\delta^{(2)}\pi = \sum_{n=1}^{\infty} \left\{ \frac{EIL}{2}\left(\frac{n\pi}{L}\right)^4 - \frac{PL}{2}\left(\frac{n\pi}{L}\right)^2 \right\} (\delta A_n)^2$$

$$= \sum_{n=1}^{\infty} \left\{ \left(\frac{n\pi}{L}\right)^2 \left(\frac{L}{2}\right) \left[EI\frac{n^2\pi^2}{L^2} - P \right] \right\} (\delta A_n)^2$$

$$= \sum_{n=1}^{\infty} \left\{ \left(\frac{n\pi}{L}\right)^2 \left(\frac{L}{2}\right) [n^2 P_E - P] \right\} (\delta A_n)^2$$

It is clear from above that when $P < P_E$ the second variation is positive-definite. And, using the theorem of Sylvester[1] we may conclude that $\delta^{(2)}\pi$ *ceases to be positive-definite when the load P equals or exceeds the Euler load. We can then say here, as in the case of the rigid-body discussion, that the buckling load is the one for which the second variation of the total potential energy ceases to be positive-definite.* We may extend this conclusion to other linear elastic stability problems wherein we have constant (not time varying) applied loads.

To get *another* energy point of view of the buckling load consider the total variation of a functional I in the following way:

$$\tilde{I} - I = \delta^{(T)}I = \int_{x_1}^{x_2} F(x, y + \epsilon\eta, y' + \epsilon\eta')\, dx - \int_{x_1}^{x_2} F(x,y,y')\, dx$$

Now expand $F(x, y + \epsilon\eta, y' + \epsilon\eta')$ as a power series in the parameter ϵ about the value $\epsilon = 0$. We get after cancellation of terms:

$$\delta^{(T)}I = \left[\frac{\partial}{\partial\epsilon}\int_{x_1}^{x_2} F(x, y + \epsilon\eta, y' + \epsilon\eta')\, dx \right]_{\epsilon=0} \epsilon$$

$$+ \frac{1}{2!}\left[\frac{\partial^2}{\partial\epsilon^2}\int_{x_1}^{x_2} F(x, y + \epsilon\eta, y' + \epsilon\eta')\, dx \right]_{\epsilon=0} \epsilon^2$$

$$+ \frac{1}{3!}\left[\frac{\partial^3}{\partial\epsilon^3}\int_{x_1}^{x_2} F(x, y + \epsilon\eta, y' + \epsilon\eta')\, dx \right]_{\epsilon=0} \epsilon^3 \quad + \cdots$$

[1] See J. LaSalle and S. Lefschetz: "Stability by Liapounov's Direct Method," Academic Press, New York, 1961, p. 36.

The first expression on the right side of the above equation is the well-known first variation (see Eq. (2.21)). The second variation may be taken as the coefficient of 1/2!, the third variation as the coefficient of 1/3!, etc. With this background let us now consider again the total potential energy π for the Euler column. We have from Eq. (9.21):

$$\pi = \frac{EI}{2} \int_0^L (w'')^2 \, dx - \frac{P}{2} \int_0^L (w')^2 \, dx \qquad (9.29)$$

To get $\tilde{\pi}$ now use $\tilde{w} = w + \epsilon\eta$. Then we have:

$$\tilde{\pi} = \frac{EI}{2} \int_0^L (w'' + \epsilon\eta'')^2 \, dx - \frac{P}{2} \int_0^L (w' + \epsilon\eta')^2 \, dx$$

$$= \frac{EI}{2} \int_0^L (w'')^2 \, dx - \frac{P}{2} \int_0^L (w')^2 \, dx + \left[EI \int_0^L w''\eta'' \, dx - P \int_0^L w'\eta' \, dx \right]\epsilon$$

$$+ \left[\frac{EI}{2} \int_0^L (\eta'')^2 \, dx - \frac{P}{2} \int_0^L (\eta')^2 \, dx \right]\epsilon^2 \qquad (9.30)$$

Noting Eq. (9.29) we may give $\delta^{(T)}\pi$ as follows:

$$\tilde{\pi} - \pi = \delta^{(T)}\pi = \left[EI \int_0^L w''\eta'' \, dx - P \int_0^L w'\eta' \, dx \right]\epsilon$$

$$+ \left[EI \int_0^L (\eta'')^2 \, dx - P \int_0^L (\eta')^2 \, dx \right]\frac{\epsilon^2}{2!}$$

Thus we can conclude that:

$$\delta^{(2)}\pi = \left[EI \int_0^L (\eta'')^2 \, dx - P \int_0^L (\eta')^2 \, dx \right]\epsilon^2 \qquad (9.31)$$

We shall now *extremize* the second variation with respect to functions which satisfy the end conditions of the problem. The corresponding Euler–Lagrange equation for such an extremization is then:

$$EI\frac{d^4\eta}{dx^4} + P\frac{d^2\eta}{dx^2} = 0$$

The determination of the existence of appropriate functions η then leads us to the same problem set forth earlier for determining the critical loads. That is, *setting the first variation of the second variation of the total potential energy equal to zero leads to the same eigenvalue problem presented earlier for establishing the critical loads.* Accordingly, the critical loads may be characterized as those loads permitting the establishment of the extremal function with appropriate boundary conditions for the

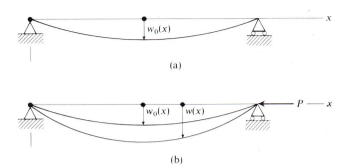

FIGURE 9.9

following condition:

$$\delta^{(1)}[\delta^{(2)}\pi] = 0 \qquad (9.32)$$

This is the criterion of *Trefftz*[1] and constitutes the second of our energy methods.
We shall make use of this criterion in the study of plates.

9.6 IMPERFECTION ANALYSIS

In the previous sections we examined a *perfectly straight* column to find that the equilibrium method, stemming from setting the first variation of π equal to zero, gave us critical loads; the extremization of the second variation or noting when π ceases to be positive-definite gave us the buckling load. We shall now set forth another criterion for the buckling load by considering a column which is initially *not* straight such as the one shown in Fig. 9.9(*a*). The initial deflection is given by the function $w_0(x)$. The total deflection with the presence of axial load P then is given by $w(x)$ as shown in Fig. 9.9(*b*).

We will give $w_0(x)$ as a Fourier series expansion in the following way

$$w_0(x) = \sum_{n=1}^{\infty} a_n \sin \frac{n\pi x}{L} \qquad (9.33)$$

so that

$$a_n = \frac{2}{L} \int_0^L w_0(x) \sin \frac{n\pi x}{L}\, dx$$

[1] E. Trefftz: "Zur Theorie der Stabilität des Elastischen Gleichgewichts," *Zeitschrift für Angewandte Mathematik und Mechanik*, Vol. 13, 1933, pp. 160–165.

The function $w(x)$ can in turn be given as follows

$$w = \sum_{n=1}^{\infty} b_n \sin \frac{n\pi x}{L} \qquad (9.34)$$

where b_n is a set of constants. Note that the end conditions are satisfied by each term in the above expansion. The total potential energy π for this case can then be given as follows:

$$\pi = \frac{1}{2} EI \int_0^L \left[\frac{d^2}{dx^2}(w - w_0) \right]^2 dx - P\varDelta \qquad (9.35)$$

We may give $\varDelta$, the contraction of the column along the x axis, as follows:

$$\varDelta = \frac{1}{2} \int_0^L \left(\frac{dw}{dx} \right)^2 dx - \frac{1}{2} \int_0^L \left(\frac{dw_0}{dx} \right)^2 dx \qquad (9.36)$$

Accordingly we have for π:

$$\pi = \frac{1}{2} EI \int_0^L \left[\frac{d^2}{dx^2}(w - w_0) \right]^2 dx - \frac{P}{2} \int_0^L \left[\left(\frac{dw}{dx} \right)^2 - \left(\frac{dw_0}{dx} \right)^2 \right] dx \qquad (9.37)$$

By using Eqs. (9.33) and (9.34) to replace w and w_0 respectively, π becomes a function of the terms b_n. Thus:

$$\pi = \frac{1}{2} EI \int_0^L \left[\sum_{1}^{\infty} (a_n - b_n)\left(\frac{n\pi}{L}\right)^2 \sin \frac{n\pi x}{L} \right]^2 dx$$

$$- \frac{P}{2} \int_0^L \left\{ \left[\sum_{1}^{\infty} b_n \frac{n\pi}{L} \cos \frac{n\pi x}{L} \right]^2 - \left[\sum_{1}^{\infty} a_n \frac{n\pi}{L} \cos \frac{n\pi x}{L} \right]^2 \right\} dx$$

To extremize π we require that:[1]

$$\frac{\partial \pi}{\partial b_m} = 0 \qquad m = 1, 2, \ldots$$

Hence:

$$- \frac{EI}{2} \int_0^L 2 \frac{\pi^2}{L^2} \left[\sum_{1}^{\infty} n^2(a_n - b_n) \sin \frac{n\pi x}{L} \right] \left[\left(\frac{m\pi}{L}\right)^2 \sin \frac{m\pi x}{L} \right] dx$$

$$- \frac{P}{2} \int_0^L 2 \left[\sum_{1}^{\infty} b_n \frac{n\pi}{L} \cos \frac{n\pi x}{L} \right] \left(\frac{m\pi}{L} \cos \frac{m\pi x}{L} \right) dx = 0 \qquad m = 1, 2, \ldots$$

[1] Remember that π is a function of the b's and is not a functional here.

Due to the orthogonality property, the terms in each integral vanish except for when $n = m$. We then get from the above equation:

$$EI\frac{\pi^4 m^4}{L^4}(a_m - b_m) + \frac{P}{L^2}m^2\pi^2 b_m = 0 \qquad m = 1, 2, \ldots \qquad (9.38)$$

We now solve for b_m to get:

$$b_m = \frac{a_m}{1 - PL^2/(\pi^2 m^2 EI)} = \frac{a_m}{1 - P/(m^2 P_E)} \qquad m = 1, 2, \ldots$$

We can then give $w(x)$ as follows:

$$w(x) = \sum_{n=1}^{\infty} \frac{a_n}{1 - (1/n^2)(P/P_E)} \sin\frac{n\pi x}{L} \qquad (9.39)$$

Unlike the results obtained for the straight beam stemming from the extremization of the total potential energy, we see here that a *non-trivial deflection exists from the very onset of the axial loading*. Note that as the load approaches the buckling load the first term in the series *blows up*.[1] We may accordingly denote here another criterion for the buckling load in that it represents the *load for which the initial deviations from straightness become amplified beyond any finite amount*. Note also that when $P = n^2 P_E$ we also get singular behavior of w, and so the imperfection analysis, by this action, yields additionally the other critical loads.

9.7 THE KINETIC METHOD

We now present another procedure toward establishing the buckling load called the *kinetic method*. In the analysis we shall pose the following query: "Is there a value of the load P for which the most general free vibration of the system, however small initially, becomes unbounded with time?"

We once again consider the pin-ended column of the previous investigations. To derive the equation of motion using Hamilton's principle we employ for the total potential energy the form given by Eq. (9.21) while for the kinetic energy we employ the expression

$$\frac{\rho A}{2}\int_0^L (\dot{w})^2 \, dx$$

---

[1] Note that if a_1 is large, the assumptions regarding linear elasticity with small rotations may be violated long before P reaches the buckling load P_E.

as explained in Sec. 7.3. Hence the Lagrangian that we shall employ in Hamilton's principle then becomes:

$$L = \int_0^L \frac{\rho A}{2}\left(\frac{\partial w}{\partial t}\right)^2 dx - \int_0^L \frac{EI}{2}\left(\frac{\partial^2 w}{\partial x^2}\right)^2 dx + \int_0^L \frac{P}{2}\left(\frac{\partial w}{\partial x}\right)^2 dx$$

$$= \int_0^L \left[\frac{\rho A}{2}(\dot{w})^2 - \frac{EI}{2}w_{xx}{}^2 + \frac{P}{2}w_x{}^2\right] dx$$

Hamilton's principle now requires that:

$$\delta^{(1)} \int_{t_1}^{t_2}\int_0^L \left[\frac{\rho A}{2}(\dot{w})^2 - \frac{EI}{2}(w_{xx})^2 + \frac{P}{2}(w_x)^2\right] dx\, dt = 0$$

Using the familiar one parameter family of varied functions $\tilde{w}(x,t) = w(x,t) + \epsilon\eta(x,t)$ where $\eta(x,t_1) = \eta(x,t_2) = 0$ we have then:

$$\frac{d}{d\epsilon}\left\{\int_{t_1}^{t_2}\int_0^L \tilde{F}\, dx\, dt\right\}_{\epsilon=0} = 0$$

$$\int_{t_1}^{t_2}\int_0^L \left(\frac{\partial F}{\partial \dot{w}}\dot{\eta} + \frac{\partial F}{\partial w_{xx}}\eta_{xx} + \frac{\partial F}{\partial w_x}\eta_x\right) dx\, dt = 0 \qquad (9.40)$$

We integrate the first expression in the integrand by parts with respect to time. Thus:

$$\int_{t_1}^{t_2}\int_0^L \frac{\partial F}{\partial \dot{w}}\dot{\eta}\, dx\, dt = \int_0^L \left[\eta\frac{\partial F}{\partial \dot{w}}\right]_{t_1}^{t_2} dx - \int_{t_1}^{t_2}\int_0^L \eta\frac{\partial}{\partial t}\left(\frac{\partial F}{\partial \dot{w}}\right) dx\, dt$$

Note that the first term on the right side is zero because of the end conditions on η. Next integrate the last expression in Eq. (9.40) by parts with respect to x. Thus:

$$\int_0^L \int_{t_1}^{t_2} \frac{\partial F}{\partial w_x}\eta_x\, dt\, dx = \int_{t_1}^{t_2}\left[\frac{\partial F}{\partial w_x}\eta\right]_0^L dt - \int_{t_1}^{t_2}\int_0^L \frac{\partial}{\partial x}\left(\frac{\partial F}{\partial w_x}\right)\eta\, dx\, dt$$

Since we require that η satisfy the end conditions of the problem, the first term on the right side of the above equation is zero. Finally we integrate by parts the second expression in the integrand of Eq. (9.40) twice with respect to x. Thus:

$$\int_0^L \int_{t_1}^{t_2} \frac{\partial F}{\partial w_{xx}}\eta_{xx}\, dt\, dx = \int_{t_1}^{t_2}\left[\frac{\partial F}{\partial w_{xx}}\eta_x\right]_0^L dt - \int_{t_1}^{t_2}\left[\frac{\partial}{\partial x}\left(\frac{\partial F}{\partial w_{xx}}\right)\eta\right]_0^L dt$$

$$+ \int_0^L \int_{t_1}^{t_2} \frac{\partial^2}{\partial x^2}\left(\frac{\partial F}{\partial w_{xx}}\right)\eta\, dx\, dt$$

The second integral on the right side of the equation clearly is zero. We then have the following result on substitution:

$$-\int_{t_1}^{t_2}\int_0^L \left\{\frac{\partial}{\partial t}\frac{\partial F}{\partial \dot{w}} - \frac{\partial^2}{\partial x^2}\frac{\partial F}{\partial w_{xx}} + \frac{\partial}{\partial x}\frac{\partial F}{\partial w_x}\right\}\eta\, dx\, dt + \int_{t_1}^{t_2}\left[\frac{\partial F}{\partial w_{xx}}\eta_x\right]_0^L dt = 0$$

Inserting for F the expression

$$\frac{\rho A}{2}(\dot{w})^2 - \frac{EI}{2}w_{xx}^2 + \frac{P}{2}w_x^2$$

we may then give the equation of motion in the following way:

$$\rho A \ddot{w} + \frac{\partial^2}{\partial x^2}\left(EI\frac{\partial^2 w}{\partial x^2}\right) + P\frac{\partial^2 w}{\partial x^2} = 0 \qquad (9.41)$$

Also we may require at the ends that $\partial F/\partial w_{xx} = EIw_{xx} = 0$ and $w = 0$.

We shall take the solution for the *freely vibrating* simply-supported column to have the following form (for uniform EI):

$$w(x,t) = \sum_{n=1}^{\infty} A_n \sin\frac{n\pi x}{L} e^{i\omega_n t} \qquad (9.42)$$

where A_n are complex constants that are determined from the initial conditions and ω_n are frequency terms that are chosen to render each term in the series a solution to the differential equation. Note that the boundary conditions $w = w_{xx} = 0$ at the ends of the column are satisfied. On substitution of the expansion into the differential equation we find that:

$$\rho A \omega_n^2 = \left(\frac{n\pi}{L}\right)^2 [n^2 P_E - P] \qquad (9.43)$$

Notice from the above result that when $P > P_E$, ω_1 becomes imaginary and, as a result, the first term in the series expansion ($n = 1$) diverges as time becomes *large*. We can conclude from this that the *buckling load is one which, when exceeded, results in a singular behavior of the vibratory motion*. This in turn means that when P is above the buckling load the equilibrium configuration is unstable since any lateral disturbance, however slight, will, within the assumptions of the theory, result in ever growing lateral motion.

9.8 GENERAL REMARKS

In reviewing the various approaches to the buckling problem of the pin-ended column we note that the *equilibrium method* gave an infinite set of eigenvalues or critical loads where nontrivial configurations could satisfy the requirements of equilibrium. The lowest critical value we called the buckling load. The *imperfection analysis* showed that at the critical loads we got infinite deflections but that between these loads theoretically finite deflection could be maintained. Using one of the so-called *energy methods* (the energy method of Trefftz) we found that the extremization of $\delta^{(2)}\pi$ again led to the set of critical loads. Furthermore, via the other energy method we found that the buckling load, P_E, marked the load where, on increasing the value of P from zero, the second variation of π ceased to be positive–definite. The implication here from extrapolation of the discussions on rigid-body systems is that up to the buckling load the equilibrium configurations are stable but at and beyond the buckling load the equilibrium configurations are not stable. This was further verified when we considered the *kinetic approach* as far as loads which exceed the critical load.

All these results stem from consideration of a particular simple problem, namely, the pin-ended column. What can we say about other *linearized* cases? *All conclusions can generally be extrapolated essentially intact.*

9.9 THE ELASTICA

Up to this time we have used simplified formulations for the study of the elastic stability of columns. From these formulations we mathematically produced such results as bifurcation points, non-trivial equilibrium configurations, etc., at the so-called critical loads. In this section we examine a more exact formulation for the column problem and we shall then be able to relate the aforementioned results of the simplified analysis with the dictates of the more exact theory. Specifically in previous sections we have in effect employed the following simplified functional for the total potential energy.

$$\pi = \frac{EI}{2} \int_0^L \left(\frac{d^2w}{dx^2}\right)^2 dx - \frac{P}{2} \int_0^L \left(\frac{dw}{dx}\right)^2 dx \qquad (9.44)$$

The expression d^2w/dx^2 used in the first integral is actually an approximation of the curvature κ of the member. You will recall from analytic geometry that κ is given exactly as:

$$\kappa = \frac{d^2w/dx^2}{[1 + (dw/dx)^2]^{3/2}}$$

Or if we use the variables θ and s at the neutral axis (see Fig. 9.10) we may also give κ exactly as:

$$\kappa = \frac{d\theta}{ds} \qquad (9.45)$$

In the second integral we have, in effect, approximated the shortening of the column Δ by the expression

$$\tfrac{1}{2} \int_0^L (dw/dx)^2 \, dx$$

We have developed this approximation in Example 3.3 and you will recall that a small slope dy/dx was a key assumption in that development. An exact evaluation of Δ is given as follows:

$$\Delta = L - \int_0^L \cos\theta\, ds = -\left\{\int_0^L \cos\theta\, ds - L\right\} \qquad (9.46)$$

Using Eqs. (9.45) and (9.46), the total potential energy then can be given in a more precise form as follows:

$$\pi = \frac{EI}{2}\int_0^L \left(\frac{d\theta}{ds}\right)^2 ds + P\left\{\int_0^L \cos\theta\, ds - L\right\} \qquad (9.47)$$

We now compute the first variation of the above functional:

$$\delta^{(1)}\pi = EI \int_0^L \frac{d\theta}{ds}\frac{d}{ds}(\delta\theta)\, ds - P\int_0^L \sin\theta\,\delta\theta\, ds$$

Integrate the first integrand by parts. We get:

$$\delta^{(1)}\pi = \left[EI\frac{d\theta}{ds}\delta\theta\right]_0^L - \int_0^L \left(EI\frac{d^2\theta}{ds^2} + P\sin\theta\right)\delta\theta\, ds$$

Setting the first variation equal to zero gives us the following Euler–Lagrange equation

$$EI\frac{d^2\theta}{ds^2} + P\sin\theta = 0 \qquad (9.48)$$

while the boundary conditions are at $s = 0, L$ are

$$\text{EITHER}\quad EI\frac{d\theta}{ds} = 0 \qquad \text{OR } \theta \text{ IS SPECIFIED} \qquad (9.49)$$

Since θ represents the local slope of the column, it is clear that the above boundary conditions represent the "*moment-slope duality.*" The differential equation is nonlinear.[1] However, we will see that it is integrable in terms of *elliptic integrals*. To carry out an immediate quadrature multiply the equation by $d\theta/ds$. Thus:

$$EI\frac{d^2\theta}{ds^2}\frac{d\theta}{ds} + P\sin\theta\frac{d\theta}{ds} = 0$$

This may be written next in the following form as you may yourself easily verify:

$$\frac{d}{ds}\left[\frac{EI}{2}\left(\frac{d\theta}{ds}\right)^2 - P\cos\theta\right] = 0$$

---

[1] It is actually the same form as the exact equation for the simple pendulum.

We may integrate to get:

$$\frac{EI}{2}\left(\frac{d\theta}{ds}\right)^2 = P \cos \theta + C_1 \tag{9.50}$$

For a simply-supported column, θ is not specified at the ends and so we use the natural boundary condition $d\theta/ds = 0$ at the ends. Denoting θ at $x = 0$ as α, and θ at $x = L$ as $-\alpha$, we can then find on applying the natural boundary conditions to Eq. (9.50) that:

$$C_1 = P \cos \alpha$$

Hence:

$$\frac{d\theta}{ds} = \left[\frac{2P}{EI}(\cos \theta - \cos \alpha)\right]^{1/2} \tag{a}$$

$$= \left[\frac{4P}{EI}\left(\sin^2 \frac{\alpha}{2} - \sin^2 \frac{\theta}{2}\right)\right]^{1/2} \tag{b} \tag{9.51}$$

where we have used familiar trigonometric identities to alter the right side of the equation. Separating variables we may next set up the following quadrature:

$$\int_0^L ds = L = \sqrt{\frac{EI}{4P}} \int_{-\alpha}^{+\alpha} \frac{d\theta}{[\sin^2 \alpha/2 - \sin^2 \theta/2]^{1/2}}$$

$$= \sqrt{\frac{EI}{4P}} \int_0^\alpha \frac{2\, d\theta}{[\sin^2 \alpha/2 - \sin^2 \theta/2]^{1/2}} \tag{9.52}$$

(In the last step above we made use of the symmetry of the deflection curve about the line $x = L/2$.) We now institute a change of variable as follows

$$\sin \frac{\theta}{2} = p \sin \phi$$

where

$$p = \sin \frac{\alpha}{2}$$

With some algebraic manipulation we may rewrite Eq. (9.52) in the following manner:

$$L = 2\sqrt{\frac{EI}{P}} \int_0^{\pi/2} \frac{d\phi}{[1 - p^2 \sin^2 \phi]^{1/2}} \tag{9.53}$$

Introducing the Euler load P_E, the above equation becomes:

$$\frac{P}{P_E} = \frac{2}{\pi} \int_0^{\pi/2} \frac{d\phi}{[1 - p^2 \sin^2 \phi]^{1/2}} \tag{9.54}$$

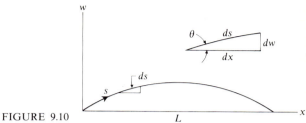

FIGURE 9.10

Using tables of elliptic integrals of the first kind[1] we could get P/P_E as a function of p and hence of α, the slope at the ends. This is done by choosing a value of P/P_E and then finding from the tables the proper value of p to satisfy the above equation.

We shall next relate p to the maximum deflection w_{max} of the column so that we will be able to plot P/P_E vs. w_{max}. For this purpose note from Fig. 9.10 that:

$$\sin \theta = \frac{dw}{ds} \qquad (9.55)$$

Accordingly, the Euler–Lagrange equation (9.48) can be written as follows on replacing $\sin \theta$:

$$EI\frac{d^2\theta}{ds^2} + P\frac{dw}{ds} = 0 \qquad (9.56)$$

We may immediately perform a quadrature to solve for w. Thus:

$$w = -\frac{EI}{P}\frac{d\theta}{ds} + C_2 \qquad (9.57)$$

Clearly $w = 0$ at the ends where, as a result of the natural boundary condition, $d\theta/ds$ must also be zero, rendering the constant of integration also equal to zero. Employing Eq. (9.51(a)) to replace $d\theta/ds$ above we then have for w:

$$w = -\sqrt{\frac{2EI}{P}}(\cos \theta - \cos \alpha)^{1/2} \qquad (9.58)$$

The maximum displacement occurs at $\theta = 0$ and so we have for w_{max}:

$$w_{max} = -\sqrt{\frac{2EI}{P}}(1 - \cos \alpha)^{1/2} \qquad (9.59)$$

[1] M. Abramowitz and I. A. Stegun, (eds.): "Handbook of Mathematical Functions," U.S. Department of Commerce, National Bureau of Standards, Vol. 55, Applied Mathematics Series, 1965.

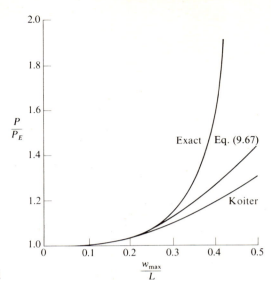

FIGURE 9.11

This may be rewritten in the following form on introducing P_E and employing a trigonometric identity for the bracketed expression:

$$\frac{w_{max}}{L} = \frac{2}{\pi} \frac{\sin \alpha/2}{\sqrt{P/P_E}} = \frac{2}{\pi} \frac{p}{\sqrt{P/P_E}} \qquad (9.60)$$

Thus with p and P/P_E appropriately related by Eq. (9.54) we can determine the corresponding value of w_{max}/L from the preceding equation. A plot of P/P_E versus w_{max}/L is shown in Fig. 9.11. At $P/P_E = 1$, the exact curve is tangent to the line $P/P_E = 1$. We see from the exact analysis that we still get the bifurcation point as we did in the linearized theory but now we no longer have the indeterminacy of load versus deflection at the buckling load. Indeed we see that ever-increasing loading is needed for an increase in deflection. We call the regime of deformation beyond the bifurcation point the *postbuckling regime*. Since the column can withstand even higher loads in the vicinity of the bifurcation point we say that we have postbuckling *stability*. (If after the bifurcation point, the structure could withstand only decreased loads then obviously we would have a situation of postbuckling *instability*.) Clearly in practical problems the question of postbuckling behavior is of significance. We shall return to this question briefly in Sec. 9.11 when we discuss the work of Koiter.

In the following section we shall consider an "intermediate" theory that will reproduce the exact results up to about $w_{max}/L = 0.3$ and which nevertheless is a considerably simpler computation than that of the elastica.

9.10 AN INTERMEDIATE THEORY

We again consider the pin-connected column of the previous sections. Our procedure will be to use certain power series expansions to form a more accurate expression for π than was used for the elementary case. We now express $\cos \theta$ as follows making use of Eq. (9.55). Thus:

$$\cos \theta = \sqrt{1 - \sin^2 \theta} = \sqrt{1 - \left(\frac{dw}{ds}\right)^2} \qquad (9.61)$$

Next we use the binomial expansion to obtain the following approximation:

$$\cos \theta \cong 1 - \frac{1}{2}\left(\frac{dw}{ds}\right)^2 - \frac{1}{8}\left(\frac{dw}{ds}\right)^4 \qquad (9.62)$$

Similarly, to obtain an approximate curvature expression noting that $\theta = \sin^{-1}(dw/ds)$, we have:

$$\frac{d\theta}{ds} = \frac{d}{ds}\left[\sin^{-1}\frac{dw}{ds}\right] = \left[1 - \left(\frac{dw}{ds}\right)^2\right]^{-1/2}\left(\frac{d^2w}{ds^2}\right)$$

Using the binomial expansion for the root we have:

$$\frac{d\theta}{ds} \cong \left[1 + \frac{1}{2}\left(\frac{dw}{ds}\right)^2\right]\left(\frac{d^2w}{ds^2}\right) \qquad (9.63)$$

We may now express the total potential energy π more accurately than in the simple case by utilizing Eqs. (9.61) and (9.63) in Eq. (9.47) as follows:

$$\begin{aligned}
\pi &= \frac{EI}{2}\int_0^L \left(\frac{d\theta}{ds}\right)^2 ds + P\left\{\int_0^L \cos\theta \, ds - L\right\} \\
&= \frac{EI}{2}\int_0^L \left[1 + \frac{1}{2}\left(\frac{dw}{ds}\right)^2\right]^2\left(\frac{d^2w}{ds^2}\right)^2 ds - \frac{P}{2}\int_0^L \left[1 + \frac{1}{4}\left(\frac{dw}{ds}\right)^2\right]\left(\frac{dw}{ds}\right)^2 ds
\end{aligned} \qquad (9.64)$$

(Note that if we delete $(dw/ds)^2$ in the brackets as being small compared with unity and replace s by x as the integration variable we then revert to classical theory presented at the outset.) We shall proceed by using the Rayleigh–Ritz method here employing as a single coordinate function the eigenfunction of the classical simple theory with s as the variable. That is:

$$w = A \sin \frac{\pi s}{L} \qquad (9.65)$$

Clearly such a function satisfies the geometric and the natural boundary conditions of the simply-supported column. Substituting Eq. (9.65) into (9.64) and using $a = A\pi/L$ we get after integration and algebraic manipulations:

$$\pi = \frac{EI\pi^2}{4L}\left\{a^2 + \tfrac{1}{4}a^4 + \tfrac{1}{32}a^6 - \frac{P}{P_E}(a^2 + \tfrac{3}{16}a^4)\right\} \qquad (9.66)$$

Extremizing π with respect to "a" then yields the following equation:

$$\tfrac{3}{16}a^4 + \left(1 - \frac{3}{4}\frac{P}{P_E}\right)a^2 + 2\left(1 - \frac{P}{P_E}\right) = 0 \qquad (9.67)$$

We may solve for "a" for each value of P/P_E. Noting that w_{max}/L equals a/π, we may then plot (P/P_E) vs. (w_{max}/L) forming the so-called intermediate theory in Fig. 9.11. Notice that the agreement between exact and intermediate theorems is quite good for $w_{max}/L < 0.3$. (In most situations of practical interest this agreement is more than satisfactory, for it is not likely that the assumption of linear elasticity would be valid much beyond that large a deflection.)

By the technique presented here for the pin-ended column we can often obtain reasonably accurate load-deflection curves without an inordinate amount of work.

9.11 A NOTE ON KOITER'S THEORY OF ELASTIC STABILITY

The bulk of the buckling analyses that we have considered thus far has been limited to the linearized eigenvalue problems that define the buckling load. In Sec. 9.9 on the "Elastica" and Sec. 9.10 on "An Intermediate Theory" we extended the results to include deflections beyond the bifurcation point and we were able to discuss *postbuckling stability*.[1]

We now present and illustrate the work of Koiter to predict postbuckling stability. The theory was developed in 1946 but has received increased attention in recent years. Like the intermediate theory presented for the pin-ended column, Koiter expresses the total potential energy as a sequence of homogeneous functionals of increasing order n. The extremization of the second-order functional leads to the eigenvalue giving the buckling load and eigenfunction ψ_1. The value of the next nonzero higher-order functional taken for the latter extremal functional ψ_1 then is significant in establishing the postbuckling behavior of the system in the vicinity of the bifurcation point and thus gives information on the important question of post-buckling stability. We illustrate these comments by again considering as a special simple case the pin-ended column, after which we shall discuss additional contributions of Koiter relating to the role of imperfections in postbuckling behavior.

We consider u to be the axial displacement of the centerline elements of the column and w to be the transverse displacement of this centerline. The new deformed geometry, which we indicate with primes, is then given as:

$$x'_1 = x_1 + u$$
$$x'_3 = w$$

Hence the new length squared of the centerline element dl' becomes:

$$(dl')^2 = \left\{ \left(1 + \frac{\partial u}{\partial x}\right)^2 + \left(\frac{\partial w}{\partial x}\right)^2 \right\} dx^2 \qquad (9.68)$$

If we assume the centerline is *inextensible*, then $dl' = dx$ and we conclude from above that:

$$\left(1 + \frac{\partial u}{\partial x}\right)^2 + \left(\frac{\partial w}{\partial x}\right)^2 = 1 \qquad (9.69)$$

---

[1] It is important to remember that the Koiter theory (as will be developed) is applicable only to bifurcation point buckling and not to limit point buckling.

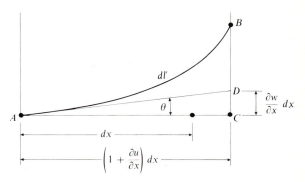

FIGURE 9.12

Now consider an element of the centerline dx of the column in a deformed configuration AB as shown in Fig. 9.12. We can say for this diagram

$$\sin \theta = \frac{(\partial w/\partial x)\,dx}{AD} = \frac{(\partial w/\partial x)\,dx}{\sqrt{(1 + \partial u/\partial x)^2 + (\partial w/\partial x)^2\,(dx)}} \qquad (9.70)$$

where we replace AD using the Pythagorean theorem. Noting Eq. (9.69) resulting from the inextensibility assumption, we simplify the above equation to:

$$\sin \theta = \frac{\partial w}{\partial x} \qquad (9.71)$$

Now differentiate with respect to x above. We get:

$$\cos \theta \frac{d\theta}{dx} = \frac{\partial^2 w}{\partial x^2}$$

Or noting that we can take $d\theta/dx = 1/\rho$, one over the radius of curvature, we have:

$$\frac{1}{\rho} = \frac{\partial^2 w/\partial x^2}{\cos \theta} = \frac{\partial^2 w/\partial x^2}{\sqrt{1 - \sin^2 \theta}} \qquad (9.72)$$

Replacing $\sin \theta$ using Eq. (9.71), the above equation becomes:

$$\frac{1}{\rho} = \frac{\partial^2 w/\partial x^2}{\sqrt{1 - (\partial w/\partial x)^2}} \qquad (9.73)$$

This is the radius of curvature that we shall henceforth use. Thus for the strain energy including only bending we have:

$$U = \frac{EI}{2} \int_0^L \left(\frac{1}{\rho}\right)^2 dx = \frac{EI}{2} \int_0^L \frac{(\partial^2 w/\partial x^2)^2}{1 - (\partial w/\partial x)^2} dx$$

And the potential of the loads using axial force N as positive for compression is

$$V = N \int_0^L \frac{\partial u}{\partial x} dx = N \int_0^L \left[\sqrt{1 - \left(\frac{\partial w}{\partial x}\right)^2} - 1 \right] dx$$

where we have used Eq. (9.69) to replace $\partial u/\partial x$. The total potential energy then becomes:

$$U + V = \frac{EI}{2} \int_0^L \frac{(\partial^2 w/\partial x^2)^2}{1 - (\partial w/\partial x)^2} dx + N \int_0^L \left[\sqrt{1 - \left(\frac{\partial w}{\partial x}\right)^2} - 1 \right] dx \qquad (9.74)$$

Next we introduce the following dimensionless variables:

$$\xi = \frac{x}{L} \qquad (a)$$

$$\psi = \frac{w}{L} \qquad (b)$$

$$\lambda = N \bigg/ \frac{\pi^2 EI}{L^2} = N/N_{\text{Euler}} \qquad (c) \quad (9.75)$$

The total potential energy functional then will be given as:

$$P^\lambda(\psi) = \frac{2L}{EI}[U + V] = \int_0^1 \frac{[\psi''(\xi)]^2}{1 - [\psi'(\xi)]^2} d\xi$$

$$+ 2\pi^2 \lambda \int_0^1 \left\{ \sqrt{1 - [\psi'(\xi)]^2} - 1 \right\} d\xi$$

Expand the denominator of the first integral and the radical of the second integral as power series:

$$P^\lambda(\psi) = \int_0^1 [\psi''(\xi)]^2 (1 + [\psi'(\xi)]^2 + \cdots) d\xi$$

$$+ 2\pi^2 \lambda \int_0^1 \left\{ [1 - \tfrac{1}{2}[\psi'(\xi)]^2 - \tfrac{1}{8}[\psi'(\xi)]^4 + \cdots] - 1 \right\} d\xi$$

Rearranging the terms we get:

$$P^\lambda(\psi) = \int_0^1 [\psi''(\xi)]^2 d\xi - \pi^2 \lambda \int_0^1 [\psi'(\xi)]^2 d\xi$$

$$+ \int_0^1 [\psi''(\xi)]^2 [\psi'(\xi)]^2 d\xi - \tfrac{1}{4}\pi^2 \lambda \int_0^1 [\psi'(\xi)]^4 d\xi$$

$$+ \cdots$$

Thus we see that the total potential energy can be expressed as the sum of a functional of order 2 plus one of order 4 and higher-order functionals that we now neglect. We express this as follows:

$$P^{\lambda}(\psi) = P_2{}^{\lambda}(\psi) + P_4{}^{\lambda}(\psi)$$

where:

$$P_2{}^{\lambda}(\psi) = \int_0^1 \{[\psi''(\xi)]^2 - \pi^2\lambda[\psi'(\xi)]^2\}\, d\xi \qquad (a)$$

$$P_4{}^{\lambda}(\psi) = \int_0^1 \{[\psi''(\xi)]^2[\psi'(\xi)]^2 - \frac{\pi^2}{4}\lambda[\psi'(\xi)]^4\}\, d\xi \qquad (b) \quad (9.76)$$

We now proceed to extremize $P_2{}^{(\lambda)}(\psi)$ with respect to a one-parameter family of varied functionals $\tilde{\psi} = \psi + \epsilon\eta$. Thus we have:

$$\left(\frac{dP_2{}^{\lambda}(\tilde{\psi})}{d\epsilon}\right)_{\epsilon=0} = 0$$

Hence:

$$2\int_0^1 (\psi''\eta'' - \pi^2\lambda\psi'\eta')\, d\xi = 0$$

Integrating by parts:

$$2\int_0^1 (\psi^{\text{IV}} + \pi^2\lambda\psi'')\eta\, d\xi + 2\psi''\eta'\Big|_0^1 - 2(\psi''' + \pi^2\lambda\psi')\eta\Big|_0^1 = 0$$

We thus are led to the following conclusion:

$$\psi^{\text{IV}} + \pi^2\lambda\psi'' = 0 \qquad (9.77)$$

Also we may use the conditions $\psi''(0) = \psi''(1) = 0$ to satisfy the above natural boundary conditions and $\psi(0) = \psi(1) = 0$ (therefore $\eta = 0$) to satisfy the kinematic boundary conditions of the problem. The general solution to Eq. (9.77) is:

$$\psi = A + B\xi + C\cos(\pi\sqrt{\lambda}\xi) + D\sin(\pi\sqrt{\lambda}\xi)$$

In satisfying the boundary conditions we arrive at the following condition

$$\sin(\pi\sqrt{\lambda}) = 0$$

so that the smallest value of λ is unity thus giving us the expected buckling load (see Eq. 9.75(c)). The eigenfunction for this load clearly is:

$$\psi_1 = \sin\pi\xi$$

Koiter now shows in his general theory[1] that the amplitude of this eigenfunction, made dimensionless here with respect to beam length, for loads λ in the neighborhood of λ_1, the critical load, is given as:

$$a = \pm \left(\frac{2}{n}\frac{A'_2}{A_n}\right)^{1/(n-2)} (\lambda_1 - \lambda)^{1/(n-2)} \qquad (a)$$

or is given as

$$a = \pm \left(-\frac{2}{n}\frac{A'_2}{A_n}\right)^{1/(n-2)} (\lambda - \lambda_1)^{1/(n-2)} \qquad (b)$$

where

$$A'_2 = \frac{d}{d\lambda}\{P_2{}^\lambda(\psi_1)\} \qquad (c)$$

and where

$$A_n = P_n^{\lambda_1}(\psi_1) \qquad (d) \quad (9.78)$$

The value of A'_2 can be assumed as negative and the choice of formula for a depends on the sign of A_n. If A_n is negative we employ the first formula (a); if it is positive we employ the second formula (b). The value of n above is the value of order of the next nonzero higher functional in the series above $n = 2$. For the problem at hand we have accordingly:

$$A'_2 = \frac{d}{d\lambda}\int_0^1 [(\psi''_1)^2 - \pi^2 \lambda(\psi'_1)^2]\,d\xi = -\pi^2 \int_0^1 (\psi'_1)^2\,d\xi = -\frac{\pi^4}{2}$$

Also

$$A_4 = P_4{}^{\lambda_1}(\psi_1) = \int_0^1 [(\psi'_1)^2(\psi''_1)^2 - \frac{\pi^2}{4}\lambda_1(\psi'_1)^4]\,d\xi = \frac{\pi^6}{32}$$

Hence we employ Eq. (9.78(b)) for a as follows:

$$a = \pm \left[\frac{2}{4}\frac{-\pi^4/2}{\pi^6/32}(\lambda - \lambda_1)\right]^{1/2}$$

Solving for λ we get:

$$\lambda = \lambda_1 + \frac{\pi^2}{8}a^2 = 1 + \frac{\pi^2}{8}a^2$$

[1] A translation of the "Stability of Elastic Equilibrium," W. T. Koiter: AFFDL-TR-70-25. Air Force Flight Dynamics Laboratory, Wright-Patterson Air Force Base, Ohio, February, 1970; or NASA Report TTF-10, 833, 1967.

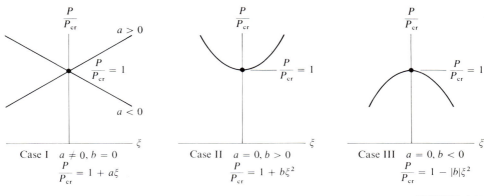

FIGURE 9.13

We see that λ increases after buckling as the amplitude a increases and so we have as already pointed out in the elastica discussion a case of postbuckling stability. (The λ vs $a = w_{max}/L$ curve (Koiter curve) is also plotted in Fig. 9.11.)

Actually there is additional information that may be gleaned from the previous analysis which we shall now discuss. It has been known since the thirties that for certain situations the buckling load was not the single significant parameter in stability considerations. Thus about this time, experimental testing of cylindrical shells revealed drastically lower *actual* buckling loads than those predicted by theory. In earlier tests on columns and plates no such serious discrepancies had been observed.

To explain this, Koiter pointed out the sensitivity of certain structures to certain types of imperfections. For background in this discussion recall that the linearized buckling analysis generally produces multiple forms of equilibrium configurations. A nonlinear analysis would show, as we have done in one specific case in Sec. 9.9, that in the neighborhood of these bifurcation points three types of postbuckling behavior are possible. In Fig. 9.13 we have plotted P/P_{cr} versus an appropriately normalized buckling displacement ξ for the aforementioned three cases. The analytical representation of these cases is given as follows:

$$\frac{P}{P_{cr}} = 1 + a\xi + b\xi^2 + \cdots \qquad (9.79)$$

where for

Case I	$a \neq 0$	and	$b = 0$
Case II	$a = 0$		$b > 0$
Case III	$a = 0$		$b < 0$

The results presented are accurate in an *asymptotic sense* and so $\xi \ll 1$. Hence only terms up to $b\xi^2$ need be retained in the equation. Note that in all cases as $\xi \to 0$, then $P \to P_{cr}$, the bifurcation load. In Case I, it is clear that the postbuckling behavior may be stable or unstable depending on the sign of "a." In practical situations, therefore, such a case would be considered unstable. In Case II the postbuckling behavior is stable, while in Case III it is unstable.

Koiter argued that certain types of geometrical imperfections on certain structures could produce loads, P_s, much lower than P_{cr}, the buckling load for the perfect structure, at which

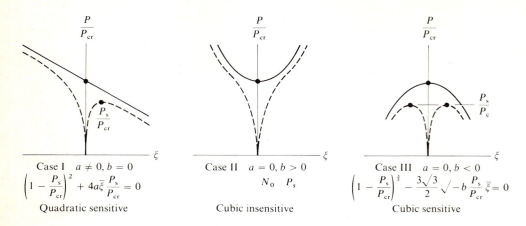

Case I $a \neq 0, b = 0$
$$\left(1 - \frac{P_s}{P_{cr}}\right)^2 + 4a\bar{\xi}\frac{P_s}{P_{cr}} = 0$$
Quadratic sensitive

Case II $a = 0, b > 0$
$$N_0 \quad P_s$$
Cubic insensitive

Case III $a = 0, b < 0$
$$\left(1 - \frac{P_s}{P_{cr}}\right)^{\frac{3}{2}} - \frac{3\sqrt{3}}{2}\sqrt{-b}\,\frac{P_s}{P_{cr}}\bar{\xi} = 0$$
Cubic sensitive

FIGURE 9.14

loads (P_s) *actual* buckling in a test would take place. And the imperfection that he included was that of the buckling mode itself for the lowest eigenvalue.[1] If $\bar{\xi}$ represents the amplitude of such an imperfection the equation corresponding to Eq. (9.79) can be shown to be of the form:

$$\left(1 - \frac{P}{P_{cr}}\right)\xi + a\xi^2 + b\xi^3 + \cdots = \frac{P}{P_{cr}}\bar{\xi} \tag{9.80}$$

Again we can distinguish three cases which we show in Fig. 9.14. The full lines correspond to those of perfect structures and therefore correspond to Fig. 9.13. The dashed lines include the geometric imperfections. In Cases I and III we have shown a maximum load possible, P_s, at which the postbuckling behavior becomes unstable and this load is considerably smaller than P_{cr}. In other words P_s is in fact the maximum (limit) load that the structures with imperfection $\bar{\xi}$ can carry. This explains the drastic decrease in buckling load of an axially compressed cylinder, for example, where an imperfection whose magnitude is only 1/10 the wall thickness reduces P_s to about 60% of P_{cr}.

It may be deduced on examining Fig. 9.14 that the *sensitivity to imperfections is directly related to the stability of the postbuckling behavior of the corresponding perfect structure.* We have already seen earlier that perfect columns are stable in postbuckling behavior (Case II) and so the tests, involving nearly perfect specimens, check the theoretical buckling load P_{cr}. This is also true for plates.

In closing we point out that considerable research is still under way to explain postbuckling behavior. We have presented only a few salient features of this work that may be of interest to the reader.[2]

---

[1] A very important assumption is involved here, i.e., that a unique mode shape corresponds to the (unique) buckling load.

[2] For a recent survey of the field see Hutchinson, J. W. and Koiter, W. T.: "Postbuckling Theory," *Applied Mechanics Reviews*, 1970, p. 1353.

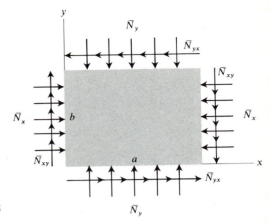

FIGURE 9.15

Part C
ELASTIC STABILITY OF PLATES

9.12 THE BUCKLING EQUATION FOR RECTANGULAR PLATES

In this section we shall present the linearized equation leading to an eigenvalue problem for the study of buckling of plates. Specifically we shall present means of determining critical values of constant applied edge loads $\bar{N}_x$, $\bar{N}_y$, $\bar{N}_{xy}$, and $\bar{N}_{yx}$ for the rectangular plate (see Fig. 9.15). We assume here that $\bar{N}_{xy} = \bar{N}_{yx}$. In doing so we shall follow a procedure analogous to that taken in Sec. 9.4, where we first found the equation of equilibrium of a column by extremizing the total potential energy. We shall then linearize the equation while noting what restrictions are actually developed in this process. In the present study we already have equations of equilibrium of plates with given edge loads from our work in the previous chapter. We accordingly now rewrite the von Kármán plate equations as follows in terms of rectangular components:

$$DV^4w = \frac{\partial^2 F}{\partial y^2}\frac{\partial^2 w}{\partial x^2} - 2\frac{\partial^2 F}{\partial x\,\partial y}\frac{\partial^2 w}{\partial x\,\partial y} + \frac{\partial^2 F}{\partial x^2}\frac{\partial^2 w}{\partial y^2} \quad (a)$$

$$\nabla^4 F = Eh\left[\left(\frac{\partial^2 w}{\partial x\,\partial y}\right)^2 - \frac{\partial^2 w}{\partial x^2}\frac{\partial^2 w}{\partial y^2}\right] \quad (b) \quad (9.81)$$

where:

$$\frac{\partial^2 F}{\partial y^2} = N_x$$

$$\frac{\partial^2 F}{\partial x^2} = N_y$$

$$-\frac{\partial^2 F}{\partial x\, \partial y} = N_{xy} \qquad (9.82)$$

In the column analysis, we linearized the equation of equilibrium (9.19) by noting that we could set $N = -P$ throughout the column. This meant effectively that we could neglect extensions along the neutral axis of the column. Here we shall set the terms analogous to N, namely $N_x(= \partial^2 F/\partial y^2)$, $N_y(= \partial^2 F/\partial x^2)$ and $N_{xy}(= -\partial^2 F/\partial x\, \partial y)$ equal to the corresponding edge loads plus a small perturbation quantity. That is, take:[1]

$$\frac{\partial^2 F}{\partial y^2} = -\overline{N}_x + \frac{\partial^2 \widetilde{F}}{\partial y^2}$$

$$\frac{\partial^2 F}{\partial x^2} = -\overline{N}_y + \frac{\partial^2 \widetilde{F}}{\partial x^2}$$

$$\frac{\partial^2 F}{\partial x\, \partial y} = \overline{N}_{xy} + \frac{\partial^2 \widetilde{F}}{\partial x\, \partial y} \qquad (9.83)$$

where the notation $\tilde{}$ indicates the perturbation quantity. We shall agree to keep only linear terms in the perturbation quantities and in $\tilde{w}$ and its derivatives. That is, we shall neglect all products and squares of these quantities. Accordingly when we substitute Eqs. (9.83) into Eqs. (9.81) we obtain after such a linearization process:

$$D\nabla^4 \tilde{w} = -\overline{N}_x \frac{\partial^2 \tilde{w}}{\partial x^2} - 2\overline{N}_{xy}\frac{\partial^2 \tilde{w}}{\partial x\, \partial y} - \overline{N}_y \frac{\partial^2 \tilde{w}}{\partial y^2} \qquad (a)$$

$$\nabla^4 \widetilde{F} = 0 \qquad\qquad\qquad (b) \quad (9.84)$$

Note that the *equations above are now uncoupled* and we need now only concern ourselves with the first of the pair.

To gain insight as to what simplifications have effectively been incorporated in the study by the previous rather formal mathematical steps, we turn (as we did for the column) to considerations of the total potential energy at this time. In particular let us consider the potential of the loading for a plate of general shape as shown in

---

[1] Note we have taken compression as positive for the applied loads and have reversed the direction of $\overline{N}_{xy}$ as positive.

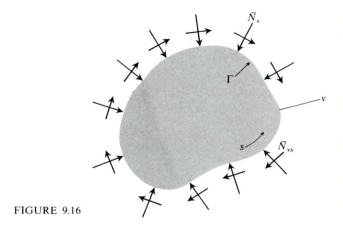

FIGURE 9.16

Fig. 9.16. We can say using negative edge loads as positive here:

$$V = -\left[\oint_\Gamma (-\bar{N}_v)(u_v)\, ds + \oint_\Gamma (-\bar{N}_{vs})(u_s)\, ds\right] \qquad (9.85)$$

In the case of a rectangular plate (see Fig. 9.15) the above equation becomes:

$$V = -\int_0^a [\bar{N}_y u_y + \bar{N}_{yx} u_x]_{y=0}\, dx - \int_0^b [(-\bar{N}_x)u_x + (-\bar{N}_{xy})u_y]_{x=a}\, dy$$

$$-\int_a^0 [(-\bar{N}_y)u_y + (-\bar{N}_{yx})u_x]_{y=b}(-dx) - \int_b^0 [(\bar{N}_x)u_x + \bar{N}_{xy}u_y]_{x=0}(-dy)$$

Collecting terms we get:

$$V = \int_0^a \bar{N}_y [(u_y)_{y=b} - (u_y)_{y=0}]\, dx + \int_0^b \bar{N}_x [(u_x)_{x=a} - (u_x)_{x=0}]\, dy$$

$$+ \int_0^a \bar{N}_{yx}[(u_x)_{y=b} - (u_x)_{y=0}]\, dx + \int_0^b \bar{N}_{xy}[(u_y)_{x=a} - (u_y)_{x=0}]\, dy$$

Now such expressions as $[(u_y)_{y=b} - (u_y)_{y=0}]$ can be given as $\int_0^b (\partial u_y/\partial y)\, dy$, etc., and accordingly noting that $\bar{N}_{xy} = \bar{N}_{yx}$, the above equation can be given as follows:

$$V = \int_0^a \int_0^b \left[\bar{N}_y \frac{\partial u_y}{\partial y} + \bar{N}_{xy}\left(\frac{\partial u_x}{\partial y} + \frac{\partial u_y}{\partial x} \right) + \bar{N}_x \frac{\partial u_x}{\partial x} \right] dx\, dy \qquad (9.86)$$

As a next step let us go back to Eq. (8.95) giving strains in terms of displacements to be used in the von Kármán plate theory. Note from this listing that in the midplane of the plate

$$\epsilon_{xx} = \frac{\partial u_x}{\partial x} + \frac{1}{2}\left(\frac{\partial w}{\partial x}\right)^2$$

$$\epsilon_{yy} = \frac{\partial u_y}{\partial y} + \frac{1}{2}\left(\frac{\partial w}{\partial y}\right)^2$$

$$\epsilon_{xy} = \frac{1}{2}\left(\frac{\partial u_y}{\partial x} + \frac{\partial u_x}{\partial y}\right) + \frac{1}{2}\left(\frac{\partial w}{\partial x}\right)\left(\frac{\partial w}{\partial y}\right)$$

If we set the strains equal to zero *in the midplane of the plate* for the ensuing analysis then we can say from above

$$\frac{\partial u_x}{\partial x} = -\frac{1}{2}\left(\frac{\partial w}{\partial x}\right)^2$$

$$\frac{\partial u_y}{\partial y} = -\frac{1}{2}\left(\frac{\partial w}{\partial y}\right)^2$$

$$\frac{\partial u_y}{\partial x} + \frac{\partial u_x}{\partial y} = -\left(\frac{\partial w}{\partial x}\right)\left(\frac{\partial w}{\partial y}\right) \tag{9.87}$$

Substituting these results into Eq. (9.86) we then have for V:

$$V = -\frac{1}{2}\int_0^a \int_0^b \left[\bar{N}_x\left(\frac{\partial w}{\partial x}\right)^2 + 2\bar{N}_{xy}\left(\frac{\partial w}{\partial x}\right)\left(\frac{\partial w}{\partial y}\right) + \bar{N}_y\left(\frac{\partial w}{\partial y}\right)^2\right] dx\, dy$$

We may now give a total potential energy functional using the above result for the external loads and using Eq. (6.27) with $u_s = v_s = q = 0$ for the strain energy of bending. We then get:

$$\pi = \frac{D}{2}\int_0^a \int_0^b \left\{(\nabla^2 w)^2 + 2(1 - v)\left[\left(\frac{\partial^2 w}{\partial x\, \partial y}\right)^2 - \frac{\partial^2 w}{\partial x^2}\frac{\partial^2 w}{\partial y^2}\right]\right\} dx\, dy$$

$$- \frac{1}{2}\int_0^a \int_0^b \left\{\bar{N}_x\left(\frac{\partial w}{\partial x}\right)^2 + 2\bar{N}_{xy}\frac{\partial w}{\partial x}\frac{\partial w}{\partial y} + \bar{N}_y\left(\frac{\partial w}{\partial y}\right)^2\right\} dx\, dy \tag{9.88}$$

It will now be left as an exercise to show that extremization of the above functional gives us the Euler–Lagrange equation, Eq. (9.84(a))—the very equation reached by the linearization process. And so we see that *this equation disregards the extensions of the midsurface.*

We can now determine the boundary conditions for the linearized buckling equation from the extremization of functional (9.88). Or more easily, they may be deduced from the von Kármán plate theory (see Eq. (8.109) by replacing N_v and N_{vs}

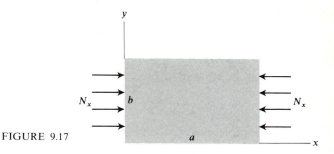

FIGURE 9.17

respectively by the applied loads $-\overline{N}_v$ and $-\overline{N}_{vs}$. Thus we have:

$$\text{EITHER } M_v = 0 \quad \text{OR} \quad \frac{\partial w}{\partial v} \text{ IS SPECIFIED} \qquad (a)$$

$$\text{EITHER} \quad Q_v + \frac{\partial M_{vs}}{\partial s} - \overline{N}_v \frac{\partial w}{\partial v} - \overline{N}_{vs} \frac{\partial w}{\partial s} = 0 \quad \text{OR} \quad w \text{ IS SPECIFIED} \qquad (b)$$

And at discontinuities on the boundary:

$$[M_{vs}\, \delta w] = 0 \qquad\qquad (c) \quad (9.89)$$

We have thus formulated the linearized buckling equation for rectangular plates.

We may now ask what loadings $\overline{N}_x$, $\overline{N}_y$, and $\overline{N}_{xy}$ or combinations of these loads permit non-trivial solutions $w \neq 0$ of the buckling equation. This is the so-called *equilibrium method* set forth in Part B. Later we shall consider the *energy method* for plates.

9.13 THE EQUILIBRIUM METHOD—AN EXAMPLE

We shall now consider the case of a simply-supported rectangular plate $a \times b \times h$ in uniaxial uniform compression (see Fig. 9.17). This means that $\overline{N}_{xy} = \overline{N}_y = 0$ and the governing equation is given as:

$$D\nabla^4 w + \overline{N}_x \frac{\partial^2 w}{\partial x^2} = 0 \qquad (9.90)$$

The boundary conditions for this problem are[1]

$$\text{at } x = 0, a; \qquad w = \frac{\partial^2 w}{\partial x^2} = 0$$

$$\text{at } y = 0, b; \qquad w = \frac{\partial^2 w}{\partial y^2} = 0 \qquad (9.91)$$

---

[1] Note that for $x = a$, and $x = 0$, M_v becomes M_x and from Eq. (6.12(a)) we see that for a simply-supported straight edge the requirement for $M_x = 0$ leads to the requirement that $\partial^2 w/\partial x^2 = 0$.

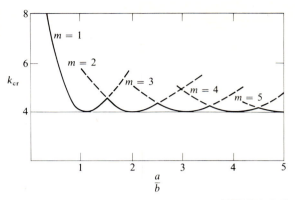

FIGURE 9.18

From our work in linear plate theory we shall propose the following expression as a solution to Eq. (9.90):

$$w = A_{mn} \sin \frac{m\pi x}{a} \sin \frac{n\pi y}{b} \qquad (9.92)$$

To satisfy the boundary conditions (9.91), clearly m and n must be integers. To avoid a trivial solution here we take m and n as nonzero. Substituting into Eq. (9.90) we have:

$$\left\{ D\left[\left(\frac{m\pi}{a}\right)^2 + \left(\frac{n\pi}{b}\right)^2 \right]^2 - \bar{N}_x \left(\frac{m\pi}{a}\right)^2 \right\} A_{mn} \sin \frac{m\pi x}{a} \sin \frac{n\pi y}{b} = 0$$

To get a nontrivial solution for the equilibrium equation it is clear from above that critical loadings $\bar{N}_x$ exist and may be given as follows.

$$(\bar{N}_x)_{\text{cr}} = D\left(\frac{a}{m\pi}\right)^2 \left[\left(\frac{m\pi}{a}\right)^2 + \left(\frac{n\pi}{b}\right)^2 \right]^2 \qquad (9.93)$$

We may introduce a so-called critical stress τ_{cr} by dividing through by h in the above equation. Replacing D by its basic definition (see Eq. (6.13)) we then have:

$$\tau_{\text{cr}} = \frac{\pi^2 E}{12(1 - v^2)} \left(\frac{h}{b}\right)^2 \left[\left(\frac{mb}{a}\right)^2 + 2n^2 + n^4\left(\frac{a}{mb}\right)^2 \right] \qquad (9.94)$$

We have here a double infinity of discrete values of $(\bar{N}_x)_{\text{cr}}$ or τ_{cr}. Whereas for columns it was easy to ascertain by inspecting which critical load would be lowest and thus become the buckling load, this is no longer the case here. It is clear immediately on inspection of Eq. (9.94) that τ_{cr} increases with n for any value of a/b, which we shall now term the *aspect ratio*. This is not so for m. Accordingly as a first step in our effort to find the minimum buckling stress we set n equal to its smallest value, (i.e.,

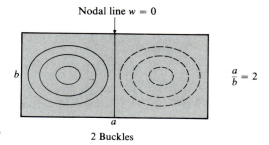

Nodal line $w = 0$

b

$\dfrac{a}{b} = 2$

a

FIGURE 9.19   2 Buckles

unity). We can then give Eq. (9.94) in the following form

$$\tau_{cr} = \frac{Ek_{cr}\pi^2}{12(1 - v^2)}\left(\frac{h}{b}\right)^2 \tag{9.95}$$

where k_{cr}, termed the *buckling coefficient*, has the form:

$$k_{cr} = \left(m\frac{b}{a} + \frac{1}{m}\frac{a}{b}\right)^2 \tag{9.96}$$

We have shown k_{cr} plotted against the aspect ratio a/b for different values of m in Fig. 9.18. It is clear on inspecting the diagram that the value of m giving the smallest buckling coefficient depends on the aspect ratio. Thus for $a/b = 1$ we see that $m = 1$, but, for $a/b = 2$, m must be 2. As a/b gets larger, notice that the critical buckling coefficient approaches the value 4. Establishing m and n for the lowest critical loads for a given ratio a/b thus establishes the buckling mode for the plate. For instance with $a/b = 2$ and $m = 2$ there must be a nodal line at $x = a/2$ (see Fig. 9.19) and we say that the plate has two buckles. The curved lines in the diagram may be thought of as contour lines of the buckled shape, the full lines indicating an upward deflection, the dashed lines indicating a downward deflection. For $m = 3$ there will be three such buckles, etc. Finally notice from Fig. 9.18 that when a/b is an integer, then $m = a/b$ for achieving the minimum value of k_{cr}.[1] Thus for such cases the length of the plate (x direction) is divided into m buckles, i.e., m half waves of length b.

---

[1] We can reach the same conclusion by treating m as a continuous variable and we then minimize k_{cr} with respect to m. Thus:

$$\frac{\partial k_{cr}}{\partial m} = 0$$

This gives us $m = a/b$.

9.14 THE RECTANGULAR PLATE VIA THE ENERGY METHOD

We now consider again the same simply-supported rectangular plate examined in the previous section (see Fig. 9.17). We will be concerned with the positive definiteness of the *second variation of the total potential energy*, $\delta^{(2)}\pi$, in this method.

As shown in Sec. 9.5 in dealing with a single parameter approach, the second variation of a functional of η may be considered to have the same form as the functional itself and so we may state for $\delta^{(2)}\pi$:

$$\delta^{(2)}\pi = \int_0^a \int_0^b \left\{ D\left[(\nabla^2\eta)^2 + 2(1-v)\left[\left(\frac{\partial^2\eta}{\partial x \partial y}\right)^2 - \left(\frac{\partial^2\eta}{\partial x^2}\right)\left(\frac{\partial^2\eta}{\partial y^2}\right) \right] \right] \right\} dx\,dy$$
$$- \int_0^a \int_0^b \left\{ \bar{N}_x \left(\frac{\partial\eta}{\partial y}\right)^2 \right\} dx\,dy \tag{9.97}$$

We may represent η now as a double sine series as follows:

$$\eta = \sum_{n=1}^{\infty} \sum_{m=1}^{\infty} A_{mn} \sin\frac{m\pi x}{a} \sin\frac{n\pi y}{b} \tag{9.98}$$

Note that the boundary conditions for η must be the same as for w and so the above series satisfies the boundary conditions. Now substitute the above result into Eq. (9.97). We get, on noting the orthogonality properties of the sine functions:

$$\delta^{(2)}\pi = \frac{\pi^2}{4} \sum_{m=1}^{\infty} \sum_{n=1}^{\infty} \left[\left\{ D\pi^2 ab\left(\frac{m^2}{a^2} + \frac{n^2}{b^2}\right)^2 - \frac{b\bar{N}_x}{a}m^2 \right\} A_{mn}{}^2 \right] \tag{9.99}$$

Since the constants A_{mn} are independent of each other it should be clear that in order for $\delta^{(2)}\pi$ to be positive-definite then *each coefficient of the A_{mn} must be positive.* Accordingly *when any coefficient becomes zero $\delta^{(2)}\pi$ is no longer positive-definite.* This leads us to the same formulation $(\bar{N}_x)_{cr}$ arrived at in the previous section (see Eq. (9.93)).

9.15 THE CIRCULAR PLATE VIA THE ENERGY METHOD

We now consider the axisymmetric buckling of a circular plate (see Fig. 9.20). A uniform loading intensity $\bar{N}_v = N$ is applied normal to the periphery as has been shown in the diagram.

We may express the total potential energy with the aid of Eq. (6.114) as follows:

$$\pi = (U+V) = \int_0^{2\pi} \int_0^a \frac{D}{2} \left\{ \left(\frac{\partial^2 w}{\partial r^2} + \frac{1}{r}\frac{\partial w}{\partial r}\right)^2 - 2(1-v)\frac{\partial^2 w}{\partial r^2}\frac{1}{r}\frac{\partial w}{\partial r} \right\} dr\, r\, d\theta$$
$$+ \int_0^{2\pi} N(u_r r)_{r=a}\, d\theta \tag{9.100}$$

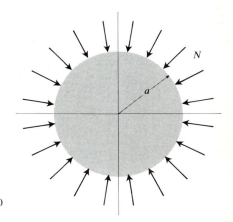

FIGURE 9.20

On integrating with respect to θ we may reformulate this expression as follows

$$(U + V) = \pi D \int_0^a \left[r \frac{\partial^2 w}{\partial r^2} + \frac{1}{r}\left(\frac{\partial w}{\partial r}\right)^2 + 2v\frac{\partial w}{\partial r}\frac{\partial^2 w}{\partial r^2} \right] dr + 2\pi D\alpha^2(ru_r)_{r=a} \qquad (9.101)$$

where $\alpha^2 = N/D$. We now examine the expression $(u_r r)_{r=a}$. This may be written as follows:

$$(ru_r)_{r=a} = \int_0^a \frac{d}{dr}(ru_r)\, dr$$

$$= \int_0^a r\frac{du_r}{dr}\, dr + \int_0^a u_r\, dr$$

$$= \int_0^a r\frac{du_r}{dr}\, dr + \int_0^a r\frac{u_r}{r}\, dr \qquad (9.102)$$

The in-plane strain displacement relations for cylindrical coordinates comparable to the relations used here for rectangular coordinates may be given as follows:

$$\epsilon_{rr} = \frac{du_r}{dr} + \frac{1}{2}\left(\frac{dw}{dr}\right)^2$$

$$\epsilon_{\theta\theta} = \frac{u_r}{r} \qquad (9.103)$$

If we set the in-plane strains equal to zero as in the previous undertaking then we see that:

$$\frac{du_r}{dr} = -\frac{1}{2}\left(\frac{dw}{dr}\right)^2$$

$$\frac{u_r}{r} = 0 \qquad (9.104)$$

Substituting these results into Eq. (9.102) we get:

$$(ru_r)_{r=a} = -\tfrac{1}{2}\int_0^a r\left(\frac{dw}{dr}\right)^2 dr$$

The total potential energy then becomes:

$$(U + V) = \pi D \int_0^a \left[rw_{rr}^2 + \frac{1}{r}(w_r)^2 + 2vw_r w_{rr} - \alpha^2 rw_r^2 \right] dr$$

The second variation then can be given as:

$$\delta^{(2)}\pi = 2\pi D \int_0^a \left[r\eta_{rr}^2 + \frac{1}{r}\eta_r^2 + 2v\eta_r\eta_{rr} - \alpha^2 r\eta_r^2 \right] dr$$

Employing the *Trefftz criterion* we extremize the above functional. The Euler–Lagrange equation for this case will be established with the aid of Eq. (2.62). Thus

$$-\frac{d^2}{dr^2}\frac{\partial F}{\partial \eta_{rr}} + \frac{d}{dr}\frac{\partial F}{\partial \eta_r} = 0 \qquad (9.105)$$

where:

$$F = r\eta_{rr}^2 + \frac{1}{r}\eta_r^2 + 2v\eta_r\eta_{rr} - \alpha^2 r\eta_r^2 \qquad (9.106)$$

We may perform one quadrature immediately on Eq. (9.105) to get:

$$-\frac{d}{dr}\frac{\partial F}{\partial \eta_{rr}} + \frac{\partial F}{\partial \eta_r} = C_1 \qquad (9.107)$$

Now substituting for F using Eq. (9.106) we get for the above equation:

$$r\eta_{rrr} + \eta_{rr} + \left(\alpha^2 r - \frac{1}{r}\right)\eta_r = C_1 \qquad (9.108)$$

The boundary conditions on η may be deduced from Eq. (2.63) to be at $r = 0$ and $r = a$:

$$\frac{\partial F}{\partial \eta_{rr}} = 0 \quad \text{OR} \quad \eta_r \text{ SPECIFIED} \qquad (a)$$

$$-\frac{d}{dr}\left(\frac{\partial F}{\partial \eta_{rr}}\right) + \frac{\partial F}{\partial \eta_r} = 0 \quad \text{OR} \quad \eta \text{ SPECIFIED} \qquad (b) \quad (9.109)$$

At this time evaluate the natural boundary condition (b) above using Eq. (9.106) for F. We get:

$$r\eta_{rrr} + \eta_{rr} + \left(\alpha^2 r - \frac{1}{r}\right)\eta_r = 0$$

at the end points. We see immediately on examining Eq. (9.108) that the constant C_1 must be zero as a result of the above boundary condition applied at $r = a$. Now consider the resulting differential equation (9.108) further. Denoting η_r as G and multiplying through by r we have:

$$r^2 G_{rr} + rG_r + (\alpha^2 r^2 - 1)G = 0 \qquad (9.110)$$

This is *Bessel's equation* of the first order. The solution which is regular at the origin for the above equation is then:

$$G = C_2 J_1(\alpha r) \qquad (9.111)$$

Since $G(=\eta_r) = 0$ at $r = 0$, we see that the kinematic boundary condition (9.109(a)) is now also satisfied at $r = 0$.[1] In getting to η itself we now consider the boundary conditions further.

For a *clamped plate* we know that $\eta = \eta_r = 0$ at $r = a$. We may readily ensure that $\eta = 0$ at $r = a$ by choosing the integration constant properly when we integrate Eq. (9.111) to get η. As for $\eta_r = 0$ at $r = a$ we require that:

$$G(a) = J_1(\alpha a) = 0$$

Choosing the first positive root of J_1 we find that:

$$\alpha a = 3.832$$

Accordingly the buckling loading N_{cr} is given as:

$$N_{cr} = D\alpha^2 = \frac{14.65D}{a^2} \qquad (9.112)$$

Now consider the *simply-supported* plate. The condition that $\eta = 0$ at $r = a$ can again be satisfied by choosing the integration constant properly on integrating Eq. (9.111) for η. The other condition that must be satisfied is the natural boundary condition (9.109(a)). That is, η must be chosen so that:

$$\frac{\partial F}{\partial \eta_{rr}} = 0 \qquad \text{at } r = a$$

Substituting for F in the above equation we get:

$$2r\eta_{rr} + 2v\eta_r = 0 \qquad \text{at } r = a$$

Substituting for η_r using Eq. (9.111) we get:

$$r\frac{dJ_1(\alpha r)}{dr} + vJ_1(\alpha r) = 0 \qquad \text{at } r = a$$

---

[1] This results from axial symmetry and the exclusion of discontinuities of slope.

Differentiating the Bessel function[1] we get:

$$\alpha a J_0(\alpha a) - J_1(\alpha a) + \nu J_1(\alpha a) = 0$$

therefore

$$J_1(\alpha a)(1 - \nu) = \alpha a J_0(\alpha a)$$

Taking $\nu = 0.3$, the smallest root for the above equation is $\alpha a = 2.049$. Hence, the buckling load in this situation becomes:

$$N_{\mathrm{cr}} = 4.20 \frac{D}{a^2} \qquad (9.113)$$

Part D
APPROXIMATION METHODS

9.16   COMMENT

We shall now consider certain aspects of the approximate solution of buckling loads and buckling modes via variational methods. Primarily we shall extend the techniques formulated in Chap. 7 for finding approximate natural frequencies (eigenvalues for the vibration problem) to that of finding approximate critical loads (eigenvalues of the elastic stability problem). We shall first establish the Rayleigh quotient for column problems and plate problems and with the background material of Chap. 7 as a basis we shall then employ the Rayleigh and the Rayleigh–Ritz methods of approximation. After this we shall consider the Kantorovich method.

9.17   THE RAYLEIGH QUOTIENT FOR BEAM-COLUMNS

We shall first formulate the Rayleigh quotient for the column following procedures analogous to those set forth in Part C of Chap. 7. We accordingly begin with the differential equation (Eq. (9.20)) for the column:

$$\frac{d^2}{dx^2}\left(EI\frac{d^2w}{dx^2}\right) = -P\frac{d^2w}{dx^2} \qquad (9.114)$$

Using the form for a differential equation given by Eq. (7.144) namely:

$$L(w) = \omega^2 M(w) \qquad (9.115)$$

---

[1] See any standard advanced mathematics text such as "Advanced Engineering Mathematics," C. R. Wylie, McGraw-Hill Book Co., Chap. 8 for a discussion of Bessel functions including differentiation properties.

it is clear that for the column:

$$L \equiv \frac{d^2}{dx^2} EI \frac{d^2}{dx^2}$$

$$M \equiv -\frac{d^2}{dx^2}$$

$$\omega^2 = P \qquad (9.116)$$

We have already shown in Sec. 3.11 that $(d^2/dx^2)EI(d^2/dx^2)$ must be a self-adjoint, positive-definite operator for the usual end supports of beams (and hence columns) and we leave it to you as an exercise to show that $-d^2/dx^2$ similarly is a self-adjoint, positive-definite operator. We can then conclude (see Sec. 7.13) that the eigenfunctions of the boundary-value problem involving Eq. (9.114) and the usual end conditions must be orthogonal. The Rayleigh quotient R for a problem having *constant product EI* is then given as:

$$R = \frac{\int_0^L wL(w)\,dx}{\int_0^L wM(w)\,dx} = -\frac{EI \int_0^L w\frac{d^4w}{dx^4}\,dx}{\int_0^L w\frac{d^2w}{dx^2}\,dx} \qquad (9.117)$$

We may rewrite the Rayleigh quotient to have better form if we integrate the expression in the denominator by parts. Thus:

$$-\int_0^L w\frac{d^2w}{dx^2}\,dx = \int_0^L \left(\frac{dw}{dx}\right)^2 dx - w\frac{dw}{dx}\Big|_0^L \qquad (9.118)$$

The last expression vanishes for fixed or simple supports at the ends of the columns. Also we shall integrate the numerator of Eq. (9.117) twice by parts. Thus:

$$\int_0^L w\frac{d^4w}{dx^4}\,dx = \int_0^L \left(\frac{d^2w}{dx^2}\right)^2 dx + w\frac{d^3w}{dx^3}\Big|_0^L - \frac{dw}{dx}\frac{d^2w}{dx^2}\Big|_0^L \qquad (9.119)$$

Again for fixed or simply-supported end conditions the last two expressions on the right side of the above equation vanish. We thus have upon using Eqs. (9.118) and (9.119) to replace the denominator and numerator respectively for the Rayleigh quotient of Eq. (9.117):

$$R = \frac{EI \int_0^L (d^2w/dx^2)^2\,dx}{\int_0^L (dw/dx)^2\,dx} \qquad (9.120)$$

We can then say for P:

$$P = \frac{EI \int_0^L (d^2w/dx^2)^2\,dx}{\int_0^L (dw/dx)^2\,dx} \qquad (9.121)$$

9.18   THE RAYLEIGH AND THE RAYLEIGH–RITZ METHODS APPLIED TO COLUMNS

In accordance with Sec. 7.15 we may find approximations from above of the buckling load P_{cr} using the Rayleigh method on the Rayleigh quotient functional. As a first step we illustrate the use of the Rayleigh method for the case of the column clamped at $x = 0$ and pinned at $x = L$.

We choose as the approximate function

$$w_{app} = a_1\left[\left(\frac{x}{L}\right)^2 - \left(\frac{x}{L}\right)^3\right] \qquad (9.122)$$

Note that this function satisfies the *boundary* conditions $w(0) = w'(0) = w(L) = 0$ *but not the natural boundary condition* $w''(L) = 0$ of the problem. Substituting the above approximation into Eq. (9.121) we get:

$$P_1 = \frac{EI(4a^2/L^3)}{(0.133a^2/L)} = 30\frac{EI}{L^2} \qquad (9.123)$$

The exact buckling load is $20.19EI/L^2$ and we have accordingly poor correlation. We can improve the accuracy of the result by choosing a function that satisfies the complete set of boundary conditions of the problem. (Recall we reached this same conclusion in Chap. 4 in our discussion of the Ritz method applied to beams.) We choose for w_1 now the function:

$$w_1 = a_1\left[\left(\frac{x}{L}\right)^2 - \frac{5}{3}\left(\frac{x}{L}\right)^3 + \frac{2}{3}\left(\frac{x}{L}\right)^4\right] \qquad (9.124)$$

The complete set of boundary conditions (including $w''(L) = 0$) is now satisfied. Substituting the above function into Eq. (9.121) we now get:

$$P_{cr} = 21.0EI/L^2$$

The agreement is now good.

Next we employ the Rayleigh–Ritz method to find a particularly good approximation of the buckling load and to find approximations *from above* of critical loads above the first. The basis for this procedure has been established in Sec. 7.6 where for an approximate eigenfunction w_{app} we employed a parameter-laden sum of coordinate functions:

$$w_{app} = A_1\phi_1 + A_2\phi_2 + \cdots + A_n\phi_n$$

We thus have for the approximate Rayleigh quotient:

$$R_{app} = \Lambda^2 = \frac{EI\displaystyle\int_0^L \left(\sum_{i=1}^n A_i\phi_i''\right)^2 dx}{\displaystyle\int_0^L \left(\sum_{i=1}^n A_i\phi_i'\right)^2 dx}$$

We introduce the notation

$$a_{ij} = \int_0^L EI\phi_i'' \phi_j'' \, dx$$

$$b_{ij} = \int_0^L \phi_i' \phi_j' \, dx \qquad (9.125)$$

so that Λ^2 becomes:

$$\Lambda^2 = \frac{\sum\limits_{i=1}^{n} \sum\limits_{j=1}^{n} a_{ij} A_i A_j}{\sum\limits_{i=1}^{n} \sum\limits_{j=1}^{n} b_{ij} A_i A_j}$$

In extremizing Λ^2 with respect to the constants A_i we have shown in Sec. 7.6 that we reach the following requirement:

$$|a_{ij} - \Lambda^2 b_{ij}| = 0 \qquad (9.126)$$

The roots $\Lambda_1^2, \ldots, \Lambda_n^2$ are the approximations from above of the first n critical loads.

We illustrate this calculation for a simply-supported column by considering for w_n the following functions

$$w_2 = A_1(\cos \xi - \cos 3\xi) + A_2(\cos 3\xi - \cos 5\xi)$$

$$w_3 = A_1(\cos \xi - \cos 3\xi) + A_2(\cos 3\xi - \cos 5\xi) + A_3(\cos 5\xi - \cos 7\xi)$$

$$\vdots$$

$$w_n = A_1(\cos \xi - \cos 3\xi) + \cdots + A_n[\cos(2n - 1)\xi - \cos(2n + 1)\xi] \qquad (9.127)$$

where $\xi = \pi x/2L$. Notice that only the kinematic boundary conditions $w = 0$ at the ends are satisfied by the above functions. Substituting the above functions into (9.125) and then using these results in Eq. (9.126) we get the results given by Table 9.1. Notice that as n goes up the approximation for the lower critical values improves. Note also *that the approximations are from above.*

We now turn our attention to plates, in particular the rectangular plate.

Table 9.1

n	$P_1 L^2/EI$	$P_2 L^2/EI$	$P_3 L^2/EI$
1	20.22		
2	20.202	59.726	
3	20.195	59.695	118.95
Exact	20.19	59.67	118.90

9.19 RAYLEIGH QUOTIENT FOR RECTANGULAR PLATES

The differential equation for buckling of a plate was shown in Sec. 9.12 to be

$$D\nabla^4 w = -\bar{N}_x \frac{\partial^2 w}{\partial x^2} - 2\bar{N}_{xy} \frac{\partial^2 w}{\partial x\,\partial y} - \bar{N}_y \frac{\partial^2 w}{\partial y^2} \qquad (9.128)$$

where the barred quantities are the edge loads which ultimately cause buckling. If these loads are developed so as to have at all times *fixed ratios between each other* and if the loads are constant on each edge then we can give the above equations in the following form:

$$D\nabla^4 w = -P^* \left[\alpha \frac{\partial^2 w}{\partial x^2} + 2\beta \frac{\partial^2 w}{\partial x\,\partial y} + \gamma \frac{\partial^2 w}{\partial y^2} \right] \qquad (9.129)$$

where P^* will be called the loading factor, a *variable* parameter, and α, β, and γ are constants. Then we can say:

$$L \equiv D\nabla^4$$

$$M \equiv -\left[\alpha \frac{\partial^2}{\partial x^2} + 2\beta \frac{\partial^2}{\partial x\,\partial y} + \gamma \frac{\partial^2}{\partial y^2} \right]$$

$$\omega^2 = P^* \qquad (9.130)$$

We may then state as a result of our discussion in Sec. 7.14:

$$P^* = \frac{D \iint w\nabla^4 w \, dx\, dy}{-\iint \left(\alpha \dfrac{\partial^2 w}{\partial x^2} + 2\beta \dfrac{\partial^2 w}{\partial x\,\partial y} + \gamma \dfrac{\partial^2 w}{\partial y^2} \right) w \, dx\, dy} \qquad (9.131)$$

We shall now consider the use of Rayleigh's method for solving the buckling problem of a simply-supported plate $a \times b$ in uniaxial compression (see Fig. 9.17). The Rayleigh quotient for this case may be given as follows, noting that $P^* = \bar{N}$, $\alpha = 1$, and $\beta = \gamma = 0$:

$$\bar{N} = \frac{D \iint w\nabla^4 w \, dx\, dy}{-\iint w(\partial^2 w/\partial x^2) \, dx\, dy}$$

By using Green's theorem to integrate by parts both integrals in the above equation (see Problem 6.13 for numerator) we find on considering the boundary conditions that:

$$\bar{N} = \frac{D \iint (\nabla^2 w)^2 \, dx\, dy}{\iint (\partial w/\partial x)^2 \, dx\, dy} \qquad (9.132)$$

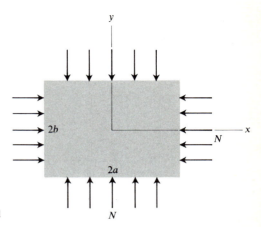

FIGURE 9.21

We consider the case where $a/b = 1$, i.e., the square plate. As an approximate buckling shape we shall use the following function

$$w = C_1 xy(x - a)(y - b)$$

so the *kinematic boundary condition* $w = 0$ is satisfied. Substituting into Eq. (9.132) we have for $\overline{N}_{cr}$:

$$\overline{N}_{cr} = \frac{44.0D}{a^2}$$

The exact value from $\overline{N}_{cr}$ as developed in Sec. 9.13 is given as:

$$(\overline{N}_{cr})_{Ex} = \frac{39.5D}{a^2}$$

The selected function does not satisfy the natural boundary conditions and an error of 11 % results.

9.20   THE KANTOROVICH METHOD

We next consider a rectangular plate $a \times b \times h$ clamped on all edges and loaded by a uniform force distribution N normal to the edges (see Fig. 9.21). For the study of the elastic stability of such a problem we may take P^* in Eq. (9.129) as N with $\alpha = \gamma = 1$ and $\beta = 0$. We then have for the appropriate differential equation the following result with x and y as the independent variables

$$\nabla^4 w + \lambda^2 \nabla^2 w = 0 \qquad (9.133)$$

where $\lambda^2 = N/D$. The boundary conditions are easily seen to be:

$$w = \frac{\partial w}{\partial x} = 0 \qquad \text{at } x = \pm a \qquad (a)$$

$$w = \frac{\partial w}{\partial y} = 0 \qquad \text{at } y = \pm b \qquad (b) \quad (9.134)$$

As pointed out in Sec. 6.9 on the static deflection of a rectangular plate, the *Kantorovich method is equivalent to rendering the Galerkin integral equal to zero.* Thus for a one term approximation with $g_0(y)$ as a known function satisfying boundary conditions (9.134(b)) and f_1 as the unknown function we have

$$w_{10}(x,y) = f_1(x)g_0(y) \qquad (9.135)$$

We then have on setting the Galerkin integral equal to zero:

$$\int_{-a}^{a} \int_{-b}^{b} (\nabla^4 w_{10} + \lambda^2 \nabla^2 w_{10}) g_0 \, dx \, dy = 0 \qquad (9.136)$$

As a choice for w_{10} we have:

$$w_{10} = f_1(x)(y^2 - b^2)^2 \qquad (9.137)$$

Substituting Eq. (9.137) into the Galerkin integral we get after integrating and dividing through by b^5:

$$\left(b^4 \frac{256}{315}\right) \frac{d^4 f_1}{dx^4} + b^2 \left[(\lambda b)^2 \frac{256}{315} - \frac{512}{105}\right] \frac{d^2 f_1}{dx^2} - \left[(\lambda b)^2 \frac{256}{105} - \frac{128}{5}\right] f_1 = 0 \qquad (9.138)$$

We have here an ordinary fourth-order differential equation with constant coefficients. The procedure is to consider a trial solution e^{mx} to form a characteristic equation for m. In substitution we get:

$$b^4 m^4 + b^2 [(\lambda b)^2 - 6] m^2 - \left[3(\lambda b)^2 - \frac{63}{2}\right] = 0$$

Since the magnitude of (λb) is not known, the roots may be real, imaginary, or complex. Consider the roots m^2 using the quadratic formula. We get:

$$m^2 = \frac{-[(\lambda b)^2 - 6] \pm \sqrt{[(\lambda b)^2 - 6]^2 + [3(\lambda b)^2 - 63/2]4}}{2b^2} \qquad (9.139)$$

If $(\lambda b)^2 = 21/2$ the second bracketed expression in the above root is zero and the magnitude of the square root equals $(\lambda b)^2 - 6$. We then have as roots for m^2:

$$m^2 = 0 \qquad m^2 = -\left[\frac{(\lambda b)^2 - 6}{b^2}\right]$$

The values of m are then either zero or are pure imaginary numbers. If now

$$(\lambda b)^2 > \frac{21}{2} \qquad (9.140)$$

the square root in Eq. (9.139) exceeds the first term $[(\lambda b)^2 - 6]$ in the numerator and we have accordingly for possible roots m, two real numbers and two imaginary numbers. For cases where $(\lambda b)^2 < 21/2$ the magnitude of the root, if it is real, is less than $(\lambda b)^2 - 6$ and so we get four pure imaginary values of m. Also there is the possibility of getting four complex values of m if the root is imaginary. Now the only kinds of roots that yield critical values of N are those corresponding to conditions (9.140)—i.e., two real roots and two imaginary roots for m. The roots for m will accordingly be denoted as follows

$$m_1 = m$$
$$m_2 = -m$$
$$m_3 = i\Gamma$$
$$m_4 = -i\Gamma$$

where using positive roots we have from Eq. (9.139):

$$m = \frac{1}{b}\left[\left\{\left[\frac{(\lambda b)^2}{2} - 3\right]^2 + \left[3(\lambda b)^2 - \frac{63}{2}\right]\right\}^{1/2} - \left\{\left[\frac{(\lambda b)^2}{2} - 3\right]\right\}\right]^{1/2}$$

$$\Gamma = \frac{1}{b}\left[\left\{\left[\frac{(\lambda b)^2}{2} - 3\right]^2 + \left[3(\lambda b)^2 - \frac{63}{2}\right]\right\}^{1/2} + \left\{\left[\frac{(\lambda b)^2}{2} - 3\right]\right\}\right]^{1/2} \qquad (9.141)$$

We thus have as a solution to the differential equation (9.138):

$$f_1(x) = A_1 \sinh mx + A_2 \cosh mx + A_3 \sin \Gamma x + A_4 \cos \Gamma x$$

To non-dimensionalize the above result replace mx by $(am)(x/a)$ and denote am as ρ_1, and replace Γx by $(a\Gamma)(x/a)$ and denote $a\Gamma$ simply as κ_1. Thus we have for $f_1(x)$

$$f_1(x) = A_1 \sinh \rho_1 \frac{x}{a} + A_2 \cosh \rho_1 \frac{x}{a} + A_3 \sin \kappa_1 \frac{x}{a} + A_4 \cos \kappa_1 \frac{x}{a}$$

where on considering Eq. (9.141) we have

$$\left\{\begin{matrix} \rho_1 \\ \kappa_1 \end{matrix}\right\} = \frac{a}{b}\left[\left\{\left[\frac{(\lambda b)^2}{2} - 3\right]^2 + \left[3(\lambda b)^2 - \frac{63}{2}\right]\right\}^{1/2} \mp \left\{\left[\frac{(\lambda b)^2}{2} - 3\right]\right\}\right]^{1/2} \qquad (9.142)$$

For critical values of N corresponding to mode shapes which are *symmetric* with respect to both the x and y axes we now set A_1 and A_3 equal to zero. Thus we have:

$$f_1(x) = A_2 \cosh \rho_1 \frac{x}{a} + A_4 \cos \kappa_1 \frac{x}{a} \qquad (9.143)$$

The boundary conditions in the x direction now require that at $x = \pm a$:

$$f_1 = 0 \qquad (a)$$

$$\frac{df_1}{dx} = 0 \qquad (b) \ (9.144)$$

We obtain:

$$A_2 \cosh \rho_1 + A_4 \cos \kappa_1 = 0 \qquad (a)$$

$$A_2 \rho_1 \sinh \rho_1 - A_4 \kappa_1 \sin \kappa_1 = 0 \qquad (b) \ (9.145)$$

For a nontrivial solution the determinant of the constants must be zero. This results in the following requirement:

$$\kappa_1 \cosh \rho_1 \sin \kappa_1 = -\rho_1 \sinh \rho_1 \cos \kappa_1$$

therefore

$$\kappa_1 \tan \kappa_1 = -\rho_1 \tanh \rho_1 \qquad (9.146)$$

For the case of a square plate ($a = b$) we next find from Eqs. (9.142) the values of $(\lambda b)^2$ that give proper κ_1 and ρ_1 which permit the satisfaction of the above equation. We may show with the aid of a computer or by graphical representation that

$$(\lambda b)^2 = 13.29, \quad 40.00, \quad 88.70, \quad \ldots$$

The buckling load N_{cr} then becomes:

$$N_{cr} = \lambda^2 D = \frac{13.29}{a^2} D \qquad (9.147)$$

As for obtaining an approximation of the *buckling mode shape* corresponding to the above result, note that the approximate values of ρ_1 and κ_1 for $(\lambda b)^2 = 13.29$ are

$$\rho_1 = 1.0046$$

$$\kappa_1 = 2.8814 \qquad (9.148)$$

Now substituting these values into Eq. (9.145(a)) we can find A_4 in terms of A_2. We can then say for w_{10}

$$w_{10} = A_2 \left[\cos (\kappa_1) \cosh \left(\rho_1 \frac{x}{a} \right) - \cosh (\rho_1) \cos \left(\kappa_1 \frac{x}{a} \right) \right] (y^2 - b^2)^2$$

As for the *extended Kantorovich method* the procedure is as follows. Start with a function $w(x,y)$ in the following form:

$$w_{10} = f_1(x) g_0(y) \qquad (9.149)$$

where $g_0(y)$ satisfies the boundary conditions and is known a priori. Solve for $f_1(x)$ exactly as we have just done in the regular Kantorovich method. Now using this function as a known function, consider a new deflection surface w_{11} defined as follows

$$w_{11} = f_1(x)g_1(y)$$

where $g_1(y)$ is now undetermined. Solve for $g_1(y)$ by the usual Kantorovich method. Now using $g_1(y)$ as known we form

$$w_{21} = f_2(x)g_1(y)$$

where $f_2(x)$ is to be determined, etc.[1]

READING

BLEICH, F.: "Buckling Strength of Metal Structures," McGraw-Hill Book Co., N.Y., 1952.

BOLOTIN, V. V.: "Dynamic Stability of Elastic Systems," Holden-Day, Inc., San Francisco, 1964.

BOLOTIN, V. V.: "Nonconservative Problems of the Theory of Elastic Systems," Pergamon Press, 1963.

COX, H. L.: "The Buckling of Plates and Shells," Pergamon Press, 1961.

FLÜGGE, W.: "Handbook of Engineering Mechanics", "Elastic Stability" by C. Libove, 44-1, McGraw-Hill Book Co., N.Y., 1962.

GERARD, G.: "Introduction to Structural Stability Theory," McGraw-Hill Book Co., N.Y.

HERRMANN, G. (Ed.): "Dynamics Stability of Structures," Pergamon Press, N.Y., 1963.

LANGHAAR, H. L.: "Energy Methods in Applied Mechanics," John Wiley and Sons, N.Y., 1962.

TIMOSHENKO, S. P. AND GERE, J. M.: "Theory of Elastic Stability," McGraw-Hill Book Co., N.Y., 1961.

ZIEGLER, H.: "Principles of Structural Stability," Blaisdell Publishers, Waltham, Mass., 1968.

PROBLEMS

9.1 Show that the buckling load of a column clamped at both ends is $P = 4P_E = 4\pi^2 EI/L^2$.

9.2 Obtain an approximate (Rayleigh–Ritz) solution for the buckling load of a clamped–clamped column.

9.3 Show that the buckling load of a column clamped at $x = 0$ and free at $x = L$ is $P = P_E/4 = \pi^2 EI/4L^2$.

9.4 Obtain an approximate (Rayleigh–Ritz) solution for the buckling load of a clamped–free column.

9.5 Consider the clamped–free column loaded as shown in Fig. 9.22. Here we take the load P not as the usual axial force, but directed along a line through $x = 0$. Explain why the buckling load of this problem is equal to that of a pinned–pinned column.

9.6 Consider now a clamped–free column subjected to a load P constrained to remain tangential to the free end of the column as it deforms (Fig. 9.23). You may verify that this is a nonconservative problem, as the loading is path-dependent. What are the boundary conditions at $x = L$? Show that the solution for the Euler equation subject to the boundary

---

[1] For a thorough discussion of the Kantorovich method and its extension for this problem see "An Extended Kantorovich Method for the Solution of Eigenvalue Problems," by Arnold D. Kerr, *International Journal of Solids and Structures*, **5**, 559–572 (1969).

FIGURE 9.22

conditions for this problem leads to the characteristic equation

$$\begin{vmatrix} k & 0 & 1 \\ \sin kl & \cos kl & 0 \\ \cos kl & -\sin kl & 0 \end{vmatrix} = 0$$

What does this imply about the buckling load? Why do we obtain this result?

9.7   Solve problem 9.6 using the kinetic method. This leads to the characteristic equation

$$0 = k^4 + 2\omega^2 + 2\omega^2 \cos \lambda_1 L \cosh \lambda_2 L + \omega k^2 \sin \lambda_1 L \sinh \lambda_2 L$$

where

$$\lambda_{1,2}{}^2 = \pm \frac{k^2}{2} + \sqrt{\left(\frac{k^2}{2}\right)^2 + \omega^2}$$

with

$$k^2 = P/EI \quad \text{and} \quad \omega^2 = \rho A \Omega^2 / EI,$$

Ω being the actual frequency of the vibrating column. The lowest root may be obtained as $P_{cr} \cong 2\pi^2 EI/L^2$. Thus the usefulness of the kinetic method for nonconservative problems has been demonstrated.

Also note that for $P \to 0$, i.e., $k^2 \to 0$, the characteristic equation is that for the frequency of an unloaded clamped–free bar, i.e.,

$$\cos \sqrt{\omega} L \cosh \sqrt{\omega} L = -1$$

9.8   Find the buckling load of a clamped–free column (standing vertically!) acted upon by its own weight. Show that the differential equation can be reduced to

$$EI \frac{d^3 w}{dx^3} = -q(L - x)\frac{dw}{dx}$$

where q is a uniformly distributed load. The calculation can be simplified by introducing the variables

$$z = \tfrac{2}{3}\sqrt{q(L - x)^3 / EI}$$

and $u = dw/dz$ so that the governing equation becomes a Bessel equation, i.e.,

$$\frac{d^2 u}{dz^2} + \frac{1}{z}\frac{du}{dz} + \left(1 - \frac{1}{9z^2}\right)u = 0$$

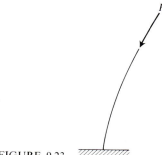

FIGURE 9.23

Then using appropriate tables of Bessel functions, satisfying appropriate boundary conditions, find the critical load

$$(qL)_{cr} = 7.837EI/L^2$$

9.9   Show that the energy functional for the previous problem (a column buckling under its own weight) is

$$U + V = \frac{EI}{2} \int_0^L \left(\frac{d^2w}{dx^2}\right)^2 dx - \frac{q}{2} \int_0^L (L - x)\frac{dw}{dx} dx$$

9.10   Obtain an approximate buckling load via the Rayleigh–Ritz method for the buckling of the clamped–free column under its own weight. (The exact answer has been obtained in Problem 9.8.)

9.11   Obtain the buckling load for a pinned–pinned column supported by a uniform elastic foundation with modulus K. The critical load is given by

$$P_{cr} = P_E\left(m^2 + \frac{KL^4}{m^2\pi^4EI}\right)$$

where m is the axial wave number. Show that for $KL^4/\pi^4EI < 4$, that the column buckles with only 1 wave ($m = 1$) and $P = P_E$; for $KL^4/\pi^4EI > 4$, show that 2 waves occur.

9.12   Use a Galerkin integral to obtain the buckling load for a simply-supported rectangular plate, under loading $\overline{N}_x = N_x{}^0(y/b)$, buckling into m waves in the x direction, and 1 wave in the y direction. What is the minimum value of the buckling coefficient

$$k_{x_\Delta} = N_x{}^0 b^2/\pi^2 D \ ?$$

9.13   For a square plate, simply supported, extend the above results (Problem 9.12) to get a two-term solution for $m = n = 1$, and $m = 2$, $n = 1$. Does this improve the accuracy of the results, or point to another possible buckling load and mode?

9.14   Attempt to improve the result of Problem 9.12 by utilizing a two-term expansion that includes $m = 1$, $n = 1$ and $m = 1$, $n = 2$. Does this change the result of problem 9.12 significantly? (Do this calculation for the rectangular plate, i.e., $a \neq b$.)

9.15   Derive an interaction formula for bi-axial buckling under uniform loads N_x^{0} and N_y^{0} for a rectangular plate that is simply supported. Construct the interaction curve for a square plate.

9.16   Compute the shear buckling coefficient for a long, simply-supported, rectangular plate. As an approximation use

$$w = w_{m1} \sin\frac{m\pi x}{a} \sin\frac{\pi y}{b} + w_{m2} \cos\frac{m\pi x}{a} \sin\frac{2\pi y}{b}$$

in conjunction with the Galerkin integral approach. Why must such a pattern be chosen, instead of the simpler forms we use for compressive loading? Why do we stipulate a long plate? An exact answer for the infinite plate, minimized with respect to an aspect ratio $\Lambda = b/(a/m) = $ (plate width)(axial buckle length) is $k_{\text{crit}} = 5.35$.

9.17   Obtain the buckling coefficient for a supported rectangular plate, under loading $\overline{N}_x = N_x^{0}$, that is *clamped* at the loaded edges. The exact answer, for a square plate, is $(k_x)_{\text{cr}} = 6.7432$. Note that a function of the form $\sin 2\pi x/a$ satisfies useful boundary conditions (for this problem) at $x = 0, a$.

The following sequence of problems is an examination of an approximate solution to the nonlinear problem of the buckling and postbuckling behavior of square plates. Two alternative loading situations will be described.

9.18   Using the plane–stress formulation of the strain energy, and the nonlinear strain–displacement relations of von Kármán plate theory (Eqs. (8.95)) derive the strain energy in rectangular coordinates for the stretching and bending of the von Kármán theory. How does this expression simplify for the case where a deflection function $w(x,y)$ is chosen such that, for a rectangular plate, $w = 0$ along all the edges?

9.19   For the assumed deflection pattern given below, where $\bar{u}$ is taken as a *prescribed* displacement in the x direction, calculate the total potential energy for a square plate including the nonlinear effects in the midplane strain–displacement relations. The assumed displacement functions are

$$u = -\bar{u}x + u_{20} \sin\frac{2\pi x}{a} + u_{22} \sin\frac{2\pi x}{a} \cos\frac{2\pi y}{a}$$

$$v = \bar{v}y + v_{02} \sin\frac{2\pi y}{a} + v_{22} \cos\frac{2\pi x}{a} \sin\frac{2\pi y}{a}$$

$$w = w_{11} \sin\frac{\pi x}{a} \sin\frac{\pi y}{a}$$

Note that the lateral deflection function is the usual one for a square, simply-supported plate. Thus it might be possible to simplify the bending strain energy of the plate by deleting the Gaussian curvature.

Further, it might be pointed out that, as we shall verify later, the in-plane displacement harmonics were chosen so as to approximately satisfy the in-plane equations of equilibrium.

9.20   By appropriate minimization of the functional (obtained in Problem 9.19 as an algebraic equation in its final form), obtain expressions for the in-plane displacement parameters in terms of the dimensionless buckling displacement w_{11}/a. Show that this quantity must be zero for $\bar{u} \le \bar{u}_{cr} = \pi^2 h^2/[3(1 - v^2)a^2]$, and that in this regime $\bar{v} = v\bar{u}$. What is the meaning of this last result?

9.21   Using the result of Problem 9.20, show that for $\bar{u} \le \bar{u}_{cr}$, that $N_x = Eh\bar{u}$, and thus that $(N_x)_{cr} = -Eh\bar{u}_{cr}$ is the critical load obtained from the linearized analysis. Further, for $\bar{u} > \bar{u}_{cr}$, show that

$$\frac{1}{h}(N_x)_{\text{av.}} = (\tau_{xx})_{\text{av.}} = \frac{1}{a}\int_0^a \tau_{xx}\,dy = -\frac{E}{2}(\bar{u} + \bar{u}_{cr})$$

Then by plotting $-\tau_{xx}$ vs. $\bar{u}$, show that the plate may continue to carry increasing load, although the slope of the load deflection curve after buckling is one half of its value prior to buckling.

9.22   Show that the result of Problem 9.21 for the buckling displacement can be written in the form

$$\left(\frac{w_{11}}{h}\right)^2 = \frac{4}{3(1 - v^2)}\left(\frac{\bar{u}}{\bar{u}_{cr}} - 1\right)$$

If the problem described in Problems 9.19–9.21 above had been repeated with an applied compressive edge loading $N_x{}^0$ (note that $\bar{u}$ would then be arbitrary, and that a potential term V would have to be added to the strain energy of Problem 9.20)

$$\left(\frac{w_{11}}{h}\right)^2 = \frac{4}{3(1 - v^2)}\left(\frac{N_x{}^0}{(N_x{}^0)_{cr}} - 1\right)$$

Plot $N_x{}^0/(N_x{}^0)_{cr}$ vs. w_{11}/h from this result, and comment upon the load-carrying capacity with increased buckling displacement and on the proclivity of the plate to buckle upwards or downwards, i.e., the sign-dependency of w_{11}/h. In light of the Koiter theory is the post-buckling behavior stable? Would you guess that a plate would be imperfection sensitive or not?

APPENDIX I

CARTESIAN TENSORS

I.1 INTRODUCTION

In this Appendix we shall consider elements of Cartesian tensors. We shall employ tensor concepts in much of this text and the accompanying notation will be used where it is most meaningful. (It will accordingly not be used exclusively.) The material of this Appendix has considerably greater application than simply for the study of solid mechanics.

I.2 VECTORS AND TENSORS

In elementary mechanics courses we dealt primarily with two classes of quantities in describing physical phenomena, namely, scalars and vectors. The scalar quantity, you will recall, required a single value for its specification and had no direction associated with it. On the other hand,

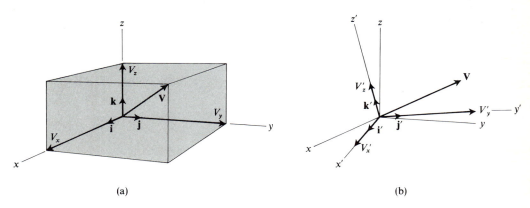

(a)   (b)

FIGURE I.1

we also worked with quantities that had both magnitude and direction. Accordingly, they could be represented as *directed line segments*. If we could combine these quantities in accordance with the parallelogram law we called them vectors. Finally you will recall we could specify a vector by giving three scalar components of the vector for a given reference frame.

In this Appendix we shall view scalars and vectors in a somewhat different manner so as to enable us properly to understand more complex quantities that we shall need for this text—namely, tensors. First note that a scalar quantity does not depend for its value on any particular reference. Stated another way, *the scalar is invariant with respect to a rotation of the coordinate axes that may be associated with the problem*. A vector, on the other hand, does require some reference for purposes of conveying its full meaning. Thus, for a velocity $\mathbf{V}$ we may refer to reference xyz (see Fig. I.1(a)) in conveying its properties in the form:

$$\mathbf{V} = V_x \mathbf{i} + V_y \mathbf{j} + V_z \mathbf{k}$$

or we may employ reference $x'y'z'$ as shown in Fig. I.1(b) to convey the same information in the form:

$$\mathbf{V} = V_{x'} \mathbf{i}' + V_{y'} \mathbf{j}' + V_{z'} \mathbf{k}'$$

Despite the need for a reference, *certain aspects* of vectors are independent of the reference. Thus, for example, the magnitude of the vector is the same for any reference that may be used. In other words we can say that the sum of the squares of the vector components is invariant for a rotation of reference axes. Stated mathematically:

$$V_x^2 + V_y^2 + V_z^2 = V_{x'}^2 + V_{y'}^2 + V_{z'}^2 \equiv \mathbf{V} \cdot \mathbf{V}$$

Also, when considering two vectors it is clear that the relative orientation of the vectors, as given by an angle θ, is independent of the coordinate system. A number of useful operations can accordingly be derived because of these invariances. These are the familiar operations of vector algebra. Vector equations involving such operations then can be stated independent of references.

There are more complex quantities that have certain characteristics independent of co-ordinates. Furthermore, certain very useful operations can be developed for such quantities

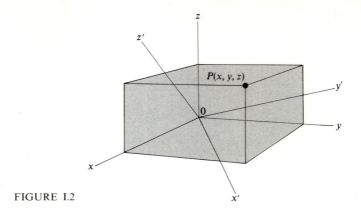

FIGURE I.2

which do not depend on a reference and which can convey certain physically important meanings. We will soon define tensor quantities as one such class of quantities. And just as we have vector equations that represent certain physical laws, so shall we be able to present certain tensor equations that play a similar role. To investigate such quantities we shall first go back to scalars and vectors to establish more useful definitions for these quantities which will permit us to proceed in an orderly way to the more complex quantities. As has been intimated in this discussion, the relation between a quantity and a reference is of critical importance. Accordingly as a first step we shall examine the transformation of coordinates under a rotation of axes. In the process we shall begin our introductory work on tensor notation.

I.3 TRANSFORMATION OF COORDINATES AND INTRODUCTION TO INDEX NOTATION

As pointed out earlier, at the heart of the discussion is the idea of a coordinate transformation. We have shown accordingly two references xyz and $x'y'z'$ rotated arbitrarily relative to each other in Fig. I.2. An arbitrary point P with coordinates x, y, z is shown. To obtain the primed coordinates $x'y'z'$ in terms of xyz coordinates, we can proceed by projecting the directed line segments x, y, and z, whose vector sum is the directed line segment $\overline{OP}$, along the primed coordinate axes. Thus for the x' axis we get:

$$x' = x \cos(x',x) + y \cos(x',y) + z \cos(x',z) \qquad \text{(I.1.)}$$

where $\cos(x',x)$ is the cosine between the x' and the x axes, etc. We shall now use another expression for the cosines in the above equations. We shall employ the letter a with two subscripts. The first subscript identifies the primed axis while the second identifies the unprimed axis. Thus we have

$$\cos(x',x) = a_{x'x}$$

$$\cos(y',z) = a_{y'z}$$

the summation process over j, only when j equals k do we get other than a zero contribution. Accordingly we can say:

$$\tau_{ij}\delta_{jk} = \tau_{ik}$$

Notice the integrity of the free indices is maintained in each equation.

I.5 OPERATIONS WITH INDICES

In the previous sections we showed how we could advantageously employ subscript notation. We shall now develop additional useful operations using indices.

Consider now the perfectly general sets $A_{q_1q_2\cdots q_r}$ and $B_{s_1s_2\cdots s_p}$ having respectively r free indices and p free indices. We define the *outer product* of these sets as a new set having $(r + p)$ free indices wherein the members of the set are found by listing all possible products between individual members of one set with individual members of the other set. This means that the outer product of $A_{q_1q_2\cdots q_r}$ and $B_{s_1s_2\cdots s_p}$ will have for three-dimensional space 3^{r+p} terms. We may denote the outer product between these sets as follows

$$A_{q_1q_2\cdots q_r}B_{s_1s_2\cdots s_p} = \mathcal{C}_{q_1q_2\cdots q_r s_1 s_2 \cdots s_p} \tag{I.14}$$

where the q's and s's are all free indices.

As a simple case consider the outer product of A_i and B_{jk}. To find all the members of the set A_iB_{jk} multiply separately A_1, A_2, and A_3 with each and every member of the set B_{jk}. Clearly we will have 27 members of the set.

A second useful operation is the so-called *inner product*. Consider two sets $A_{q_1q_2\cdots q_r}$ and $B_{s_1s_2\cdots s_p}$ which we may denote as $\{A\}$ and $\{B\}$ respectively. If we express a pair of the indices between the two sets using the same letter in order to form dummy indices, and then form a set of all possible products for the remaining free indices wherein each member of the set involves a sum over the aforestated dummy indices, we have what is called an inner product for the two sets over the aforementioned pair of indices. Clearly, the inner product will have $p + r - 2$ free indices thus having 3^{p+r-2} separate members in the set (in a three dimensional space) with each member consisting of a sum over the chosen dummy indices. Thus if we let s_1 and q_1 be represented by k we have as the inner product of $A_{q_1q_2\cdots q_n}$ and $B_{s_1s_2\cdots s_p}$ over the indices q_1 and s_1 the result:

$$A_{kq_2\cdots q_n}B_{ks_2\cdots s_p} = C_{q_2q_3\cdots q_n s_2 s_3 \cdots s_p}$$

Each term in the set $\{C\}$ is a summation over k according to the rules we have established for handling the dummy index.

The following is another kind of inner product of $\{A\}$ and $\{B\}$.

$$A_{iq_2jq_4\cdots q_r}B_{ijs_3\cdots s_p} = C_{q_2q_4\cdots q_r s_3 s_4 \cdots s_p}$$

where the inner product has been formed over pairs of indices (q_1s_1) and (q_3s_2) respectively.

If A_i and B_i represent components of vectors $\mathbf{A}$ and $\mathbf{B}$ then the inner product A_iB_i is clearly the ordinary dot product of vector analysis. Thus:

$$A_iB_i = \mathbf{A} \cdot \mathbf{B}$$

Also the inner product of A_i with itself, $A_i A_i$, is simply the square of the magnitude of the vector **A**. Thus:

$$A_i A_i = |\mathbf{A}|^2$$

As a final operation, consider a set of quantities having r free indices $A_{q_1 q_2 \cdots q_r}$. By making two of the indices the same to form a pair of dummy indices we decrease the number of different members of the set from 3^r to 3^{r-2}. We call this operation *contraction*. Thus the contraction of $\{A\}$ over indices $q_2 q_3$ is given as follows:

$$A_{q_1 j j q_4 \cdots q_r} = C_{q_1 q_4 \cdots q_r}$$

We could contract over two or more sets of indices. Thus for two sets of indices we have an example:

$$A_{j i i j q_5 \cdots q_r} = C_{q_5 \cdots q_r}$$

We shall find the preceding operations of considerable use in the sections to follow.

EXAMPLE I.1   Given the following sets:

$$A_i = \begin{pmatrix} 2 \\ 1 \\ 3 \end{pmatrix} \qquad B_{jk} = \begin{pmatrix} 2 & 1 & 0 \\ -1 & 3 & 2 \\ 1 & 4 & -1 \end{pmatrix}$$

(a) What are the terms C_{112} and C_{213} of the outer product between A_i and B_{jk}?
The outer product of the above sets may be given as

$$A_i B_{jk} = C_{ijk}$$

Here for C_{112} we have $(A_1)(B_{12}) = (2)(1) = 2$, and for C_{213} we have $A_2 B_{13} = (1)(0) = 0$.

(b) What is the set representing the inner product over indices i and j?
We have for this computation:

$$A_i B_{ik} = A_1 B_{1k} + A_2 B_{2k} + A_3 B_{3k}$$

The set then is given as follows:

$$A_i B_{ik} = \begin{pmatrix} [A_1 B_{11} + A_2 B_{21} + A_3 B_{31}] \\ [A_1 B_{12} + A_2 B_{22} + A_3 B_{32}] \\ [A_1 B_{13} + A_2 B_{23} + A_3 B_{33}] \end{pmatrix}$$

Inserting numbers we have for the set:

$$A_i B_{ik} = \begin{pmatrix} [(2)(2) + (1)(-1) + (3)(1)] \\ [(2)(1) + (1)(3) \quad + (3)(4)] \\ [(2)(0) + (1)(2) \quad + (3)(-1)] \end{pmatrix} = \begin{pmatrix} 6 \\ 17 \\ -1 \end{pmatrix}$$

(c) What are B_{22} and B_{jj}?

B_{22} is simply one term of the set B_{jk} namely (3). On the other hand B_{jj} represents a contraction. Thus:

$$B_{jj} = B_{11} + B_{22} + B_{33} = 2 + 3 - 1 = 4 \qquad ////$$

I.6 SCALARS AND VECTORS

We shall now reconsider scalar and vector quantities in terms of rotation of axes and in terms of the notation that we have just presented. The new perspective that we will achieve will help in setting forth tensor concepts later.

First note that since a scalar has only a single value we have no need for subscripts. And since there is no change in a scalar when there is a rotation of the reference axes we have:

$$M' = M \qquad (I.15)$$

as the relation between primed and unprimed axes.

As for a vector we have learned that the specifications of three rectangular scalar components is sufficient to specify the vector. We indicated briefly in the previous section that we can accordingly express the vector using one subscript—a free index—with the understanding that this subscript takes on separately the integers 1, 2, and 3 to form an ordered array of components. That is, V_i represents the set (V_1, V_2, V_3). However, it is only when a specific reference is indicated (usually at the end of computations) that we concern ourselves with specific values of the triplet for the reference of interest. In other words, V_i may be used in a manner similar to **V** during mathematical formulations with certain rules regulating the handling of the subscripts to yield the vector algebra familiar from vector mechanics. There are times, however, when V_i is meant to represent *any one* element of the set and not the whole set. The context of the discussion should make clear at any time how V_i is being used.

Now consider a vector **V** (or V_i) shown in Fig. I.3 as a directed line segment. In accord with the parallelogram law, we have for the rectangular components of **V**:

$$V_x = V \cos(V, x)$$
$$V_y = V \cos(V, y)$$
$$V_z = V \cos(V, z)$$

When the reference is rotated to a primed position as shown in the diagram we will have a new set of components which again can be computed from the parallelogram law as follows:

$$V'_x = V_x \cos(x', x) + V_y \cos(x', y) + V_z \cos(x', z)$$
$$V'_y = V_x \cos(y', x) + V_y \cos(y', y) + V_z \cos(y', z)$$
$$V'_z = V_x \cos(z', x) + V_y \cos(z', y) + V_z \cos(z', z)$$

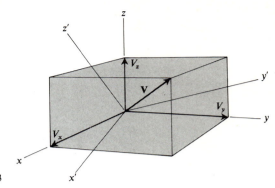

FIGURE I.3

Using the double index notation we can express these equations as follows:

$$\boxed{V'_i = a_{ij}V_j} \qquad (\text{I.16})$$

We can conclude that a *necessary condition* for a quantity to be a vector is that the *components transform according to the above equation under a rotation of axes.*

We can also readily show that the above transformation relation is a *sufficient* condition for identifying vector quantities. Thus, consider two ordered sets of numbers $(A_1, A_2,$ and $A_3)$ and $(B_1, B_2,$ and $B_3)$ associated with a reference $x_1, x_2,$ and x_3. Suppose that each set transforms under a rotation of axes to new triplets of numbers in accordance with Eq. (I.16). That is for axes $x'_1, x'_2,$ and x'_3:

$$A'_i = a_{ij}A_j$$
$$B'_i = a_{ij}B_j \qquad (\text{I.17})$$

Now add Eq. (I.17) in the following way:[1]

$$A'_i + B'_i = a_{ij}(A_j + B_j) \qquad (\text{I.18})$$

Denoting the sum on the left side of the above equation as C'_i we have:

$$C'_i = a_{ij}(A_j + B_j)$$

Now carry out an inner product using C'_i on the left side of the equation and $a_{ik}(A_k + B_k)$ which equals C'_i, on the right side of the equation. We get:

$$C'_i C'_i = a_{ij}(A_j + B_j)a_{ik}(A_k + B_k)$$
$$= a_{ij}a_{ik}(A_jA_k + B_jB_k + A_jB_k + B_jA_k) \qquad (\text{I.19})$$

---

[1] It should be clear that we are adding corresponding components of the vector equations.

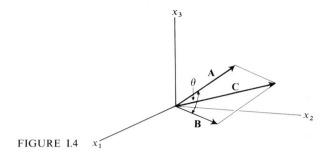

FIGURE I.4 x_1

Noting Eq. (I.13) we have:

$$C_i'C_i' = \delta_{jk}(A_jA_k + B_jB_k + A_jB_k + B_jA_k)$$
$$= A_jA_j + B_jB_j + 2A_jB_j \qquad (I.20)$$

Now consider two directed line segments respectively having components (A_1, A_2, and A_3) and (B_1, B_2, and B_3) in reference x_1, x_2, and x_3. These are shown in Fig. I.4. The parallelogram law permits us to find the quantity D by the law of cosines as follows:

$$D^2 = A^2 + B^2 + 2AB \cos \theta$$

Using the Pythagorean Theorem and double index notation we can express the above equation as follows:

$$D_iD_i = A_kA_k + B_kB_k + 2\mathbf{A} \cdot \mathbf{B}$$
$$= A_kA_k + B_kB_k + 2A_kB_k$$

Since the length of D is the same for any reference we can replace D_iD_i by $D_i'D_i'$ to get:

$$D_i'D_i' = A_kA_k + B_kB_k + 2A_kB_k$$

The right side of the above expression is identical to that of Eq. (I.20) and so we see that $C_i'C_i' = D_i'D_i'$. Thus the sets A_i and B_i, as a result of transformation equation (I.17), *satisfy the parallelogram law when directed line segments are associated with these triplets.* These sets thus are vectors and accordingly we can conclude that transformation equation (I.16) is sufficient for establishing vector quantities. We shall henceforth use this transformation law as the defining criterion for the identification of vectors.

I.7 TENSORS: SYMMETRY AND SKEW-SYMMETRY

In the previous section we have redefined vectors and scalars in terms of certain transformation equations for the components of these quantities. Thus considering a primed and an unprimed reference rotated arbitrarily relative to each other about a common origin we have the following defining transformation equations:

SCALARS: $M = M'$
VECTORS: $A_i' = a_{ij}A_j$

We shall now generalize these transformation equations so as to define other more complex quantities. Thus we will define a *second-order tensor* as a set of nine components (consequently denoted with two free subscripts, as, for example, A_{ij}) which transforms under a rotation from an unprimed to a primed set of axes according to the following equation

$$A'_{ij} = a_{ik}a_{jl}A_{kl} \qquad (\text{I.21})$$

where the a's are the familiar direction cosines discussed earlier.[1] Many important quantities in the engineering sciences are second-order tensors. And as a result of the above transformation equations, second-order tensors have certain distinct and useful properties. Notice the way the defining transformation builds up from scalars, with no free subscripts, to vectors, with one free subscript, to the second-order tensor with two free subscripts. For this reason we call the scalar a *zeroth order tensor* and the vector a *first-order tensor*. Continuing on in this process, we can define a *p*th order tensor as follows:

$$A'_{q_1 q_2 \cdots q_p} = a_{q_1 j_1}a_{q_2 j_2} \cdots a_{q_p j_p}A_{j_1 j_2 \cdots j_p} \qquad (\text{I.22})$$

wherein we have sets of quantities requiring *p* free indices.

We shall say that a tensor such as T_{ijkl} is *symmetric* with respect to any two of its indices—say *jk*—if the values of the tensor components corresponding to these indices are equal to the tensor components corresponding to the reverse of these indices. That is:

$$T_{ijkl} = T_{ikjl} \qquad (\text{I.23})$$

In the case of a second-order tensor we have for symmetry the single possibility:

$$A_{ij} = A_{ji} \qquad (\text{I.24})$$

In the representation of this set as an array it means that the terms on one side of the main diagonal may be considered as mirror image values of the terms on the other side of the main diagonal. This has been shown in the following representation:

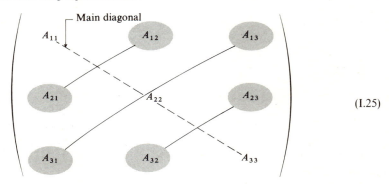

$$(\text{I.25})$$

[1] Using Eq. (I.12) we leave it for you to show for the transformation from primed to unprimed coordinates that for second-order tensors we have:

$$A_{ij} = a_{ki}a_{lj}A'_{kl}$$

Furthermore, a tensor such as T_{ijkl} is said to be *skew-symmetric* in any two of its indices, say jk, if the components of the tensors with indices reversed are negatives of each other. Thus we have as a definition of skew-symmetry:

$$T_{ijkl} = -T_{ikjl} \qquad \text{skew-symmetric in } jk \quad (a)$$
$$T_{ijkl} = -T_{ilkj} \qquad \text{skew-symmetric in } jl \quad (b) \qquad (\text{I}.26)$$

Consider next those components of a skew-symmetric tensor wherein the indices involved in the skew-symmetry have identical numbers. For T_{ijkl} in Eq. (I.26(a)) for instance, we are referring to T_{i11l}, T_{i22l}, and T_{i33l}, and for T_{ijkl} in Eq. (I.26(b)) we are referring to T_{i1k1}, T_{i2k2}, and T_{i3k3}. It is clear that such terms must be zero. Thus, considering T_{i11l} we see from the skew-symmetry requirement:

$$T_{i11l} = -T_{i11l} \qquad (\text{I}.27)$$

Clearly for each i and l we must have $T_{i11l} = 0$ in order to satisfy the above condition.

For a second-order tensor we have for skew-symmetry the single possibility:

$$B_{ij} = -B_{ji} \qquad (\text{I}.28)$$

And in the array representation we have

$$\begin{pmatrix} 0 & B_{12} & B_{13} \\ -B_{12} & 0 & B_{23} \\ -B_{13} & -B_{23} & 0 \end{pmatrix}$$

where the main diagonal must be composed only of zeros.

Now suppose we have a tensor T_{ijkl} which is *symmetric* in i, j and a tensor B_{srtp} which is *skew-symmetric* in s, r. It is useful to note that the inner product of these tensors for indices i and s, and j and r give a set all of whose terms are zero. Thus:

$$T_{ijkl} B_{ijtp} = C_{kltp} = 0$$

To show this most simply, expand the expression on the left side of the above equation as follows:

$$(T_{11kl} B_{11tp} + T_{12kl} B_{12tp} + T_{13kl} B_{13tp} +$$
$$T_{21kl} B_{21tp} + T_{22kl} B_{22tp} + T_{23kl} B_{23tp} +$$
$$T_{31kl} B_{31tp} + T_{32kl} B_{32tp} + T_{33kl} B_{33tp})$$

Because of the skew-symmetry of B_{ijtp}, the terms B_{11tp}, B_{22tp}, and B_{33tp} along the "diagonal" of the arrangement are clearly zero, and the terms at "image" positions about the diagonal are negatives of each other because of the symmetry in ij of T_{ijkl} and the skew-symmetry in ij of B_{ijtp}. The sum of the terms for any value of the free indices $kltp$ accordingly is zero and so all terms for the set C_{kltp} are zero.

The converse to the above conclusion can be very useful. Suppose A_{ijkl} is a tensor *skew-symmetric* in ij but otherwise *perfectly arbitrary* and it is known that:

$$A_{ijkl} B_{ijps} = C_{klps} = 0$$

Clearly we can conclude that B_{ijps} must be *symmetric* in the indices i, j. A similar statement can be developed for establishing the skew-symmetry of a tensor using an arbitrary symmetric tensor.

We will next show that it is not always necessary to employ Eq. (I.22) to identify a set of quantities as a tensor. That is, we will show by the so-called *quotient rule* that it is not always necessary to go to the defining transformation equations in order to establish the tensor character of a set of quantities.

Suppose that A_i represents an *arbitrary vector* and that the inner product of A_i and a set of terms C_{irs} forms a *second-order tensor* B_{rs}. We can show that for such a circumstance C_{irs} is a *third-order tensor*. That is, if:

$$B_{rs} = C_{irs}A_i \qquad \text{(I.29)}$$

where A_i is an arbitrary vector and B_{rs} is a second-order tensor then C_{irs} is a third-order tensor. To show this, transform B_{rs} to a new set of coordinates according to the rule of second-order tensors as follows

$$B'_{rs} = a_{rm}a_{sl}B_{ml} = a_{rm}a_{sl}C_{iml}A_i \qquad \text{(I.30)}$$

wherein we have substituted from Eq. (I.29) in the last step. But A_i being a vector, can be transformed as follows:

$$A_i = a_{ji}A'_j$$

Substituting into Eq. (I.30) we get:

$$B'_{rs} = a_{rm}a_{sl}a_{ji}C_{iml}A'_j \qquad \text{(I.31)}$$

We now replace the left side of the above equation using the right side of Eq. (I.29) in primed coordinates as follows:

$$C'_{irs}A'_i = a_{rm}a_{sl}a_{ji}C_{iml}A'_j$$

Now replacing the dummy index i on the left side of the equation by j, we get on rearranging of terms:

$$[C'_{jrs} - a_{rm}a_{ji}a_{sl}C_{iml}]A'_j = 0$$

Since A'_j is arbitrary we can conclude that:

$$C'_{jrs} = a_{ji}a_{rm}a_{sl}C_{iml}$$

You will recognize this equation as the defining relation for third-order tensors and so we have shown that C_{ijk} is a third-order tensor. This is one of many quotient laws. It is a simple matter to extend the above result as follows:

If $C_{q_1 q_2 \cdots q_n}$ is a set of 3^n quantities which form an $(n-1)$th order tensor when multiplied as an inner product over any one of its indices with an arbitrary vector A_i, then $C_{q_1 q_2 \cdots q_n}$ is an nth order tensor.

Another quotient law that will prove useful is given as follows:

If A_{ij} is a set of 9 quantities which forms a second-order tensor when multiplied as an inner product with arbitrary second-order tensors, B_{km}, over indices i and k, or over indices j and m, then A_{ij} is a second-order tensor.

Finally, we will demonstrate that taking the partial derivative of an nth order tensor $A_{p_1 p_2 \cdots p_n}$ with respect to x_i, where i is a free index, results in a tensor of rank $(n + 1)$. We shall find it convenient here to adopt the convention of denoting partial derivatives of a tensor by making use of commas and indices as follows:

$$\frac{\partial (\ )}{\partial x_i} \equiv (\ )_{,i}; \qquad \frac{\partial^2 (\ )}{\partial x_i \partial x_j} \equiv (\ )_{,ij}$$

Accordingly we wish to show at this time $A_{j_1 \cdots j_n, i}$ is an $(n + 1)$th-order tensor. Suppose we start with a primed reference x_i'. Then we may say:

$$A'_{j_1 \cdots j_n, i} \equiv \partial (A'_{j_1 \cdots j_n})/\partial x_i'$$

Since $A'_{j_1 \cdots j_n}$ is an nth order tensor we use its transformed form in the unprimed reference in the above equation:

$$A'_{j_1 \cdots j_n, i} = \frac{\partial}{\partial x_i'} (a_{j_1 k_1} a_{j_2 k_2} \cdots a_{j_n k_n} A_{k_1 \cdots k_n})$$

Also we note that:

$$\frac{\partial}{\partial x_i'} = \frac{\partial}{\partial x_l} \frac{\partial x_l}{\partial x_i'}$$

But $\partial x_l / \partial x_i' = a_{il}$. Hence we have:

$$\frac{\partial}{\partial x_i'} = a_{il} \frac{\partial}{\partial x_l}$$

Thus we have:

$$A'_{j_1 \cdots j_n, i} = a_{j_1 k_1} \cdots a_{j_n k_n} a_{il} \frac{\partial}{\partial x_l} (A_{k_1 \cdots k_n})$$

$$= a_{j_1 k_1} \cdots a_{j_n k_n} a_{il} A_{k_1 \cdots k_n, l}$$

The above statement is the defining one for a tensor of order $(n + 1)$ and so we have demonstrated that $A_{j_1 \cdots j_n, i}$ is an $(n + 1)$th order tensor.

Since we can perform the operation of taking partials with respect to other coordinates x_j, x_k, etc., we can conclude that:

$$A_{p_1 \cdots p_n, jk} = \frac{\partial^2}{\partial x_j \partial x_k} (A_{p_1 \cdots p_n})$$

is an $(n + 2)$th-order tensor while

$$A_{p_1 \cdots p_n, jkl} = \frac{\partial^3}{\partial x_j \partial x_k \partial x_l} (A_{p_1 \cdots p_n})$$

is an $(n + 3)$th-order tensor.

As exercises at the end of this Appendix we shall ask you to prove the following statements:

1 If each term of an nth-order tensor is multiplied by a constant, then the resulting set is also an nth-order tensor.

2 The outer product of an *m*th-order tensor and an *n*th-order tensor results in an $(m + n)$th-order tensor.

3 The inner product of an *m*th-order tensor and an *n*th-order tensor over *k* indices results in a tensor of order $(m + n - 2k)$.

I.8 VECTOR OPERATIONS USING TENSOR NOTATION: THE ALTERNATING TENSOR

We have already shown that the dot product $\mathbf{A} \cdot \mathbf{B}$ can be expressed as $A_i B_i$ using index notation. Another operation that we have used extensively is the cross product $\mathbf{A} \times \mathbf{B} = \mathbf{C}$ where,

$$C_1 = A_2 B_3 - A_3 B_2$$
$$C_2 = A_3 B_1 - A_1 B_3 \quad\quad (\text{I}.32)$$
$$C_3 = A_1 B_2 - A_2 B_1$$

In order to be able to get this result using index notation we introduce the *alternating tensor* ϵ_{ijk} defined as follows:[1]

$\epsilon_{ijk} = 0$ for those terms of the set for which *i, j, k* do not form some permutation of 1,2,3. (Example; if any two of the subscripts are equal, then such terms are zero.)

$\epsilon_{ijk} = 1$ for those terms of the set having indices that form the sequence 1, 2, 3 or that can be arranged by an *even* number of permutations to form this sequence.

$\epsilon_{ijk} = -1$ for those terms in the set that require an *odd* number of permutations to reach the sequence 1, 2, 3.

Thus we have:

$$\epsilon_{112} = \epsilon_{212} = \epsilon_{331} = \cdots = 0$$
$$\epsilon_{123} = \epsilon_{231} = \epsilon_{312} = 1$$
$$\epsilon_{213} = \epsilon_{321} = \epsilon_{132} = -1$$

Now if we go back to Eq. (I.32), we can readily demonstrate that the following expression represents the terms comprising the set of the cross product.

$$C_i = \epsilon_{ijk} A_j B_k \quad\quad (\text{I}.33)$$

Thus carrying out the double summation over dummy indices *j* and *k*, we get for $i = 1$:

$$C_1 = \epsilon_{1jk} A_j B_k = \epsilon_{111} A_1 B_1 + \epsilon_{121} A_2 B_1 + \epsilon_{131} A_3 B_1 + \epsilon_{112} A_1 B_2 + \epsilon_{122} A_2 B_2$$
$$+ \epsilon_{132} A_3 B_2 + \epsilon_{113} A_1 B_3 + \epsilon_{123} A_2 B_3 + \epsilon_{133} A_3 B_3$$

[1] One can show that ϵ_{ijk} is a third-order cartesian tensor. This is not, however, a simple step. See Hodge: "Continuum Mechanics," McGraw-Hill Book Co., Chap. 4.

Now employing the definition of the alternating tensor we find

$$C_1 = A_2 B_3 - A_3 B_2$$

as you yourself may verify.

Now consider the *triple scalar product* $(\mathbf{A} \times \mathbf{B}) \cdot \mathbf{C}$. Using double index notation we get:

$$(\mathbf{A} \times \mathbf{B}) \cdot \mathbf{C} = \epsilon_{ijk} A_j B_k C_i \qquad (I.34)$$

Furthermore, you will recall from your mechanics courses that:

$$(\mathbf{A} \times \mathbf{B}) \cdot \mathbf{C} = \begin{vmatrix} A_1 & A_2 & A_3 \\ B_1 & B_2 & B_3 \\ C_1 & C_2 & C_3 \end{vmatrix} \qquad (I.35)$$

and so comparing Eqs. (I.35) and (I.34) we get the following expression for a determinant.

$$\begin{vmatrix} A_1 & A_2 & A_3 \\ B_1 & B_2 & B_3 \\ C_1 & C_2 & C_3 \end{vmatrix} = \epsilon_{ijk} A_j B_k C_i \qquad (I.36)$$

We may put the above relation in a more convenient form from an index point of view by noting that

$$\epsilon_{ijk} A_j B_k C_i = \epsilon_{kij} A_i B_j C_k \qquad (I.37)$$

where we have changed the dummy indices. Now ϵ_{kij} can be made into ϵ_{ijk} by two interchanges of indices and so we have:

$$\epsilon_{kij} = \epsilon_{ijk}$$

Accordingly, Eq. (1.37) can be written as:

$$\epsilon_{ijk} A_j B_k C_i = \epsilon_{ijk} A_i B_j C_k$$

We can then write Eq. (I.36) in the following manner:

$$\begin{vmatrix} A_1 & A_2 & A_3 \\ B_1 & B_2 & B_3 \\ C_1 & C_2 & C_3 \end{vmatrix} = \epsilon_{ijk} A_i B_j C_k$$

Let us next consider the *field operators* of vector analysis using double index notation. First, we have the *gradient operator* which acting on a function ϕ is given as:

$$\nabla \phi = \frac{\partial \phi}{\partial x} \mathbf{i} + \frac{\partial \phi}{\partial y} \mathbf{j} + \frac{\partial \phi}{\partial z} \mathbf{k}$$

To write $\nabla\phi$ in index notation we can express the Cartesian unit vectors as follows:

$$\mathbf{i} = \mathbf{i}_1$$
$$\mathbf{j} = \mathbf{i}_2$$
$$\mathbf{k} = \mathbf{i}_3$$

Accordingly we can then say:

$$\boxed{\nabla\phi = \phi_{,j}\mathbf{i}_j} \qquad (\text{I}.38)$$

Also, the expression $\phi_{,i}$ with i as a free index, is the index counterpart of $\nabla\phi$.

The *divergence* of a vector, you will recall, is given as follows in vector notation:

$$\text{div } \mathbf{V} = \nabla \cdot \mathbf{V} = \frac{\partial V_x}{\partial x} + \frac{\partial V_y}{\partial y} + \frac{\partial V_z}{\partial z}$$

Using index notation, we get:

$$\text{div } \mathbf{V} = \frac{\partial V_i}{\partial x_i} = V_{i,i} \qquad (\text{I}.39)$$

Next we consider the *curl* operator. Again from vector analysis, we have:

$$\text{curl } \mathbf{V} = \nabla \times \mathbf{V} = \begin{vmatrix} \mathbf{i} & \mathbf{j} & \mathbf{k} \\ \dfrac{\partial}{\partial x_1} & \dfrac{\partial}{\partial x_2} & \dfrac{\partial}{\partial x_3} \\ V_x & V_y & V_z \end{vmatrix}$$

$$= \begin{vmatrix} \mathbf{i}_1 & \mathbf{i}_2 & \mathbf{i}_3 \\ \dfrac{\partial}{\partial x_1} & \dfrac{\partial}{\partial x_2} & \dfrac{\partial}{\partial x_3} \\ V_1 & V_2 & V_3 \end{vmatrix}$$

Then by direct analogy to the determinant expansion above we can say:

$$\boxed{\nabla \times \mathbf{V} = \epsilon_{ijk}\mathbf{i}_i\frac{\partial V_k}{\partial x_j} = \epsilon_{ijk}V_{k,j}\mathbf{i}_i} \qquad (\text{I}.40)$$

Or considering the index i to be a free index, we can express $\nabla \times \mathbf{V}$ indicially simply as $\epsilon_{ijk}V_{k,j}$.

Finally, we consider the *Laplacian* operator ∇^2. From vector analysis we have:

$$\nabla^2 = \frac{\partial^2}{\partial x_1{}^2} + \frac{\partial^2}{\partial x_2{}^2} + \frac{\partial^2}{\partial x_3{}^2}$$

Hence, using index notation, we get:

$$\nabla^2 \equiv {}_{,ii}$$

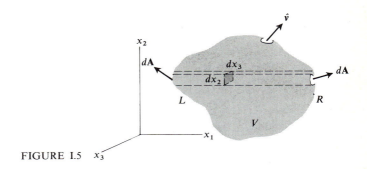

FIGURE I.5

And so:

$$\boxed{\nabla^2\phi \equiv \phi_{,ii}} \qquad (\text{I.41})$$

I.9   GAUSS' THEOREM

In the previous section we considered the so-called differential field operators of divergence, curl, etc., in tensor notation. Equally important for us are certain integral theorems. At this time we present Gauss' theorem in a reasonably general form.

Assume we have an nth order tensor $T_{jk...}$, defined at each point in space (i.e., it is an nth order tensor *field*). Consider a domain V in space of such a shape that in the direction x_1 lines parallel to the x_1 axis can pierce the boundary only twice (see Fig. I.5). Imagine now that this volume is composed of infinitesimal prisms having sides dx_2 and dx_3 as has been shown in the diagram. Consider one prism and compute the following integral over the volume δV of this prism.

$$\iiint_{\delta V} \frac{\partial}{\partial x_1}(T_{jk...})\, dx_1\, dx_2\, dx_3$$

Carrying out integration with respect to x_1 we get

$$\iiint_{\delta V} \frac{\partial}{\partial x_1}(T_{jk...})\, dx_1\, dx_2\, dx_3 = \iint (T_{jk...}\, dx_2\, dx_3)_R - \iint (T_{jk...}\, dx_2\, dx_3)_L$$

where the first expression on the right side of the above equation is evaluated at the right end of the prism while the second expression is evaluated at the left end of the prism. Using $\mathbf{v}$ to represent the unit vector normal to the boundary surface of V, and considering v_1, v_2, and v_3 to be the direction cosines of $\mathbf{v}$,[1] we can replace dx_2, dx_3 at the right end of the prism by $(+v_1\, dA)$ and by $(-v_1\, dA)$ on

---

[1] When we use xyz notation, we generally write v_1, v_2, and v_3 as a_{nx}, a_{ny}, and a_{nz} with n being indicative of the unit normal.

the left end of the prism.[1] We thus have:

$$\iiint_{\delta V} \frac{\partial}{\partial x_1}(T_{jk\cdots})\, dx_1\, dx_2\, dx_3 = \iint (T_{jk\cdots}v_1\, dA)_R + \iint (T_{jk\cdots}v_1\, dA)_L$$

Now integrating over all the prisms comprising the volume V we get:[2]

$$\iiint_V \frac{\partial}{\partial x_1}(T_{jk\cdots})\, dx_1\, dx_2\, dx_3 = \iint_R (T_{jk\cdots})v_1\, dA + \iint_L (T_{jk\cdots})v_1\, dA$$

The right side of the above equation clearly covers the entire surface of volume V and is accordingly replaceable by a closed-surface integral. We have then the following statement:

$$\iiint_V \frac{\partial}{\partial x_1}(T_{jk\cdots})\, dx_1\, dx_2\, dx_3 = \oiint_S (T_{jk\cdots})v_1\, dA \tag{I.42}$$

We developed Eq. (I.42) for the x_1 direction. We could have proceeded in a similar manner in any direction x_i. Accordingly, the generalization of the above statement is given as follows:

$$\boxed{\iiint_V \frac{\partial T_{ik\cdots}}{\partial x_i}\, dV = \oiint_S (T_{jk\cdots})v_i\, dA} \tag{I.43}$$

where i is a free index. This is *Gauss' theorem* in a generalized form.[3] Using the notation presented for a spatial partial derivative, we can also give this equation as follows:

$$\boxed{\iiint_V (T_{jk\cdots})_{,i}\, dv = \oiint_S (T_{jk\cdots})v_i\, dA} \tag{I.44}$$

Suppose that $T_{jk\cdots}$ is the zeroth-order tensor ϕ (i.e., a scalar). We then have:

$$\iiint_V \phi_{,i}\, dv = \oiint_S \phi v_i\, dA$$

If we wish to revert to vector notation, this form of Gauss' law becomes

$$\iiint_V \nabla\phi\, dv = \oiint_S \phi\, d\mathbf{A} \tag{I.45}$$

[1] The appearance of the minus sign here is a result of the use of the *outward-normal* convention for area vectors. Thus, v_1 on the left side of the prism is clearly negative for the kind of domain we have chosen to work with and a minus sign must be included so that the product $(-v_1\, dA)$ be the positive number needed to replace $dx_2\, dx_3$ for that side.

[2] We are assuming tacitly here, in order to be able to carry out the integration, that the surface of the domain V can be split up into a finite number of parts such that there is a continuously varying tangent plane on each piece. That is, the surface should be *piecewise smooth*.

[3] This theorem can be extended to bodies whose shape is such that lines parallel to $x_1x_2x_3$ axes cut the surface more than twice, provided such a body can be decomposed into contiguous composite bodies that do have the property specified in the development.

where $d\mathbf{A} = \mathbf{v}\,dA$. Suppose next that $T_{jk\ldots}$ is a first-order tensor (i.e., a vector) V_j. We then get:

$$\iiint_V V_{j,i}\,dv = \oiint_S V_j v_i\,dA \qquad (1.46)$$

Next perform a contraction operation on the free indices i and j in the above equation. We get:

$$\iiint_V V_{j,j}\,dv = \oiint_S V_j v_j\,dA \qquad (1.47)$$

You should have no difficulty in writing this equation in terms of vectors to arrive at:

$$\boxed{\iiint_V \operatorname{div}\mathbf{V}\,dv = \oiint_S \mathbf{V}\cdot d\mathbf{A}} \qquad (1.48)$$

We will have much use of this form of Gauss' theorem. In this form it is called the *divergence theorem*.[1]

Let us consider next that the vector field $\mathbf{V}$ is given as follows:

$$\mathbf{V} = \phi\mathbf{i} + \psi\mathbf{j} + \kappa\mathbf{k}$$

Then Eq. (I.48) can be given in the following way:

$$\iiint_V \left(\frac{\partial\phi}{\partial x} + \frac{\partial\psi}{\partial y} + \frac{\partial\kappa}{\partial z}\right)dv = \oiint_S (\phi a_{vx} + \psi a_{vy} + \kappa a_{vz})\,dA \qquad (1.49)$$

where $a_{vx} = v_1$, etc., are the direction cosines of the outer normal to the boundary. In this form the equation is called *Green's Theorem*.[2] If we consider a two-dimensional simplification of Green's theorem we get the following result:

$$\iint_S \left(\frac{\partial\phi}{\partial x} + \frac{\partial\psi}{\partial y}\right)dA = \oint_\Gamma (\phi a_{vx} + \psi a_{vy})\,dl \qquad (1.50)$$

We now set $\psi = 0$ and $\kappa = 0$ in Eq. (I.49) and let ϕ be the product of two functions u and w. We get on substitution:

$$\iiint_V \frac{\partial(uw)}{\partial x}\,dv = \iiint_V u\frac{\partial w}{\partial x}\,dv + \iiint_V w\frac{\partial u}{\partial x}\,dv = \oiint_S (uw)a_{vx}\,dA$$

Rearranging, we reach the following formula representing an *integration by parts*:

$$\iiint_V u\frac{\partial w}{\partial x}\,dv = \oiint_S (uw)a_{vx}\,dA - \iiint_V w\frac{\partial u}{\partial x}\,dv$$

---

[1] The divergence theorem has a simple physical interpretation. You will recall from your earlier work in mechanics, that div $\mathbf{B}$ represents the net efflux of the flux of vector field $\mathbf{B}$ per unit volume at a point. And accordingly, the volume integration on the left side of the above equations represents the net efflux of flux of vector field $\mathbf{B}$ from volume V. This net flux is then equated in this equation to the flux piercing the bounding surface S of the volume V. We have here a statement concerning the conservation of flux for a volume V for the vector $\mathbf{B}$.

[2] Green's theorem may be proven without vectorial considerations for any three functions ϕ, ψ, and κ that are piecewise continuous and differentiable. See Kaplan, W.: "Advanced Calculus," Addison-Wesley Co., Chap. 5.

Generalizing for any coordinate x_i we get:

$$\iiint_V u \frac{\partial w}{\partial x_i} \, dv = \oiint_S (uw) a_{vx_i} \, dA - \iiint_V w \frac{\partial u}{\partial x_i} \, dv \qquad (I.51)$$

The corresponding integration by parts formula for two dimensions is:

$$\iint_S u \frac{\partial w}{\partial x_i} \, dA = \oint_\Gamma (uw) a_{vx_i} \, dl - \iint_S w \frac{\partial u}{\partial x_i} \, dA \qquad (I.52)$$

We have much occasion in this text to utilize the above formulas.

I.10   GREEN'S FORMULA

Another formula that is useful to us in the text is Green's formula and its various special forms. For this purpose consider the following integral

$$I = \iiint_V \phi_{,i} \psi_{,i} \, dv \qquad (I.53)$$

wherein, you will note, we have as an integrand the inner product of two gradients. The functions ϕ and ψ are assumed to be continuous in the domain V with continuous first and second partial derivatives in this domain. The domain V has all the restriction needed for the development of Gauss' theorem. Note now that:

$$\left(\frac{\partial \phi}{\partial x_1} \right) \left(\frac{\partial \psi}{\partial x_1} \right) = \frac{\partial}{\partial x_1} \left(\phi \frac{\partial \psi}{\partial x_1} \right) - \phi \frac{\partial^2 \psi}{\partial x_1{}^2}$$

Similar relations may be written for x_2 and x_3 so that Eq. (I.53) may be expressed as follows:

$$\iiint_V \phi_{,i} \psi_{,i} \, dv = - \iiint_V \phi \psi_{,jj} \, dv + \iiint_V \frac{\partial}{\partial x_j} \left(\phi \frac{\partial \psi}{\partial x_j} \right) dv \qquad (I.54)$$

We may now use Gauss' theorem for the last integral. That is:

$$\iiint_V \frac{\partial}{\partial x_j} \left(\phi \frac{\partial \psi}{\partial x_j} \right) dv = \oiint_S \left(\phi \frac{\partial \psi}{\partial x_j} \right) v_j \, dA$$

But on expanding $(\partial \psi / \partial x_j) v_j$ we realize that this is simply the directional derivative of ψ in the direction $\mathbf{v}$ normal to the surface of the domain V. Thus we can say

$$\iiint_V \frac{\partial}{\partial x_j} \left(\phi \frac{\partial \psi}{\partial x_j} \right) dv = \oiint_S \phi \frac{\partial \psi}{\partial v} \, dA$$

We can then express Eq. (I.54) in the following manner:

$$\iiint_V \phi_{,i} \psi_{,i} \, dv = - \iiint_V \phi \psi_{,jj} \, dv + \oiint_S \phi \frac{\partial \psi}{\partial v} \, dA \qquad (I.55)$$

This is a preliminary form of Green's formula which, using vector notation, should be readily recognized to have the following form:

$$\iiint_V \nabla\phi \cdot \nabla\psi \, dv = -\iiint_V \phi\nabla^2\psi \, dv + \oiint_S \phi\frac{\partial\psi}{\partial v}\, dA \qquad (I.56)$$

Now write Eq. (I.55) with ϕ and ψ interchanged. Subtracting equations we see that the left sides cancel leaving the following result:

$$\iiint_V (\phi\psi_{,jj} - \psi\phi_{,jj}) \, dv = \iint_S \left(\phi\frac{\partial\psi}{\partial v} - \psi\frac{\partial\phi}{\partial v}\right) dA \qquad (I.57)$$

This is the well-known *Green's formula*. In *vector notation* we get:

$$\iiint_V (\phi\nabla^2\psi - \psi\nabla^2\phi) \, dv = \iint_S \left(\phi\frac{\partial\psi}{\partial v} - \psi\frac{\partial\phi}{\partial v}\right) dA \qquad (I.58)$$

In two dimensions the volume integral becomes a surface integral while the closed surface integral becomes a line integral.

I.11 CLOSURE

We have introduced certain basic definitions and concepts for the use of Cartesian tensors. Hopefully, this material has been linked with your earlier work on vectors. As should be clear from the development, Cartesian tensors are valid only for Cartesian references. Actually they represent a special case of general curvilinear tensors valid for all references. In such studies we distinguish between contravariant tensors which transform from one curvilinear reference x_i at a point to another curvilinear reference x'_i at the point via the rule:

$$A'^{ij} = \frac{\partial x'^i}{\partial x^k}\frac{\partial x'^j}{\partial x^l}A^{kl} \qquad (I.59)$$

and covariant tensors for which

$$A'_{ij} = \frac{\partial x^k}{\partial x'^i}\frac{\partial x^l}{\partial x'^j}A_{kl} \qquad (I.60)$$

Note that the position of the subscripts (down or up) indicates covariance or contravariance. For Cartesian references rotated relative to each other Eqs. (I.59) and (I.60) become identical in form, with the derivatives becoming the familiar direction cosines. There is no longer a need to distinguish between covariance and contravariance.

We do not restrict ourselves in this text to one particular notation. Rather we use vector notation, tensor notation, or make use of summation operators to yield what (to the authors) seems to be the clearest presentation for the situation at hand.

READING

FREDERICK, D. AND CHANG, T. S.: "Continuum Mechanics," Allyn and Bacon Co., Boston, 1965.

FUNG, Y. C.: "A First Course in Continuum Mechanics," Prentice-Hall Inc., N.J., 1969.

HODGE, P. G.: "Continuum Mechanics," McGraw-Hill Book Co., N.Y., 1970.

PRAGER, W.: "Introduction to Mechanics of Continua," Ginn & Co., Mass., 1961.

SHAMES, I. H.: "Engineering Mechanics—Statics," Prentice-Hall, Inc., N.J., 1966, Chaps. 1, 2.

SHAMES, I. H.: "Mechanics of Deformable Solids," Prentice-Hall, Inc., N.J., 1964, Chap. 2.

SMIRNOV, V. I.: "Advanced Calculus," Pergamon Press, N.Y., 1964.

SOKOLNIKOFF, I. S.: "Tensor Analysis," John Wiley and Sons, N.Y., 1967.

PROBLEMS

I.1 What are the transformation equations between reference triads which are rotated about the x and x' axes (which are coincident) through an angle of 30° in the $yz(y'z')$ plane?

What are the primed coordinates of position $(2, 0, 3)$ in the unprimed reference?

I.2 Given the following values:

$$A_1 = 2 \qquad A_2 = 3 \qquad A_3 = 4$$
$$B_{11} = 0 \qquad B_{12} = 2 \qquad B_{13} = -2$$
$$B_{21} = -3 \qquad B_{22} = 1 \qquad B_{23} = -1$$
$$B_{31} = 6 \qquad B_{32} = -3 \qquad B_{33} = 0$$
$$C_1 = 1 \qquad C_2 = 3 \qquad C_3 = 8$$

Compute or denote the array for the following:

(*a*) $A_i B_{2i}$
(*b*) $B_{j2} C_j$
(*c*) $A_i B_{22} C_j$

I.3 Given the following information:
(*a*) The direction cosines between a vector **V** and a set of axes x_i as $\cos(\mathbf{V},x_i)$;
(*b*) The rotation matrix between the x_i axes and a set of primed axes x_i';

Show that the vector **V** has the following direction cosines in the x_i' system:

$$\cos(\mathbf{V},x_i') = \cos(x_i',x_j)\cos(\mathbf{V},x_j)$$

(Note the sum over j.)

I.4 Derive Eq. (I.13), i.e., $a_{ji}a_{jk} = \delta_{ik}$.

I.5 Using data from Problem I.2, compute:
(*a*) $B_{ij}\delta_{kj}$ \qquad for $i = 1, k = 3$
(*b*) $A_i C_j \delta_{ij} B_{km}$ for $k = 1, m = 2$
(*c*) $A_m C_j B_{km} \delta_{jk}$
(*d*) $(A_i B_{kr} C_m)(\delta_{ik}\delta_{rm})$

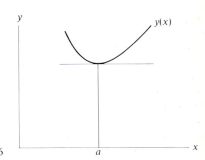

FIGURE I.6

I.6   (a) Using data of Problem I.2, what is the representation of the outer product of the sets A_i and B_{2j}? What is the value of the inner product of these sets?

    (b) What is the representation of the outer product of C_i, B_{21} and A_j? What is the inner product?

    (c) What is the value of $C_i B_{sj}$ contracted for indices i, j with $s = 2$, and contracted for indices i, s for $j = 1$?

I.7   Show that if λ is a constant and A_{ij} is a second-order tensor, then λA_{ij} is a second-order tensor.

I.8   Show that if A_{ij} and B_{ij} are second-order tensors, then $(A_{ij} + B_{ij})$ is a second-order tensor C_{ij}, wherein we add respective components of A_{ij} and B_{ij} to form C_{ij}.

I.9   Show that the outer product of an rth-order tensor and an nth-order tensor is an $(r + n)$th-order tensor.

I.10   Show that the inner product of an nth-order tensor with a kth-order tensor over c sets of indices results in a tensor of rank $(n + k - 2c)$.

I.11   If A_{ijkl} is a fourth-order tensor show that A_{iikl} is a second-order tensor.

I.12   Prove for a set A_{ij} of nine quantities that when we take an inner product with arbitrary second-order tensors B_{km} over indices i, k, that A_{ij} is a second-order tensor.

I.13   Evaluate the following terms:

    (a) ϵ_{112} (b) ϵ_{132} (c) ϵ_{231} (d) ϵ_{131}

I.14   Show that $\epsilon_{ijk}\epsilon_{ijp} = 0$ for those terms where the index k does not equal the index p. Also show that $\epsilon_{ijk}\epsilon_{ijp} = 2$ for those terms where the index k equals the index p. Thus we can say for the representation of $\epsilon_{ijk}\epsilon_{ijp}$

$$\epsilon_{ijk}\epsilon_{ijp} = C_{kp} = \begin{pmatrix} 2 & 0 & 0 \\ 0 & 2 & 0 \\ 0 & 0 & 2 \end{pmatrix}$$

Thus show that

$$\epsilon_{ijk}\epsilon_{ijp} = 2\delta_{kp}$$

Finally show that the contraction of the above over kp gives 6.

I.15   Show that $\delta_{ik}\epsilon_{ikm} = 0$.

I.16 Compute the set of terms corresponding to ($i = 1$, $i = 2$, $i = 3$):

$$(x_1{}^2 + 2x_2{}^3 + 3x_3)_{,i}$$

I.17 If $\mathbf{V} = 3x\mathbf{i} + 2y^2\mathbf{j} + 10z^3\mathbf{k}$, compute:

    (*a*) $V_{i,i}$ (*b*) $V_{i,i2}$ (*c*) $V_{j,3j}$

I.18 If $\phi = x^2 + 2\sin y + xz^3$, compute:
    (*a*) $\phi_{,i}$
    (*b*) $\phi_{,ii}$
    (*c*) $\phi_{,iij}$
    (*d*) $\phi_{,ii2}$

I.19 Write indicially:
    (*a*) $(\nabla^2\phi)\mathbf{A} \cdot \mathbf{B}$
    (*b*) $[(\mathbf{A} \times \mathbf{B}) \cdot \mathbf{C}]\mathbf{D}$
    (*c*) $\nabla^4\phi$

I.20 Show that:

$$\epsilon_{ijk}\epsilon_{imn} = \delta_{jm}\delta_{kn} - \delta_{jn}\delta_{km}$$

I.21 Utilizing the result of Problem 20 prove indicially:
    (*a*) $(\mathbf{A} \times \mathbf{B}) \cdot (\mathbf{C} \times \mathbf{D}) = (\mathbf{A} \cdot \mathbf{C})(\mathbf{B} \cdot \mathbf{D}) - (\mathbf{A} \cdot \mathbf{D})(\mathbf{B} \cdot \mathbf{C})$
    (*b*) $\mathbf{U} \times (\mathbf{V} \times \mathbf{W}) = (\mathbf{U} \cdot \mathbf{W})\mathbf{V} - (\mathbf{U} \cdot \mathbf{V})\mathbf{W}$
    (*c*) $\operatorname{curl}(\operatorname{grad} \phi) = \mathbf{0}$
    (*d*) $\operatorname{div}(\operatorname{curl} \mathbf{A}) = 0$

I.22 On a boundary curve C around an area A show that $a_{vx} = dy/ds$ and $a_{vy} = -dx/ds$, where $\mathbf{s}$ is tangent to the curve and where $\mathbf{v}$ is the outward normal unit vector. Now use Eq. (I.50) to derive Green's transformation.

$$\oint (V_x \, dx + V_y \, dy) = \iint \left(\frac{\partial V_y}{\partial x} - \frac{\partial V_x}{\partial y}\right) dx \, dy$$

for a vector field $\mathbf{V}$.

I.23 Show that Eq. (I.51) can also be derived directly from Gauss' theorem (Eq. I.44).

I.24 Generalize the result of Problem I.22 into the three-dimensional transformation of Kelvin (often called Stokes theorem):

$$\oint_C \mathbf{V} \cdot d\mathbf{l} = \iint_A \boldsymbol{v} \cdot (\nabla \times \mathbf{V}) \, dA$$

I.25 If $\partial P/\partial x = \partial Q/\partial y$ for the functions $P(x,y)$ and $Q(x,y)$, show that the line integral

$$\oint (Q \, dx + P \, dy)$$

is independent of path. In three dimensions, for a vector field $\mathbf{V}$, show that the analogous condition is:

$$\nabla \times \mathbf{V} = \mathbf{0}.$$

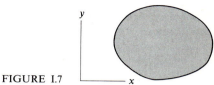

FIGURE I.7

I.26 Show that the area enclosed in a curve of the form shown in Fig. I.7 (note a straight line can cut the curve no more than twice) is:

$$A = \tfrac{1}{2} \oint (x\,dy - y\,dx)$$

I.27 Use the divergence theorem to show that for a domain of volume V enclosed by a surface S

$$\iint_S \mathbf{r} \cdot \mathbf{v}\,dA = 3V$$

where $\mathbf{r} = x\mathbf{i} + y\mathbf{j} + z\mathbf{k}$.

I.28 From Green's formula (Eq. (I.58)), show that a special case is:

$$\iint_A \nabla^2 \phi\,dA = \oint_C \frac{\partial \phi}{\partial v}\,ds$$

Next show that the condition for ϕ to be harmonic is the path-independence of the line integral:

$$\oint \left(\frac{\partial \phi}{\partial x}\,dy - \frac{\partial \phi}{\partial y}\,dx \right)$$

APPENDIX II

ROTATION TENSOR FOR A DEFORMING ELEMENT

We shall now demonstrate that for a deforming element each nonzero component of the rotation tensor represents the average of the angle of rotation for all line segments in the element about a coordinate axis. For this purpose consider the neighborhood of point P in a body (Fig. II.1). A reference xyz has been set up at point P and an arbitrary infinitesimal line segment $\overline{PQ}$ has been shown there in the undeformed geometry with polar angles θ and ϕ. We wish to consider the rotation of this line segment and so we must consider the displacement vectors of points P and Q. We may give these as $(\mathbf{u})_P$ and $(\mathbf{u})_P + (d\mathbf{u})$ respectively. Of these displacements only $d\mathbf{u}$ gives rise to possible rotation of the element. Hence in Fig. II.2 we have disregarded the translation of the element from $(\mathbf{u})_P$ and have shown only $d\mathbf{u}$. We shall consider here only the component of rotation about the z axis and so we have shown in the diagram the projection of $\overline{PQ}$ onto the xy plane as $\overline{PQ'}$. Components du_x and du_y of $d\mathbf{u}$ are shown. We may then express the angle of

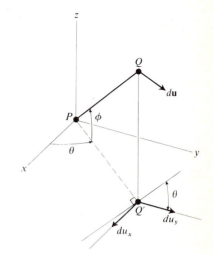

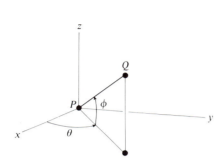

FIGURE II.1

FIGURE II.2

rotation of $\overline{PQ}$ about the z axis as follows

$$\frac{du_y \cos \theta}{\overline{PQ'}} - \frac{du_x \sin \theta}{\overline{PQ'}} \tag{II.1}$$

Now the differentials of displacement can be expressed as follows, using components of the directed line segment $\overline{PQ}$

$$du_x = \left(\frac{\partial u_x}{\partial x}\right)_P (\overline{PQ})_x + \left(\frac{\partial u_x}{\partial y}\right)_P (\overline{PQ})_y + \left(\frac{\partial u_x}{\partial z}\right)_P (\overline{PQ})_z$$

$$du_y = \left(\frac{\partial u_y}{\partial x}\right)_P (\overline{PQ})_x + \left(\frac{\partial u_y}{\partial y}\right)_P (\overline{PQ})_y + \left(\frac{\partial u_y}{\partial z}\right)_P (\overline{PQ})_z$$

Using the polar angles to get the components of $\overline{PQ}$ in terms of $\overline{PQ}$, the above equations become

$$du_x = \left(\frac{\partial u_x}{\partial x}\right)_P (\overline{PQ}) \cos \phi \cos \theta + \left(\frac{\partial u_x}{\partial y}\right)_P \overline{PQ} \cos \phi \sin \theta + \left(\frac{\partial u_x}{\partial z}\right)_P \overline{PQ} \sin \phi$$

$$du_y = \left(\frac{\partial u_y}{\partial x}\right)_P (\overline{PQ}) \cos \phi \cos \theta + \left(\frac{\partial u_y}{\partial y}\right)_P \overline{PQ} \cos \phi \sin \theta + \left(\frac{\partial u_y}{\partial z}\right)_P \overline{PQ} \sin \phi \tag{II.2}$$

Employing Eqs. (II.2) in Eq. (II.1) and noting that $\overline{PQ'}$ is simply $\overline{PQ} \cos \phi$, we get for the rotation of $\overline{PQ}$ about the z axis

$$\left[\left(\frac{\partial u_y}{\partial x}\right)_P \cos^2 \theta + \left(\frac{\partial u_y}{\partial y}\right)_P \sin \theta \cos \theta + \left(\frac{\partial u_y}{\partial z}\right)_P \tan \phi \cos \theta\right]$$

$$- \left[\left(\frac{\partial u_x}{\partial x}\right)_P \sin \theta \cos \theta + \left(\frac{\partial u_x}{\partial y}\right)_P \sin^2 \theta + \left(\frac{\partial u_x}{\partial z}\right)_P \tan \phi \sin \theta\right] \tag{II.3}$$

Now keep ϕ constant and find the average value of the above quantity as θ goes from zero to 2π. Note that the average value of $\sin^2 \theta$ and $\cos^2 \theta$ is $\frac{1}{2}$ over this interval while the average value of $\sin \theta$, $\cos \theta$ and $(\sin \theta \cos \theta)$ is zero. Accordingly we get the following result for this averaging process:

$$\frac{1}{2}\left[\left(\frac{\partial u_y}{\partial x}\right)_P - \left(\frac{\partial u_x}{\partial y}\right)_P\right] \qquad \text{(II.4)}$$

Note that since expression (II.3) is independent of the length of $\overline{PQ}$ the subsequent expression (II.4) is the average rotation for *all* line segments at a fixed angle ϕ. But on inspecting expression (II.4) we see that this expression is not dependent on angle ϕ (that is, it is valid for any ϕ) and so we may conclude that the expression (II.4) represents the average angular rotation of *all* line segments in the neighborhood of P about the z axis. We may delete the subscript P now to express (II.4) as follows

$$(\phi_z)_{\text{av}} = \frac{1}{2}\left(\frac{\partial u_y}{\partial x} - \frac{\partial u_x}{\partial y}\right)$$

Similarly we can show

$$(\phi_x)_{\text{av}} = \frac{1}{2}\left(\frac{\partial u_z}{\partial y} - \frac{\partial u_y}{\partial z}\right)$$

$$(\phi_y)_{\text{av}} = \frac{1}{2}\left(\frac{\partial u_x}{\partial z} - \frac{\partial u_z}{\partial x}\right)$$

This proves the assertion made at the outset of this appendix.

INTEGRATION OF $\int_0^1 \xi^a [1-\xi^g]^h \, d\xi$

$a \rightarrow$	$\downarrow h$	$\frac{1}{2}$	1	2	3	4	5
$g = \frac{1}{2}$	$\frac{1}{2}$	3.041×10^{-1}	2.031×10^{-1}	1.136×10^{-1}	7.481×10^{-2}	5.398×10^{-2}	4.129×10^{-2}
	1	1.660×10^{-1}	9.995×10^{-2}	4.762×10^{-2}	2.778×10^{-2}	1.818×10^{-2}	1.282×10^{-2}
	2	6.610×10^{-2}	3.329×10^{-2}	1.190×10^{-2}	5.556×10^{-3}	3.030×10^{-3}	1.832×10^{-3}
	3	3.280×10^{-2}	1.425×10^{-2}	3.968×10^{-3}	1.515×10^{-3}	6.993×10^{-4}	3.663×10^{-4}
	4	1.856×10^{-2}	7.107×10^{-3}	1.587×10^{-3}	5.050×10^{-4}	1.998×10^{-4}	9.158×10^{-5}
	5	1.145×10^{-2}	3.935×10^{-3}	7.213×10^{-4}	1.942×10^{-4}	6.660×10^{-5}	2.693×10^{-5}
$g = 1$	$\frac{1}{2}$	3.919×10^{-1}	2.665×10^{-1}	1.523×10^{-1}	1.015×10^{-1}	7.378×10^{-2}	5.673×10^{-2}
	1	2.660×10^{-1}	1.666×10^{-1}	8.333×10^{-2}	5.000×10^{-2}	3.333×10^{-2}	2.381×10^{-2}
	2	1.517×10^{-1}	8.328×10^{-2}	3.333×10^{-2}	1.667×10^{-2}	9.524×10^{-3}	5.952×10^{-3}
	3	1.009×10^{-1}	4.995×10^{-2}	1.667×10^{-2}	7.143×10^{-3}	3.571×10^{-3}	1.984×10^{-3}
	4	7.323×10^{-2}	3.328×10^{-2}	9.523×10^{-3}	3.571×10^{-3}	1.587×10^{-3}	7.937×10^{-4}
	5	5.618×10^{-2}	2.376×10^{-2}	5.952×10^{-3}	1.984×10^{-3}	7.937×10^{-4}	3.608×10^{-4}
$g = 2$	$\frac{1}{2}$	4.784×10^{-1}	3.331×10^{-1}	1.962×10^{-1}	1.332×10^{-1}	9.803×10^{-2}	7.605×10^{-2}
	1	3.803×10^{-1}	2.500×10^{-1}	1.333×10^{-1}	8.333×10^{-2}	5.714×10^{-2}	4.167×10^{-2}
	2	2.764×10^{-1}	1.666×10^{-1}	7.619×10^{-2}	4.167×10^{-2}	2.540×10^{-2}	1.667×10^{-2}
	3	2.210×10^{-1}	1.250×10^{-1}	5.079×10^{-2}	2.500×10^{-2}	1.385×10^{-2}	8.333×10^{-3}
	4	1.860×10^{-1}	9.995×10^{-2}	3.694×10^{-2}	1.667×10^{-2}	8.525×10^{-3}	4.762×10^{-3}
	5	1.616×10^{-1}	8.328×10^{-2}	2.842×10^{-2}	1.190×10^{-2}	5.683×10^{-3}	2.976×10^{-3}
$g = 3$	$\frac{1}{2}$	5.228×10^{-1}	3.694×10^{-1}	2.221×10^{-1}	1.528×10^{-1}	1.135×10^{-1}	8.872×10^{-2}
	1	4.438×10^{-1}	2.999×10^{-1}	1.667×10^{-1}	1.070×10^{-1}	7.500×10^{-2}	5.556×10^{-2}
	2	3.549×10^{-1}	2.250×10^{-1}	1.111×10^{-1}	6.429×10^{-2}	4.091×10^{-2}	2.778×10^{-2}
	3	3.041×10^{-1}	1.840×10^{-1}	8.333×10^{-2}	4.451×10^{-2}	2.630×10^{-2}	1.667×10^{-2}
	4	2.702×10^{-1}	1.577×10^{-1}	6.667×10^{-2}	3.338×10^{-2}	1.856×10^{-2}	1.111×10^{-2}
	5	2.456×10^{-1}	1.392×10^{-1}	5.556×10^{-2}	2.635×10^{-2}	1.392×10^{-2}	7.937×10^{-3}

APPENDIX IV

TO SHOW THAT LAGRANGE MULTIPLIERS ARE ZERO FOR DEVELOPMENT OF THE RAYLEIGH-RITZ METHOD

We must now verify the assertion in Sec. 7.16 that the $k - 1$ Lagrange multipliers are zero. We note first that for the case where $k = 1$, there is no constraint for the function W as to orthogonality and so the last expression in Eq. (7.167) will not be present. (Note more formally, for $k = 1$ the summation is empty for this case.) *Then Eq. (7.167) can be given as Eq. (7.168).* Now suppose that Eq. (7.168) holds for $S = 1, 2, \ldots, (k - 1)$. We will now *prove* that $\lambda^{(1)} = \lambda^{(2)} = \cdots = \lambda^{(k-1)} = 0$ and hence from our discussion of Sec. (7.16), Eq. (7.168) then holds *also* for $S = k$. On this basis knowing that Eq. (7.168) holds for $S = 1$ we could then deduce that it holds also for $S = 2$ and so forth for all values of S, thereby justifying the step of setting all the Lagrange multipliers equal to zero.

To show this, we go back to Eqs. (7.167) which we now express using C's associated with the kth approximate eigenfunction as follows:

$$\sum_{j=1}^{n} (G_{ij} - \Lambda^2 E_{ij})C_j^{(k)} - \tfrac{1}{2}\left[\sum_{S=1}^{k-1} \lambda^{(S)} \sum_{j=1}^{n} C_j^{(S)}E_{ij}\right]\left[\sum_{v=1}^{n} \sum_{p=1}^{n} C_v^{(k)}C_p^{(k)}E_{vp}\right] = 0$$

$$i = 1, 2, \ldots n$$

Now multiply by $C_i^{(f)}$ where f is *any integer less than* k and sum over i from 1 to n. We thus get:

$$\sum_{i=1}^{n}\sum_{j=1}^{n}(G_{ij} - \Lambda^2 E_{ij})C_j^{(k)}C_i^{(f)} - \tfrac{1}{2}\left[\sum_{S=1}^{k-1}\lambda^{(S)}\sum_{j=1}^{n}\sum_{i=1}^{n}C_i^{(f)}C_j^{(S)}E_{ij}\right]$$

$$\times\left[\sum_{v=1}^{n}\sum_{p=1}^{n}C_v^{(k)}C_p^{(k)}E_{vp}\right] = 0$$

But from the orthogonality conditions we have imposed on the approximate eigenfunctions we can conclude (see Eq. (7.166)) that

$$\sum_{j=1}^{n}\sum_{i=1}^{n}C_i^{(k)}C_j^{(f)}E_{ij} = 0$$

in the first expression and that only when $S = f$ do we get other than zero in the second expression. Thus we have:

$$\sum_{i=1}^{n}\sum_{j=1}^{n}G_{ij}C_i^{(f)}C_j^{(k)} - \tfrac{1}{2}\lambda^{(f)}\left\{\sum_{j=1}^{n}\sum_{i=1}^{n}C_i^{(f)}C_j^{(f)}E_{ij}\right\}\left\{\sum_{v=1}^{n}\sum_{p=1}^{n}C_v^{(k)}C_p^{(k)}E_{vp}\right\} = 0$$

$$(IV.1)$$

Now let us focus on the first expression in the above equation. It may be rewritten as follows:

$$\sum_{i=1}^{n}\sum_{j=1}^{n}G_{ij}C_i^{(f)}C_j^{(k)} = \sum_{j=1}^{n}C_j^{(k)}\left(\sum_{i=1}^{n}C_i^{(f)}G_{ij}\right) \qquad (IV.2)$$

We can now go back to Eq. (7.168) which we now assume is valid in this discussion for all values of f. Thus we have

$$\sum_{j=1}^{n}C_j^{(f)}G_{ij} = \Lambda^2\sum_{j=1}^{n}C_j^{(f)}E_{ij} \qquad i = 1, 2, \ldots, n$$

Using the fact that G_{ij} and E_{ij} are symmetric as a result of the self-adjoint property of the operators L and M (see Eqs. (7.165)) we can change indices in the above equations as follows:

$$\sum_{i=1}^{n}C_i^{(f)}G_{ij} = \Lambda^2\sum_{i=1}^{n}C_i^{(f)}E_{ij} \qquad j = 1, 2, \ldots, n$$

Now use the above equation to replace the bracketed expression on the right side of Eq. (IV.2) so that

$$\sum_{i=1}^{n}\sum_{j=1}^{n}G_{ij}C_i^{(f)}C_j^{(k)} = \sum_{j=1}^{n}C_j^{(k)}\left(\Lambda^2\sum_{i=1}^{n}C_i^{(f)}E_{ij}\right)$$

$$= \Lambda^2\sum_{i=1}^{n}\sum_{j=1}^{n}C_j^{(k)}C_i^{(f)}E_{ij}$$

Again we invoke the orthogonality conditions (Eq. (7.166)) for the approximate eigenfunctions (note $f \neq k$) to show that the last expression above is zero. Thus in returning to Eq. (IV.1) we must conclude

$$\tfrac{1}{2}\lambda^{(f)}\left\{\sum_{j=1}^{n}\sum_{i=1}^{n}C_i^{(f)}C_j^{(f)}E_{ij}\right\}\left\{\sum_{v=1}^{n}\sum_{p=1}^{n}C_v^{(k)}C_p^{(k)}E_{vp}\right\}=0$$

Since $\iiint WM(W)\,dv$, which is the origin of each of the bracketed expressions above, is positive-definite, we can conclude that $\lambda^{(f)}=0$. Now since f can be any integer up to and including $k-1$, we can conclude from above that $\lambda^{(1)}=\lambda^{(2)}\cdots=\lambda^{(k-1)}=0$. Thus having assumed that

$$\sum_{j=1}^{n}(G_{ij}-\Lambda^2 E_{ij})C_j^{(S)}=0\qquad i=1,2,\ldots,n\qquad\text{(IV.3)}$$

for all values of S up to $k-1$, we see that the Lagrange multipliers up to $\lambda^{(k-1)}$ are zero. But now going to Eq. (7.167) we see that this is the *precise condition for making the above equation valid for S equal to k also*. We could then similarly prove with Eq. (IV.3) valid for S up to k that $\lambda^{(1)}=\lambda^{(2)}=\cdots=\lambda^{(k)}=0$ and thus verify that Eq. (7.168) is valid also for $S=k+1$.

The process can then be continued. It needs only a starting point and as pointed out at the outset we do know that (Eq. IV.3) is valid for $S=1$. With this as the beginning step we can conclude that all the Lagrange multipliers are indeed zero for all values of k.

INDEX